Modelling and Control in Solid Mechanics

A.M. Khludnev
J. Sokolowski

Birkhäuser Verlag
Basel · Boston · Berlin

Authors:

A.M. Khludnev
Lavrentyev Institute of Hydrodynamics
of the Russian Academy of Sciences
Novosibirsk 630090
Russia

J. Sokolowski
Institut Elie Cartan
Laboratoire de Mathématiques
Université Henri Poincaré Nancy I
B.P. 239
54506 Vandoeuvre-Les-Nancy
France

1991 Mathematics Subject Classification 73KXX, 73T05, 93B35, 35Q72

A CIP catalogue record for this book is available from the Library of Congress, Washington D.C., USA

Deutsche Bibliothek Cataloging-in-Publication Data
Chludnev, A. M.:
Modelling and control in solid mechanics / A. M. Khludnev ;
J. Sokolowski. - Basel ; Boston ; Berlin : Birkhäuser, 1997
(International series of numerical mathematics ; Vol. 122)

NE: Sokolowski, Jan:; GT

Softcover reprint of the hardcover 1st edition 1997
Printed on acid-free paper produced from chlorine-free pulp. TCF ∞
Cover design: Heinz Hiltbrunner, Basel

ISBN-13: 978-3-0348-9855-3 e-ISBN-13: 978-3-0348-8984-1
DOI: 10.1007/978-3-0348-8984-1

9 8 7 6 5 4 3 2 1

Contents

Preface

New trends in free boundary problems and new mathematical tools together with broadening areas of applications have led to attempts at presenting the state of art of the field in a unified way. In this monograph we focus on formal models representing contact problems for elastic and elastoplastic plates and shells.

New approaches open up new fields for research. For example, in crack theory a systematic treatment of mathematical modelling and optimization of problems with cracks is required. Similarly, sensitivity analysis of solutions to problems subjected to perturbations, which forms an important part of the problem solving process, is the source of many open questions. Two aspects of sensitivity analysis, namely the behaviour of solutions under deformations of the domain of integration and perturbations of surfaces seem to be particularly demanding in this context.

On writing this book we aimed at providing the reader with a self-contained study of the mathematical modelling in mechanics. Much attention is given to modelling of typical constructions applied in many different areas. Plates and shallow shells which are widely used in the aerospace industry provide good examples. Allied optimization problems consist in finding the constructions which are of maximal strength (endurance) and satisfy some other requirements, eg. weight limitations.

Mathematical modelling of plates and shells always requires a reasonable compromise between two principal needs. One of them is the accuracy of the description of a physical phenomenon (as required by the principles of mechanics). The other need is the sufficiency of model simplicity which should ensure, within the formalism employed, meaningful quantitative features of solutions. That compromise is essential for the applicability of models to practical optimal design problems.

The monograph covers the following topics.

1. Qualitative properties of solutions to contact problems for elastic plates and shells with rigid bodies and allied optimization problems.
2. Determination of extreme crack shapes in plates by optimization methods.
3. Existence theorems for exact formulations of contact problems for elastoplastic beams and plates.
4. Sensitivity analysis of solutions to contact problems for plates and shells and its application to solving optimal design and shape optimization problems.

In the models used the existence of free boundaries is a consequence of restrictions imposed on solutions. In contact problems there occur local restrictions. Geometrically, those restrictions represent the nonpenetration condition of the two bodies in contact. The nonpenetration condition is imposed on solutions to contact problems regardless of the form of the mechanical model adopted. Other restrictions defined by the state equation are related to the physical nature of the problem in question. Such restrictions, also being local, are typical in elastoplastic models for example.

Problems of the elasticity theory are often expressed as variational inequalities of the form

$$u \in K: \quad \langle \Pi'(u), \bar{u} - u \rangle \geq 0 \quad \forall\, \bar{u} \in K,$$

where Π is the energy functional related to a plate or a shell, and $\Pi'(u)$ is the derivative of Π evaluated at the point u. The convex set K defines a specific type of restrictions, eg. the nonpenetration condition, imposed on admissible displacements. Models of this type are analysed in Chapter 2.

The nonpenetration condition for the shallow shell-rigid punch system is a typical geometric restriction imposed on the displacement vector $\omega = (U, w)$ of the shallow shell middle-surface and has the form

$$w(x) - U(x) \cdot \nabla \Phi(x) \geq \Phi(x), \quad x \in \Omega,$$

where the function $y = \Phi(x)$ describes the punch shape. A mathematical model with the above nonpenetration condition is formulated in Section 3 of Chapter 2.

Solutions to contact elastoplastic problems are saddle points of the Lagrangian

$$L(u, \bar{v}) \leq L(u, v) \leq L(\bar{u}, v) \quad \forall\, (\bar{u}, \bar{v}) \in K \times B.$$

Restrictions are described by the set $K \times B$. Models of this type are considered in Section 8 of Chapter 3.

Regularity of solutions to variational inequalities depends heavily on the restrictions imposed and cannot be substantially improved by increasing the regularity of data. For instance, for an elliptic variational inequality with a differential operator of the $2m$ order we can expect at most that solutions belong to the space $H^{m+1}_{loc}(\Omega)$.

In contact problems of a shallow shell with a rigid punch this lack of regularity of solutions is physically justified. In these problems the interaction forces, unknown in advance, are not regular. The interaction between the shell and the punch is described by a nonnegative measure μ with the support contained in the contact domain Ω. The character of the interaction forces represented by this measure depends on the gradient of the function Φ which describes the punch.

As shown in Section 3.3 of Chapter 2 the measure μ has the representation

$$\langle \Pi'(\omega), \bar{\omega} \rangle = \int_\Omega \frac{\bar{w} - \bar{U} \cdot \nabla \Phi}{\sqrt{1 + |\nabla \Phi|^2}} d\mu$$

which holds for all sufficiently smooth functions $\bar{\omega} = (\bar{U}, \bar{w})$ with compact supports.

A regularity result can be formulated as follows. If $|\nabla\Phi(x)| > 0$ for $x \in \Omega_0, \Omega_0 \subset \Omega$, then

$$(U, w) \in H^2_{loc}(\Omega_0) \times H^4_{loc}(\Omega_0)$$

and the density of the measure μ with respect to the Lebesgue measure belongs to the space $L^2_{loc}(\Omega_0)$. When $\nabla\Phi(x) = 0$ the above regularity result is no longer valid.

The majority of optimization and optimal control problems encountered in this book belong to the shape optimization field. This includes not only problems of optimal choice of the domain of integration for systems of partial differential equations, but also problems of finding a form of a punch or an external restriction so that to get the desired properties of solutions to mathematical models. For instance, in contact problems it is desirable to have sufficient regularity of solutions in the domain of contact.

In its classical formulation a shape optimization problem is one of minimization or maximization of a quality functional (which involves function variables defining the geometry of the given structure) over all admissible geometries of the structure in question. The existence of solutions to these problems is discussed in Chapter 4, and in Chapter 5 the necessary optimality conditions are derived. Examples of shape optimization problems are provided by the problem of the optimal choice of the punch shape in contact problems and the problem of finding extremal shapes of cracks in elastic plates.

We have made an extensive study of a variational inequality depending upon a small parameter $\varepsilon > 0$, with the fourth-order nonlinear operator involving the given function φ, namely

$$\int_\Omega \big(\varepsilon\Delta^2 w - \varphi(\|\nabla w\|^2)\Delta w - f\big)(\bar{w} - w)dx \geq 0 \quad \forall\, \bar{w} \geq \psi\,.$$

For fixed $\varepsilon > 0$ and the given punch shape $\psi \in \Psi$, the function $w = w_\varepsilon(\psi)$ solves the variational inequality provided that $w \geq \psi$ in Ω. For $\varepsilon > 0$ the function $w_\varepsilon(\psi)$ describes the contact of a nonlinear plate with a punch. For $\varepsilon = 0$ the above model with a nonlinear elliptic operator corresponds to a membrane with properties depending upon the choice of the function φ.

In Section 6 of Chapter 4, for the problem

$$\inf_{\psi\in\Psi} J_\varepsilon(\psi) \quad \text{with the quality functional} \quad J_\varepsilon(\psi) = \|w_\varepsilon(\psi) - w_*\|$$

we prove the convergence of the optimal solutions ψ^ε as $\varepsilon \to 0$. which yields the convergence

$$J_\varepsilon(\psi^\varepsilon) \to J_0(\psi^0), \quad \varepsilon \to 0.$$

We propose a new approach to the crack theory. This allows us to eliminate physical contradictions appearing in the classical approach.

In the classical approach to the modelling of cracks in elastic bodies, conditions on crack faces are given in the form of equations. Examples exist showing that solutions satisfying those equations may not satisfy the nonpenetration condition, i.e. that the crack faces may penetrate each other, which is physically impossible.

The novelty of our approach is that we admit inequalities in conditions on crack faces. These inequalities correspond to the nonpenetration conditions of crack faces.

The models with cracks are regarded as contact problems. For the plate with a vertical crack given as $y = \psi(x)$ in the plane x, y, the displacement U of the plate around crack faces in the plane x, y satisfies the condition

$$[\, U \,]\nu \geq 0 \quad \text{on} \quad \Gamma_\psi .$$

By $[\, U \,] = U^+ - U^-$ we denote the jump of the function U on the graph Γ_ψ of the function $y = \psi(x)$. The extremal crack shape is a solution to the optimization problem

$$\sup_{\psi \in \Psi} \|U - U_*\|.$$

In Section 10.3 a more detailed condition is imposed on crack shapes involving normal displacements w, tangential displacements U, and the distance z to the middle surface of the plate, namely

$$[U - z \cdot \nabla w]\nu \geq 0 \quad \text{on} \quad \Gamma_\psi, \quad |\, z \,| \leq \Lambda.$$

By 2Λ we denote the thickness of the plate.

In contact elastoplastic problems with the nonpenetration condition the solution M should also satisfy conditions of the type

$$\Phi(M) \leq 0$$

where Φ is the continuous and convex function. It is a frequently applied trick in the analysis of mathematical models of these problems that the restrictions are expressed exclusively as equality and inequality type relations. These relations, however, do not reflect accurately all the restrictions imposed on solutions in the original formulation of the problem and a part of the boundary conditions is usually lost. We propose another approach based on regularization and penalization which overcomes this drawback.

Differentiability of solutions to certain of the models analysed in this book is treated in a detailed way in Chapter 5. We concentrate on perturbations of the right-hand side and the coefficients of the differential operator, as well as on deformations of the domain of integration.

In Section 3.3.1 of Chapter 5 dealing with the variational inequality for the Kirchhoff plate we study the differentiability of solutions with respect to deformations of the domain of integration. The variational inequality is of the form

$$w \in K(\Omega) : \int_\Omega \Delta w \Delta(\varphi - w) dx \geq \int_\Omega f(\varphi - w) dx \quad \forall\, \varphi \in K(\Omega)$$

and the set $K(\Omega)$ of admissible displacements is given as

$$K(\Omega) = \{\varphi \in H_0^2(\Omega) \mid \varphi(x) \geq \psi(x) \text{ in } \Omega\} .$$

The so-called shape derivative $w'(\Omega; V)$ (defined in Section 2 of Chapter 5) of solutions to this inequality is given as a solution to the following variational inequality

$$w' \in S : \quad \int_\Omega \Delta w' \Delta(\varphi - w') dx \geq 0 \quad \forall \varphi \in S .$$

In proving results such as this we make an extensive use of the results on the conical differentiability of metric projections onto closed convex sets in Sobolev spaces. For instance, we use this approach in Section 6 to analyse sensitivity of solutions to some shape optimization problems. In Section 8 for shallow shells we propose an approach to the sensitivity analysis of solutions to systems of equations under deformations of the middle surface.

The content of the book is divided into five chapters. Chapter 1 covers basic facts on functional spaces, variational inequalities, and measure theory used in the next chapters. Chapter 2 is devoted to the contact problems in elasticity, and in particular to qualitative properties of solutions such as regularity, and to constructions of measures describing interaction forces in the domain of contact. Elastoplastic models are treated in Chapter 3 with a detailed presentation of existence results for static, quasi-static and dynamic problems. The optimization and optimal control problems are investigated in Chapter 4. Existence results are proved for selected shape optimization and optimal control problems. In Chapter 5 the sensitivity analysis to some models investigated in this book is presented. For example, we prove the conical differentiability of solutions to the fourth-order elliptic variational inequalities. Optimality conditions for a number of the shape optimization and optimal design problems are also discussed.

The authors express their gratitude to all persons who contributed to this book in many different ways. We are grateful to Professor Karl-Heinz Hoffmann from Munich for his initiative to publish this book in the series International Series of Numerical Mathematics of which he is one of the editors. For some years the cooperation with him has been a great pleasure for us. We are indebted to Professor Murali Rao from Gainesville for valuable collaboration. We would also like to express sincere thanks to our Russian, Polish and French friends, especially to Professor Michel Bernadou from INRIA-Rocqencourt.

A part of the book was translated by T.G. Pritchina. The remaining translation work and the revision of the entire text are due to Dr. Ewa Bednarczuk. We thank them both for their efforts.

In their work on this book the authors were supported by the Lavrentyev Institute of Hydrodynamics of the Russian Academy of Sciences, the State Committee for Scientific Research of the Republic of Poland under grant no. 2 1207 9101 and by the Systems Research Institute of the Polish Academy of Sciences.

Chapter 1
Introduction

1 Elements of mathematical analysis and calculus of variations

1.1 Functional spaces. Elementary properties

Let $x = (x_1, \dots, x_N)$ be a point of the N-dimensional Euclidean space $\mathbb{R}^N$, and let $\Omega \subset \mathbb{R}^N$ be a bounded domain. By $L^p(\Omega)$, $1 \le p < \infty$, we denote the Banach space of all functions whose p-th power is integrable, with the norm

$$\|f\|_{L^p(\Omega)} = \left(\int_\Omega |f|^p dx \right)^{\frac{1}{p}} .$$

When $p = 2$ the resulting space is a Hilbert space. The inner product of two functions in this space is equal to the integral of the product of those functions over the domain. For $p = \infty$ a norm is defined as

$$\|f\|_{L^\infty(\Omega)} = \sup_{x \in \Omega} \text{ess} \, |f(x)| .$$

Let $s \ge 0$ be an integer and $1 \le p \le \infty$. The Sobolev space $W_p^s(\Omega)$ contains all functions which together with all their derivatives up to order s belong to $L^p(\Omega)$. A norm in this space is given by the formula

$$\|f\|_{W_p^s(\Omega)} = \sum_{|\alpha|=0}^{s} \sum_{\alpha} \|D^\alpha f\|_{L^p(\Omega)} ,$$

where the second summation is taken over all $\alpha_1, \dots, \alpha_N$ such that $|\alpha| = \alpha_1 + \dots + \alpha_N$. In the sequel we shall often consider the Hilbert space $W_2^s(\Omega)$ denoted by $H^s(\Omega)$. The Banach space of all functions which are continuous in the closure $\overline{\Omega}$, endowed with the norm

$$\|f\|_{C(\overline{\Omega})} = \sup_{x \in \Omega} |f(x)|$$

is denoted by $C(\overline{\Omega})$. By $C^k(\overline{\Omega})$ we denote the space of all functions having the first k derivatives continuous in $\overline{\Omega}$, equipped with the norm

$$\|f\|_{C^k(\overline{\Omega})} = \sum_{|\alpha|=0}^{k} \sum_{\alpha} \|D^\alpha f\|_{C(\overline{\Omega})} .$$

Elements of a given functional space may belong to another space. Imbedding theorems describe properties of functions regarded as elements of different functional spaces. Below we formulate one of those theorems.

Theorem 1.1. *Let $\Omega \subset \mathbb{R}^N$ be a bounded domain with piecewise smooth boundary Γ and let Γ_r be the intersection of Ω with an r-dimensional smooth hyperplane, $r \leq N$. Then for every function $f \in W_p^s(\Omega)$, where $s \geq 1$, $p > 1$, $N > sp$, $r > N - sp$, there exists a trace $f\big|_{\Gamma_r}$ on Γ_r such that*

$$f\big|_{\Gamma_r} \in L^q(\Gamma_r)\,, \quad q \leq \frac{pr}{N - sp}\,,$$

$$\big\|f\big|_{\Gamma_r}\big\|_{L^q(\Gamma_r)} \leq c\,\|f\|_{W_p^s(\Omega)}\,.$$

For $N = sp$, q can take any value less than $+\infty$. If $sp > N$, then $f \in C^k(\overline{\Omega})$, $k = s - 1 - \left[\dfrac{N}{p}\right]$, and

$$\|f\|_{C^k(\overline{\Omega})} \leq c\,\|f\|_{W_p^s(\Omega)}\,.$$

In the above inequalities c is a positive constant independent of f, and $\left[\dfrac{N}{p}\right]$ denotes the integer part of $\dfrac{N}{p}$. Moreover, the imbedding of $W_p^s(\Omega)$ in $L^q(\Gamma_r)$ is compact for $N \geq sp$, $q < \dfrac{pr}{N - sp}$. When $sp > N$, the imbedding of $W_p^s(\Omega)$ in $C^k(\overline{\Omega})$ is compact.

In the sequel we shall also make use of some other imbedding theorems. Let us now formulate several well-known inequalities for functions defined on Ω.

1. Hölder inequality. For any functions $f \in L^p(\Omega)$, $g \in L^q(\Omega)$, the inequality

$$\|fg\|_{L^1(\Omega)} \leq \|f\|_{L^p(\Omega)}\,\|g\|_{L^q(\Omega)}\,, \quad \frac{1}{p} + \frac{1}{q} = 1\,, \quad p \geq 1$$

holds.

2. Poincaré-Friedrichs inequality. For every function $f \in H_0^1(\Omega)$, the inequality

$$\|f\|_{L^2(\Omega)} \leq c\,\|\nabla f\|_{L^2(\Omega)}$$

holds, where c is a constant independent of f, and $H_0^1(\Omega)$ is the completion in the norm of $H^1(\Omega)$ of smooth functions with compact supports.

3. First Korn inequality. Let $\Gamma_0 \subset \Gamma$ be a subset of the boundary Γ with positive measure, and Γ be smooth. Let us denote by $H_{\Gamma_0}^1(\Omega)$ the closure in $H^1(\Omega)$ of smooth functions equal to zero near Γ_0, and let $u = (u_1, u_2, \ldots, u_N) \in H_{\Gamma_0}^1(\Omega)$,

$\varepsilon_{ij}(u) = \dfrac{1}{2}\left(\dfrac{\partial u_i}{\partial x_j} + \dfrac{\partial u_j}{\partial x_i}\right)$. In particular, when $\Gamma_0 = \Gamma$, the space $H^1_{\Gamma_0}(\Omega)$ coincides with $H^1_0(\Omega)$. For every function $u \in H^1_{\Gamma_0}(\Omega)$, the inequality

$$\sum_{i,j=1}^{N} \|\varepsilon_{ij}(u)\|^2_{L^2(\Omega)} \geq c\,\|u\|^2_{H^1_{\Gamma_0}(\Omega)}$$

holds with a constant c independent of u.

Let us also recall the formula of integration by parts

$$\int_\Omega f g_{x_i} dx = \int_\Gamma f g \nu_i d\Gamma - \int_\Omega g f_{x_i} dx$$

which is valid for all $f, g \in H^1(\Omega)$.

Here $\nu = (\nu_1, \ldots, \nu_N)$ is the unit outward normal to the boundary Γ. In Chapter 5 the unit outward normal to $\Gamma = \partial\Omega$ is denoted by $n = (n_1, \ldots, n_N)$.

When dealing with dynamic problems it is necessary to consider spaces of functions defined on $(0,T)$ and taking values in a Banach space B. The Banach space of measurable functions defined on $(0,T)$ with values in B with the norm

$$\|f\|_{L^p(0,T;B)} = \left(\int_0^T \|f(t)\|^p_B dt\right)^{\frac{1}{p}}, \quad 1 \leq p < \infty\,,$$

will be denoted by $L^p(0,T;B)$. For $p = \infty$ we put

$$\|f\|_{L^\infty(0,T;B)} = \sup_{t\in(0,T)} \text{ess}\, \|f(t)\|_B\,.$$

Let V be a normed space and let $J : V \mapsto \mathbb{R}$ be any functional. A continuous and linear functional δJ_u, such that for every $v \in V$

$$\delta J_u(v) = \lim_{\lambda\to 0} \frac{J(u+\lambda v) - J(u)}{\lambda}\,, \tag{1.1}$$

is called the derivative (Gateaux derivative) of J at u. The set of all continuous linear functionals defined on V is called the dual space of V and is denoted by V^*. The dual space is also a normed space with the norm

$$\|u^*\|_{V^*} = \sup_{u\neq 0} \frac{u^*(u)}{\|u\|_V}\,.$$

The definition of the derivative gives rise to the existence of the operator $V \mapsto V^*$ assigning to every $u \in V$ the derivative δJ_u . A functional J is called differentiable if it has the derivative δJ_u at every $u \in V$.

1.2 Variational inequalities

A set $K \subset V$ is said to be convex if $\lambda u_1 + (1-\lambda)u_2 \in K$ for every $u_1, u_2 \in K$, $\lambda \in (0,1)$. Let $K \subset V$ be a convex set and let J be a differentiable functional on V. We consider the minimization problem

$$\inf_{u \in K} J(u) \ . \tag{1.2}$$

An element $u_0 \in K$ is called a solution to problem (1.2) if

$$J(u_0) = \inf_{u \in K} J(u) \ .$$

Let $u_0 \in K$ be a solution to problem (1.2). Then

$$J(u) - J(u_0) \geq 0 \quad \forall\, u \in K \ .$$

By putting $u = \lambda u_1 + (1-\lambda)u_0$, $u_1 \in K$, $\lambda \in (0,1)$, we get

$$\frac{J\big(u_0 + \lambda(u_1 - u_0)\big) - J(u_0)}{\lambda} \geq 0 \ .$$

Passing to the limit as $\lambda \to 0$, one easily finds that

$$u_0 \in K \ : \ \delta J_{u_0}(u_1 - u_0) \geq 0 \quad \forall\, u_1 \in K \ . \tag{1.3}$$

Inequality (1.3) is what is called a variational inequality. It has been obtained from the problem of minimizing the functional J over the set K. In the sequel we shall study more closely relationships between minimization problems and variational inequalities but now we want to indicate one essential issue. It is obvious that problem (1.2) is more general than that of minimization over the whole space. It is a well-known fact that in the latter problem the Euler equation provides a necessary optimality condition. Variational inequality (1.3) generalizes the Euler equation, ie. for $K = V$ the Euler equation follows from (1.3). Indeed, by taking $u_1 = u_0 + u$, where $u \in V$ is an arbitrary element, and substituting it into (1.3), we obtain the inequality

$$\delta J_{u_0}(u) \geq 0 \ .$$

Next, by replacing u with $-u$, we get

$$\delta J_{u_0}(u) = 0 \quad \forall\, u \in V$$

which is the Euler equation.

1.3 Minimization problems for convex functionals

A functional $J : V \mapsto \mathbb{R}$ is said to be convex if

$$J\big(\lambda u_0 + (1-\lambda)u_1\big) \leq \lambda J(u_0) + (1-\lambda)J(u_1) \tag{1.4}$$

for all $u_0, u_1 \in V$, $\lambda \in (0,1)$. A functional J is strictly convex if it is convex and the equality in (1.4) does not occur for $u_0 \neq u_1$.

Let J be a convex and differentiable functional. We shall prove that

$$J(u_0) - J(u_1) \geq \delta J_{u_1}(u_0 - u_1) \quad \forall\, u_0, u_1 \in V \ . \tag{1.5}$$

In fact, it follows from (1.4) that

$$J\big(u_1 + \lambda(u_0 - u_1)\big) - J(u_1) \leq \lambda\big(J(u_0) - J(u_1)\big) \ .$$

Dividing both sides of the above inequality by λ and passing to the limit as $\lambda \to 0$ we get

$$\delta J_{u_1}(u_0 - u_1) \leq J(u_0) - J(u_1)$$

which is the required inequality (1.5).

As was mentioned before, if $K \subset V$ is a convex set, then the variational inequality

$$u_0 \in K \ : \ \delta J_{u_0}(u_1 - u_0) \geq 0 \quad \forall\, u_1 \in K \tag{1.6}$$

is a necessary condition for the functional J to attain its minimum over the set K at $u_0 \in K$. Now we prove that condition (1.6) is also sufficient provided J is convex . In fact, let us assume that (1.6) holds. Then it follows from (1.5) that

$$J(u_1) - J(u_0) \geq \delta J_{u_0}(u_1 - u_0) \geq 0 \quad \forall\, u_1 \in K \ .$$

This means that

$$J(u_1) - J(u_0) \geq 0 \quad \forall\, u_1 \in K \ .$$

Thus, we have proved the following result.

Theorem 1.2. *Inequality (1.6) is a necessary and sufficient condition for a convex and differentiable functional J to attain its minimum over K at $u_0 \in K$.*

We have shown that (1.5) holds for convex and differentiable functionals. Now we shall prove the converse, namely, that (1.5) implies the convexity of J. To this end, let us take any $u, u_0 \in V$ and substitute $u_1 = (1-\lambda)u + \lambda u_0$ into inequality (1.5). This gives

$$J(u_0) - J\big((1-\lambda)u + \lambda u_0\big) \geq \delta J_{(1-\lambda)u+\lambda u_0}\big[(1-\lambda)(u_0 - u)\big] \ . \tag{1.7}$$

In the same way we get

$$J(u) - J\big((1-\lambda)u + \lambda u_0\big) \geq \delta J_{(1-\lambda)u+\lambda u_0}\big[-\lambda(u_0 - u)\big] \ . \tag{1.8}$$

Multiplying (1.7) and (1.8) by λ and $(1-\lambda)$, respectively, and summing up the resulting inequalities we obtain

$$\lambda J(u_0) + (1-\lambda)J(u) - J\big((1-\lambda)u + \lambda u_0\big) \geq 0 \ . \tag{1.9}$$

This means that functional J is convex.

An analogous result can be proved for J being strictly convex. Namely, if J is a strictly convex functional, then

$$J(u_0) - J(u_1) > \delta J_{u_1}(u_0 - u_1) \ , \quad u_0 \neq u_1 \ , \tag{1.10}$$

and the converse is also true. To prove this we need the following lemma. Its proof will be given below.

Lemma 1.1. *For every convex and differentiable functional* J *the function* $\lambda^{-1}[J(u+\lambda u_0) - J(u)]$ *is nondecreasing with respect to* λ.

Now let us suppose that inequality (1.10) holds. Hence, inequalities (1.7) and (1.8) will be strict whenever $u_0 \neq u$. This means that inequality (1.9) will also be strict. To prove the converse implication let us take a strictly convex functional J. Then

$$J\big(u + \lambda(u_0 - u)\big) < (1-\lambda)J(u) + \lambda J(u_0) \ , \quad u \neq u_0 \ .$$

And consequently,

$$\frac{J\big(u + \lambda(u_0 - u)\big) - J(u)}{\lambda} < J(u_0) - J(u) \ .$$

In view of the above lemma, we infer that the left-hand side of the above inequality converges from above to $\delta J_u(u_0 - u)$. Thus,

$$\delta J_u(u_0 - u) < J(u_0) - J(u) \ ,$$

which proves the result.

Proof of Lemma 1.1. Let us consider the function

$$\varphi(\lambda) = \frac{J(u+\lambda u_0) - J(u)}{\lambda} \ .$$

We shall prove that $\varphi'(\lambda) \geq 0$. We have

$$\varphi'(\lambda) = \frac{\lambda \frac{d}{d\lambda} J(u+\lambda u_0) - \big[J(u+\lambda u_0) - J(u)\big]}{\lambda^2} \ . \tag{1.11}$$

It can easily be found that

$$\frac{d}{d\lambda} J(u+\lambda u_0) = \delta J_{u+\lambda u_0}(u_0) \ .$$

Thus, it follows from (1.11) that

$$\varphi'(\lambda) = \frac{\lambda \delta J_{u+\lambda u_0}(u_0) - J(u+\lambda u_0) + J(u)}{\lambda^2} \ .$$

Due to (1.5), the numerator in the above formula is nonnegative, and consequently $\varphi'(\lambda) \geq 0$, which is the required conclusion. □

1.4 Derivative of a convex functional

Let V be a normed space and let V^* be its dual. An operator $A : V \mapsto V^*$ is said to be monotone if

$$(Au - Au_1)(u - u_1) \geq 0 \quad \forall\, u, u_1 \in V \ .$$

As pointed out in Section 1.1, one may consider the operator $\delta J : V \mapsto V^*$ which to every element $u \in V$ assigns the derivative δJ_u of J at u.

Theorem 1.3. *A differentiable functional J is convex if and only if δJ is a monotone operator.*

Proof. **Necessity.** In the previous section we have proved that

$$\begin{aligned} J(u) - J(u_1) &\geq \delta J_{u_1}(u - u_1) \ , \quad \text{and} \\ J(u_1) - J(u) &\geq \delta J_u(u_1 - u) \ . \end{aligned}$$

Summing up the above inequalities we obtain

$$(\delta J_u - \delta J_{u_1})(u - u_1) \geq 0 \quad \forall\, u, u_1 \in V \ , \tag{1.12}$$

which proves the assertion.

Sufficiency. Suppose that inequality (1.12) holds. Let us consider the function

$$\varphi(\lambda) = J\big(u_1 + \lambda(u - u_1)\big) \ , \quad \lambda \in (0,1) \ ,$$

for fixed elements $u, u_1 \in V$. We shall prove that

$$\Delta \equiv \lambda J(u) + (1 - \lambda)J(u_1) - J\big(\lambda u + (1 - \lambda)u_1\big)$$

is nonnegative. In view of the well-known result on finite differences we obtain

$$\begin{aligned} \Delta &= \lambda\big[\varphi(1) - \varphi(\lambda)\big] + (1 - \lambda)\big[\varphi(0) - \varphi(\lambda)\big] \\ &= \lambda(1 - \lambda)\varphi'(\xi_1) - \lambda(1 - \lambda)\varphi'(\xi_2) \ , \end{aligned}$$

where $\xi_1 \in (\lambda, 1)$, $\xi_2 \in (0, \lambda)$; and therefore $\xi_1 > \xi_2$. Thus

$$\Delta = \lambda(1 - \lambda)\big[\varphi'(\xi_1) - \varphi'(\xi_2)\big] \ .$$

On the other hand,

$$\varphi'(\xi_i) = \delta J_{u_1 + \xi_i(u - u_1)}(u - u_1) \ , \quad i = 1, 2 \ .$$

Hence, by (1.12), we can write

$$\varphi'(\xi_1) - \varphi'(\xi_2) = \big(\delta J_{u_1 + \xi_1(u - u_1)} - \delta J_{u_1 + \xi_2(u - u_1)}\big)(u - u_1) \geq 0 \ .$$

This means that $\Delta \geq 0$, which proves the convexity of J. □

Let $K \subset V$ be a convex set and let J be a convex and differentiable functional. It has been shown in Section 1.3 that the problem of minimization of J over K is equivalent to the following variational inequality

$$u \in K : \quad \delta J_u(u_1 - u) \geq 0 \quad \forall\, u_1 \in K \; . \tag{1.13}$$

Variational inequality (1.13) can be expressed in another equivalent way. Namely, the following theorem is valid.

Theorem 1.4. *Let $\delta J : V \mapsto V^*$ be a continuous mapping. Inequality (1.13) is fulfilled if and only if*

$$u \in K : \quad \delta J_{u_1}(u_1 - u) \geq 0 \quad \forall\, u_1 \in K \; . \tag{1.14}$$

Proof. By the convexity of J, the operator δJ is monotone. Thus,

$$(\delta J_u - \delta J_{u_1})(u - u_1) \geq 0 \quad \forall\, u, u_1 \in K \; . \tag{1.15}$$

If an element u satisfies the variational inequality (1.13), then by summing up (1.13) and (1.15) we obtain (1.14). To prove the converse implication we assume (1.14). Choosing $u_1 = (1-\lambda)u + \lambda u_0$, $u_0 \in K$, it is easily seen that

$$\lambda \delta J_{(1-\lambda)u + \lambda u_0}(u_0 - u) \geq 0 \; .$$

After dividing both sides of the above inequality by λ and passing to the limit as $\lambda \to 0$ it follows from the continuity of δJ that

$$\delta J_u(u_0 - u) \geq 0 \quad \forall\, u_0 \in K$$

which completes the proof. □

1.5 Minimization problems for nonsmooth functionals

As before, let $K \subset V$ be a convex set and let $J : V \mapsto \mathbb{R}$ be a convex functional. Furthermore, assume that J is represented as the sum of a differentiable functional and a nondifferentiable one, ie. $J = J_1 + J_2$, where J_1, J_2 are convex, J_1 is differentiable and $\delta J_1 : V \mapsto V^*$ is continuous. We study the problem

$$\inf_{u \in K} J(u) \; . \tag{1.16}$$

Theorem 1.5. *The following three conditions are equivalent:*

$$u \in K \quad \text{is a solution to problem (1.16)} \; , \tag{1.17}$$

$$u \in K \; : \quad \delta J_{1u}(u_0 - u) + J_2(u_0) - J_2(u) \geq 0 \quad \forall\, u_0 \in K \; , \tag{1.18}$$

$$u \in K \; : \quad \delta J_{1u_0}(u_0 - u) + J_2(u_0) - J_2(u) \geq 0 \quad \forall\, u_0 \in K \; . \tag{1.19}$$

Proof. First we prove that (1.17) is equivalent to (1.18). Let $u \in K$ be a solution to problem (1.16). Then

$$J(u) \le J\big((1-\lambda)u + \lambda u_0\big) \quad \forall\, u_0 \in K \ , \quad \lambda \in (0,1) \ .$$

In view of the convexity of J_2, this implies that

$$J_1(u) + J_2(u) \le J_1\big((1-\lambda)u + \lambda u_0\big) + (1-\lambda)J_2(u) + \lambda J_2(u_0) \ .$$

Dividing the above inequality by λ we get

$$\frac{1}{\lambda}\big[J_1\big((1-\lambda)u + \lambda u_0\big) - J_1(u)\big] + J_2(u_0) - J_2(u) \ge 0$$

and after passing to the limit as $\lambda \to 0$ (1.18) follows. To prove the converse, let us assume that u satisfies the variational inequality (1.18). Then, by the convexity of J_1, one has

$$J_1(u_0) - J_1(u) \ge \delta J_{1u}(u_0 - u) \quad \forall\, u_0 \in K \ .$$

By summing up the above inequality and (1.18) we find that

$$J(u_0) - J(u) \ge 0 \quad \forall\, u_0 \in K \ .$$

This means that u solves problem (1.16).

Let us now prove that (1.18) is equivalent to (1.19). Assume that (1.18) holds. It follows from the monotonicity of δJ_1 that

$$(\delta J_{1u_0} - \delta J_{1u})(u_0 - u) \ge 0 \ .$$

Now, in view of (1.18), inequality (1.19) follows. To prove the converse, suppose that (1.19) is satisfied. By taking $(1-\lambda)u + \lambda u_1 = u_0$, $u_1 \in K$, we get

$$\lambda \delta J_{1[(1-\lambda)u+\lambda u_1]}(u_1 - u) + J_2\big((1-\lambda)u + \lambda u_1\big) - J_2(u) \ge 0 \ .$$

By the convexity of J_2, the left-hand side of the above inequality can be estimated from above. After dividing the resulting inequality by λ and passing to the limit as $\lambda \to 0$ we obtain (1.18), which completes the proof. □

1.6 Weak convergence. Compactness principles

Let V be a normed space and let V^* be its dual. A sequence of elements $u_n \in V$ weakly converges to an element u if for each $u^* \in V^*$

$$u^*(u_n) \to u^*(u) \ , \quad n \to \infty \ .$$

One may consider also the dual space of V^*, denoted by V^{**} . For any element $u \in V$, the functional $u^* \mapsto u^*(u)$, where $u^* \in V^*$, is linear and continuous on V^* and thus is an element of the space V^{**}. For every element $u \in V$, the functional $u^{**} \in V^{**}$

$$u^{**}(u^*) = u^*(u) \quad \forall\, u^* \in V$$

establishes an imbedding of the space V into its second dual V^{**}. The imbedding operator will be denoted by π. If $\pi V = V^{**}$, the space V is called reflexive. The spaces $L^p(\Omega)$, $1 < p < \infty$, possess the reflexivity property, ie.

$$\left(L^p(\Omega)\right)^* = L^q(\Omega) \ , \quad \frac{1}{p} + \frac{1}{q} = 1 \ .$$

The spaces $L^1(\Omega)$ and $L^\infty(\Omega)$ are not reflexive; the dual space of $L^1(\Omega)$ is $L^\infty(\Omega)$ while the dual of $L^\infty(\Omega)$ is larger than $L^1(\Omega)$. The notion of weak-(∗) convergence introduced below is used for spaces which are not reflexive. A sequence of elements $u_n^* \in V^*$ is called weakly-(∗) convergent to an element $u^* \in V^*$ if for each $u \in V$

$$u_n^*(u) \to u^*(u) \ , \quad n \to \infty \ .$$

Some compactness properties are related to reflexivity. We recall two of them.

Theorem 1.6. *A bounded set of a reflexive Banach space is weakly compact.*

Theorem 1.7. *For a separable normed space, a bounded set of the dual space is weakly-(∗) compact.*

The term "weakly compact" ("weakly-(∗) compact") means that every bounded sequence contains a weakly (weakly-(∗)) convergent subsequence.

In particular, balls in the space $L^\infty(\Omega)$ are weakly-(∗) compact.

1.7 Weak semicontinuity of functionals

A functional $J : V \mapsto \mathbb{R}$ is called weakly lower semicontinuous at u if the condition

$$u_n \to u \quad \text{weakly}$$

implies

$$\underline{\lim}\, J(u_n) \geq J(u) \ .$$

A simple example of a weakly lower semicontinuous functional on V is provided by the norm. In fact, taking $J(u) = \|u\|$ we get that if $u_n \to u$ weakly in V, then $\underline{\lim}\ \|u_n\| \geq \|u\|$. When comparing weak convergence with the strong convergence (convergence in norm) it is evident that if $u_n \to u$ strongly in V, then $u_n \to u$ weakly in V. It is also worth noting here that in general the continuity of a functional does not imply its weak lower semicontinuity. Clearly, weak lower semicontinuity does not imply continuity. But any weakly lower semicontinuous functional is strongly lower semicontinuous, ie. the condition

$$u_n \to u \quad \text{strongly in} \quad V$$

implies

$$\underline{\lim}\ J(u_n) \geq J(u) \ .$$

This property follows from the fact that every strongly convergent sequence u_n is weakly convergent. Now we prove two theorems providing sufficient conditions for weak lower semicontinuity of functionals.

Theorem 1.8. *Suppose that*

$$J(u) - J(u_0) \geq \delta J_{u_0}(u - u_0) \quad \forall\, u \in V \tag{1.20}$$

holds at $u_0 \in V$. Then the functional J is weakly lower semicontinuous at u_0. If (1.20) is fulfilled for every $u, u_0 \in V$, then J is weakly lower semicontinuous at every $u_0 \in V$.

Proof. Let $u_n \to u_0$ weakly in V. Since $\delta J_{u_0} \in V^*$, we have

$$\delta J_{u_0}(u_n - u_0) \to 0 \ .$$

It follows from (1.20) that

$$\underline{\lim}\ \big[J(u_n) - J(u_0)\big] \geq \lim\ \delta J_{u_0}(u_n - u_0) = 0 \ .$$

Hence,

$$\underline{\lim}\ J(u_n) \geq J(u_0)$$

which proves the theorem. □

Theorem 1.9. *Let J satisfy the condition*

$$(\delta J_u - \delta J_{u_0})(u - u_0) \geq 0 \quad \forall\, u, u_0 \in V \ . \tag{1.21}$$

Then J is weakly lower semicontinuous on V.

Proof. It was proved in Section 1.4 that inequality (1.21) is equivalent to the convexity of J. On the other hand, the convexity of J is equivalent to (1.21). Hence, the result follows from the above theorem. □

1.8 Existence of solutions to the minimization problem

A functional $J : V \mapsto \mathbb{R}$ is coercive if

$$J(u) \to +\infty \quad \text{when} \quad \|u\| \to +\infty \ .$$

A set $M \subset V$ is called weakly closed, if the conditions $u_n \to u$ weakly, $u_n \in M$, imply $u \in M$.

Theorem 1.10. *A closed convex set of a reflexive Banach space is weakly closed.*

Now, let us assume that V is a reflexive space and $K \subset V$ is a closed convex set. Below we prove the existence result for solutions to the minimization problem.

Theorem 1.11. *Let a functional J be coercive and weakly lower semicontinuous. Then the problem*

$$\inf_{u \in K} J(u) \tag{1.22}$$

has a solution.

Proof. Let us take a minimizing sequence u_n, ie. a sequence possessing the property

$$J(u_n) \to \inf_{u \in K} J(u) \ .$$

Since J is coercive, the sequence u_n is bounded, ie.

$$\|u_n\| \leq c \ ,$$

where c is independent of n. For, if not, the sequence u_n would not be minimizing. By the reflexivity of V, one may choose a subsequence u_i of the sequence u_n such that $u_i \to u$ weakly in V. In view of Theorem 1.10, and since $u_i \in K$, we have $u \in K$. Putting

$$d = \inf_{u \in K} J(u)$$

we obtain

$$d = \underline{\lim}\, J(u_n) = \underline{\lim}\, J(u_i) \geq J(u) \geq d \ .$$

The first inequality here follows from the weak lower semicontinuity of J. Thus, we have found an element u such that

$$d = J(u) \ , \quad u \in K$$

which means that u solves problem (1.22). This proves the assertion of the theorem. □

1.9 The case of a Hilbert space

Let V be a Hilbert space with the norm defined by the inner product

$$\|u\|^2 = (u, u) \ .$$

Consider a bilinear continuous functional $B : V \times V \mapsto \mathbb{R}$, $B(u, u_1) = B(u_1, u)$, $B(u, u) \geq 0$ for all $u \in V$. Let $F : V \mapsto \mathbb{R}$ be a linear continuous functional and let $K \subset V$ be a closed convex set. We consider the functional

$$J(u) = \frac{1}{2}\, B(u, u) - F(u)$$

and we investigate the minimization problem

$$\inf_{u \in K} \ J(u) \ . \tag{1.23}$$

Theorem 1.12. *A solution to problem (1.23) exists if and only if there exists a solution of the variational inequality*

$$u \in K \ : \quad B(u, \bar{u} - u) \geq F(\bar{u} - u) \quad \forall\, \bar{u} \in K \ . \tag{1.24}$$

Proof. First, we find the derivative δJ_u of J. It is easy to see that, by the symmetry of B,

$$\lim_{\lambda \to 0} \ \frac{J(u + \lambda \bar{u}) - J(u)}{\lambda} = B(u, \bar{u}) - F(\bar{u}) \ .$$

Thus,

$$\delta J_u(\bar{u}) = B(u, \bar{u}) - F(\bar{u}) \ .$$

Hence, the mapping $u \mapsto \delta J_u$ is monotone which entails that the functional J is convex. As was proved in Section 1.3, this, in turn, ensures the solvability of (1.23) if and only if the inequality

$$u \in K \ : \quad \delta J_u(\bar{u} - u) \geq 0 \quad \forall\, \bar{u} \in K \tag{1.25}$$

is solvable. Taking into account the above formula for δJ_u, we get the conclusion. □

Let us now formulate sufficient conditions for the solvability of (1.24). We assume that there exists a constant $c > 0$ such that

$$B(u, u) \geq c \|u\|^2 \ .$$

We shall prove that all the assumptions of Theorem 1.11 are satisfied which will allow us to deduce the solvability of problem (1.23) and, consequently, that of (1.24). To this end we recall the Riesz theorem.

Theorem 1.13. *Every linear and continuous functional G defined on a Hilbert space V has a unique representation*

$$G(u) = (g, u) \quad \forall\, u \in V \ , \tag{1.26}$$

and, moreover, $\|G\|_* = \|g\|$.

According to our assumptitions, for each $u \in V$ the function $B(u, u_0)$ is linear and continuous at u_0. By the Riesz theorem, there exists an element Tu such that

$$B(u, u_0) = (Tu, u_0) \quad \forall\, u_0 \in V \ .$$

Let us take any sequence u_n, $u_n \to u$ weakly. Then

$$\begin{aligned} B(u_n, u_n) &= B(u, u) + 2B(u, u_n - u) + B(u_n - u, u_n - u) \\ &= B(u, u) + 2(Tu, u_n - u) + B(u_n - u, u_n - u) \ . \end{aligned}$$

The second term of the latter sum converges to zero because of the weak convergence of the sequence u_n, the third term is nonnegative. Hence

$$\underline{\lim}\, B(u_n, u_n) \geq B(u, u) \ . \tag{1.27}$$

Moreover, there exists an element $f \in V$ such that

$$F(u) = (f, u) \ . \tag{1.28}$$

It follows also from (1.27) that the functional J is weakly lower semicontinuous. By the Cauchy inequality,

$$|F(u)| \leq \|f\| \cdot \|u\| \ ,$$

the functional J is coercive since

$$J(u) \geq \frac{c}{2} \|u\|^2 - \|f\| \cdot \|u\| \to +\infty \ , \quad \text{whenever} \quad \|u\| \to +\infty \ .$$

The Riesz theorem guarantees the reflexivity of the Hilbert space V. In this way we have proved that all the assumptions of Theorem 1.11 are satisfied and thus we obtain the existence of a solution to problem (1.24). It is easily seen that a solution is unique.

1.10 Elements of measure theory

Let $\Omega \subset \mathbb{R}^N$ be a bounded domain. The least $\mathfrak{S}$-algebra containing all compacts from Ω is called the Borel $\mathfrak{S}$-algebra. A nonnegative $\mathfrak{S}$-additive real-valued function defined on the Borel $\mathfrak{S}$-algebra which is finite for all compacts $K \subset \Omega$ will be called a nonnegative measure. If not misleading, the term "nonnegative" will be sometimes omitted. If ν is a nonnegative measure and $\nu(\Omega) < +\infty$, then ν is called finite. Let Ω_ν be the largest open set such that $\nu(\Omega_\nu) = 0$. The set $\text{spt}\nu = B_\nu = \Omega \backslash \Omega_\nu$ (closed with respect to Ω) is called the support of the nonnegative measure ν. Obviously, Ω_ν may be empty. A measure ν is finite if B_ν is compact. A nonnegative measure ν is concentrated on a Borel set B if $\nu(\Omega \backslash B) = 0$. In particular, a measure ν may be concentrated on a subset of the support B_ν.

Any $\mathfrak{S}$-additive real-valued function defined on the Borel $\mathfrak{S}$-algebra and finite for all compacts $K \subset \Omega$ is called a measure on Ω.

For every measure ν, there exists a decomposition of the domain Ω into the union of two disjoined Borel sets Ω^+ and Ω^- such that $\nu(B) \geq 0$ for all $B \subset \Omega^+$, $\nu(B) \leq 0$ for all $B \subset \Omega^-$. In general, the decomposition $\Omega = \Omega^+ \cup \Omega^-$ is not unique, but the nonnegative measures ν^+, ν^- defined by the formulae

$$\nu^+(B) = \nu(B \cap \Omega^+) \ , \quad \nu^-(B) = -\nu(B \cap \Omega^-)$$

do not depend upon the choice of Ω^+, Ω^-. The nonnegative measure $|\nu| = \nu^+ + \nu^-$ is called the full variation of the measure ν.

Let $f(x)$ be a function locally integrable with respect to the Lebesgue measure. This function defines the measure

$$\nu(B) = \int_B f\,(x)dx \ ,$$

which is called absolutely continuous with respect to the Lebesgue measure. In particular, we have $\nu(B) = 0$ for every set B of zero Lebesgue measure. The function $f(x)$ is called the density of the measure ν.

Denote by $C_0(\Omega)$ the space of all continuous functions with compact supports defined on Ω. In this space convergence is defined as follows. A sequence $\varphi_n \in C_0(\Omega)$ converges to $\varphi \in C_0(\Omega)$ if the supports of the functions φ_n belong to a certain compact subset of Ω and, moreover, φ_n converges uniformly to φ. The set of all measures defined on Ω will be denoted by $\mathcal{M}_\Omega$. It can be proved that the formula

$$F(\varphi) = \int_\Omega \varphi(x) d\nu(x) \ , \quad \nu \in \mathcal{M}_\Omega \ , \tag{1.29}$$

defines a linear continuous functional on $C_0(\Omega)$. In fact, let $\varphi_n \to \varphi$. Then

$$\begin{aligned} |F(\varphi_n) - F(\varphi)| &\leq \int_K |\varphi_n(x) - \varphi(x)| \, d\,|\nu|(x) \\ &\leq |\nu|(K) \max_K |\varphi_n(x) - \varphi(x)| \ . \end{aligned}$$

Moreover, formula (1.29) gives a general form of linear continuous functionals on $C_0(\Omega)$. It means that for every linear and continuous functional F on $C_0(\Omega)$ there exists a unique measure $\nu \in \mathcal{M}_\Omega$ such that (1.29) holds. Hence, functionals can be identified with measures and represented in the form

$$\nu(\varphi) = \int_\Omega \varphi(x)d\nu(x) \ , \quad \nu \in \mathcal{M}_\Omega \ .$$

A linear and continuous functional F on $C_0(\Omega)$ is said to be positive, if $F(\varphi) \geq 0$ for all $\varphi \in C_0(\Omega)$, $\varphi(x) \geq 0$. Although in this case the term "nonnegative" would be more proper we accept the term "positive" which is more common. A linear and continuous functional is positive if and only if in formula (1.29) the measure ν is nonnegative. Positive functionals on $C_0(\Omega)$ possess an important property formulated below.

Theorem 1.14. *Let a functional F be linear and positive on $C_0(\Omega)$. Then F is continuous and can be represented in the form (1.29) with a measure ν being nonnegative.*

Proof. Since F is linear, it suffices to prove its continuity at zero . Let us choose a sequence φ_n converging to zero in $C_0(\Omega)$. Let $K \subset \Omega$ be a compact subset of Ω containing all supports of φ_n and let $\xi \in C_0(\Omega)$ be a function such that $0 \leq \xi \leq 1$, $\xi \equiv 1$ on K. Denote

$$\delta_n = \max_K |\varphi_n(x)| \ .$$

Then

$$-\delta_n \xi(x) \leq \varphi_n(x) \leq \delta_n \xi(x) \ .$$

Hence

$$-\delta_n F(\xi) \leq F(\varphi_n) \leq \delta_n F(\xi) \ .$$

This means that

$$|F(\varphi_n)| \leq \delta_n F(\xi) \ .$$

Since $\delta_n \to 0$, we get the assertion of the theorem. □

In the space of measures there exist different concepts of convergence. We shall use, mainly, the concept of weak-(*) convergence. Namely, a sequence of measures ν_n weakly-(*) converges to a measure ν, if for each function $\varphi \in C_0(\Omega)$

$$\int_\Omega \varphi(x)d\nu_n(x) \to \int_\Omega \varphi(x)d\nu(x) \ .$$

This convergence is sometimes called weak convergence. From our point of view the term weak-(*) convergence is more justified.

It is worth noting that the weak-(*) limit of nonnegative measures is a nonnegative measure.

Let us now formulate some results characterizing weak-(*) convergence of nonnegative measures. Assume that a set S is dense in the space $C_0(\Omega)$ in the following sense: for every element $\varphi \in C_0(\Omega)$ there exists a sequence $\varphi_n \in S$ such that $\varphi_n \to \varphi$ in $C_0(\Omega)$. An important example of a dense set in $C_0(\Omega)$ is the space $C_0^\infty(\Omega)$ of infinitely differentiable functions with compact supports. An important fact stated below says that it is enough to check weak-$(*)$ convergence on dense subsets of $C_0(\Omega)$ only.

Theorem 1.15. *Let a set S be dense in $C_0(\Omega)$ and for every element $\varphi \in S$*

$$\nu_n(\varphi) \to \nu(\varphi) \ ,$$

where ν_n, ν are nonnegative measures. Then

$$\nu_n \to \nu \quad \text{weakly-}(*) \ .$$

Theorem 1.16. *Let a set S be dense in $C_0(\Omega)$ and let, for every element $\varphi \in S$, the sequence $\{\nu_n(\varphi)\}$ be fundamental for some nonnegative measures ν_n. Then, the nonnegative measures ν_n converge weakly-$(*)$ to some nonnegative measure ν.*

Below we formulate two theorems characterizing semicontinuity of measures.

Theorem 1.17. *Let a sequence of nonnegative measures ν_n weakly-$(*)$ converge to ν and let $K \subset \Omega$ be compact. Then*

$$\overline{\lim}\, \nu_n(K) \le \nu(K) \ .$$

Theorem 1.18. *Let a sequence of nonnegative measures ν_n weakly-$(*)$ converge to ν and let $G \subset \Omega$ be an open set. Then*

$$\underline{\lim}\, \nu_n(G) \ge \nu(G) \ .$$

A relatively compact set B is normal with respect to a nonnegative measure ν if the ν-measure of its boundary ∂B is zero.

Theorem 1.19. *Let a sequence of nonnegative measures ν_n converge weakly-$(*)$ to ν. Then, for every normal set B,*

$$\lim\ \nu_n(B) = \nu(B) \ .$$

Proof. Define $\overline{B} = B \cup \partial B$, $\overset{\circ}{B} = B \backslash \partial B$. By the normality of B, we have $\nu(B) = \nu(\overline{B}) = \nu(\overset{\circ}{B})$. It follows from the above theorems that

$$\underline{\lim}\, \nu_n(B) \ge \underline{\lim}\, \nu_n(\overset{\circ}{B}) \ge \nu(\overset{\circ}{B}) = \nu(B) \ ,$$
$$\overline{\lim}\, \nu_n(B) \le \overline{\lim}\, \nu_n(\overline{B}) \le \nu(\overline{B}) = \nu(B) \ .$$

Hence the result follows. □

Now we formulate a criterion for weak-(*) compactness of nonnegative measures. A set L of nonnegative measures is called weakly-(*) bounded if for every element $\varphi \in C_0(\Omega)$ there exists a constant $c(\varphi)$ such that

$$\left|\nu(\varphi)\right| \le c(\varphi)$$

for all $\nu \in L$.

Theorem 1.20. *Every weakly-(*) bounded set L of nonnegative measures is weakly-(*) compact.*

Let $K \subset \mathbb{R}^N$ be compact. Consider the Banach space $C(K)$ of all continuous functions on K with the norm

$$\|\varphi\| = \sup_K \left|\varphi(x)\right| .$$

Denote by $\mathcal{M}_K$ the set of all measures defined on the $\mathfrak{S}$-algebra of Borel subsets of K satisfying the condition $|\nu|(K) < +\infty$. The following theorem provides us with a representation for elements of the dual space of $C(K)$.

Theorem 1.21. *Any linear continuous functional F defined on $C(K)$ can be represented in the form*

$$F(\varphi) = \int_K \varphi(x) d\nu(x) , \quad \nu \in \mathcal{M}_K ,$$

and, moreover, the representation is unique, $\|F\| = |\nu|(K)$.

Let us now introduce the notion of a subharmonic function. The domain $\Omega \subset \mathbb{R}^N$ is assumed to be bounded. We denote by $\rho(x)$ the distance from x to the boundary $\partial\Omega$. Let $u(x)$ be a function locally integrable in Ω and taking values in the extended reals $\overline{\mathbb{R}}$. For $x \in \Omega$ and $r > 0$, $r < \rho(x)$, we define

$$(\pi u)(x,r) = \frac{1}{|B_r(x)|} \int_{B_r(x)} u(y) dy ,$$

where $B_r(x)$ is a ball of radius r and center x. It follows from the rules of differentiability of integrals that, for almost all $x \in \Omega$, $\lim_{r \to 0} (\pi u)(x,r)$ exists and is equal to $u(x)$ almost everywhere in Ω.

A function $u : \Omega \mapsto \overline{\mathbb{R}}$ is called subharmonic if it is locally integrable in Ω and, for every $x \in \Omega$, the function $(\pi u)(x,r)$ is nondecreasing with respect to r on the segment $\big(0, \rho(x)\big)$ and

$$\lim_{r \to 0} (\pi u)(x,r) = u(x) , \quad x \in \Omega .$$

It is clear that any subharmonic function satisfies the inequalities $-\infty \le u(x) < +\infty$ for all $x \in \Omega$.

It follows from the definition that, if for every $x \in \Omega$, the function $(\pi u)(x,r)$ is nondecreasing with respect to r, then the function $\bar{u}(x) = \lim_{r\to 0}(\pi u)(x,r)$ has the property $\bar{u}(x) = u(x)$ for almost all $x \in \Omega$. Hence, $(\pi\bar{u})(x,r) = (\pi u)(x,r)$ for all x, r. Obviously, the function $\bar{u}$ is subharmonic. Thus, if for locally integrable function $u : \Omega \mapsto \overline{\mathbb{R}}$ and every $x \in \Omega$, the formula $(\pi u)(x,r)$ leads to a function which is nondecreasing with respect to r, then u is equivalent to a subharmonic function.

A function $u : \Omega \mapsto \overline{\mathbb{R}}$ is upper semicontinuous at $x \in \Omega$, if for every sequence x_n converging to x, one has

$$\overline{\lim}\, u(x_n) \leq u(x) \ .$$

It should be noted here that any subharmonic function $u(x)$ is upper semicontinuous on Ω. This is a consequence of the following theorem.

Theorem 1.22. *If $u_n(x)$ is a nonincreasing sequence of upper semicontinuous functions in Ω, then the pointwise limit $u(x) = \lim u_n(x)$ is an upper semicontinuous function.*

A function u is called superharmonic if $-u$ is subharmonic.

It can be proved that the inequality $\Delta u(x) \geq 0$ holds in Ω for any subharmonic and sufficiently smooth function $u(x)$, Δ the Laplace operator. The converse is also true, namely, if $u(x)$ is a sufficiently smooth function such that $\Delta u(x) \geq 0$ in Ω, then u is a subharmonic function. When u is not sufficiently smooth one can use the following property of subharmonic functions.

Theorem 1.23. *Let $u : \Omega \mapsto \overline{\mathbb{R}}$ be a subharmonic function. Then there exists a nonnegative measure ν on Ω such that*

$$\int_\Omega \Delta\varphi(x)u(x)dx = \int_\Omega \varphi(x)d\nu(x)$$

holds for all $\varphi \in C_0^\infty(\Omega)$.

The measure mentioned in the above theorem can be given by the formula $\nu = \Delta u$, where the Laplace operator is understood in the sense of distributions. This formula, however, requires some explanation. Usually, distributions are defined as linear and continuous functionals on the space $C_0^\infty(\Omega)$. A distribution g is called positive if its values $g(\varphi)$ are nonnegative for all $\varphi \in C_0^\infty(\Omega)$, $\varphi(x) \geq 0$. A positive distribution can be regarded as a nonnegative measure. We shall prove this fact. In view of Theorem 1.14 we need to show that the function g can be extended to a linear positive functional on $C_0(\Omega)$. By the density of $C_0^\infty(\Omega)$ in $C_0(\Omega)$, for any $\varphi \in C_0(\Omega)$, there exists a sequence $\varphi_n \in C_0^\infty(\Omega)$ such that $\varphi_n \to \varphi$ in $C_0(\Omega)$. Moreover, $\{g(\varphi_n)\}$ is a convergent sequence. Indeed, let us assume that the supports of all φ_n belong to a compact $K \subset \Omega$ and a function $\xi \in C_0^\infty(\Omega)$ is given such that $0 \le \xi \le 1$, $\xi \equiv 1$ on K. Then

$$\left|\varphi_n(x) - \varphi_m(x)\right| \le \delta_{n,m}\xi(x) \ ,$$

where

$$\delta_{n,m} = \max_{\Omega} \left|\varphi_n(x) - \varphi_m(x)\right| \to 0 \ , \quad n, m \to \infty \ .$$

And consequently,

$$\left|g(\varphi_n) - g(\varphi_m)\right| \le \delta_{n,m} g(\xi) \ .$$

Therefore, the limit of $g(\varphi_n)$ exists. It is easily seen that this limit does not depend upon the choice of the approximate sequence φ_n. We accept this limit as $g(\varphi)$. Obviously, the constructed functional g is linear and positive on $C_0(\Omega)$ which completes the proof of the assertion. It allows us to give an exact meaning to the aforementioned equality $\nu = \Delta u$.

2 Mathematical models of elastic bodies. Contact problems

2.1 Linear elastic bodies and shallow shells

Let $\omega = (u_1, \ldots, u_N)$ be the displacement vector of material points of an elastic body, $\varepsilon_{ij} = \frac{1}{2}\left(\frac{\partial u_i}{\partial x_j} + \frac{\partial u_j}{\partial x_i}\right)$ be components of the strain tensor, and σ_{ij} be components of the stress tensor. A linear elastic body in the state of static equilibrium (for $N = 3$) is described by equations shown below. In the domain $\Omega \subset \mathbb{R}^3$ we seek the displacement vector ω and components of the stress tensor σ_{ij} satisfying the equilibrium equations

$$-\frac{\partial \sigma_{ij}}{\partial x_j} = f_i \ , \quad i = 1, 2, 3 \ , \tag{2.1}$$

and Hooke's law

$$\sigma_{ij} = a_{ijkl}\varepsilon_{kl} \ , \quad i, j = 1, 2, 3 \ . \tag{2.2}$$

Here f_i are given volume forces, a_{ijkl} are bounded functions which satisfy the conditions of symmetry and positive definiteness, ie.

$$a_{ijkl} = a_{jikl} = a_{klij} \ , \quad a_{ijkl}\xi_{kl}\xi_{ij} \ge c\xi_{ij}\xi_{ij} \ , \quad c > 0 \ .$$

From now on we use the summation convention over repeated indices. It is easily seen that the substitution of (2.2) into (2.1) leads to a system of linear equations with respect to the components of the displacement vector. This system is called the Lamé equations. The displacement vector can be given on the boundary Γ,

$$\omega = \omega_0 \ . \tag{2.3}$$

Also the surface force vector can be given on Γ,

$$\sigma_{ij}\nu_j = g_i \ , \quad i = 1,2,3 \ . \tag{2.4}$$

Here by (ν_1, ν_2, ν_3) we denote the components of the unit outward normal to Γ. We can consider also the mixed boundary value problem. In this problem we impose a condition like (2.3) on a part of the boundary. For the remaining part of the boundary a condition of the type (2.4) holds. Those problems can also be formulated in variational form. For instance, the problem with condition (2.3) corresponds to the minimization of the energy functional

$$J(\omega) = \frac{1}{2}\int_\Omega \sigma_{ij}\varepsilon_{ij}dx - \int_\Omega f_i u_i dx \tag{2.5}$$

over the set of functions ω whose first derivatives are square integrable in Ω and which satisfy (2.3).

A mathematical model of a linear shallow shell is derived from the three-dimensional linear model of an elastic body on the basis of some hypotheses. In the domain $\Omega \subset \mathbb{R}^2$ we want to find functions $u, v, w, N_{ij}, \varepsilon_{ij}$ satisfying the equations

$$\Delta^2 w + k_{11}N_{11} + k_{22}N_{22} = f \ , \tag{2.6}$$

$$\frac{\partial N_{ij}}{\partial x_j} = -f_i \ , \quad i = 1,2 \ , \tag{2.7}$$

$$\begin{gathered} N_{11} = \varepsilon_{11} + \sigma\varepsilon_{22} \ , \quad N_{22} = \varepsilon_{22} + \sigma\varepsilon_{11} \ , \\ N_{12} = \frac{1}{2}\left(1-\sigma\right)\varepsilon_{12} \ , \end{gathered} \tag{2.8}$$

$$\begin{gathered} \varepsilon_{11} = \frac{\partial u}{\partial x_1} + k_{11}w \ , \quad \varepsilon_{22} = \frac{\partial v}{\partial x_2} + k_{22}w \ , \\ \varepsilon_{12} = \frac{\partial u}{\partial x_2} + \frac{\partial v}{\partial x_1} \ . \end{gathered} \tag{2.9}$$

The functions u, v describe displacements of the middle surface point along tangential directions x_1, x_2, w is the normal displacement, N_{ij} are integrated stresses, Δ is the Laplace operator over x_1, x_2, σ is a constant, $0 < \sigma < \frac{1}{2}$, f, f_i are given exterior forces, and k_{11}, k_{22} are curvatures along axes x_1, x_2, respectively. For simplicity, in the above equations some coefficients are assumed to be equal one. Only essential coefficients are distinguished. When $k_{11} = k_{22} = 0$, the equations (2.6)–(2.9) correspond to the plate. The simplest boundary conditions describing clamping along Γ have the form

$$u = v = w = \frac{\partial w}{\partial \nu} = 0 \ .$$

Other boundary conditions can also be studied. As in the three-dimensional case, the above-formulated problem can be written in variational form. In such a case (2.6)–(2.9) are the Euler equations for the energy functional

$$J(\omega) = \frac{1}{2}\int_\Omega (\Delta w)^2 dx + \frac{1}{2}\int_\Omega \Big\{ \varepsilon_{11}^2 + \varepsilon_{22}^2 + 2\sigma\varepsilon_{11}\varepsilon_{22} + \frac{1}{2}(1-\sigma)\varepsilon_{12}^2 \Big\} dx - \int_\Omega F\omega dx \,, \qquad (2.10)$$

$$\omega = (u, v, w) \,, \quad F = (f_1, f_2, f) \,.$$

2.2 Mathematical models of contact problems

As was shown in Section 1, if $J : V \mapsto \mathbb{R}$ is a convex and differentiable functional and $K \subset V$ is a convex set, then the problems

$$\inf_{\omega \in K} J(\omega)$$

and

$$\omega \in K \ : \quad \delta J_\omega(\overline{\omega} - \omega) \geq 0 \quad \forall\, \overline{\omega} \in K \qquad (2.11)$$

are equivalent. For most contact problems considered in this book this equivalence holds.

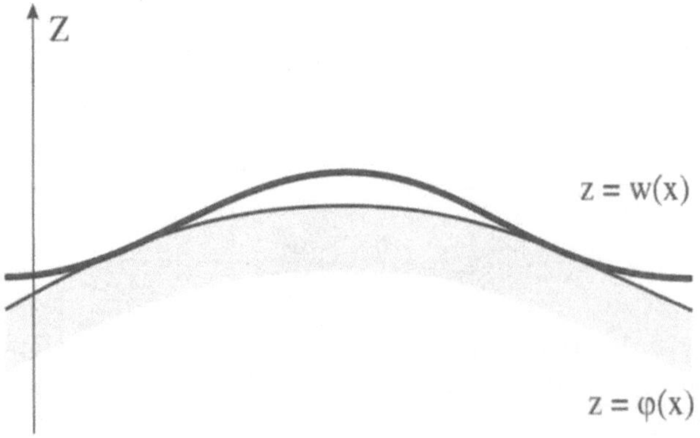

Fig. 1. Contact problem for a plate.

To illustrate this we formulate a contact problem between an elastic plate and a rigid punch in absence of tangential displacements of the plate. The choice of the set of admissible displacements K depends upon nonpenetration conditions which are of geometrical nature. In this case J denotes the energy functional for the plate.

Let $\Omega \subset \mathbb{R}^2$ be a bounded domain occupied by the middle surface of the plate. The axis z is orthogonal to the middle surface. Denote by w displacements of the plate along the axis z. Let the punch shape be described by the equation $z = \varphi(x)$, $x \in \Omega$. Displacements have to satisfy the nonpenetration condition $w(x) \geq \varphi(x)$. We introduce the set of admissible displacements

$$K = \left\{ w \in H_0^2(\Omega) \middle| w(x) \geq \varphi(x), \quad x \in \Omega \right\} .$$

The space $H_0^2(\Omega)$ is obtained as the completion in the norm of the space $H^2(\Omega)$ of infinitely differentiable functions with compact supports. Boundary conditions for w correspond to clamping along Γ. The inequality $w \geq \varphi$ holds for functions belonging to K. It has a geometrical character and can be interpreted as a mutual nonpenetration condition of the punch-plate system. It follows that displacements w satisfy the restriction $w \geq \varphi$. The situation is illustrated in Fig. 1, where the punch shape is represented by the shadowed area. The energy functional for the plate is of the form

$$J(w) = \frac{1}{2} \int_\Omega (\Delta w)^2 \, dx - \int_\Omega f w dx \ .$$

Then, inequality (2.11) can be written as

$$w \in K \ : \int_\Omega \Delta w (\Delta \overline{w} - \Delta w) \, dx \geq \int_\Omega f(\overline{w} - w) \, dx \quad \forall \, \overline{w} \in K \ .$$

Assume that a solution w is sufficiently smooth. Then this inequality is equivalent to the following relations

$$w - \varphi \geq 0 \ , \quad \Delta^2 w - f \geq 0 \ , \quad (w - \varphi)\left(\Delta^2 w - f\right) = 0 \ .$$

Thus, if the strong inequality $w(x) - \varphi(x) > 0$ is satisfied at x, then the functional J fulfills the Euler equation

$$\Delta^2 w - f = 0 \ .$$

The inequality $w(x) - \varphi(x) > 0$ represents the lack of contact at x. On the other hand, the relation $\Delta^2 w(x) - f(x) > 0$ implies that $w(x) - \varphi(x) = 0$. The latter signifies the presence of contact at x.

One should note that in the case of a shallow shell the restriction would contain all components of the displacement vector. Namely, let (u, v, w) be the displacement vector and let $z = \varphi(x)$ be the equation describing a punch shape. The nonpenetration condition can be written in the form

$$w(x) \geq \varphi\big(x + U(x)\big) \ , \quad U = (u, v) \ .$$

We assume that the displacements U are small enough to use the Taylor expansion formula. Taking into account the linear terms only we get

$$w(x) - U(x) \cdot \nabla \varphi(x) \geq \varphi(x) \ , \quad x \in \Omega \ . \tag{2.12}$$

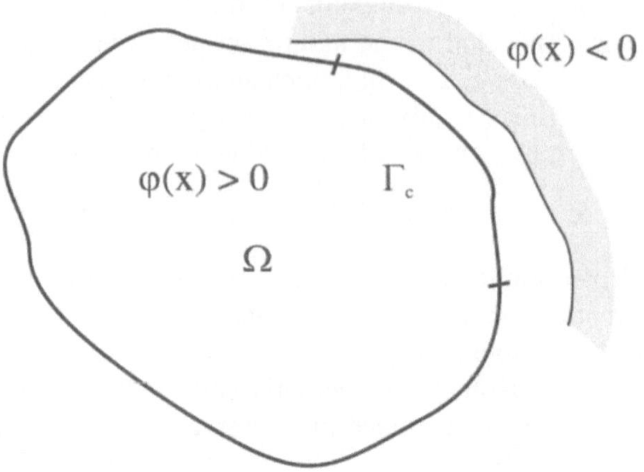

Fig. 2. The contact problem in $\mathbb{R}^3$.

In this way we find that the set of admissible displacements contains the functions (u, v, w) satisfying (2.12) in the domain Ω. The strict inequality in (2.12) represents the absence of contact at $x \in \Omega$, while the equality signifies the presence of contact.

In the examples just considered restrictions are imposed on the sought functions in the domain and boundary conditions can be fixed in advance. In the three-dimensional contact problem the situation is quite different, namely the equilibrium equations are fulfilled in the domain, while the restriction is imposed on the boundary. For example, let an elastic body occupy a bounded domain $\Omega \subset \mathbb{R}^3$. A punch shape is described by the equation $\varphi(x) = 0$ (see Fig. 2). Assume that the surface $\varphi(x)$ divides the space into two domains. The first one is characterized by the inequality $\varphi(x) > 0$, while in the second one $\varphi(x) < 0$. At $x \in \Gamma_c$ the restriction has the form

$$\varphi\big(x + \omega(x)\big) \geq 0 \ ,$$

where $\omega(x)$ is the displacement vector. The linearized condition has the form

$$\omega(x) \cdot \nabla\varphi(x) \geq -\varphi(x) \ , \quad x \in \Gamma_c \ . \tag{2.13}$$

This inequality defines admissible functions in the three-dimensional problem. An analogous condition can be obtained in contact problems for two elastic bodies. Then the energy functional is of the form (2.5). The variational inequality corresponding to the minimization of this energy functional over the set of admissible displacements can be easily derived. The same reasoning applies to the shallow shell with the energy functional determined by formula (2.11), and the set of admissible displacements given by (2.12).

As was mentioned in the preface the main peculiarity of problems with restrictions is that, in general, solutions are not smooth. A simple illustrative example is provided by the variational inequality arising in the contact problem for a beam

$$w \in K \ : \quad \int_a^b w_{xx}(\overline{w}_{xx} - w_{xx})dx \geq \int_a^b f(\overline{w} - w)dx \quad \forall\, \overline{w} \in K \ ,$$

$$K = \big\{ w \in H_0^2(a,b) \big| w(x) \geq \varphi(x) \ , \quad x \in (a,b) \big\} \ .$$

A large set of exact solutions to this problem was found in (Cimatti, 1973) for a variety of functions φ describing the punch shape. It was proved that solutions w satisfy equations of the type

$$w_{xxxx} = \sum_{i=1}^{m} \delta_i + f \ ,$$

where δ_i are Dirac measures. This example exhibits the possibility of the occurrence of concentrated interactions between a beam and a punch. In this book we investigate the problem of concentrated interactions between an elastic body and a punch in its full generality. As seen in the examples contact points can be found only after the corresponding problem is fully solved. This is the main drawback when coping with free boundary problems. In the book we investigate problems similar to those described above as well as other mathematical models of this sort.

Chapter 2
Variational Inequalities in Contact Problems of Elasticity

Solutions to the classical Signorini problem describe states of equilibrium of a linearly elastic body, a part of its boundary being supported by a nondeformable friction free surface. On the contact surface boundary conditions are expressed by the Signorini type system of inequalities. This distinguishes the Signorini problem from other formulations of contact problems where boundary conditions are of bilateral type. Those inequalities allow the boundary points to be placed at the admissible side of the boundary surface (or, in the limit case, to reach this surface). Then, tangent and normal components of the contact force are nonpositive, with the tangent component equal to zero.

A characteristic feature of the Signorini problem as well as the main difficulty when dealing with it is that the contact domain is not known a priori. This means that pointwise boundary conditions can be determined only after displacement fields are found.

From the mathematical standpoint the Signorini problem is the problem of minimizing the potential energy functional of the body over the set of admissible displacement fields. This set consists of displacement fields satisfying the contact condition that the scalar products of displacement vectors and inward normal vectors to the contact surface are nonnegative on the contact surface.

The present chapter provides a detailed analysis of the variational inequalities arising in contact problems of elasticity. For three-dimensional bodies contact problems are analyzed in the first two sections. In the next sections we deal with contact problems concerning plates and shells.

1 Contact between an elastic body and a rigid body

1.1 Problem formulation

In this section we formulate the contact problem involving an elastic body and a rigid body. This problem is a generalization of the classical Signorini problem.

We start with some preliminary assumptions and notations. By Ω, $\Omega \subset \mathbb{R}^3$, we denote the three-dimensional bounded domain occupied by the body, its boundary surface Γ being smooth. Components of the strain and the stress associated with a displacement field $\omega = (u_1, u_2, u_3)$ are given by

$$\varepsilon_{ij}(\omega) = \frac{1}{2}\left(\frac{\partial u_i}{\partial x_j} + \frac{\partial u_j}{\partial x_i}\right),$$

$$\sigma_{ij}(\omega) = c_{ijkl}\varepsilon_{kl}(\omega), \quad i,j = 1,2,3,$$

respectively. Here c_{ijkl} are components of the elasticity tensor. The symmetry conditions

$$c_{ijkl} = c_{ijlk} = c_{klij}, \quad c_{ijkl} \in C^\infty(\bar{\Omega})$$

and the positive definiteness condition

$$c_{ijkl}\sigma_{kl}\sigma_{ij} \geq c\sigma_{ij}\sigma_{ij}, \quad c > 0$$

are satisfied with a positive constant c. We have made use of the summation convention over repeated indices $i, j, k, l = 1, 2, 3$.

We restrict our considerations to the case where the elastic moduli are infinitely many times differentiable fields, ie. $c_{ijkl} \in C^\infty(\bar{\Omega})$. We assume that the boundary Γ is composed of three closed parts Γ_ω, Γ_σ, Γ_c, $\Gamma = \Gamma_\omega \cup \Gamma_\sigma \cup \Gamma_c$.

The potential energy of the body is defined by

$$\Pi(\omega) = \frac{1}{2}\int_\Omega \sigma_{ij}(\omega)\varepsilon_{ij}(\omega)dx - \int_\Omega f_i u_i dx - \int_{\Gamma_\sigma} g_i u_i d\Gamma \tag{1.1}$$

for displacement fields ω from the Sobolev space $H^1_\omega(\Omega)$ of functions whose first generalized derivatives in Ω belong to $L^2(\Omega)$ and whose traces vanish on Γ_ω. The derivative of Π at ω is denoted by $\Pi'(\omega)$. In (1.1), the functions $f_i \in L^2(\Omega)$, and $g_i \in L^2(\Gamma_\sigma)$ are given.

The shape of the nondeformable surface that supports the body (ie. the surface of the punch) is given by the equation $\Phi(x) = 0$, and the admissible region is defined by the inequality $\Phi(x) \geq 0$. We assume that $|\nabla\Phi(x)| \neq 0$ for $x \in \Gamma_c$.

Let K be the set of functions belonging to $H^1_\omega(\Omega)$ that satisfy the inequality

$$\omega(x) \cdot \nabla\Phi(x) \geq -\Phi(x), \quad x \in \Gamma_c \ . \tag{1.2}$$

Assume that K is nonempty. A solution to the minimization problem of functional (1.1) over the set K satisfies the following variational inequality

$$\omega \in K: \quad \langle \Pi'(\omega), \chi - \omega\rangle \geq 0 \quad \forall\, \chi \in K \ . \tag{1.3}$$

Let $\nu = (\nu_1, \nu_2, \nu_3)$ be the outward normal vector to the boundary Γ. The angle between $\nabla\Phi$ and $-\nu$ is assumed to be small on Γ_c. If the shape of the nondeformable surface coincides with Γ_c, then inequality (1.2) reduces to $u_i\nu_i \leq 0$ and (1.3) takes the form of the Signorini problem in its classical formulation (cf. Fichera, 1972).

Assume now that mes$\Gamma_\omega > 0$. According to the first Korn inequality, the quadratic part of the energy functional turns out to be the squared norm of ω in

$H^1_\omega(\Omega)$. Hence, the functional Π is coercive and the existence of a solution to our minimization problem follows from the lower semicontinuity of Π on the space $H^1_\omega(\Omega)$. The solution is unique.

1.2 Regularity of solutions. Construction of measures

Solution regularity at interior points of Γ_c (relative to Γ) is a corner stone of the analysis of the problem. This subject will be dealt with in the present subsection.

With further applications in mind we start with Theorem 1.1, being actually a particular case of a result of (Nečas,1975) which applies to a larger class of elliptic operators of second order.

Theorem 1.1. *Let Φ be a function of class C^3. If $\omega \in K$ solves variational inequality (1.3), then for each $x \in \Gamma_c \backslash \partial\Gamma_c$*

$$\omega \in H^2(\mathcal{O}(x) \cap \Omega) \ ,$$

where $\mathcal{O}(x)$ is a neighbourhood of x.

Now, on subsets of Γ_c, we shall construct a measure that characterizes the interaction between the body and the contact surface. As we shall see, local properties of this measure strongly depend upon the solution regularity on Γ_c.

To get our results we need an additional requirement on Γ_c. Let $\widetilde{\Gamma}_c$ be a part of the boundary Γ, $\Gamma_c \subset \widetilde{\Gamma}_c$, and $\partial\widetilde{\Gamma}_c$ has no points in common with $\partial\Gamma_c$. Assume that $\widetilde{\gamma}_c \subset \mathbb{R}^2$ is a bounded domain with boundary $\partial\widetilde{\gamma}_c$. We require the existence of a one-to-one mapping of class C^∞ of $\widetilde{\gamma}_c \cup \partial\widetilde{\gamma}_c$ into $\widetilde{\Gamma}_c$ which transforms $\gamma_c \cup \partial\gamma_c$ onto Γ_c, where $\gamma_c \subset \widetilde{\gamma}_c$, and the boundary $\partial\gamma_c$ is of class C^1. Moreover, denoting by $x = x(y_1, y_2)$ a representation of this mapping, we require that the Jacobian matrix $\frac{\partial x}{\partial y}$ be of rank 2 at each point of $\widetilde{\gamma}_c \cup \partial\widetilde{\gamma}_c$.

The following two cases will be considered separately.

(i) $\partial\Gamma_c \subset \partial\Gamma_\sigma$.
(ii) $\partial\Gamma_c$ and $\partial\Gamma_\omega$ have a nonempty intersection.

Consider first the case where for each $x_0 \in \partial\Gamma_c$ there exists a neighbourhood $\mathcal{O}(x_0)$ such that $\mathcal{O}(x_0) \cap \Gamma \subset \Gamma_c \cup \Gamma_\sigma$. Then the following theorem is true.

Theorem 1.2. *On the $\mathfrak{S}$-algebra of Borel subsets of the boundary Γ_c there exists a nonnegative measure μ such that for each $\chi \in H^1_\omega(\Omega) \cap C(\Gamma_c)$ the representation*

$$\langle \Pi'(\omega), \chi \rangle = \int_{\Gamma_c} \frac{\chi \cdot \nabla\Phi}{|\nabla\Phi|} d\mu \tag{1.4}$$

holds.

Proof. Note first that if a vector function $\chi \in H^1_\omega(\Omega)$ fulfills the inequality $\chi\nabla\Phi \geq 0$ on Γ_c, then

$$\langle \Pi'(\omega), \chi \rangle \geq 0 \ . \tag{1.5}$$

Indeed, for $\varepsilon > 0$, we have $\omega + \varepsilon\chi \in K$, and hence by substituting $\omega + \varepsilon\chi$ as a test function into (1.3) and dividing both sides by ε we get inequality (1.5). By $\mathcal{V}$ we denote the linear manifold of functions χ_* defined on Γ_c,

$$\chi_* = \frac{\chi \cdot \nabla\Phi}{|\nabla\Phi|}, \quad \chi \in H^1_\omega(\Omega) \cap C(\Gamma_c) \ . \tag{1.6}$$

The linear functional

$$\Psi(\chi_*) = \langle \Pi'(\omega), \chi \rangle$$

is well-defined on $\mathcal{V}$. To see this, consider $\chi^1_* = \chi^2_*$ on Γ_c. According to (1.5), $\langle \Pi'(\omega), \chi^1 - \chi^2 \rangle \geq 0$. Since the opposite inequality also holds we get $\Psi(\chi^1_*) = \Psi(\chi^2_*)$, which is the required equality.

Note that by inequality (1.5), the functional Ψ is positive.

Let Ω_ε be a domain with the boundary parametrized by $x = y - \varepsilon\nu(y)$, $y \in \Gamma$, $0 \leq \varepsilon \leq \varepsilon_0$, where $\nu(y)$ is the outward normal vector at y. For sufficiently small ε_0, there exists a one-to-one correspondence between Γ and the boundary of Ω_ε.

Let γ be a function in $C^1(\bar{\Omega})$ which equals one outside $\Omega_{\frac{1}{2}\varepsilon_0}$ and vanishes on $\Omega_{\frac{3}{4}\varepsilon_0}$. Since Γ_c possesses common points with Γ_σ there exists a part Γ^0_c of the boundary Γ such that $\Gamma^0_c \subset \Gamma_c \cup \Gamma_\sigma$, $\Gamma_c \subset \Gamma^0_c$, and $\partial\Gamma_c$ is disjoint with $\partial\Gamma^0_c$. We assume that $\Gamma^0_c \subset \widetilde{\Gamma}_c$.

We construct a function $n \in C^1(\bar{\Omega}\backslash\Omega_{\varepsilon_0})$ which equals one for $x = y - \varepsilon\nu(y)$, $y \in \Gamma_c$, $0 \leq \varepsilon \leq \varepsilon_0$, and vanishes for $x = y - \varepsilon\nu(y)$, $y \in \Gamma\backslash\Gamma^0_c$, $0 \leq \varepsilon \leq \varepsilon_0$. To this end, consider a function which is identically equal to one on γ_c. Since $\partial\gamma_c$ is of class C^1, this function can be extended beyond γ_c in such way that the resulting function is bounded on a domain which contains γ_c (see Mikhailov, 1976).

Now, using the fact that there exists a one-to-one mapping of $\widetilde{\gamma}_c \cup \partial\widetilde{\gamma}_c$ onto $\widetilde{\Gamma}_c$, we can construct a function which is bounded on Γ^0_c and equals one on Γ_c. This function can be extended to the whole boundary Γ by putting its values equal to zero at respective points and then an extension can be constructed along the arcs $x = y - \varepsilon\nu(y)$ on $\bar{\Omega}\backslash\Omega_{\varepsilon_0}$. The function obtained in this way is denoted by n.

The fact that n equals one on Γ_c is not essential for the existence of the extension. Thus, for an arbitrary function belonging to $C^1(\Gamma_c)$, its extension to the whole boundary Γ such that the extended function vanishes on $\Gamma\backslash\Gamma^0_c$ can be obtained in a similar way.

We define the vector function

$$h(x) = \begin{cases} 0, & x \in \Omega_{\varepsilon_0} \\ -\nu(y)\gamma(x)n(x), & x = y - \varepsilon\nu(y), \quad y \in \Gamma, \quad 0 \leq \varepsilon \leq \varepsilon_0 \end{cases} \tag{1.7}$$

on $\bar{\Omega}$. Clearly, $h \in H^1_\omega(\Omega) \cap C(\Gamma_c)$.

Note that

$$h\frac{\nabla\Phi}{|\nabla\Phi|} \geq \delta > 0 \quad \text{on} \quad \Gamma_c, \quad \delta = \text{const} \ ,$$

since the angle between the inward normal $-\nu(y)$ and $\frac{\nabla\Phi}{|\nabla\Phi|}$ is smaller than $\frac{\pi}{2}$. In the sequel we shall also make use of other properties of the function h.

Let $\varphi \in C^1(\Gamma_c)$. Define

$$p(y) = -|\nabla\Phi|^{-1}\nu(y)\nabla\Phi(y), \quad y \in \Gamma_c \ .$$

The function φp^{-1} can be extended to the whole boundary in such a way that the resulting function is of class C^1 and is equal to zero on $\Gamma \setminus \Gamma_c^0$. Again, the latter function can be extended to $\bar{\Omega} \setminus \Omega_{\varepsilon_0}$ assuming a constant value along the arcs $x = y - \varepsilon\nu(y), \quad y \in \Gamma, \ 0 \leq \varepsilon \leq \varepsilon_0$. The function obtained will be denoted by $P(x), x \in \bar{\Omega} \setminus \Omega_{\varepsilon_0}$.

The vector function

$$\chi(x) = \begin{cases} 0, & x \in \Omega_{\varepsilon_0} \\ -\nu(y)\gamma(x)P(x), & x = y - \varepsilon\nu(y), \quad y \in \Gamma, \ 0 \leq \varepsilon \leq \varepsilon_0 \end{cases}$$

has the property that

$$\chi \in H^1_\omega(\Omega) \cap C(\Gamma_c), \quad \chi\frac{\nabla\Phi}{|\nabla\Phi|} = \varphi \quad \text{on} \quad \Gamma_c \ .$$

Thus it is readily seen that the manifold $\mathcal{V}$ encompasses all functions from the space $C^1(\Gamma_c)$, and hence its completion with respect to the sup norm $\|\cdot\|_{C(\Gamma_c)} = \max_{\Gamma_c}|\cdot|$ coincides with the space $C(\Gamma_c)$.

Consequently,

$$|\chi_*| \leq \|\chi_*\|_{C(\Gamma_c)}\frac{h \cdot \nabla\Phi}{\delta|\nabla\Phi|} \ ,$$

and, since Ψ is positive definite,

$$|\Psi(\chi_*)| \leq c\|\chi_*\|_{C(\Gamma_c)} \ , \quad c = \delta^{-1}\langle\Pi'(\omega), h\rangle \ .$$

Thus, Ψ is a linear fuctional on $C(\Gamma_c)$. Since this functional is also positive there exists a nonnegative measure μ such that (Kantorovich, Akilov, 1984)

$$\Psi(\chi_*) = \int_{C(\Gamma_c)} \chi_* d\mu \quad \forall\, \chi_* \in C(\Gamma_c)$$

and, moreover,

$$\mu(\Gamma_c) = \|\Psi\| \leq \delta^{-1}\langle\Pi'(\omega), h\rangle \ .$$

Therefore, (1.4) holds for the function χ_* defined by (1.6), where $\chi \in H^1_\omega(\Omega) \cap C(\Gamma_c)$ and the proof of Theorem 1.2 is completed. □

Now we show that the measure constructed above is absolutely continuous on $\Gamma_c \setminus \partial\Gamma_c$ with respect to the Lebesgue measure.

Let $x_0 \in \Gamma_c \setminus \partial\Gamma_c$. There exists a sufficiently small neighbourhood $\mathcal{O}(x_0)$ such that $\mathcal{O}(x_0) \cap \partial\Gamma_c = \emptyset$. Let $\xi = (\xi_1, \xi_2, \xi_3) \in H^1_\omega(\Omega) \cap C(\Gamma_c)$ be a vector field such that the support of ξ satisfies $\mathrm{spt}\xi \subset \mathcal{O}(x_0) \cap \bar{\Omega}$. By Theorem 1.2,

$$\langle \Pi'(\omega), \xi \rangle = \int_{\Gamma_c} \frac{\xi \cdot \nabla\Phi}{\mid \nabla\Phi \mid} d\mu \ . \tag{1.8}$$

Taking into account the smoothness of the solution in a vicinity of x_0, we conclude that the left-hand side of (1.8) is equal to

$$\int_\Omega \{\sigma_{kj} \frac{\partial \xi_k}{\partial x_j} - f_k \xi_k\} dx = \int_{\Gamma_c} \sigma_{kj} \nu_j \xi_k d\Gamma \ . \tag{1.9}$$

Recall that a measure γ is singular with respect to the Lebesgue measure if γ is concentrated on a set of zero Lebesgue measure. For any measure defined on the $\mathfrak{S}$-algebra of Borel subsets of Γ_c there exists a unique decomposition into the absolutely continuous and the singular parts. Thus,

$$\mu(B) = \int_B q d\Gamma + \gamma(B) \ .$$

Here q is an integrable function with respect to the Lebesgue measure, called the density of the measure μ, $B \subset \Gamma_c$ is a Borel set. The first integral is the absolutely continuous part of μ with respect to the Lebesgue measure on Γ_c.

We show that (1.8) and (1.9) imply that $\gamma \equiv 0$ on $\Gamma_c \setminus \partial\Gamma_c$. By our assumptions, $|\nabla\Phi| \neq 0$ on Γ_c. Assume that the coordinate system is chosen so that $\Phi_{x_i} \equiv \frac{\partial\Phi}{\partial x_i}(x) \neq 0$ for $x \in \mathcal{O}(x_0) \cap \Gamma_c$, $i = 1, 2, 3$. Furthermore, we can assume that ξ_i is the only nonzero component of the vector function ξ. Therefore, the right hand sides of (1.8) and (1.9) lead to the following equation, without summation over repeated indices i,

$$\int_{\Gamma_c} \xi_i \frac{\Phi_{x_i}}{\mid \nabla\Phi \mid} d\mu = \int_{\Gamma_c} \sigma_{ij} \nu_j \xi_i d\Gamma \ .$$

Hence, the singular part of the measure μ vanishes on $\mathcal{O}(x_0) \cap \Gamma_c$, and the density q is given as

$$q = \frac{\sigma_{ij} \nu_j \mid \nabla\Phi \mid}{\Phi_{x_i}} \ , \quad i = 1, 2, 3 \ . \tag{1.10}$$

Denoting $(\sigma \cdot \nu)_i = \sigma_{ij} \nu_j$ we get

$$q = \frac{\nabla\Phi \cdot \sigma \cdot \nu}{\mid \nabla\Phi \mid} \ .$$

Indeed, in view of (1.10),

$$q \equiv q \frac{\Phi^2_{x_1}}{\mid \nabla\Phi \mid^2} + q \frac{\Phi^2_{x_2}}{\mid \nabla\Phi \mid^2} + q \frac{\Phi^2_{x_3}}{\mid \nabla\Phi \mid^2} = \frac{\nabla\Phi \cdot \sigma \cdot \nu}{\mid \nabla\Phi \mid} \ .$$

Let us consider the case (ii), where $\partial\Gamma_c \cap \partial\Gamma_\omega \neq \emptyset$.

Theorem 1.3. *There exists a nonnegative measure μ defined on the $\mathfrak{S}$-algebra of Borel subsets of $\Gamma_c \setminus \partial\Gamma_c$ such that (1.4) holds for all $\chi \in H^1_\omega(\Omega) \cap C_0(\Gamma_c)$.*

By $H^1_\omega(\Omega) \cap C_0(\Gamma_c)$ we denote the subspace of the space $H^1_\omega(\Omega)$ consisting of functions whose traces on Γ_c are continuous and which have compact supports in $\Gamma_c \setminus \partial\Gamma_c$. Convergence $\varphi_n \to \varphi$ of a sequence φ_n in $C_0(\Gamma_c)$ is defined in a standard way, ie. $\mathrm{spt}\varphi_n \subset \mathcal{K}$, $\mathcal{K} \subset \Gamma_c \setminus \partial\Gamma_c$ is compact, $\varphi_n(x) \to \varphi(x)$ uniformly on $\mathcal{K}$.

The proof of Theorem 1.3 is analogous to that of Theorem 1.2 but now our functional space is different. The linear variety $\mathcal{V}$ consists of functions such that

$$\chi_* = \frac{\chi \cdot \nabla\Phi}{|\nabla\Phi|}, \quad \chi \in H^1_\omega(\Omega) \cap C_0(\Gamma_c) .$$

In particular, $C^1_0(\Gamma_c) \subset \mathcal{V}$. To show this we are to construct, for a given function $\varphi \in C^1_0(\Gamma_c)$, a function $\chi \in H^1_\omega(\Omega) \cap C_0(\Gamma_c)$ such that $\frac{\chi \cdot \nabla\Phi}{|\nabla\Phi|} = \varphi$ on Γ_c.

The construction is simpler than that in the proof of Theorem 1.2 since now φ can be extended to the whole boundary Γ by assigning it the value zero.

Thus, the linear functional Ψ defined on $\mathcal{V}$, $\Psi(\chi_*) = \langle \Pi'(\omega), \chi \rangle$ is extended to the space $C_0(\Gamma_c)$ by continuity. The representation (1.4) follows from the well-known formula for nonnegative linear functionals on the space of finite (ie. with compact supports) and continuous functions. The density of the measure μ is $\frac{\nabla\Phi \cdot \sigma \cdot \nu}{|\nabla\Phi|}$, and $\mu(\mathcal{K}) < \infty$ for any compact $\mathcal{K} \subset \Gamma_c \setminus \partial\Gamma_c$.

2 Contact between two elastic bodies

2.1 Formulation of the problem. Regularity of solutions

Consider two elastic bodies occupying in a natural state bounded domains Ω and Ω' in $\mathbb{R}^3$ with boundaries Γ, Γ', respectively. A common part of the boundaries is Γ_c, and ω, ω' are displacement vectors of the bodies. The nonpenetration condition on Γ_c has the form

$$\omega(x)\nu(x) - \omega'(x)\nu(x) \le 0 , \quad x \in \Gamma_c . \tag{2.1}$$

Here $\nu = (\nu_1, \nu_2, \nu_3)$ is the unit outward normal on Γ. All the quantities corresponding to the second body are indicated by primes. We assume that $\Gamma = \Gamma_c \cup \Gamma_\omega \cup \Gamma_\sigma$, $\Gamma' = \Gamma_c \cup \Gamma_{\omega'} \cup \Gamma_{\sigma'}$, $\Gamma, \Gamma' \in C^\infty$, and the Lebesgue measures of Γ_ω and $\Gamma_{\omega'}$ are positive.

By $H^1_\omega(\Omega)$ we denote the Sobolev space of vector-functions as defined in the previous section. The same notation is used for functions defined on Ω', $H(\Omega) = H^1_\omega(\Omega) \times H^1_{\omega'}(\Omega')$.

The energy functional for two elastic bodies is the sum of the energy functionals defined in Ω and Ω', respectively. The energy functional in Ω has the form

$$\Pi(\omega) = \frac{1}{2}\int_{\Omega} \sigma_{ij}(\omega)\varepsilon_{ij}(\omega)dx - \int_{\Omega} f_i u_i dx - \int_{\Gamma_\sigma} g_i u_i d\Gamma \ ,$$

$$\sigma_{ij}(\omega) = c_{ijkl}\varepsilon_{kl}(\omega) \ , \quad \varepsilon_{ij}(\omega) = \frac{1}{2}\left(\frac{\partial u_i}{\partial x_j} + \frac{\partial u_j}{\partial x_i}\right) \ ,$$

and the energy functional in Ω' is obtained by replacing $\omega = (u_1, u_2, u_3)$, f_i, g_i, c_{ijkl} with $\omega' = (u_1', u_2', u_3')$, f_i', g_i', c_{ijkl}', respectively.

Denote by K the subset of $H(\Omega)$ such that for any $(\omega, \omega') \in K$, inequality (2.1) is satisfied almost everywhere on Γ_c, and assume that K is nonempty. Denote $E(\omega, \omega') = \Pi(\omega) + \Pi'(\omega')$. Solutions to the contact problem minimize the energy functional $E(\omega, \omega')$ over the set K. Therefore, any solution $\psi = (\omega, \omega')$ satisfies the variational inequality

$$\psi \in K \ : \quad \langle\, E'(\psi),\ \chi - \psi \,\rangle \geq 0 \quad \forall\, \chi \in K \ , \tag{2.2}$$

where $E'(\psi)$ is the derivative of the functional E at $\psi = (\omega, \omega')$.

Problem (2.2) can also be formulated in a another way. Namely, the equilibrium equations

$$\frac{\partial \sigma_{ij}}{\partial x_j} = -f_i \ , \quad i = 1,2,3 \ ; \quad \frac{\partial \sigma_{ij}'}{\partial x_j} = -f_i' \ , \quad i = 1,2,3 \ , \tag{2.3}$$

are satisfied in Ω and in Ω' and the following boundary conditions hold

$$\omega = 0 \quad \text{on} \quad \Gamma_\omega \ ; \quad \omega' = 0 \quad \text{on} \quad \Gamma_{\omega'} \ ;$$

$$\sigma_{ij}\nu_j = g_i \quad \text{on} \quad \Gamma_\sigma \ , \ i = 1,2,3 \ ; \ \sigma_{ij}'\nu_j' = g_i' \quad \text{on} \quad \Gamma_{\sigma'} \ , \ i = 1,2,3 \ , \tag{2.4}$$

$$\begin{cases} \omega\nu - \omega'\nu < 0 \\ \sigma_{kj}\nu_j\nu_k = 0 \\ \sigma_\tau = \sigma_\tau' = 0 \end{cases} \quad \text{or} \quad \begin{cases} \omega\nu - \omega'\nu = 0 \\ \sigma_{kj}\nu_j\nu_k \leq 0 \\ \sigma_\tau = \sigma_\tau' = 0 \end{cases} \quad \text{on} \quad \Gamma_c \ , \tag{2.5}$$

$$\sigma_{kj}(\omega)\nu_j\nu_k = \sigma_{kj}'(\omega')\nu_j\nu_k \quad \text{on} \quad \Gamma_c \tag{2.6}$$

where

$$(\sigma_\tau)_i = \sigma_{ij}\nu_j - (\sigma_{kj}\nu_j\nu_k)\nu_i \ , \quad i = 1,2,3 \ ,$$

and $\sigma_{ij} = \sigma_{ij}(\omega)$, $\sigma_{ij}' = \sigma_{ij}'(\omega')$.

To verify that the variational solution satisfies (2.3) it is enough to substitute the test function

$$\psi + \psi_1 \ , \quad \psi_1 = (\varphi, \varphi') \ , \quad \varphi \in C_0^\infty(\Omega) \ , \quad \varphi' \in C_0^\infty(\Omega')$$

into (2.2). Now the boundary conditions (2.4)–(2.6) are given in a formal way. This means that a weak solution is assumed to be sufficiently smooth. Integrating by parts (2.2) and taking into account (2.3) with $\chi = (\varphi, \varphi')$ we get

$$\int_{\Gamma_\sigma \cup \Gamma_c} \sigma_{ij}\nu_j(\varphi_i - \omega_i)d\Gamma + \int_{\Gamma_{\sigma'} \cup \Gamma_c} \sigma'_{ij}\nu'_j(\varphi'_i - \omega'_i)d\Gamma$$
$$- \int_{\Gamma_\sigma} g_i(\varphi_i - \omega_i)d\Gamma - \int_{\Gamma_{\sigma'}} g'_i(\varphi'_i - \omega'_i)d\Gamma \geq 0 \ .$$

Since φ and φ' are arbitrary functions defined on Γ_σ, $\Gamma_{\sigma'}$, respectively, conditions (2.4) follow, and hence

$$\int_{\Gamma_c} \sigma_{ij}\nu_j(\varphi_i - \omega_i)d\Gamma + \int_{\Gamma_c} \sigma'_{ij}\nu'_j(\varphi'_i - \omega'_i)d\Gamma \geq 0 \ . \tag{2.7}$$

Moreover, $\nu = -\nu'$ on Γ_c, and the representation

$$\sigma_{ij}\nu_j(\varphi_i - \omega_i) = (\sigma_{ij}\nu_j\nu_i)(\varphi\nu - \omega\nu) + \sigma_\tau(\varphi_\tau - \omega_\tau)$$

follows by standard decomposition of the vector $(\sigma_{1j}\nu_j,\ \sigma_{2j}\nu_j,\ \sigma_{3j}\nu_j)$ into the sum of the normal $(\sigma_{ij}\nu_j\nu_i)\nu$ and the tangential σ_τ components. Thus, from (2.7), we get

$$\begin{aligned}\int_{\Gamma_c} (\sigma_{ij}\nu_j\nu_i)(\varphi\nu - \omega\nu)d\Gamma - \int_{\Gamma_c} (\sigma'_{ij}\nu_j\nu_i)(\varphi'\nu - \omega'\nu)d\Gamma \\ + \int_{\Gamma_c} \sigma_\tau(\varphi_\tau - \omega_\tau)d\Gamma - \int_{\Gamma_c} \sigma'_\tau(\varphi'_\tau - \omega'_\tau)d\Gamma \geq 0 \ .\end{aligned} \tag{2.8}$$

This inequality is satisfied for all $(\varphi, \varphi') \in K$. The tangential components of φ, and φ' are arbitrary on Γ_c, and hence, from (2.8) it follows that $\sigma_\tau = 0$, $\sigma'_\tau = 0$ on Γ_c.

If, in a neighbourhood of a certain point x of Γ_c, it would be $\sigma_{ij}\nu_j\nu_i > 0$, then we would immediately get a contradiction. Indeed, by letting $\varphi' = 0$, $\varphi\nu \leq 0$ on Γ_c, $\varphi \in H^1_\omega(\Omega)$ (its support being compact in Γ_c), $\varphi' \in H^1_{\omega'}(\Omega')$, and assuming that $\varphi\nu < 0$ in a neighbourhood of x, we would get $(t\varphi)\nu \leq 0$ on Γ_c for $t > 0$, and therefore, the function $(t\varphi, \varphi')$ would be admissible which, for sufficiently large t, would contradict (2.8).

Thus, it must be $\sigma_{ij}\nu_j\nu_i \leq 0$ on Γ_c. By similar arguments, we get (2.6) and the inequality $\sigma'_{ij}\nu_j\nu_i \leq 0$ on Γ_c.

Taking into account (2.6) we can rewrite inequality (2.8) in the form

$$\int_{\Gamma_c} (\sigma_{ij}\nu_j\nu_i)\big[(\varphi\nu - \varphi'\nu) - (\omega\nu - \omega'\nu)\big]d\Gamma \geq 0 \ .$$

and (2.5) result.

Let $f_i \in L^2(\Omega)$, $g_i \in L^2(\Gamma_\sigma)$, $f'_i \in L^2(\Omega')$, $g'_i \in L^2(\Gamma_{\sigma'})$, $i = 1, 2, 3$. Then the problem of minimization of the functional E over the set K is uniquely solvable.

Since K is closed and convex in $H(\Omega)$, it is weakly closed. Therefore, the existence of solutions follows from the coercivity and the weak lower semicontinuity of the functional E on $H(\Omega)$. Uniqueness follows directly from (2.2) by standard arguments.

A solution to problem (2.2) has second order derivatives at interior points of Γ_c relative to Γ. This fact is a direct consequence of the work (Yakunina, 1981). More precisely, we have the following result.

Theorem 2.1. *For each $x \in \Gamma_c \backslash \partial \Gamma_c$ there exists a neighbourhood $\mathcal{O}(x) \subset \Omega \cup \Omega' \cup \Gamma_c$ such that*

$$\omega \in H^2\big(\mathcal{O}(x) \cap \Omega\big) \ , \quad \omega' \in H^2\big(\mathcal{O}(x) \cap \Omega'\big) \ .$$

2.2 Construction of a measure

Let Γ_c satisfy the requirement described in Section 1.2. Assume that there exists a mapping $x'(y_1, y_2)$ which coincides with $x(y_1, y_2)$ on $\gamma_c \cup \partial\gamma_c$. Moreover, assume that this mapping possesses properties analogous to those of $x(y_1, y_2)$ listed at the beginning of Section 1.2 and establishes a one-to-one correspondence between $\tilde{\gamma}'_c \cup \partial\tilde{\gamma}'_c$ and $\tilde{\Gamma}'_c$, where $\gamma_c \subset \tilde{\gamma}'_c$, $\Gamma_c \subset \tilde{\Gamma}'_c$, $\tilde{\Gamma}'_c \subset \Gamma'$, and $\partial\Gamma_c$ has no points in common with $\partial\tilde{\Gamma}'_c$.

Let $\varphi = (\gamma, \gamma') \in H(\Omega)$ satisfy the inequality $\gamma\nu - \gamma'\nu \le 0$ almost everywhere on Γ_c. Then, for $\psi = (\omega, \omega')$ which solves (2.2) we have

$$\langle\, E'(\psi), \varphi \,\rangle \ge 0 \ . \tag{2.9}$$

This can easily be verified as follows. Since the function ψ belongs to K and φ satisfies the above inequality on Γ_c, we have $\psi + \varepsilon\varphi \in K$. The functional E attains its minimum over K at ψ. In consequence, $E(\psi + \varepsilon\varphi) - E(\psi) \ge 0$. Dividing the latter inequality by ε and passing to the limit as $\varepsilon \to 0$ we get inequality (2.9).

Now we construct a measure on subsets of Γ_c which allows us to characterize the interaction between the two bodies. As in the previous section, we have to consider several cases corresponding to how the sets $\Gamma_c, \Gamma_\sigma, \Gamma_\omega, \Gamma_{\sigma'}, \Gamma_{\omega'}$ are placed with respect to one another.

First we consider the case where for every $x_0 \in \partial\Gamma_c$ there exists a neighbourhood $\mathcal{O}(x_0)$ such that $\mathcal{O}(x_0) \cap \Gamma \subset \Gamma_c \cup \Gamma_\sigma$ and $\mathcal{O}(x_0) \cap \Gamma' \subset \Gamma_c \cup \Gamma_{\sigma'}$.

Theorem 2.2. *On the $\mathfrak{S}$-algebra of Borel subsets of Γ_c there exists a nonnegative measure μ such that for every $\varphi = (\gamma, \gamma') \in H(\Omega) \cap C(\Gamma_c)$*

$$\langle\, E'(\psi), \varphi \,\rangle = -\int_{\Gamma_c} (\gamma\nu - \gamma'\nu)\, d\mu \ . \tag{2.10}$$

Proof. For every function $\varphi = (\gamma, \gamma') \in H(\Omega) \cap C(\Gamma_c)$ we put

$$\varphi_* = \gamma\nu - \gamma'\nu \quad \text{on} \quad \Gamma_c \ .$$

The linear space of all these functions on Γ_c will be denoted by W. Let F be a linear functional defined on W by the formula

$$F(\varphi_*) = -\langle\, E'(\psi), \varphi \,\rangle \ . \tag{2.11}$$

It is easily seen that F is well-defined. In fact, if $\varphi_*^{(1)} = \varphi_*^{(2)}$, then, by (2.9),

$$\langle\, E'(\psi), \varphi^{(1)} - \varphi^{(2)} \,\rangle = 0 \ .$$

Consequently, $F(\varphi_*^{(1)}) = F(\varphi_*^{(2)})$. It follows also that the functional F is positive.

Denote by $\mathcal{W}$ the Banach space obtained as the completion of W in the norm $\|\cdot\|_{C(\Gamma_c)} = \max\limits_{\Gamma_c} |\cdot|$. As in the proof of Theorem 1.2, we can show that W contains all the functions from $C^1(\Gamma_c)$. Hence, $\mathcal{W}$ coincides with $C(\Gamma_c)$.

Now we prove that there exists a function $\varphi_0 = (\beta, \beta') \in H(\Omega)$ such that, for a certain constant δ, $\beta\nu - \beta'\nu \le \delta < 0$ almost everywhere on Γ_c. The function $h \in H^1_\omega(\Omega)$ defined by (1.7) satisfies the equality

$$-h\nu = 1 \quad \text{on} \quad \Gamma_c \ .$$

On the other hand, $\nu = -\nu'$ on Γ_c. Hence, as before, we can construct a function $h' \in H^1_{\omega'}(\Omega')$ such that

$$h'\nu = 1 \quad \text{on} \quad \Gamma_c \ .$$

Thus, the function $\varphi_0 = (\beta, \beta')$ with $(\beta, \beta') = (h, h')$ has the required property.

For an arbitrary $\varphi_* \in W$,

$$|\varphi_*| \le \delta^{-1}(\beta\nu - \beta'\nu) \cdot \|\varphi_*\|_{C(\Gamma_c)} \ .$$

Since F is positive

$$|F(\varphi_*)| \le c\|\varphi_*\|_{C(\Gamma_c)} \ , \qquad c = -\delta^{-1}\langle\, E'(\psi) \ , \ \varphi_0 \,\rangle \ . \tag{2.12}$$

This implies that F is a linear and continuous functional on $\mathcal{W}$, which together with its positivity yields the existence of a nonnegative measure μ such that for every $\varphi_* \in C(\Gamma_c)$,

$$F(\varphi_*) = \int_{\Gamma_c} \varphi_* d\mu \ ,$$

and

$$\mu(\Gamma_c) \le -\delta^{-1}\langle\, E'(\psi) \ , \ \varphi_0 \,\rangle \ .$$

This completes the proof. □

The measure constructed above can be decomposed into the singular part and the absolutely continuous part, $\mu = \mu_0 + \mu_1$.

Now we shall prove that the singular part μ_0 can be nonzero only on $\partial\Gamma_c$. This, in turn, will imply that a singular interaction between the two bodies cannot occur at $x \in \Gamma_c \backslash \partial\Gamma_c$.

Let us take any $x_0 \in \Gamma_c \backslash \partial\Gamma_c$. There exists a neighbourhood $\mathcal{O}(x_0)$ of x_0 which is disjoint with $\partial\Gamma_c$. Choose $\varphi = (\gamma, \gamma') \in H(\Omega) \cap C(\Gamma_c)$ with the support contained in $\mathcal{O}(x_0)$. By (2.10),

$$\langle E'(\psi), \varphi \rangle = -\int_{\Gamma_c} (\gamma\nu - \gamma'\nu) d\mu \ . \tag{2.13}$$

In virtue of Theorem 2.1, the solution ψ has the derivatives up to order two in $\mathcal{O}(x_0) \cap \Omega$ and $\mathcal{O}(x_0) \cap \Omega'$. Integrating by parts the left-hand side of (2.13) we get

$$\int_{\mathcal{O}(x_0)\cap\Omega} \left\{ \sigma_{ij}(\omega) \frac{\partial \gamma_i}{\partial x_j} - f_i \gamma_i \right\} dx + \int_{\mathcal{O}(x_0)\cap\Omega'} \left\{ \sigma'_{ij}(\omega') \frac{\partial \gamma'_i}{\partial x_j} - f'_i \gamma'_i \right\} dx$$

$$= \int_{\Gamma_c} \left\{ \sigma_{ij}(\omega)\nu_j\gamma_i - \sigma'_{ij}(\omega')\nu_j\gamma'_i \right\} d\Gamma \ .$$

Using the equalities $\sigma_{ij}(\omega)\nu_j\nu_i = \sigma'_{ij}(\omega')\nu_j\nu_i$ and $\sigma_\tau = \sigma'_\tau$ on Γ_c, the left-hand side of (2.13) can be rewritten as

$$\begin{aligned} &\int_{\Gamma_c} \left[(\sigma_{ij}(\omega)\nu_j\nu_i)\gamma_k\nu_k - (\sigma'_{ij}(\omega')\nu_j\nu_i)\gamma'_k\nu_k \right] d\Gamma \\ &\qquad = \int_{\Gamma_c} (\sigma_{ij}(\omega)\nu_j\nu_i)(\gamma\nu - \gamma'\nu) d\Gamma \ . \end{aligned} \tag{2.14}$$

Thus, from (2.13), (2.14) it follows that

$$-\int_{\Gamma_c} (\gamma\nu - \gamma'\nu) d\mu = \int_{\Gamma_c} (\sigma_{ij}(\omega)\nu_j\nu_i)(\gamma\nu - \gamma'\nu) d\Gamma \ .$$

Therefore, the singular part μ_0 of the measure μ equals zero on $\Gamma_c \backslash \partial\Gamma_c$. Note that we have actually proved that the density of the measure μ is equal to $-\sigma_{ij}(\omega)\nu_j\nu_i$.

Finally, we state a theorem concerning the situation when $\partial\Gamma_c$ has common points with $\partial\Gamma_\omega$.

Theorem 2.3. *On the $\mathfrak{S}$-algebra of Borel subsets of $\Gamma_c \backslash \partial\Gamma_c$ there exists a nonnegative measure μ such that (2.10) holds for all $\varphi = (\gamma, \gamma') \in H(\Omega) \cap C_0(\Gamma_c)$. The density of this measure is $-\sigma_{kj}(\omega)\nu_j\nu_k$ and its singular part equals zero. Moreover, $\mu(B) < +\infty$ for each compact $B \subset \Gamma_c \backslash \partial\Gamma_c$.*

This theorem can be proved in much the same way as the previous one, so here we give only the main ideas of the proof. By W we denote the linear variety of functions φ_* defined on Γ_c such that

$$\varphi_* = \gamma\nu - \gamma'\nu \ , \quad \varphi = (\gamma, \gamma') \in H(\Omega) \cap C_0(\Gamma_c) \ .$$

We claim that this variety contains $C_0^1(\Gamma_c)$. To see this, we have to construct, for any $\alpha \in C_0^1(\Gamma_c)$, a function $\varphi = (\gamma, \gamma') \in H(\Omega) \cap C_0(\Gamma_c)$ such that $\varphi_* = \gamma\nu - \gamma'\nu$ coincides with α on Γ_c. To this end, we extend the function α to the boundaries Γ and Γ', by putting its value equal to zero, and next we extend the resulting function to the domains Ω and Ω'. Now, in view of the positivity of the functional F defined on W by the formula (2.11), F can be extended to $C_0(\Gamma_c)$.

For functions $\varphi = (\gamma, \gamma') \in H(\Omega) \cap C_0(\Gamma_c)$, formula (2.10) follows from the representation of linear positive functionals on the space of continuous functions with compact supports.

In this case a continuity of functionals follows from their positivity and linearity (see Chapter 1, Section 1.10), an estimate of the type (2.12) is not needed.

If Γ_c does not satisfy the requirement formulated at the beginning of Section 2.2, a local variant of Theorem 2.3 can be obtained. In this case a representation of the type (2.10) can be derived by using test functions with compact supports.

3 Contact between a shallow shell and a rigid punch

In this section we study the contact problem for a shallow shell. The problem is to minimize the energy functional over the set of admissible displacements. Solutions to this kind of problems are defined by variational inequalities. Special attention is paid to the smoothness of solutions. The results obtained will allow us to formulate conditions under which concentrated interaction forces appear.

3.1 Existence of solutions

Equilibrium equations for a shallow shell have the form

$$\Delta^2 w + k_{11}N_{11} + k_{22}N_{22} = f \ , \tag{3.1}$$

$$\frac{\partial N_{ij}}{\partial x_j} = -f_i \ , \quad i = 1, 2 \ , \tag{3.2}$$

$$N_{11} = \varepsilon_{11} + \sigma\varepsilon_{22} \ , \quad N_{22} = \varepsilon_{22} + \sigma\varepsilon_{11} \ , \quad N_{12} = \frac{1}{2}(1 - \sigma)\varepsilon_{12} \ , \tag{3.3}$$

$$\varepsilon_{11} = u_{x_1} + k_{11}w \ , \quad \varepsilon_{22} = v_{x_2} + k_{22}w \ , \quad \varepsilon_{12} = u_{x_2} + v_{x_1} \ . \tag{3.4}$$

Here σ, $0 < \sigma < \frac{1}{2}$, is a constant, f, f_i, k_{11}, k_{22} are given functions. We use the summation convention over repeated indices, $\Delta = \frac{\partial^2}{\partial x_1^2} + \frac{\partial^2}{\partial x_2^2}$, x_1, x_2 are orthogonal coordinates on the middle surface of the shell, $(x_1, x_2) \in \Omega$, and we assume the symmetry with respect to the indices of the functions in question.

In (3.1)–(3.4), the functions $u, v, w, N_{ij}, \varepsilon_{ij}$ are unknown. If we substitute N_{ij}, ε_{ij} according to (3.3), (3.4) into system (3.1), (3.2), then this system becomes complete over u, v, w.

The energy functional for the shell is of the form

$$\begin{aligned}\Pi(\omega) = &\int_\Omega (\Delta w)^2 dx + \int_\Omega \left[\varepsilon_{11}^2 + \varepsilon_{22}^2 + 2\sigma\varepsilon_{11}\varepsilon_{22} + \frac{1}{2}(1-\sigma)\varepsilon_{12}^2\right] dx \\ &- 2\int_\Omega (f_1 u + f_2 v + f w) dx \ , \quad \omega = (u, v, w) \ .\end{aligned}$$

This functional is defined on the space $H(\Omega) = H_0^1(\Omega) \times H_0^1(\Omega) \times H_0^2(\Omega)$. In particular, the boundary conditions for u, v, w are fixed. Notice that relations (3.1), (3.2) are the Euler equations for the functional $\Pi(\omega)$. In this case ε_{ij}, N_{ij} are related to u, v, w by formulae (3.3), (3.4).

Let the z-axis be directed orthogonally to the middle surface of the shell and $z = \Phi(x_1, x_2)$ be the equation of the punch shape. The displacements u, v, w of middle surface points (along axes x_1, x_2, z) satisfy the restriction

$$w(x) - U(x) \cdot \nabla\Phi(x) \geq \Phi(x) \ , \quad x \in \Omega \ , \tag{3.5}$$

where $U = (u, v)$, $\nabla\Phi = (\Phi_{x_1}, \Phi_{x_2})$. In deriving (3.5) we apply the linear approach.

Denote by K the set of functions $\omega = (u, v, w)$ belonging to $H(\Omega)$, which satisfy inequality (3.5) almost everywhere in Ω. The contact problem between the shell and the punch can be formulated as the minimization problem

$$\inf_{\omega \in K} \Pi(\omega) \ . \tag{3.6}$$

The set K is convex and closed in $H(\Omega)$. In fact, let $\omega^n \in K$ and $\omega^n \to \omega$ strongly in $H(\Omega)$. We can choose a subsequence of ω^n converging almost everywhere in Ω to the function ω. The functions $\omega^n = (u^n, v^n, w^n)$ satisfy inequality (3.5) almost everywhere. Thus, the function $\omega = (u, v, w)$ also satisfies this inequality.

We assume that the set K is nonempty. The nonemptiness of K depends on the function Φ and, in particular, on the boundary condition imposed on Φ.

Suppose that $f, f_1, f_2 \in L^2(\Omega)$, $\Phi \in C^\infty(\bar{\Omega})$, $k_{11}, k_{22} \in H^{1+\varepsilon}(\Omega)$, $\varepsilon > 0$, and the boundary $\partial\Omega$ is of class C^∞.

Now we are going to show that the functional Π is coercive on $H(\Omega)$, ie.,

$$\Pi(\omega) \to +\infty \quad \text{when} \quad \|\omega\|_{H(\Omega)} \to +\infty \ .$$

Let Π_1 be the nonlinear part of the functional Π, ie.,

$$\Pi_1(\omega) = \int_\Omega (\Delta w)^2 dx + \int_\Omega \left[\varepsilon_{11}^2 + \varepsilon_{22}^2 + 2\sigma\varepsilon_{11}\varepsilon_{22} + \frac{1}{2}(1-\sigma)\varepsilon_{12}^2\right] dx \ .$$

For $r > 0$, consider a transformation of the unit sphere S of $H(\Omega)$,

$$\tilde{U} \mapsto \delta r \tilde{U} \ , \quad \tilde{w} \mapsto r\tilde{w} \ , \quad \tilde{U} = (\tilde{u}, \tilde{v}) \ , \quad (\tilde{U}, \tilde{w}) \in S \ ,$$

where δ is a positive constant to be chosen later.

We assume that in the space $H_0^1(\Omega) \times H_0^1(\Omega)$ the squared norm is of the form

$$\|U\|_1^2 = \int_\Omega \left[u_{x_1}^2 + v_{x_2}^2 + 2\sigma u_{x_1} v_{x_2} + \frac{1}{2}(1-\sigma)\left(u_{x_2} + v_{x_1}\right)^2\right] dx \ .$$

Moreover, we assume that for functions w in the space $H_0^2(\Omega)$ the squared norm $\|w\|_2^2$ coincides with the first integral in the formula for $\Pi_1(\omega)$ and

$$\|\omega\|_{H(\Omega)}^2 = \|U\|_1^2 + \|w\|_2^2 \ .$$

Let $(\tilde{U}, \tilde{w}) \in S$, $\|\tilde{w}\|_2^2 \geq \frac{1}{4}$, and let ω be the image of $(\tilde{U}, \tilde{w})$ under the above transformation. Then $\Pi_1(\omega) \geq \frac{r^2}{4}$. Moreover, for $\|\tilde{U}\|_1^2 \geq \frac{3}{4}$ we have $\Pi_1(\omega) \geq \frac{3}{4}\delta^2 r^2 - c_1 \delta r^2$, where c_1 bounds from above the integral

$$2\int_\Omega \left\{k_{11} w u_{x_1} + k_{22} w v_{x_2} + \sigma\left(k_{11} w v_{x_2} + k_{22} w u_{x_1}\right)\right\} dx$$

on S. By choosing δ such that $\frac{3}{4}\delta^2 - c_1\delta = 1$, we ensure that the images of elements from the unit sphere satisfy the inequality $\Pi_1(\omega) \geq \frac{r^2}{4}$. The linear part of $\Pi(\omega)$ can be bounded by cr, $r > 0$. Hence,

$$\Pi(\omega) \geq \frac{r^2}{4} - cr \ ,$$

which proves the coercivity of the functional $\Pi(\omega)$.

Now we prove the weak lower semicontinuity of the functional $\Pi(\omega)$ on the space $H(\Omega)$.

Let us consider the part of $\Pi(\omega)$ which depends on $\varepsilon_{ij} = \varepsilon_{ij}(\omega)$. This part is convex with respect to ε_{ij}. Let $\omega^n \to \omega$ weakly in $H(\Omega)$, $\omega^n = (u^n, v^n, w^n)$. Then $\varepsilon_{ij}(\omega^n) \to \varepsilon_{ij}(\omega)$ weakly in $L^2(\Omega)$ and therefore this part of the functional Π is weakly lower semicontinuous. The first term in the representation of $\Pi(\omega)$ is also weakly lower semicontinuous.

The properties of the functional $\Pi(\omega)$ guarantee the existence of solutions to problem (3.6). A solution satisfies the inequality

$$\omega \in K \ : \ \langle\, \Pi'(\omega),\ \tilde{\omega} - \omega \,\rangle \geq 0 \quad \forall\, \tilde{\omega} \in K \ , \tag{3.7}$$

where $\Pi'(\omega)$ is the derivative of the functional Π at ω.

It should be noted that variational inequality (3.7) is uniquely solvable. In fact, suppose that ω^1, ω^2 solve inequality (3.7). Let us substitute $\tilde{\omega} = \omega^2$, $\omega = \omega^1$ and $\tilde{\omega} = \omega^1$, $\omega = \omega^2$ into (3.7). Summing up the resulting inequalities we obtain

$$\langle\, \Pi_1'(\omega^1 - \omega^2), \omega^1 - \omega^2 \,\rangle \leq 0 \ .$$

Denote by $\omega = (u, v, w)$ the difference $\omega^1 - \omega^2$. From the above inequality it follows that

$$\int_\Omega (\Delta w)^2 dx + \int_\Omega \left[\varepsilon_{11}^2 + \varepsilon_{22}^2 + 2\sigma \varepsilon_{11} \varepsilon_{22} + \frac{1}{2}(1-\sigma)\varepsilon_{12}^2 \right] dx \leq 0 \ ,$$

where ε_{ij} are related to u, v, w by formulae (3.4). Hence, $\omega \equiv 0$.

Inequality (3.7) is the main object of our investigation in this section.

3.2 Regularity of solutions

We show now that the local regularity of the solution is by one higher than the local regularity of the variational solution provided that $|\nabla \Phi(x)| > 0$. We next sharpen this result by showing that w has derivatives of order four in a neighbourhood of the points satisfying the above condition.

Theorem 3.1. *Let $x_0 \in \Omega$ and $|\nabla \Phi(x_0)| > 0$. There exists a neighbourhood Ω_0 of x_0 such that*

$$(u, v, w) \in H^2(\Omega_0) \times H^2(\Omega_0) \times H^3(\Omega_0) \ . \tag{3.8}$$

Proof. Without loss of generality we assume that $\Phi_{x_1}(x_0) \neq 0$, $\Phi_{x_2}(x_0) \neq 0$. Take a neighbourhood Ω_3 of x_0 such that $|\Phi_{x_1}(x)| > 0$, $|\Phi_{x_2}(x)| > 0$, $x \in \Omega_3$. Denote by e_i the basis vectors, $i = 1, 2$, and take $|\tau| > 0$,

$$d_{\pm i\tau} h(x) = \tau^{-1}\big[h(x \pm \tau e_i) - h(x)\big], \quad \Delta_{i\tau} = -d_{-i\tau} d_{i\tau} \ .$$

It is possible to choose domains $\Omega_0 \subset \Omega_1 \subset \Omega_2 \subset \Omega_3$ in such a way that dist $(\partial\Omega_1, \partial\Omega_2) \geq q > 0$, for a certain constant q, and $\bar{\Omega}_0 \subset \Omega_1$, $\bar{\Omega}_2 \subset \Omega_3$.

We begin by proving that for a function $\varphi \in C_0^\infty(\Omega_1)$, $\varphi \equiv 1$ on Ω_0, $0 \leq \varphi \leq 1$ everywhere, the vector $\omega_\tau = (u_\tau, v_\tau, w_\tau)$ with components

$$\begin{gathered} u_\tau = u + \frac{\tau^2}{2}\varphi^2 \Phi_{x_1}^{-1} \Delta_{i\tau}(u\Phi_{x_1}) \ , \quad v_\tau = v + \frac{\tau^2}{2}\varphi^2 \Phi_{x_2}^{-1} \Delta_{i\tau}(v\Phi_{x_2}) \ , \\ w_\tau = w + \frac{\tau^2}{2}\varphi^2 \Delta_{i\tau}(w - \Phi) \end{gathered} \tag{3.9}$$

belongs to the set K for $0 < |\tau| < q$.

We start by noting that if, for functions h and H,

$$h(x) \geq H(x) \ , \quad x \in \Omega \ ,$$

then

$$h_\tau(x) \geq H(x) \ , \quad x \in \Omega \ ,$$

for $h_\tau = h + \frac{\tau^2}{2}\varphi^2 \Delta_{i\tau}(h - H)$, φ being a function on Ω with compact support, $|\varphi| \leq 1$, and $|\tau| > 0$ sufficiently small.

In fact,

$$\begin{aligned} h_\tau(x) &= h(x) + \frac{\tau^2}{2}\varphi^2 \Delta_{i\tau}(h - H)(x) = (1 - \varphi^2)h(x) \\ &+ \frac{\varphi^2}{2}\big[h(x + \tau e_i) + h(x - \tau e_i)\big] - \frac{\tau^2}{2}\varphi^2 \Delta_{i\tau} H(x) \\ &\geq (1 - \varphi^2)H(x) + \frac{\varphi^2}{2}\big[H(x + \tau e_i) + H(x - \tau e_i)\big] \\ &\qquad - \frac{\tau^2}{2}\varphi^2 \Delta_{i\tau} H(x) = H(x) \ . \end{aligned} \tag{3.10}$$

Denote $f_0 = f - k_{11}N_{11} - k_{22}N_{22}$ and substitute ω_τ as a test function to inequality (3.7). This gives

$$\begin{aligned} &\int_\Omega \Delta w \Delta\big(\varphi^2 \Delta_{i\tau}(w - \Phi)\big)dx + \int_\Omega N_{11}\big(\varphi^2 \Phi_{x_1}^{-1} \Delta_{i\tau} u \Phi_{x_1}\big)_{x_1} dx \\ &+ \int_\Omega N_{12}\big(\varphi^2 \Phi_{x_1}^{-1} \Delta_{i\tau} u \Phi_{x_1}\big)_{x_2} dx + \int_\Omega N_{12}\big(\varphi^2 \Phi_{x_2}^{-1} \Delta_{i\tau} v \Phi_{x_2}\big)_{x_1} dx \\ &\qquad + \int_\Omega N_{22}\big(\varphi^2 \Phi_{x_2}^{-1} \Delta_{i\tau} v \Phi_{x_2}\big)_{x_2} dx \\ &\geq \frac{2}{\tau^2}\int_\Omega \big[f_0(w_\tau - w) + f_1(u_\tau - u) + f_2(v_\tau - v)\big]dx \ . \end{aligned} \tag{3.11}$$

The sequence of formulae given below has the property that the difference between any two subsequent terms can be bounded from above by the second unbracketed term of the right-hand side of inequality (3.14):

$$\begin{aligned} &\int_\Omega \Delta w \Delta(\varphi^2 \Delta_{i\tau} w)dx \to \int_\Omega \Delta(\varphi w)\Delta\big(\Delta_{i\tau}(\varphi w)\big)dx \to \\ &\to - \int_\Omega \Delta(\varphi w) d_{-i\tau} d_{i\tau} \Delta(\varphi w)dx \to \\ &\to - \int_\Omega \Delta\big(d_{i\tau}(\varphi w)\big)\Delta\big(d_{i\tau}(\varphi w)\big)dx \ . \end{aligned} \tag{3.12}$$

The same property is manifested also by other terms of the left-hand side of (3.11). Namely, let us denote the values $N_{11}, N_{12}, \ldots, \varepsilon_{11}, \varepsilon_{12} \ldots$, by $\sigma_{11}(u, v)$,

$\sigma_{12}(u,v),\ldots,e_{11}(u,v),e_{12}(u,v,),\ldots$, respectively, assuming that those values are obtained from formulae (3.3), (3.4) for $k_{11}=k_{22}=0$. In this way we get

$$N_{11}=\sigma_{11}(u,v)+k_{11}w+\sigma k_{22}w\ ,\quad N_{12}=\sigma_{12}(u,v)\ ,$$

$$N_{22}=\sigma_{22}(u,v)+k_{22}w+\sigma k_{11}w\ .$$

Then, from the second term of the left-hand side of (3.11) we take the "main" part containing the highest order derivatives. As a result we obtain the following sequence of formulae

$$\begin{aligned}\int_\Omega \sigma_{11}(u,v)\big(\varphi^2\Phi_{x_1}^{-1}\Delta_{i\tau}u\Phi_{x_1}\big)_{x_1}dx \to \int_\Omega \sigma_{11}(u,v)\big(\varphi^2\Delta_{i\tau}u\big)_{x_1}dx \to\\ \to \int_\Omega \sigma_{11}(\varphi u,\varphi v)\big(\Delta_{i\tau}(\varphi u)\big)_{x_1}dx \to\\ \to -\int_\Omega \sigma_{11}\big(d_{i\tau}(\varphi u),d_{i\tau}(\varphi v)\big)\big(d_{i\tau}(\varphi u)\big)_{x_1}dx\ .\end{aligned} \tag{3.13}$$

Here we do not sum up over i. The difference of any two neighbouring terms of this sequence either equals zero or can be bounded from above by the first unbracketed term of the right-hand side of (3.14). The last term of (3.12) can be bounded as follows

$$\int_\Omega \Delta\big(d_{i\tau}(\varphi w)\big)\Delta\big(d_{i\tau}(\varphi w)\big)dx \ge c\|d_{i\tau}(\varphi w)\|_2^2\ .$$

Moreover, by the first Korn inequality, the sum of terms which are obtained in the same way as the last term in (3.13), can be bounded as follows

$$\int_\Omega \sigma_{kj}(d_{i\tau}(\varphi U))e_{kj}(d_{i\tau}(\varphi U))dx \ge c\|d_{i\tau}(\varphi U)\|_1^2\ .$$

Here c is a constant which does not depend on U. By using the Cauchy inequality, the right-hand side of (3.11) can be easily estimated. Thus, we obtain the inequality

$$\begin{aligned}&\|d_{i\tau}(\varphi w)\|_2^2+\|d_{i\tau}(\varphi U)\|_1^2\\ &\le c\Big[\|f_1\|_0^2+\|f_2\|_0^2+\|u\|_1^2+\|v\|_1^2+\big(\|u\|_1+\|v\|_1\big)\big(\|d_{i\tau}(\varphi u)\|_1\\ &+\|d_{i\tau}(\varphi v)_1\|\big)\Big]+c\Big[\|f_0\|_0^2+\|w\|_2^2+\|w\|_2\|d_{i\tau}(\varphi w)\|_2\Big]\ ,\end{aligned} \tag{3.14}$$

where constants in the right-hand side are uniform with respect to τ. From this inequality it follows that $\|d_{i\tau}(\varphi w)\|_2+\|d_{i\tau}(\varphi U)\|_1$ are bounded uniformly with respect to τ. This completes the proof of Theorem 3.1. □

One can observe that the assertion of Theorem 3.1 remains valid if the equality $|\nabla\Phi(x_0)| = 0$ holds for every $x_0 \in \Omega_4$, where $\bar{\Omega}_0 \subset \Omega_4$.

In fact, there exists a constant $\delta \le 0$ such that (3.5) takes the form

$$w(x) \ge \delta$$

for $x \in \Omega_4$. Let $\varphi_1, \varphi_2 \in C_0^\infty(\Omega_4)$. Then, $\tilde{\omega} = (u + \varphi_1, v + \varphi_2, w)$ belongs to the set K. Substituting this vector into (3.7), one can deduce that equations (3.2) are satisfied in Ω_4 in the sense of distributions. These equations can be rewritten as

$$-\frac{\partial \sigma_{ij}(u,v)}{\partial x_j} = F_i(w) \ , \quad i = 1,2 \ , \tag{3.15}$$

where $\sigma_{ij}(u,v)$ are as in (3.13), the functions $F_i(w)$ depend only on w.

Equations (3.15) are two-dimensional equilibrium equations of linear theory of elasticity with the right-hand sides $F_i(w) \in L^2(\Omega_4)$. Hence, $u, v \in H^2_{\text{loc}}(\Omega_4)$.

As in the proof of Theorem 3.1, the function $w_\tau = w + \frac{\tau^2}{2}\varphi^2 \Delta_{i\tau} w$ has the property that $(u, v, w_\tau) \in K$. Substituting (u, v, w_τ) as a test function into (3.7) we obtain

$$\int_\Omega \Delta w \Delta(\varphi^2 \Delta_{i\tau} w) dx \ge \int_\Omega f_0 \varphi^2 \Delta_{i\tau} w dx \ .$$

and

$$\|d_{i\tau}(\varphi w)\|_2 \le c$$

with a constant c uniform with respect to τ. Thus, $w \in H^3(\Omega_0)$. $\bar{\Omega}_0 \subset \Omega_4$.

In the sequel we shall establish a stronger result. Namely, in Theorem 3.3 we shall prove that under the assumptions of Theorem 3.1

$$w \in H^4_{\text{loc}}(\Omega_0)$$

which means that the smoothness of w is even higher than that already shown. In getting this result the condition $|\nabla\Phi(x_0)| > 0$ is essential and the result is no longer true if we admit $|\nabla\Phi(x_0)| = 0$.

On the other hand, it is always true that

$$\Phi_{x_1} w \ , \quad \Phi_{x_2} w \in H^3(\Omega) \ .$$

To demonstrate this, we take $\omega = (u, v, w)$ which solves (3.7) and a function $\varphi \in C_0^\infty(\Omega)$. Then $\tilde{\omega} = (u + \varphi, v, w + \varphi\Phi_{x_1}) \in K$. By substituting the function $\tilde{\omega}$ into (3.7) and taking into account that φ is arbitrary, we obtain that the equation

$$\Phi_{x_1} \Delta^2 w = F_1 \ ,$$

is satisfied in Ω in the sense of distributions, where

$$F_1 = \Phi_{x_1}(f - k_{11}N_{11} - k_{22}N_{22}) + \left(\frac{\partial N_{11}}{\partial x_1} + \frac{\partial N_{12}}{\partial x_2} + f_1\right) \ .$$

By the same arguments,

$$\Phi_{x_2}\Delta^2 w = F_2 \ .$$

From the structure of F_1, F_2 it follows that $F_1, F_2 \in H^{-1}(\Omega)$. The above equations can be rewritten as

$$\Delta^2(\Phi_{x_1}w) = \tilde{F}_1 \ , \quad \Delta^2(\Phi_{x_2}w) = \tilde{F}_2 \ .$$

It is seen that the functions $\tilde{F}_1$, $\tilde{F}_2$ also belong to the space $H^{-1}(\Omega)$. Using the well-known regularity results for the Dirichlet problem for the biharmonic equation we conclude that $\Phi_{x_1}w, \Phi_{x_2}w \in H^3(\Omega)$.

3.3 Absence of concentrated forces

In this section we shall construct on subsets of the domain Ω a nonnegative measure which characterizes the interaction between the shell and the punch.

Note first that

$$\langle\, \Pi'(\omega),\ \omega_0 \,\rangle \geq 0 \tag{3.16}$$

for any vector $\omega_0 = (u_0, v_0, w_0) \in H(\Omega)$ such that $w_0 - U_0\nabla\Phi \geq 0$, $U_0 = (u_0, v_0)$. Indeed, $\omega + \omega_0 \in K$, and we can substitute $\omega + \omega_0$ into (3.7) as a test function, which gives (3.16).

Recall that $C_0(\Omega)$ is the space of functions with compact supports and continuous in Ω endowed with the following convergence: a sequence φ_n converges to φ in $C_0(\Omega)$ if φ_n converges uniformly to φ and supports of all φ_n belong to a fixed compact $K \subset \Omega$ (see Chapter 1, Section 1).

Theorem 3.2. *On the $\mathfrak{S}$-algebra of Borel subsets of Ω there exists a nonnegative measure μ such that the representation*

$$\langle\, \Pi'(\omega), \tilde{\omega} \,\rangle = \int_\Omega \frac{\tilde{w} - \tilde{U}\cdot\nabla\Phi}{\sqrt{1+|\nabla\Phi|^2}}\, d\mu \tag{3.17}$$

holds for any $\tilde{\omega} = (\tilde{U}, \tilde{w}) \in H(\Omega) \cap C_0(\Omega)$.

Proof. Take $\tilde{\omega} = (\tilde{U}, \tilde{w}) \in H(\Omega) \cap C_0(\Omega)$ and define

$$\tilde{\omega}_* = \frac{\tilde{w} - \tilde{U}\cdot\nabla\Phi}{\sqrt{1+|\nabla\Phi|^2}} \ . \tag{3.18}$$

The linear space of all functions $\tilde{\omega}_*$ is denoted by V. Let

$$\psi(\tilde{\omega}_*) = \langle\, \Pi'(\omega), \tilde{\omega} \,\rangle \ .$$

This is a well-defined functional on V. To show this, let us take a vector $\approx{\omega} \in H(\Omega) \cap C_0(\Omega)$ such that $\approx{\omega}_*$ coincides with $\tilde{\omega}_*$. Then $\tilde{\omega}_* - \approx{\omega}_* = 0$, and hence, in view of (3.16), one gets $\psi(\tilde{\omega}_*) = \psi(\tilde{\tilde{\omega}}_*)$. Therefore, ψ is a linear and positive functional on V.

Now we show that this functional can be extended to the space $C_0(\Omega)$. Choose $h \in C_0(\Omega)$. Suppose that functions $\varphi_n \in C_0^\infty(\Omega)$ possess the property that $\varphi_n \to h$ in $C_0(\Omega)$. We have to prove that the sequence $\psi(\varphi_n)$ converges.

The supports $S(\varphi_n)$ of the functions φ_n belong to a fixed compact $B \subset \Omega$. Let $g \in C_0^\infty(\Omega)$ be a function such that $g \equiv 1$ on B and $0 \le g \le 1$ everywhere on Ω. Then

$$|\varphi_m(x) - \varphi_n(x)| \le \delta(m,n)g(x) , \quad \delta(m,n) = \max_{\Omega} |\varphi_m(x) - \varphi_n(x)| .$$

Since ψ is positive, it follows that

$$|\psi(\varphi_m) - \psi(\varphi_n)| \le \delta(m,n)\psi(g) .$$

We have $\delta(m,n) \to 0$ as $m,n \to \infty$. Hence, the limit of $\psi(\varphi_n)$ exists and we denote it by $\psi(h)$. It is easily seen that this limit does not depend on the choice of φ_n and, consequently, ψ is a linear and positive functional. One can choose $\tilde{w}_n = \sqrt{1+|\nabla\Phi|^2}\varphi_n$, $\tilde{U}_n = 0$ as approximating sequence of vectors. In so doing all the needed conditions will be satisfied.

Finally, since every positive linear functional on the space $C_0(\Omega)$ is defined by a nonnegative measure μ, we have

$$\psi(h) = \int_\Omega h d\mu \qquad \forall\, h \in C_0(\Omega) .$$

This representation coincides with (3.17) for every function $\tilde{\omega}_*$ defined by (3.18), where $\tilde{\omega} \in H(\Omega) \cap C_0(\Omega)$. The proof is completed. □

The following observation is worth of mentioning here. Since $\omega_0 = (0, 0, \tilde{w})$ satisfies (3.16), the inequality

$$\int_\Omega \Delta w \Delta \tilde{w} dx + \int_\Omega (k_{11}N_{11} + k_{22}N_{22} - f)\tilde{w} dx \ge 0$$

holds for all $\tilde{w} \in H_0^2(\Omega)$, $\tilde{w} \ge 0$. This implies that $\nu \equiv \Delta^2 w + k_{11}N_{11} + k_{22}N_{22} - f$ is a positive distribution, and consequently, it is a nonnegative measure on the domain Ω. In particular, $\nu(B) < +\infty$ for each compact $B \subset \Omega$.

Now we decompose the measure μ into the sum of the absolutely continuous and the singular components

$$\mu(B) = \mu_0(B) + \mu_1(B) . \tag{3.19}$$

Here μ_0 is a singular measure concentrated on the set of the Lebesgue measure zero and μ_1 is an absolutely continuous measure defined by a function $q(x)$ which is locally integrable in Ω. Namely, for any Borel set $B \subset \Omega$

$$\mu_1(B) = \int_B q(x) dx .$$

The following result is true.

Theorem 3.3. *Let $x_0 \in \Omega$ and $|\nabla\Phi(x_0)| > 0$. There exists a neighbourhood Ω_0 of x_0 such that the singular component μ_0 of the measure μ equals zero on Ω_0.*

Proof. Let us take a neighbourhood Ω_0 as in the proof of Theorem 3.1. Let $\varphi \in C_0^\infty(\Omega_0)$ and $\varepsilon > 0$. From (3.5) it follows that the vector ω_ε, $(u + \varepsilon\varphi\Phi_{x_1}^{-1}, v, w + \varepsilon\varphi) \equiv \omega_\varepsilon$, belongs to the set K. Since the functional Π attains its minimum at $\omega = (u, v, w)$, we have $\Pi(\omega_\varepsilon) - \Pi(\omega) \geq 0$. Because φ is arbitrary, we can divide this inequality by ε, and passing to the limit as $\varepsilon \to 0$ we deduce that the equation

$$\Delta^2 w + k_{11}N_{11} + k_{22}N_{22} - f = \left(\frac{\partial N_{1j}}{\partial x_j} + f_1\right)\Phi_{x_1}^{-1}$$

is satisfied in Ω_0. By Theorem 3.1, the right-hand side of this equation belongs to $L^2(\Omega_0)$. Hence, $w \in H^4_{\mathrm{loc}}(\Omega_0)$.

Choose $\tilde{\omega} = (0, 0, \tilde{w})$, where $\tilde{w} \in H_0^2(\Omega) \cap C_0(\Omega_0)$. From (3.17) it follows that

$$\langle\, \Pi'(\omega), \tilde{\omega}\,\rangle = \int_\Omega \frac{\tilde{w}}{\sqrt{1 + |\nabla\Phi|^2}}\, d\mu\ .$$

Integrating by parts the left-hand side of this relation, we get

$$\int_\Omega (\Delta^2 w + k_{11}N_{11} + k_{22}N_{22} - f)\tilde{w}dx = \int_\Omega \frac{\tilde{w}}{\sqrt{1 + |\nabla\Phi|^2}}\, d\mu\ .$$

The assertion of the theorem is proved. Note that we have actually proved that the derivative of the measure μ on the set Ω_0 is equal to $(\Delta^2 w + k_{11}N_{11} + k_{22}N_{22} - f)\sqrt{1 + |\nabla\Phi|^2}$. □

3.4 Parallel punch

As mentioned before, if $\nabla\Phi(x) = 0$ for $x \in \Omega_0$, then restriction (3.5) takes the form $w \geq \delta$ in Ω_0. In this case the measure μ has the following property.

Theorem 3.4. *Let Ω_0 be a subdomain of Ω such that $\nabla\Phi \equiv 0$ in Ω_0. Then none of the isolated contact points in Ω_0 belongs to the support* $\mathrm{spt}\mu = S(\mu)$ *of the measure μ.*

This theorem has the following mechanical interpretation. The punch which is parallel to Ω_0 cannot interact with the shell at an isolated point of Ω_0. From Theorem 3.3 it follows that there are no concentrated contact forces on any subdomain Ω_1 such that $|\nabla\Phi(x)| > 0, x \in \Omega_1$. If $\nabla\Phi \equiv 0$ in Ω, then w satisfies variational inequality (3.28) (see below) and concentrated contact forces can be constructed for $k_{11} = k_{22} = 0$ (see Caffarelli, Friedman, 1979).

Lemma 3.1. *The function Δw is upper semicontinuous on Ω.*

This lemma actually means that there exists an upper semicontinuous function on Ω which is equivalent to Δw.

Proof of Lemma 3.1. Suppose that $\psi \in H^4(\Omega) \cap H_0^2(\Omega)$ solves the problem

$$\Delta^2 \psi = f_0 \ , \quad \psi\big|_{\partial\Omega} = 0 \ , \quad \frac{\partial \psi}{\partial n}\bigg|_{\partial\Omega} = 0 \ .$$

Recall that $f_0 = f - k_{11}N_{11} - k_{22}N_{22}$. Let us choose an infinitely many times differentiable function $\theta(x) \geq 0$ such that

$$\theta(x) = \theta(|x|) \ , \quad \int_0^1 \theta(\tau)\tau d\tau = 1 \ , \quad \theta(\tau) \equiv 0 \quad \text{when } \tau \geq 1 \ ,$$

and extend the functions w and ψ by putting their values equal to zero beyond Ω. Denote by $w_\varepsilon, \psi_\varepsilon$ the mollifiers of w, ψ given by the formulae

$$w_\varepsilon(x) = \frac{1}{2\pi\varepsilon^2} \int_{|x-y|\leq\varepsilon} \theta\left(\frac{x-y}{\varepsilon}\right) w(y) dy \ . \tag{3.20}$$

Let $g = w - \psi$ and let $B_\rho(x)$ be the ball of radius ρ centered at x. By Green's formula, for every $B_\rho(x) \subset \Omega$,

$$\Delta g_\varepsilon(x) = \frac{1}{2\pi\rho} \int_{\partial B_\rho(x)} \Delta g_\varepsilon(y) dy - \frac{1}{2\pi} \int_{B_\rho(x)} \Delta^2 g_\varepsilon(y) \ \ln \frac{\rho}{|x-y|} dy \ . \tag{3.21}$$

According to the observation which follows the proof of Theorem 3.2, the function $\nu \equiv \Delta^2 w - f_0 = \Delta^2 w - \Delta^2 \psi$ is positive in the sense of distributions. Thus, by Lemma 3.3 below, $\Delta^2 g_\varepsilon(x) = \Delta^2 w_\varepsilon(x) - \Delta^2 \psi_\varepsilon(x) \geq 0$ for all x. $\mathrm{dist}(x, \partial\Omega) \geq \varepsilon$. Since the logarithmic function is increasing, by (3.21) we get

$$\frac{1}{2\pi\rho} \int_{\partial B_\rho(x)} \Delta g_\varepsilon(y) dy \leq \frac{1}{2\pi\rho_1} \int_{\partial B_{\rho_1}(x)} \Delta g_\varepsilon(y) dy \ , \quad \rho_1 \geq \rho \ .$$

Multiplying this inequality by $\rho\rho_1$ and integrating first over ρ from zero to ρ and next over ρ_1 from ρ to ρ_1 we find that

$$\frac{1}{\pi\rho^2} \int_{B_\rho(x)} \Delta g_\varepsilon(y) dy \leq \frac{1}{\pi\rho_1^2} \int_{B_{\rho_1}(x)} \Delta g_\varepsilon(y) dy \ . \tag{3.22}$$

After passing to the limit as $\varepsilon \to 0$, we see that the function Δg satisfies inequality (3.22). For almost all $x \in \Omega$, we have

$$b_\rho(x) \equiv \frac{1}{\pi\rho^2} \int_{B_\rho(x)} \Delta g(y) dy \to \Delta g(x) \ , \quad \rho \to 0 \ .$$

The function b_ρ is continuous and nondecreasing, hence Δg is upper semicontinuous. Taking into account that $\Delta\psi \in C(\bar{\Omega})$, the assertion of Lemma 3.1 follows. □

Lemma 3.2. *The function Δg is subharmonic in Ω.*

This follows from the proof of Lemma 3.1.

Proof of Theorem 3.4. Suppose that $x_0 \in \Omega_0$ is an isolated contact point belonging to the support $S(\mu)$. This means that $w(x_0) = \delta$ and $w(x) > \delta$ in some neighbourhood of x_0, $x \neq x_0$. We choose a ball $B_\rho(x_0) \subset \Omega_0$ with a sufficiently small radius ρ. From Green's formula it follows that

$$w(x_0) = \frac{1}{2\pi\rho} \int_{\partial B_\rho(x_0)} w(y)dy - \frac{1}{2\pi} \int_{B_\rho(x_0)} \Delta w(y) \, \ln \rho |x_0 - y|^{-1} dy \ . \tag{3.23}$$

Since $w(x_0) = \delta$ and $w(x) \geq \delta$ in Ω_0, from (3.23) we deduce that

$$\frac{1}{2\pi} \int_{B_\rho(x_0)} \Delta w(y) \, \ln \rho |x_0 - y|^{-1} dy \geq 0 \ .$$

This inequality yields the existence of a sequence $y_i \in B_\rho(x_0)$ such that $\Delta w(y_i) \geq 0$. Assume that $y_i \to y_\rho$. By the upper semicontinuity of Δw, we clearly have $\Delta w(y_\rho) \geq 0$. Passing to the limit when ρ tends to zero and making use of the upper semicontinuity of Δw, we get the inequality

$$\Delta w(x_0) \geq 0 \ . \tag{3.24}$$

Further, let $H(x) = \dfrac{1}{2\pi} \ln |x|^{-1}$, and let $\vartheta(x)$ be the Dirac measure. The symbol $*$ stands for the convolution of two distributions. It is seen that $\mu = \nu$ on Ω_0.

Consider the potential

$$H * \nu_\rho(x) = \int_{B_\rho(x_0)} H(x-y) d\nu(y) \ ,$$

where ν_ρ is the restriction of the measure ν to $B_\rho(x_0)$. Note that ν_ρ is a finite measure, and hence, by the Fubini theorem, this potential is locally integrable.

We claim that the function

$$\alpha(x) = \Delta g(x) + H * \nu_\rho(x) \ , \quad x \in B_\rho(x_0) \ , \tag{3.25}$$

is harmonic in $B_\rho(x_0)$. In fact, we have

$$\Delta\alpha = \Delta^2 g + \Delta(H * \nu_\rho) = \nu_\rho - (\vartheta * \nu_\rho) = 0 \ .$$

This follows from the equality $\Delta H = -\vartheta$ which is satisfied in the sense of distributions.

Formula (3.25) holds almost everywhere in $B_\rho(x_0)$, and, the functions Δg, $-H * \nu_\rho$ are subharmonic in $B_\rho(x_0)$. It is a well known fact that for subharmonic functions the mean values over a ball of radius r considered at a given point converge to the value of the function as $r \to 0$ (see the formula for $b_\rho(x)$ in the proof of Lemma 3.1). Thus, equality (3.25) holds for all $x \in B_\rho(x_0)$. Taking into account (3.24) we have

$$\int_{B_\rho(x_0)} H(x_0 - y)d\nu(y) = -\Delta g(x_0) + \alpha(x_0) < +\infty \ . \tag{3.26}$$

From this it follows that $\nu(x_0) = 0$, which completes the proof. □

Finally, we turn to the case where $\nabla\Phi \equiv 0$ in Ω. Then

$$K = \left\{(\tilde{u}, \tilde{v}, \tilde{w}) \in H(\Omega) \middle| \tilde{w}(x) \geq \delta \quad \text{a.e.in} \quad \Omega\right\} \ . \tag{3.27}$$

It is easily seen that, by (3.7), equations (3.15) are satisfied. Moreover,

$$\int_\Omega \Delta w(\Delta\tilde{w} - \Delta w)dx \geq \int_\Omega f_0(\tilde{w} - w)dx \tag{3.28}$$

for every function $\tilde{w} \in H_0^2(\Omega)$ such that $\tilde{w}(x) \geq \delta$ almost everywhere in Ω.

Equations (3.15) are satisfied in Ω. Hence, by smoothness results for linear elasticity problems, $u, v \in H^2(\Omega) \cap H_0^1(\Omega)$. If, in addition, $\delta < 0$, we obtain $w \in H^3(\Omega) \cap H_0^2(\Omega)$. To prove this, we note that $w \in C(\bar{\Omega})$, and therefore, by the equality $w\big|_{\partial\Omega} = 0$, we have $w > \delta$ in some neighbourhood of the boundary. Hence, by (3.28), the equation

$$\Delta^2 w = f_0 \tag{3.29}$$

is satisfied in this neighbourhood, and in view of a well-known result for biharmonic equation, in this neighbourhood $w \in H^3$. We have already proved that $w \in H^3_{\text{loc}}(\Omega)$. Hence, $w \in H^3(\Omega)$, which is the required conclusion.

Now we prove that $\nu(\Omega) < +\infty$ provided $\delta < 0$. In fact, equation (3.29) is fulfilled in some neighbourhood of the boundary $\partial\Omega$. Hence, this neighbourhood is of measure zero. Besides, $\nu(B) < +\infty$ for every compact $B \subset \Omega$. Consequently, $\nu(\Omega) < +\infty$.

Put $\delta = 0$ in (3.27). We claim that if the solution to problem (3.7) has the property that

$$f_0 \equiv f - k_{11}N_{11} - k_{22}N_{22} \leq 0 \quad \text{in } \Omega \ , \tag{3.30}$$

then $w \equiv 0$ in Ω. To prove this, let us note that for a given f_0, finding a solution w to inequality (3.28) is equivalent to solving the minimization problem

$$\inf_{\substack{\tilde{w} \in H_0^2(\Omega) \\ \tilde{w} \geq 0}} \left\{\frac{1}{2}\int_\Omega (\Delta\tilde{w})^2 dx - \int_\Omega f_0 \tilde{w} dx\right\} \ .$$

When $f_0 \leq 0$ the minimized functional is nonnegative and attains its minimum at $\tilde{w} \equiv 0$, which proves the assertion. In particular, when $k_{11} = k_{22} = 0$, inequality (3.30) is a requirement imposed exclusively on f.

Finally, we state the result used in the proof of Lemma 3.1.

Lemma 3.3. *For a given* $h \in H_0^2(\Omega)$, *let* $\Delta^2 h$ *be a positive function in the sense of distributions and let* h_ε *be the mollifier of* h. *Then,* $\Delta^2 h_\varepsilon(x) \geq 0$ *for all* $x \in \Omega$, dist $(x, \partial\Omega) \geq \varepsilon$.

Proof. Let $\varphi \in C_0^\infty(\Omega)$, $\varphi \geq 0$. Suppose that the distance between the support of φ and $\partial\Omega$ is greater than ε. Then

$$\int_\Omega \Delta^2 h_\varepsilon(x)\varphi(x)dx = \frac{1}{2\pi\varepsilon^2}\int_\Omega dx\varphi(x)\left(\Delta_x^2 \int_{|x-y|\leq\varepsilon} \theta\left(\frac{x-y}{\varepsilon}\right)h(y)dy\right)$$

$$= \frac{1}{2\pi\varepsilon^2}\int_\Omega dx\varphi(x)\left(\int_{|x-y|\leq\varepsilon} \Delta_y\theta\left(\frac{x-y}{\varepsilon}\right)\Delta_y h(y)dy\right) \geq 0 \ . \qquad \square$$

4 Contact between two elastic plates

4.1 Problem formulation. Properties of the solution

Assume that $\Omega \subset \mathbb{R}^2$ is a simply connected bounded domain with smooth boundary, and $\delta \geq 0$ is a constant.

By K we denote a closed convex set in $H_0^2(\Omega) \times H_0^2(\Omega)$,

$$K = \left\{(u,v) \in H_0^2(\Omega) \times H_0^2(\Omega) \middle| u - v \geq -\delta \quad \text{in} \quad \Omega\right\} \ .$$

Formally, the contact problem between two plates is the problem of minimization of the energy functional

$$\Pi(u,v) = \int_\Omega \left\{a_1(\Delta u)^2 + a_2(\Delta v)^2 - 2Fu - 2Gv\right\} dx$$

over the set K. It is assumed that the plates occupy identical domains and in the natural state they are at a given distance δ from each other. The z-axis is parallel to the normal to the middle planes. Functions u and v describe displacements of the upper plate (which has the greater coordinate z) and the lower plate, respectively, $F, G \in L^2(\Omega)$ are given functions, a_1, a_2 are positive constants characterizing bending properties of the plates.

A solution $(u,v) \in K$ to the above minimization problem exists and satisfies the variational inequality

$$\begin{aligned}\int_\Omega &\left\{a_1 \Delta u(\Delta\tilde{u} - \Delta u) + a_2\Delta v(\Delta\tilde{v} - \Delta v)\right.\\ &\left.- F(\tilde{u} - u) - G(\tilde{v} - v)\right\} dx \geq 0 \qquad \forall\, (\tilde{u}, \tilde{v}) \in K \ .\end{aligned} \tag{4.1}$$

This inequality gives necessary and sufficient conditions for the minimum.

Let C be the coincidence set,

$$C = \left\{x \in \Omega \,\middle|\, u(x) - v(x) = -\delta\right\} \ .$$

Accordingly, $N = \Omega \backslash C$ is the noncoincidence domain.

The purpose of this section is to study the structure of N and C. Some solution regularity results will also be obtained. The main result is the proof of connectedness of the noncoincidence domain under some assumptions upon F and G.

For $\varphi \in H_0^2(\Omega)$, $\varphi \geq 0$ and $\varepsilon > 0$, the pair of functions $(u + \varepsilon\varphi, v)$ belongs to the set K. Hence, $\Pi(u + \varepsilon\varphi, v) \geq \Pi(u, v)$ and

$$\int_\Omega a_1 \Delta u \Delta\varphi dx + \frac{1}{2} a_1 \varepsilon \int_\Omega (\Delta\varphi)^2 dx - \int_\Omega F\varphi dx \geq 0 \ .$$

Passing to the limit as $\varepsilon \to 0$, we conclude that $\nu \equiv a_1 \Delta^2 u - F$ is a positive distribution, and hence, it is a nonnegative measure on Ω. By similar arguments, $-(a_2 \Delta^2 v - G)$ is also a measure.

It is easily seen that these two measures coincide. Indeed, for every function $\varphi \in H_0^2(\Omega)$, we have $(u+\varepsilon\varphi, v+\varepsilon\varphi) \in K$. Hence, the inequality $\Pi(u+\varepsilon\varphi, v+\varepsilon\varphi) \geq \Pi(u, v)$ implies

$$\int_\Omega (a_1 \Delta u \Delta\varphi - F\varphi) dx + \int_\Omega (a_2 \Delta v \Delta\varphi - G\varphi) dx \geq 0 \ .$$

Since φ is arbitrary,

$$a_1 \Delta^2 u - F = -a_2 \Delta^2 v + G \ , \tag{4.2}$$

which is the sought equality.

The support $S(\nu)$ of the measure ν is a subset of C. Indeed, it is easy to verify that the equations

$$a_1 \Delta^2 u = F \ , \qquad a_2 \Delta^2 v = G$$

are satisfied in the domain N.

The following property can be easily established. If $\frac{F}{a_1} - \frac{G}{a_2} > 0$ in Ω_0, $\Omega_0 \subset \Omega$, the coincidence set has no interior points in Ω_0.

To prove this, suppose that the converse is true. Then, in some neighbourhood of an interior contact point we have $\Delta^2 u = \Delta^2 v$ and, hence

$$\frac{\nu}{a_1} + \frac{\nu}{a_2} = \Delta^2 u - \frac{F}{a_1} - \Delta^2 v + \frac{G}{a_2} = \frac{G}{a_2} - \frac{F}{a_1} < 0 \ .$$

This, however, contradicts the nonnegativity of the measure ν.

In other words, if the above property holds, then none of the balls in Ω_0, even those of arbitrarily small radius, contain only contact points.

Now we are going to prove that isolated points of $S(\nu)$ cannot be contact points.

According to Lemma 4.4 below, $\Delta u - \Delta v \in L^\infty_{\text{loc}}(\Omega)$. From (4.2) it follows that $a_1 \Delta u + a_2 \Delta v \in L^\infty(\Omega)$, and hence $\Delta u \in L^\infty_{\text{loc}}(\Omega)$.

Suppose that there exists an isolated contact point x_0 belonging to $S(\nu)$. Let $B_r(x_0)$ be the ball. Its boundary will be denoted by $\partial B_r(x_0)$. Then, for sufficiently small r, the equation

$$a_1 \Delta^2 u = F \tag{4.3}$$

is satisfied in the domain $B_r^0(x_0) \equiv B_r(x_0)\backslash\{x_0\}$.

Now we show that this equation is satisfied in $B_r(x_0)$. In fact, let $F_0 \in H^2\big(B_r(x_0)\big) \cap H_0^1\big(B_r(x_0)\big)$ solve the problem in $B_r(x_0)$

$$\Delta F_0 = F \ , \qquad F_0\Big|_{\partial B_r(x_0)} = 0 \ .$$

Equation (4.3) can be written as $\Delta(a_1\Delta u - F_0) = 0$. It is satisfied in $B_r^0(x_0)$. By regularity results for biharmonic equations, $\Delta u \in H^2_{\mathrm{loc}}\big(B_r^0(x_0)\big)$, and consequently the function Δu is continuous in $B_r^0(x_0)$. In virtue of the imbedding theorems, the function F_0 is also continuous in $B_r^0(x_0)$. Moreover, the function $a_1\Delta u - F_0$ is bounded in $B_r^0(x_0)$. From the well-known removable singularity theorem for harmonic functions it follows that $\Delta(a_1\Delta u - F_0) = 0$ in $B_r(x_0)$. This equation, however, contradicts the fact that $x_0 \in S(\nu)$, which proves the assertion.

In what follows we shall prove some regularity results.

Lemma 4.1. *We have* $u, v \in H^3_{\mathrm{loc}}(\Omega)$.

Proof. Let $\Omega_2 \subset \Omega_1 \subset \Omega$ be domains such that dist $(\partial\Omega_1, \partial\Omega) \geq q > 0$, where q is a constant, $\bar{\Omega}_2 \subset \Omega_1$. Take a function $\varphi \in C_0^\infty(\Omega_1)$, $\varphi \equiv 1$ on Ω_2, $0 \leq \varphi \leq 1$ everywhere, and the operators $d_{i\tau}$, $\Delta_{i\tau}$ as defined in the preceding section.

We begin by noting that the functions

$$u_\tau = u + \frac{\tau^2}{2}\varphi^2\Delta_{i\tau}u \ , \quad v_\tau = v + \frac{\tau^2}{2}\varphi^2\Delta_{i\tau}v \ , \quad 0 < |\tau| < q \ ,$$

satisfy the inequality $u_\tau - v_\tau \geq -\delta$, or equivalently, $(u_\tau, v_\tau) \in K$.

In fact, the relations

$$\begin{aligned} u_\tau - v_\tau &= u - v + \frac{\tau^2}{2}\varphi^2\Delta_{i\tau}(u-v) = (1-\varphi^2)(u-v)(x) \\ &+ \frac{\varphi^2}{2}(u-v)(x+\tau e_i) + \frac{\varphi^2}{2}(u-v)(x-\tau e_i) \\ &\geq (1-\varphi^2)(-\delta) + \frac{\varphi^2}{2}(-\delta) + \frac{\varphi^2}{2}(-\delta) = -\delta \end{aligned} \tag{4.4}$$

hold which means that $(u_\tau, v_\tau) \in K$.

Substituting the pair (u_τ, v_τ) into (4.1) as a test function we get

$$\begin{aligned} a_1\int_\Omega \Delta u\Delta(\varphi^2\Delta_{i\tau}u)dx + a_2\int_\Omega \Delta v\Delta(\varphi^2\Delta_{i\tau}v)dx \\ \geq \int_\Omega F\varphi^2\Delta_{i\tau}u dx + \int_\Omega G\varphi^2\Delta_{i\tau}v dx \ . \end{aligned} \tag{4.5}$$

We can apply to (4.5) the technique used in Section 3 to show that

$$\|d_{i\tau}(\varphi u)\|_2^2 + \|d_{i\tau}(\varphi v)\|_2^2 \le c\{\|F\|_0^2 + \|G\|_0^2 + \|u\|_2^2$$

$$+\|v\|_2^2 + \|u\|_2\|d_{i\tau}(\varphi u)\|_2 + \|v\|_2\|d_{i\tau}(\varphi v)\|_2\}$$

with a constant c which is uniform with respect to τ. Hence, the left-hand side of this inequality is bounded and therefore $\varphi u, \varphi v \in H^3(\Omega)$, ie. $u, v \in H^3_{\text{loc}}(\Omega)$. This proves the lemma. □

In general, one cannot expect higher order regularity. To see this, one can put $a_1 = a_2$ and substitute the pair $(v + \tilde{w}, v)$ into (4.1) as a test function with $\tilde{w}$ being an element from $H_0^2(\Omega)$ such that $\tilde{w} \ge -\delta$ in Ω. By substituting the pair $(u, u - \tilde{w})$ into (4.1) and summing up the resulting relations we find that, for $w = u - v$,

$$\int_\Omega \left[a_1 \Delta w(\Delta\tilde{w} - \Delta w) - (F - G)(\tilde{w} - w)\right] dx \ge 0 .$$

Thus, the difference $u - v$ satisfies the variational inequality for the biharmonic operator, and, in general, it does not belong to the space $H^4_{\text{loc}}(\Omega)$ (see Caffarelli, Friedman, 1979). At the same time, from (4.2) it follows that $a_1 u + a_2 v \in H^4(\Omega) \cap H_0^2(\Omega)$.

Lemma 4.2. *The function Δu (resp. Δv) is upper (resp. lower) semicontinuous in Ω.*

Proof. Let n be the unit outward normal to $\partial\Omega$. Assume that $\psi \in H^4(\Omega) \cap H_0^2(\Omega)$ solves the problem

$$a_1 \Delta^2 \psi = F \ ; \quad \psi = \frac{\partial \psi}{\partial n} = 0 \quad \text{on} \quad \partial\Omega .$$

The functions u and ψ can be extended beyond Ω without loss of regularity. Then, the mollifiers can be introduced by the formulae

$$u_\varepsilon(x) = \frac{1}{2\pi\varepsilon^2} \int_{|x-y|\le\varepsilon} \theta\left(\frac{x-y}{\varepsilon}\right) u(y) dy .$$

where $\theta(x) \ge 0$ is an infinitely many times differentiable function such that

$$\theta(x) = \theta(|x|) \ , \quad \int_0^1 \theta(t) t dt = 1 \ , \quad \theta(t) = 0 \quad \text{when} \quad |t| \ge 1 .$$

Set $f = u - \psi$. The functions

$$b_r(x) \equiv \frac{1}{\pi r^2} \int_{B_r(x)} \Delta f(y) dy$$

converge to $\Delta f(x)$ for almost all $x \in \Omega$ as $r \to 0$.

Now we are going to show that $b_r(x)$ is nondecreasing as a function of r. This, in turn, will entail the upper semicontinuity of Δf, since b_r is continuous, which together with the relation $\Delta u \in C(\bar{\Omega})$ will complete the proof of this lemma.

To prove that $b_r(x)$ is nondecreasing, we use Green's formula

$$\Delta f_\varepsilon(x) = \frac{1}{2\pi r}\int_{\partial B_r(x)} \Delta f_\varepsilon(y)dy - \frac{1}{2\pi}\int_{B_r(x)} \Delta^2 f_\varepsilon(y)\ln r|x-y|^{-1}dy \ . \tag{4.6}$$

The inequality $\Delta^2 u - \Delta^2\psi \geq 0$ is satisfied in the sense of distributions. This implies that $\Delta^2 f_\varepsilon(x) = \Delta^2 u_\varepsilon(x) - \Delta^2\psi_\varepsilon(x) \geq 0$ for all x whose distance from $\partial\Omega$ is greater than ε.

Consider equality (4.6) for $r_1 \geq r$. Since $\ln\ r|x-y|^{-1}$ is increasing with respect to r, we have

$$\frac{1}{2\pi r}\int_{\partial B_r(x)} \Delta f_\varepsilon(y)dy \leq \frac{1}{2\pi r_1}\int_{\partial B_{r_1}(x)} \Delta f_\varepsilon(y)dy \ .$$

Hence,

$$\frac{1}{\pi r^2}\int_{B_r(x)} \Delta f_\varepsilon(y)dy \leq \frac{1}{\pi r_1^2}\int_{B_{r_1}(x)} \Delta f_\varepsilon(y)dy \ .$$

Passing to the limit in this inequality as $\varepsilon \to 0$ and taking into account that $\Delta f_\varepsilon \to \Delta f$ in $L^2(\Omega)$ we get the inequality $b_r(x) \leq b_{r_1}(x)$. As already mentioned, this relation completes the proof of the first part of this lemma.

The same reasoning applies to the function $g = v - \xi$, where $\xi \in H^4(\Omega) \cap H_0^2(\Omega)$ solves the problem

$$a_2\Delta^2\xi = G \ ; \quad \xi = \frac{\partial\xi}{\partial n} = 0 \quad \text{on} \quad \partial\Omega \ .$$

This yields the lower semicontinuity of Δg and Δv, since $\Delta\xi \in C(\bar{\Omega})$ and the proof of the lemma is completed. □

Note that in this proof we have actually shown that Δf is a subharmonic function, whereas Δg is superharmonic.

Lemma 4.3. *For any $x_0 \in C$, we have $\Delta u(x_0) \geq \Delta v(x_0)$.*

Proof. Put $w = u - v$. By Green's formula,

$$w(x_0) = I_1 - I_2 \ ,$$

$$I_1 = \frac{1}{2\pi r}\int_{\partial B_r(x_0)} w(y)dy \ , \quad I_2 = \frac{1}{2\pi}\int_{B_r(x_0)} \Delta w(y)\ \ln\ r|x_0 - y|^{-1}dy \ .$$

Since $w(y) \geq -\delta$ and $w(x_0) = -\delta$, we have $I_2 \geq 0$. Hence, there exists a sequence $y_i \in B_r(x_0)$ such that $\Delta w(y_i) \geq 0$. We can assume that $y_i \to y_r$. The function $\Delta w = \Delta u - \Delta v$ is upper semicontinuous. Thus, passing to the upper limit as $i \to \infty$, one gets $\Delta w(y_r) \geq 0$. Passing to the limit as $r \to 0$ and using again the upper semicontinuity of Δw we conclude that $\Delta u(x_0) \geq \Delta v(x_0)$. The lemma is proved. □

Let f and g be the functions introduced in Lemma 4.2. We can prove the following result.

Lemma 4.4. *The relation* $\Delta f - \Delta g \in L^{\infty}_{\mathrm{loc}}(\Omega)$ *holds.*

Proof. Let $H(x) = \dfrac{1}{2\pi} \ln |x|^{-1}$ and let $\vartheta(x)$ be the Dirac measure. By $*$ we denote the convolution of two distributions. We take $\bar{\Omega}_0 \subset \Omega$ and denote by $\tilde{\nu}_0$ the restriction of the measure $\tilde{\nu} = \dfrac{\nu}{a_1} + \dfrac{\nu}{a_2}$ to Ω_0.

Consider the potential

$$H * \tilde{\nu}_0(x) = \int_{\Omega_0} H(x-y) d\tilde{\nu}(y) \ .$$

In view of the Fubini theorem and since the measure $\tilde{\nu}_0$ is finite, this potential is a locally integrable function with respect to the Lebesgue measure.

We shall prove that the function

$$\gamma(x) = \Delta f(x) - \Delta g(x) + H * \tilde{\nu}_0(x) \ , \quad x \in \Omega_0 \ , \tag{4.7}$$

is harmonic in Ω_0. Let us first note that $\Delta H(x) = -\vartheta(x)$ and

$$\Delta\gamma = \Delta^2 f - \Delta^2 g + \Delta(H * \tilde{\nu}_0) = \tilde{\nu}_0 + (\Delta H * \tilde{\nu}_0) = \tilde{\nu}_0 - \vartheta * \tilde{\nu}_0 = 0 \ .$$

Relation (4.7) is fulfilled almost everywhere in Ω_0. Moreover, the functions $-H*\tilde{\nu}_0$ and $\Delta f - \Delta g$ are subharmonic in Ω. Recall that the mean values of a subharmonic function over the ball of radius r converge to the value of this function at a given point as $r \to 0$. This means that (4.7) is valid for all $x \in \Omega_0$.

Now we take into account the well-known theorem on boundedness of potentials (see Landkof, 1966). This result says that if a potential of a finite measure μ is bounded from above on the support $S(\mu)$, then it is bounded from above at every point.

By Lemma 4.3, for every $x_0 \in S(\tilde{\nu})$, one has $\Delta u(x_0) \geq \Delta v(x_0)$. Let $\bar{\Omega}_1 \subset \Omega_0$, $\bar{\Omega}_2 \subset \Omega_1$. From (4.7) it follows that for all $x_0 \in \Omega_1 \cap S(\tilde{\nu})$

$$\int_{\Omega_0} H(x_0 - y) d\tilde{\nu}(y) = \gamma(x_0) + \Delta g(x_0) - \Delta f(x_0) < c < +\infty \ .$$

Consequently, the potential of the measure $\tilde{\nu}_1$ (the restriction of $\tilde{\nu}$ to Ω_1) is bounded from above for all x. This potential is also bounded from below. Thus, taking into account the decomposition

$$H * \tilde{\nu}_0(x) = \int_{\Omega_1} H(x-y) d\tilde{\nu}(y) + \int_{\Omega_0 \setminus \Omega_1} H(x-y) d\tilde{\nu}(y) \ ,$$

from (4.7) it follows that the function $|\Delta f - \Delta g|$ is bounded on Ω_2. The lemma is proved. □

4.2 Connectedness of the noncoincidence domain

The main result of this section can be formulated as follows.

Theorem 4.1. *Let* $\delta > 0$ *and* $\dfrac{F}{a_1} \leq \dfrac{G}{a_2}$ *on* Ω. *Then, the noncoincidence domain* N *is connected.*

Proof. The functions u, v are equal to zero on $\partial\Omega$ and continuous in $\bar{\Omega}$. Hence, the noncoincidence domain contains a neighbourhood of the boundary $\partial\Omega$.

Suppose that the assertion of the theorem is not true. Then, a connected component N_1 of the noncoincidence domain N exists such that none of the curves lying in N connects N_1 with the above neighbourhood.

In view of Lemma 4.3,

$$\Delta u(x_0) \geq \Delta v(x_0) \tag{4.8}$$

on the boundary ∂N_1. Moreover, the equations

$$a_1 \Delta^2 u = F \ , \quad a_2 \Delta^2 v = G \tag{4.9}$$

are satisfied in the domain N_1.

Below we shall prove that the inequality

$$\Delta u(x) \geq \Delta v(x) \ , \quad x \in N_1 \tag{4.10}$$

is a consequence of (4.8), (4.9). Next, by noting that $u(x) - v(x) = -\delta$ on the boundary ∂N_1 and taking into account the well-known maximum principle we deduce from (4.10) that $u - v \leq -\delta$ in N_1, contradictory to the definition of N_1.

Thus, to complete the proof we have to validate (4.10). Let the domains Ω_1, Ω_2 have the properties that $\bar{N}_1 \subset \Omega_1$, $\bar{\Omega}_1 \subset \Omega_2$, $\bar{\Omega}_2 \subset \Omega$. The representation of the type (4.7) holds in the domain Ω_2 so that

$$\Delta f(x) - \Delta g(x) = \gamma(x) - H * \tilde{\nu}_2(x) \ , \quad x \in \Omega_2 \ . \tag{4.11}$$

Here $\tilde{\nu}_2$ is the restriction of the measure $\tilde{\nu}$ to Ω_2.

The potential $H * \tilde{\nu}_2(x)$ can be written as

$$H * \tilde{\nu}_2(x) = -p(x) + \frac{1}{2\pi} \int_{\Omega_2 \setminus \bar{\Omega}_1} \ln\, |x - y|^{-1} d\tilde{\nu}(y) - \frac{1}{2\pi} \tilde{\nu}(\bar{\Omega}_1)\, \ln\, m \ ,$$

$$x \in \Omega_2 \ , \quad m = \text{const} \ , \quad m > \text{diam}\, \Omega_1 \ ,$$

$$p(x) = -\frac{1}{2\pi} \int_{\bar{\Omega}_1} \ln\, m|x - y|^{-1} d\tilde{\nu}(y) \ .$$

By this, formula (4.11) can be rewritten as

$$\Delta f(x) - \Delta g(x) = p(x) + \beta(x) \ . \tag{4.12}$$

The function $\beta(x)$ is harmonic in Ω_1. By Lemma 4.4, for $x \in \bar{\Omega}_1$, the functions

$$p_r(x) \equiv -\frac{1}{2\pi}\int_{\bar{\Omega}_1\setminus B_r(x)} \ln\, m|x-y|^{-1} d\tilde{\nu}(y)$$

converge from above to $p(x)$ as $r \to 0$. By Egorov's theorem, for any $\varepsilon > 0$, there exists a closed subset $\Omega_\varepsilon \subset \bar{\Omega}_1$ such that $\tilde{\nu}(\bar{\Omega}_1\setminus\Omega_\varepsilon) < \varepsilon$ and p_r converges to p uniformly in Ω_ε.

Define

$$p_{r,\varepsilon}(x) = -\frac{1}{2\pi}\int_{\bar{\Omega}_1\setminus B_r(x)} \ln\frac{m}{|x-y|} d\tilde{\nu}_\varepsilon(y) \ ,$$

$$p_\varepsilon(x) = -\frac{1}{2\pi}\int_{\bar{\Omega}_1} \ln\frac{m}{|x-y|} d\tilde{\nu}_\varepsilon(y) \ ,$$

where $\tilde{\nu}_\varepsilon$ is the restriction of the measure $\tilde{\nu}$ to Ω_ε. Taking into account the uniform convergence of p_r to p in Ω_ε we find that

$$0 \le p_{r,\varepsilon}(x) - p_\varepsilon(x) \le \frac{1}{2\pi}\int_{\bar{\Omega}_1\cap B_r(x)} \ln\, m|x-y|^{-1} d\tilde{\nu}(y) \to 0 \ .$$

The last convergence is uniform in Ω_ε as $r \to 0$. This means that p_ε are continuous in Ω_ε. Note that $S(\tilde{\nu}_\varepsilon) \subset \Omega_\varepsilon$.

The well-known theorem on continuity of potentials states that the continuity on the support of the measure yields the continuity everywhere (see Landkof, 1966). By this theorem, p_ε are continuous in Ω_1. Next, the inequality $p_\varepsilon \ge p$ and (4.12) ensure that $p_\varepsilon + \beta \ge \Delta f - \Delta g$ in Ω_1. In particular, this inequality is valid on ∂N_1. With (4.8) we find that $p_\varepsilon + \beta + \Delta\psi - \Delta\xi \ge 0$ on ∂N_1. Since $\Delta p_\varepsilon = 0$ and $\Delta\beta = 0$ on N_1, the inequality

$$\Delta(p_\varepsilon + \beta + \Delta\psi - \Delta\xi) = \Delta^2\psi - \Delta^2\xi = \frac{F}{a_1} - \frac{G}{a_2} \le 0 \quad \text{on} \quad N_1$$

follows. The functions $p_\varepsilon + \beta + \Delta\psi - \Delta\xi$ are continuous in $\bar{N}_1$, and hence, the maximum principle implies that

$$p_\varepsilon + \beta + \Delta\psi - \Delta\xi \ge 0 \quad \text{in} \quad N_1 \ . \tag{4.13}$$

Therefore, for all $x \in N_1$,

$$0 \le p_\varepsilon(x) - p(x) \le \frac{1}{2\pi}\int_{\bar{\Omega}_1} \ln\, m|x-y|^{-1} d(\tilde{\nu} - \tilde{\nu}_\varepsilon) \to 0 \ , \quad \varepsilon \to 0 \ ,$$

since $S(\tilde{\nu}) \cap N_1 = \emptyset$. Thus, from (4.13), it can be deduced that

$$\Delta\psi(x) - \Delta\xi(x) \ge \lim_{\varepsilon\to 0}\big(-p_\varepsilon(x) - \beta(x)\big) = -p(x) - \beta(x) \ , \quad x \in N_1 \ .$$

With (4.12) we obtain (4.10) and the theorem is proved. □

5 Regularity of solutions to variational inequalities of order four

In this section some examples of variational inequalities of order four are analysed. It is shown that solutions possess square integrable derivatives up to order four. The result concerns the situations where we admit points at which the constraints are active or the bounds are attained.

5.1 The contact problem of a plate with a membrane

Assume that $\Omega \subset \mathbb{R}^2$ is a bounded domain with smooth boundary $\partial\Omega$. Let $\delta \geq 0$ be a constant and $f, g \in L^2(\Omega)$. Consider a closed and convex set in the product of Sobolev spaces,

$$K = \{(u,v) \in H_0^2(\Omega) \times H_0^1(\Omega) | u - v \geq -\delta \quad \text{a. e. in } \Omega\}.$$

There exists a unique solution (u, v) to the variational inequality

$$(u,v) \in K : \quad \int_\Omega \{\Delta u(\Delta \tilde{u} - \Delta u) + \nabla v \cdot (\nabla \tilde{v} - \nabla v)\} dx \qquad (5.1)$$
$$- \int_\Omega \{f(\tilde{u} - u) + g(\tilde{v} - v)\} dx \geq 0 \quad \forall\, (\tilde{u}, \tilde{v}) \in K\,.$$

Here Δ is the Laplace operator with respect to $x = (x_1, x_2) \in \Omega$, and ∇ stands for the gradient operator.

Inequality (5.1) describes the equilibrium state of a plate in contact with a membrane. In the stress free state both structures remain at a distance δ from each other. After the loading is applied, they touch each other and the equilibrium configuration of both structures becomes fixed. This problem is equivalent to the minimization of the energy functional over the set K,

$$\inf_{(u,v)\in K} \int_\Omega \{(\Delta u)^2 + (\nabla v)^2 - 2fu - 2gv\} dx. \qquad (5.2)$$

By the coercivity and the weak lower semicontinuity of the energy functional on the product space $H_0^2(\Omega) \times H_0^1(\Omega)$, there exists a solution to (5.2).

The following theorem gives the interior regularity of the solutions.

Theorem 5.1. *We have*

$$u \in H^4_{\text{loc}}(\Omega), \quad v \in H^2_{\text{loc}}(\Omega)\,.$$

Proof. Assume that Ω_1, Ω_2 are given subdomains of Ω, $\bar{\Omega}_2 \subset \Omega_1$, $\text{dist}(\partial\Omega_1, \partial\Omega) \geq q$, q is a constant.

Let us introduce an auxiliary function $\varphi \in C_0^\infty(\Omega_1)$, $\varphi \equiv 1$ on Ω_2 and $0 \leq \varphi \leq 1$ everywhere. As in the previous section one can show (cf. (4.4)) that a pair of functions u_τ, v_τ with the components defined by

$$\begin{aligned} u_\tau &= u - \frac{\tau^2}{2}\varphi^2 d_{-i\tau} d_{i\tau} u, \\ v_\tau &= v - \frac{\tau^2}{2}\varphi^2 d_{-i\tau} d_{i\tau} v, \quad 0 < |\tau| < q, \end{aligned}$$

satisfies the inequality $u_\tau - v_\tau \geq -\delta$ and consequently, belongs to the set K. Substituting (u_τ, v_τ) into (5.1) as a test function we get

$$\begin{aligned} &\int_\Omega \Delta u \Delta(\varphi^2 d_{-i\tau} d_{i\tau} u) dx + \int_\Omega \nabla v \cdot \nabla(\varphi^2 d_{-i\tau} d_{i\tau} v) dx \\ &- \int_\Omega f\varphi^2 d_{-i\tau} d_{i\tau} u dx - \int_\Omega g\varphi^2 d_{-i\tau} d_{i\tau} v dx \leq 0. \end{aligned} \tag{5.3}$$

By definition of the set K, for any $\varphi \in H_0^2(\Omega)$,

$$(u + \varepsilon\varphi, v + \varepsilon\varphi) \in K \ .$$

Substituting this pair of functions into (5.1) one finds that the equation

$$\Delta^2 u - f = \Delta v + g \tag{5.4}$$

is satisfied in Ω.

Since $v \in H_0^1(\Omega)$, we have $\Delta v \in H^{-1}(\Omega)$. Therefore, by the existence theorems for the Dirichlet problem for the biharmonic equation, one concludes that $u \in H^3(\Omega) \cap H_0^2(\Omega)$. Consequently, the first, third and fourth terms in inequality (5.3) are of lower order, which means that the terms are bounded from above by a constant which depends only on the $H^2 \times H^1$-norm of the solution and on φ.

Now, let us investigate the second term. Consider the chain of the formulae

$$\begin{aligned} &\int_\Omega \nabla v \cdot \nabla(\varphi^2 d_{-i\tau} d_{i\tau} v) dx \\ \to &\int_\Omega \nabla(\varphi v) \cdot \nabla(\varphi d_{-i\tau} d_{i\tau} v) dx \\ \to &\int_\Omega \nabla(\varphi v) \cdot \nabla(d_{-i\tau} d_{i\tau} \varphi v) dx \\ \to &\int_\Omega d_{i\tau} \nabla(\varphi v) \cdot d_{i\tau} \nabla(\varphi v) dx \ . \end{aligned} \tag{5.5}$$

The differences between subsequent terms of this chain can be estimated from above by the quantity

$$c(\|v\|_1^2 + \|v\|_1 \|d_{i\tau}(\varphi v)\|_1),$$

c is a constant independent of v and τ. Here $\|\cdot\|_s$ denotes the norm in the Sobolev space $H^s(\Omega)$. Hence, we obtain the following estimate

$$\|d_{i\tau}(\varphi v)\|_1^2 \le c\{\|v\|_1^2 + \|v\|_1\|d_{i\tau}(\varphi v)\|_1 + \|u\|_3^2 + \|f\|_0^2 + \|g\|_0^2\},$$

with c being independent of u, v, τ. Thus,

$$\|d_{i\tau}(\varphi v)\|_1 \le c$$

and we conclude that $\varphi v \in H^2(\Omega)$, $v \in H^2_{\text{loc}}(\Omega)$. In view of (5.4) we find that $u \in H^4_{\text{loc}}(\Omega)$ which completes the proof. □

In the sequel we deal with the solution regularity up to the boundary.

Theorem 5.2. *Let $\delta > 0$. Then*

$$u \in H^4(\Omega) \cap H_0^2(\Omega), \quad v \in H^2(\Omega) \cap H_0^1(\Omega).$$

Proof. Let Ψ solve the problem

$$\Delta\Psi = g \text{ in } \Omega, \quad \Psi = 0 \text{ on } \partial\Omega\,.$$

Then, $\Psi \in H^2(\Omega) \cap H_0^1(\Omega)$ and hence, in view of the imbedding theorem, $\Psi \in C(\bar{\Omega})$.

First, we show that if $r \to 0$, then

$$\frac{1}{r^2}\int_{B_r(x)} |w(y)|dy \to 0, \quad r = \text{dist}(x, \partial\Omega), \tag{5.6}$$

uniformly with respect to x, where $w = v + \Psi$.

Let x_0 be a fixed point in $\partial\Omega$. We shift the origin to this point. Let the curve $x_2 = \alpha(x_1)$ parametrize the boundary $\partial\Omega$ in a sufficiently small neighbourhood $\mathcal{O}$ of x_0. For any $x = (x_1, x_2) \in \mathcal{O} \cap \Omega$, we have $x_2 > \alpha(x_1)$. We transform the coordinates in $\mathcal{O}$,

$$y_1 = x_1, \quad y_2 = x_2 - \alpha(x_1)\,.$$

It is assumed that y_0 is the image of x_0 under this transformation.

For any smooth function $\tilde{h}(y) = h(x)$ vanishing at $y_2 = 0$ we have

$$\tilde{h}(y) = \int_0^{y_2} \tilde{h}_{y_2} dy_2\,.$$

Consequently, for

$$G_\varepsilon = \{|y - y_0| < \varepsilon \mid \ y_2 > 0\},$$

it follows that

$$\frac{1}{\varepsilon^2}\int_{G_\varepsilon}|\tilde{h}|dy \le \frac{\varepsilon}{\varepsilon^2}\int_{G_\varepsilon}|\tilde{h}_{y_2}|dy \le c\frac{\varepsilon^2}{\varepsilon^2}\Big(\int_{G_\varepsilon}\tilde{h}^2_{y_2}dy\Big)^{\frac{1}{2}} \le c\|\nabla h\|_{L^2(\mathcal{O}_\varepsilon)}\,. \tag{5.7}$$

Here $\mathcal{O}_\varepsilon$ is the image of G_ε and c is a constant independent of $\tilde{h}$. The right-hand side of (5.7) tends to zero as $\varepsilon \to 0$. If the function $\tilde{h}$ belongs to the Sobolev space H^1, and vanishes in mean for $y_2 = 0$ then it can be approximated by a sequence of smooth functions. Hence, (5.7) holds for $\tilde{h} \in H^1$.

Accordingly, for the function $\tilde{w}(y) = w(x)$

$$\frac{1}{\varepsilon^2}\int_{G_\varepsilon}|\tilde{w}(y)|dy \to 0\,.$$

Hence (5.6) follows.

Now, by using (5.6) we show that the equation

$$\Delta v + g = 0 \tag{5.8}$$

is satisfied in a vicinity of the boundary $\partial\Omega$.

Note that $\Delta w = \Delta v + g$ is a nonnegative distribution (generalized function). Indeed, let us take $\varphi \in H_0^2(\Omega)$, $\varphi \ge 0$. Then the pair $(u, v-\varphi)$ belongs to the set K. Substituting this pair into (5.1) we get

$$\int_\Omega(-\nabla v\cdot\nabla\varphi + g\varphi)dx \ge 0\,.$$

According to Theorem 5.1, $\Delta w \in L^2_{\mathrm{loc}}(\Omega)$ and $w \in C(\Omega)$. It follows from Green's formula that

$$w(x_0) = \frac{1}{2\pi r}\int_{\partial B_r(x_0)} w(y)dy - \frac{1}{2\pi}\int_{B_r(x_0)}\Delta w(y)\ln r|x_0-y|^{-1}dy.$$

Here $x_0 \in \Omega$ is an arbitrary point and $r < \operatorname{dist}(x_0, \partial\Omega)$. The same inequality is satisfied for $r_1 \ge r$, $r_1 < \operatorname{dist}(x_0,\partial\Omega)$. Since $\ln r_1|x_0-y|^{-1} \ge \ln r|x_0-y|^{-1}$ and $\Delta w \ge 0$, it follows that

$$\frac{1}{2\pi r}\int_{\partial B_r(x_0)} w(y)dy \le \frac{1}{2\pi r_1}\int_{\partial B_{r_1}(x_0)} w(y)dy$$

and hence

$$w_r(x_0) \equiv \frac{1}{\pi r^2}\int_{\partial B_r(x_0)} w(y)dy \le \frac{1}{\pi r_1^2}\int_{\partial B_{r_1}(x_0)} w(y)dy.$$

Thus $w_r(x_0) \downarrow w(x_0)$, with $r \to 0$. Furthermore, by the imbedding theorem $u \in C(\bar{\Omega})$, and therefore there exists a neighbourhood $\tilde{\mathcal{O}}$ of the boundary $\partial\Omega$ such that $u + \delta > \frac{\delta}{2}$ and $|\Psi| < \frac{\delta}{4}$ in $\tilde{\mathcal{O}}$.

Suppose that there exists a point x_0 in $\tilde{\mathcal{O}}$, such that $u(x_0) - v(x_0) = -\delta$, ie. x_0 is a contact point. Hence, $v(x_0) = u(x_0) + \delta > \frac{\delta}{2}$, and, according to what we have just proved, $w_r(x_0) \geq w(x_0)$. Thus,

$$w_r(x_0) \geq \frac{\delta}{2} + \Psi(x_0) .$$

Here r can be equal to the distance to the boundary. Since $|\Psi(x_0)| < \frac{\delta}{4}$, we get a contradiction with (5.6), since (5.6) holds for any point x_0 in a vicinity of $\partial\Omega$. Hence, the strict inequality $u - v > -\delta$ holds in $\tilde{\mathcal{O}}$.

In view of inequality (5.1), equation (5.8) is satisfied in $\tilde{\mathcal{O}}$, which implies that $v \in H^2(\tilde{\mathcal{O}})$. From Theorem 5.1 it follows that $v \in H^2(\Omega) \cap H_0^1(\Omega)$, and, by (5.4), we deduce that $u \in H^4(\Omega) \cap H_0^2(\Omega)$, which completes the proof. □

5.2 The contact problem for a shell

The contact problem between a shallow shell and a rigid obstacle is considered. Let $\Omega \subset \mathbb{R}^2$ be a bounded domain with smooth boundary $\partial\Omega$. Consider the product space $\mathcal{H}(\Omega) = H_0^1(\Omega) \times H_0^1(\Omega) \times H_0^2(\Omega)$ and the energy functional defined on $\mathcal{H}(\Omega)$,

$$\Pi(\omega) = \int_\Omega \{(\nabla u)^2 + (\nabla v)^2 + (\Delta w)^2 - 2F\omega\}dx, \tag{5.9}$$
$$\omega = (u, v, w), \quad F = (f_1, f_2, f) .$$

Let $z = \phi(x)$ be the boundary of the punch (a rigid body), where $\phi \in C^\infty(\bar{\Omega})$. Let K be the subset of $\mathcal{H}(\Omega)$ containing functions such that

$$w(x) - U(x) \cdot \nabla\phi(x) \geq \phi(x), \quad U = (u, v), \tag{5.10}$$

almost everywhere in Ω. The form of constraints (5.10) coincides with that considered in Section 3. These constraints describe the linear approximation of the nonpenetration condition of the unilateral contact between the shell and the punch parametrized by $z = \phi(x)$, $x \in \Omega$.

The problem of minimization of the functional Π over the set K is equivalent to the following variational inequality

$$\omega \in K \;:\; \langle \Pi'(\omega), \tilde{\omega} - \omega \rangle \geq 0 \quad \forall\, \tilde{\omega} \in K. \tag{5.11}$$

The aim of this section is to prove the following result.

Theorem 5.3. *Let* $|\nabla\phi(x)| > 0$ *for* $x \in \Omega$ *and* $\phi < 0$ *on* $\partial\Omega$. *Then* $(u, v, w) \in H^2(\Omega) \times H^2(\Omega) \times H^4(\Omega)$.

Proof. Let $\omega = (u, v, w)$ solve inequality (5.11). Choose an arbitrary function $\varphi \in C_0^\infty(\Omega)$. Then $\widetilde{\omega} = (u + \varphi, v, w + \varphi\phi_{x_1})$ belongs to K. Substitute $\widetilde{\omega}$ into inequality (5.11). Since φ is arbitrary, the following equation

$$\phi_{x_1}(\Delta^2 w - f) = \Delta u + f_1 \tag{5.12}$$

is satisfied in Ω.

Accordingly,

$$\phi_{x_2}(\Delta^2 w - f) = \Delta v + f_2 \ . \tag{5.13}$$

Now, multiplying equations (5.12) and (5.13) by ϕ_{x_1} and ϕ_{x_2}, respectively, and summing up the resulting equations we get

$$|\nabla\phi|^2(\Delta^2 w - f) = \Delta h + G,$$

where $h = U \cdot \nabla\phi$. Note that the function G depends on derivatives of u, v, ϕ, and $G \in L^2(\Omega)$.

Now, constraint (5.10) can be written as

$$w(x) - h(x) \geq \phi(x), \quad x \in \Omega. \tag{5.14}$$

By (5.11), $\Delta^2 w - f \geq 0$. Hence, $\Delta h + G \geq 0$. Let $p = h + g$, where $g \in H^2(\Omega) \cap H_0^1(\Omega)$ solves the Dirichlet problem

$$\Delta g = G \quad \text{in } \Omega, \quad g = 0 \quad \text{on } \partial\Omega.$$

Thus, in view of the previous result, it follows that $\Delta p \geq 0$.

This inequality enables us to prove that $p_r(x_0) \downarrow p(x_0)$ as $r \downarrow 0$, where

$$p_r(x) = \frac{1}{\pi r^2}\int_{B_r(x)} p(y)dy \ , \quad r < \operatorname{dist}(x, \partial\Omega) \ .$$

As in the preceeding section one can show that if $r \downarrow 0$, then

$$\frac{1}{\pi r^2}\int_{B_r(x)} |p(y)|dy \to 0 \ , \quad r = \operatorname{dist}(x, \partial\Omega) \tag{5.15}$$

uniformly with respect to x.

Now one can easily show that in the vicinity of $\partial\Omega$ there are no contact points, ie. inequality (5.14) is strict. Indeed, since $w \in C(\bar{\Omega})$, there exists a neighbourhood $\mathcal{O}$ of the boundary $\partial\Omega$ such that in $\mathcal{O}$, $\phi < -\delta$, $w - \phi > \frac{\delta}{2}$, and $\delta > 0$ is a sufficiently small constant. Assume that in the neighbourhood $\mathcal{O}$ there exist contact points satisfying (5.14) as equality. At any such point $h(x_0) = w(x_0) - \phi(x_0) > \frac{\delta}{2}$ and, consequently, $p(x_0) > \frac{\delta}{2} + g(x_0)$. Thus

$$p_r(x_0) \geq \frac{\delta}{2} + g(x_0), \tag{5.16}$$

where r can be chosen to be equal to the distance between x_0 and the boundary $\partial\Omega$. Inequality (5.16) contradicts (5.15), since for x_0 close to $\partial\Omega$ we have $|g(x_0)| < \delta/4$. Thus, in the vicinity of the boundary there are no points of contact and

$$-\Delta u = f_1 , \quad -\Delta v = f_2 , \quad \Delta^2 w = f .$$

This means that in the vicinity of the boundary we have $(u, v, w) \in H^2 \times H^2 \times H^4$, which, along with the fact that $(u, v, w) \in H^2_{\text{loc}}(\Omega) \times H^2_{\text{loc}}(\Omega) \times H^4_{\text{loc}}(\Omega)$, completes the proof of the theorem. □

The condition $|\nabla\phi(x)| > 0$ is essential in the above theorem. If $\nabla\phi \equiv 0$, then problem (5.11) reduces to the variational inequality for the biharmonic operator and therefore, in general, w fails to be in $H^4(\Omega)$.

6 Boundary value problems for nonlinear shells

6.1 General remarks

Let $\Omega \subset \mathbb{R}^2$ be a bounded domain with infinitely many times differentiable boundary Γ and let $H^{2,0}(\Omega) = H^2(\Omega) \cap H^1_0(\Omega)$. On the space $H(\Omega) = H^1_0(\Omega) \times H^1_0(\Omega) \times H^{2,0}(\Omega)$ we consider the functional $\Pi(\omega)$,

$$\begin{aligned} \Pi(\omega) = &\int_\Omega \left\{ w^2_{x_1x_1} + w^2_{x_2x_2} + 2\sigma w_{x_1x_1} w_{x_2x_2} + 2(1-\sigma) w^2_{x_1x_2} \right\} dx \\ &+ \int_\Omega \left\{ \varepsilon^2_{11} + \varepsilon^2_{22} + 2\sigma\varepsilon_{11}\varepsilon_{22} + \frac{1}{2}(1-\sigma)\varepsilon^2_{12} \right\} dx , \end{aligned} \tag{6.1}$$

$$\omega = (U, w) , \quad U = (u, v) ,$$

where σ, $0 < \sigma < \frac{1}{2}$, is a positive constant,

$$\begin{aligned} \varepsilon_{11} &= u_{x_1} + k_{11} w + \frac{1}{2} w^2_{x_1} , \\ \varepsilon_{22} &= u_{x_2} + k_{22} w + \frac{1}{2} w^2_{x_2} , \\ \varepsilon_{12} &= u_{x_2} + v_{x_1} + w_{x_1} w_{x_2} . \end{aligned} \tag{6.2}$$

The Euler equations for the functional $\Pi(\omega)$ describe equilibrium states of nonlinear shallow shells and have the form

$$\frac{\partial N_{ij}}{\partial x_j} = 0 , \quad i = 1, 2 , \tag{6.3}$$

$$\Delta^2 w - \left(N_{ij} w_{x_j}\right)_{x_i} + k_{11} N_{11} + k_{22} N_{22} = 0 , \tag{6.4}$$

$$N_{11} = \varepsilon_{11} + \sigma\varepsilon_{22} , \quad N_{22} = \varepsilon_{22} + \sigma\varepsilon_{11} , \quad N_{12} = N_{21} = \frac{1}{2}(1-\sigma)\varepsilon_{12} . \tag{6.5}$$

In this section the unilateral boundary value problems will be considered for equations (6.3)–(6.5). Dropping out nonlinear terms in (6.2)–(6.5), one easily gets the linear equations of shallow shells as considered in Section 3. Replacing the space $H^{2,0}(\Omega)$ used in the definition of $H(\Omega)$ by $H_0^2(\Omega)$ we find out that the value of functional (6.1) coincides with the quadratic part of the functional analysed in Section 3. Indeed, the formula

$$\int_\Omega (\Delta w)^2 dx = \int_\Omega \left\{ w_{x_1x_1}^2 + w_{x_2x_2}^2 + 2\sigma w_{x_1x_1} w_{x_2x_2} + 2(1-\sigma) w_{x_1x_2}^2 \right\} dx$$

is valid for all $w \in H_0^2(\Omega)$.

Suppose that $k_{11}, k_{22} \in H^{1+\varepsilon}(\Omega)$, $\varepsilon > 0$. Due to imbedding theorems, one has k_{11}, $k_{22} \in C(\bar{\Omega})$. Note that if $\omega^n = (u^n, v^n, w^n)$ weakly converges to $\omega = (u, v, w)$ in $H(\Omega)$, then we can choose a subsequence ω^k such that $\underline{\lim}\ \Pi(\omega^k) \geq \Pi(\omega)$. To see this one can choose a subsequence ω^k such that $w_{x_i}^k \to w_{x_i}$ strongly in $L^4(\Omega)$.

Now, we establish the coercivity of the functional $\Pi(\omega)$, ie., we show that

$$\Pi(\omega) \to +\infty \quad \text{when} \quad \|\omega\|_{H(\Omega)} \to +\infty \ .$$

Assume that the squared norm $\|\cdot\|_2^2$ of the space $H^{2,0}(\Omega)$ coincides with the first integral in the definition of $\Pi(\omega)$. The functional $\Pi(\omega)$ can be written as

$$\Pi(\omega) = \|w\|_2^2 + A(U,U) + b(U,w) + d(w) \ , \quad U = (u,v) \ , \tag{6.6}$$

where

$$\begin{aligned} A(V_1,V_2) &= \int_\Omega \Big\{ \varepsilon_{11}^{(1)}\varepsilon_{11}^{(2)} + \varepsilon_{22}^{(1)}\varepsilon_{22}^{(2)} + \sigma\varepsilon_{11}^{(1)}\varepsilon_{22}^{(2)} + \sigma\varepsilon_{11}^{(2)}\varepsilon_{22}^{(1)} \\ &\quad + \frac{1}{2}(1-\sigma)\varepsilon_{12}^{(1)}\varepsilon_{12}^{(2)} \Big\} dx \ , \quad V_i = \left(\varphi^{(i)}, \psi^{(i)}\right) \ , \\ b(V_1,w) &= 2\int_\Omega \Big\{ \varepsilon_{11}^{(1)}(k_{11}w + \frac{1}{2}w_{x_1}^2) + \varepsilon_{22}^{(1)}(k_{22}w + \frac{1}{2}w_{x_2}^2) \\ &\quad + 2\sigma\varepsilon_{11}^{(1)}(k_{22}w + \frac{1}{2}w_{x_2}^2) + 2\sigma\varepsilon_{22}^{(1)}(k_{11}w + \frac{1}{2}w_{x_1}^2) \\ &\quad + \frac{1}{2}(1-\sigma)\varepsilon_{12}^{(1)} w_{x_1} w_{x_2} \Big\} \ dx \ , \\ d(w) &= \int_\Omega \Big\{ (k_{11}w + \frac{1}{2}w_{x_1}^2)^2 + (k_{22}w + \frac{1}{2}w_{x_2}^2)^2 + 2\sigma(k_{11}w \\ &\quad + \frac{1}{2}w_{x_1}^2)(k_{22}w + \frac{1}{2}w_{x_2}^2) + \frac{1}{2}(1-\sigma)w_{x_1}^2 w_{x_2}^2 \Big\} \ dx \ , \\ \varepsilon_{11}^{(i)} &= \varphi_{x_1}^{(i)} \ , \quad \varepsilon_{22}^{(i)} = \psi_{x_2}^{(i)} \ , \quad \varepsilon_{12}^{(i)} = \varphi_{x_2}^{(i)} + \psi_{x_1}^{(i)} \ , \quad i = 1,2 \ . \end{aligned} \tag{6.7}$$

Consider the transformation of the unit sphere S in the space $H(\Omega)$,

$$U \mapsto \delta R^2 U \ , \quad w \mapsto Rw \ , \quad (U,w) \in S \ , \tag{6.8}$$

where $\delta > 0$ is a constant to be chosen later, R is a positive parameter.

Let $(\tilde{U}, \tilde{w}) \in S$ and $\|\tilde{w}\|_2 \geq \frac{1}{2}$. Then, obviously

$$\Pi(\omega) \geq \frac{R^2}{4} \ , \tag{6.9}$$

where ω is the image of $(\tilde{U}, \tilde{w})$. By the first Korn inequality, $\|\cdot\|_1 = A(\cdot,\cdot)^{1/2}$ can be accepted as a norm in the space $H_0^1(\Omega) \times H_0^1(\Omega)$.

If $(\tilde{U}, \tilde{w}) \in S$, $\|\tilde{U}\|_1 \geq \frac{\sqrt{3}}{2}$, from (6.6) we easily deduce that

$$\Pi(\omega) \geq \frac{3}{4}\delta^2 R^4 - c_1 \delta R^4 - c_2 \delta R^3 \ , \tag{6.10}$$

where ω is the image of $(\tilde{U}, \tilde{w})$. The constant c_1 is an upper bound on S of the integral

$$\int_\Omega \left\{ \varepsilon_{11}^{(1)} w_{x_1}^2 + \varepsilon_{22}^{(1)} w_{x_2}^2 + 2\sigma\varepsilon_{11}^{(1)} w_{x_2}^2 + 2\sigma\varepsilon_{22}^{(1)} w_{x_1}^2 + \frac{1}{2}(1-\sigma)\varepsilon_{12}^{(1)} w_{x_1} w_{x_2} \right\} dx \ .$$

The quantities $\varepsilon_{11}^{(1)}, \varepsilon_{22}^{(1)}, \varepsilon_{12}^{(1)}$ are chosen as in (6.7). The constant c_2 is an upper bound of the remaining part of $b(U, w)$ on S having the third order of homogeneity with respect to R for the transformation (6.8). By choosing δ such that $\frac{3}{4}\delta^2 - c_1\delta = 1$ we get from (6.9), (6.10) that

$$\Pi(\omega) \geq c_3 R^2 \ , \quad c_3 > 0 \ ,$$

for sufficiently large R. This shows the coercivity of $\Pi(\omega)$.

Let K be a non-empty closed convex set in $H(\Omega)$. Since it is weakly closed, the following statement is valid.

Lemma 6.1. *There exists a solution of the minimization problem*

$$\inf_{\tilde{\omega} \in K} \Pi(\tilde{\omega}) \ . \tag{6.11}$$

A minimizer of Π over K satisfies the variational inequality

$$\omega \in K \ : \ \langle \Pi'(\omega), \tilde{\omega} - \omega \rangle \geq 0 \ , \quad \forall \tilde{\omega} \in K \ ,$$

where $\Pi'(\omega)$ is the derivative of the functional Π at ω.

Note the following simple fact. Let Φ be a convex and weakly lower semicontinuous functional on $H^{2,0}(\Omega)$, $\Phi(w) \neq -\infty$, Φ not identically equal to $+\infty$. Then the problem

$$\inf_{\tilde{\omega} \in K} \{\Pi(\tilde{\omega}) + \Phi(\tilde{w})\} \ , \quad \tilde{\omega} = (\tilde{u}, \tilde{v}, \tilde{w}) \ , \tag{6.12}$$

has a solution. To check this, observe that the functional Φ can be bounded from below

$$\Phi(w) \geq F(w) + c_0 \ , \tag{6.13}$$

where $F(w)$ is a linear and continuous functional on $H^{2,0}(\Omega)$ and c_0 is a constant (Ekeland, Temam, 1976).

In the proof of the coercivity of $\Pi(\omega)$, transformation (6.8) is linear with respect to R for the coordinate w. Hence, taking into account (6.13), we get

$$\Pi(\tilde{\omega}) + \Phi(\tilde{w}) \geq cR^2 \ , \quad c > 0 \ ,$$

for large R, which proves the coercivity of functional (6.12). Hence, problem (6.12) is solvable. We can assume that Φ is convex and lower semicontinuous (not equal to $-\infty$ and not identically equal to $+\infty$). In this case the functional Φ is also weakly lower semicontinuous.

A solution $\omega \in K$ to problem (6.12) is also a solution to the variational inequality

$$\begin{gathered} \omega \in K : \langle\, \Pi'(\omega), \tilde{\omega} - \omega \,\rangle + \Phi(\tilde{w}) - \Phi(w) \geq 0 \quad \forall\, \tilde{\omega} \in K \ , \\ \omega = (u, v, w) \ , \quad \tilde{\omega} = (\tilde{u}, \tilde{v}, \tilde{w}) \ . \end{gathered} \tag{6.14}$$

This section is devoted to the analysis of variational inequality (6.14) for some particular Φ and K.

6.2 Inequalities on the boundary. Convergence of solutions

6.2.1. By $B(\cdot \ , \ \cdot)$, we denote the bilinear form appearing in the first term of the representation of $\Pi(\omega)$, namely,

$$\begin{aligned} B(\varphi, \psi) = \int_\Omega \Big\{ & \varphi_{x_1x_1}\psi_{x_1x_1} + \varphi_{x_2x_2}\psi_{x_2x_2} + \sigma\big(\varphi_{x_1x_1}\psi_{x_2x_2} \\ & + \varphi_{x_2x_2}\psi_{x_1x_1}\big) + 2(1-\sigma)\varphi_{x_1x_2}\psi_{x_1x_2} \Big\}\, dx \ . \end{aligned}$$

In this section $\langle\, \varphi, \psi \,\rangle$ denotes the integral $\int_\Omega \varphi\psi dx$. Let $\nu = (\nu_1, \nu_2)$ be the unit outward normal to the boundary Γ. We define on Γ

$$M(\varphi) = \sigma \Delta\varphi + (1-\sigma)\frac{\partial^2 \varphi}{\partial \nu^2} \ ,$$

$$T(\varphi) = -\frac{\partial}{\partial \nu}\Delta\varphi - (1-\sigma)\frac{\partial}{\partial s}\frac{\partial^2 \varphi}{\partial \nu \partial s} \ ,$$

where $\dfrac{\partial}{\partial s}$ denotes the derivative along the tangent to Γ, $s = (-\nu_2, \nu_1)$. For sufficiently smooth functions φ, ψ, Green's formula

$$B(\varphi, \psi) = \int_\Omega \Delta^2 \varphi\psi dx + \int_\Gamma T(\varphi)\psi d\Gamma + \int_\Gamma M(\varphi)\psi_\nu d\Gamma \ , \quad \psi_\nu \equiv \frac{\partial \psi}{\partial \nu} \ ,$$

is valid (Duvaut, Lions, 1972). Let $W(\Omega)$ be the space of functions

$$\{\varphi \mid \varphi \in H^2(\Omega)\ ,\quad \Delta^2\varphi \in L^1(\Omega)\}$$

with the norm

$$\|\varphi\|^2_{W(\Omega)} = \|\varphi\|^2_{H^2(\Omega)} + \|\Delta^2\varphi\|^2_{L^1(\Omega)}\ .$$

By $\langle\ \cdot,\cdot\ \rangle_p$ we denote the duality pairing between $H^{-p}(\Gamma)$ and $H^p(\Gamma)$. The following statement allows us to define $T(\varphi)$ and $M(\varphi)$ on the boundary for $\varphi \in W(\Omega)$.

Lemma 6.2. *There exists a linear and continuous operator*

$$\Lambda : W(\Omega) \mapsto H^{-3/2}(\Gamma) \times H^{-1/2}(\Gamma)\ ,\quad \Lambda(\varphi) = \big(\Lambda_0(\varphi), \Lambda_1(\varphi)\big)\ ,$$

such that

$$\Lambda(\varphi) = \big(T(\varphi), M(\varphi)\big)\quad \forall\, \varphi \in C^\infty(\bar{\Omega})\ .$$

Moreover, the generalized Green formula

$$B(\varphi,\psi) = \langle\ \Delta^2\varphi, \psi\ \rangle + \langle\ \Lambda_0(\varphi), \psi\ \rangle_{\frac{3}{2}} + \langle\ \Lambda_1(\varphi), \psi_\nu\ \rangle_{\frac{1}{2}} \qquad \forall\, \psi \in H^2(\Omega) \tag{6.15}$$

is valid for $\varphi \in W(\Omega)$.

Proof. First we note that, by imbedding theorems, $\psi \in L^\infty(\Omega)$. Hence, in (6.15) the expression $\langle\ \Delta^2\varphi, \psi\ \rangle$ is well-defined. For any function $\psi \in H^2(\Omega)$, one can define on Γ the trace $\gamma_0\psi$ and the derivative $\gamma_1\psi$ along the normal, $\gamma_0\psi \in H^{3/2}(\Gamma)$, $\gamma_1\psi \in H^{1/2}(\Gamma)$. Moreover, there exists an operator

$$l\ :\ H^{3/2}(\Gamma) \times H^{1/2}(\Gamma) \mapsto H^2(\Omega)$$

such that $\gamma_0 l(u_1,u_2) = u_1$, $\gamma_1 l(u_1,u_2) = u_2$ (see Lions, Magenes, 1968). Take any function $\psi \in H^2(\Omega)$, and choose $u_1 \in H^{3/2}(\Gamma)$, $u_2 \in H^{1/2}(\Gamma)$ such that $u_1 = \gamma_0\psi$, $u_2 = \gamma_1\psi$. For any $\varphi \in W(\Omega)$, denote

$$L_\varphi(u_1,u_2) = B(\varphi,\psi) - \langle\ \Delta^2\varphi, \psi\ \rangle\ .$$

It is easily seen that $L_\varphi(u_1,u_2)$ does not depend on ψ. Indeed, let $\psi_1, \psi_2 \in H^2(\Omega)$, $\gamma_0\psi_i = u_1$, $\gamma_1\psi_i = u_2$. We show that

$$B(\varphi,\psi) - \langle\ \Delta^2\varphi, \psi\ \rangle = 0$$

for $\psi = \psi_1 - \psi_2 \in H^2_0(\Omega)$. This is obviously true for all $\psi \in C_0^\infty(\Omega)$, and, by the density of $C_0^\infty(\Omega)$ in $H_0^2(\Omega)$, it holds for $\psi \in H_0^2(\Omega)$. The space $H^2(\Omega)$ is continuously imbedded in $L^\infty(\Omega)$ and the operator l is continuous. Hence,

$$|L_\varphi(u_1,u_2)| \le c\ \|\varphi\|_{W(\Omega)}\ \|\psi\|_{H^2(\Omega)} \le$$
$$\le c\ \|\varphi\|_{W(\Omega)}\ \big(\|u_1\|_{H^{3/2}(\Gamma)} + \|u_2\|_{H^{1/2}(\Gamma)}\big)\ .$$

This means that the functional $(u_1, u_2) \mapsto L_\varphi(u_1, u_2)$ is linear and continuous on $H^{3/2}(\Gamma) \times H^{1/2}(\Gamma)$. Consequently, there exists an element $\Lambda(\varphi) = (\Lambda_0(\varphi), \Lambda_1(\varphi)) \in H^{-3/2}(\Gamma) \times H^{-1/2}(\Gamma)$ such that

$$L_\varphi(u_1, u_2) = \langle \Lambda_0(\varphi), u_1 \rangle_{3/2} + \langle \Lambda_1(\varphi), u_2 \rangle_{1/2} \ .$$

Moreover,

$$\|\Lambda(\varphi)\|_{H^{-3/2}(\Gamma)\times H^{-1/2}(\Gamma)} \leq c\, \|\varphi\|_{W(\Omega)} \ .$$

This proves the continuity of Λ, since Λ is linear.

Let $\varphi \in C^\infty(\bar{\Omega})$ and $\psi \in C^2(\bar{\Omega})$. We have

$$L_\varphi(\gamma_0\psi, \gamma_1\psi) = B(\varphi, \psi) - \langle \Delta^2\varphi, \psi \rangle = \int_\Gamma T(\varphi)\gamma_0\psi d\Gamma$$
$$+ \int_\Gamma M(\varphi)\gamma_1\psi d\Gamma = \langle T(\varphi), \gamma_0\psi \rangle_{3/2} + \langle M(\varphi), \gamma_1\psi \rangle_{1/2} \ .$$

In view of the density of $(\gamma_0\psi, \gamma_1\psi)$ in $H^{3/2}(\Gamma)\times H^{1/2}(\Gamma)$, the same representation holds for $L_\varphi(u_1, u_2)$, $(u_1, u_2) \in H^{3/2}(\Gamma) \times H^{1/2}(\Gamma)$. In view of this representation we infer that

$$(\Lambda_0(\varphi), \Lambda_1(\varphi)) = (T(\varphi), M(\varphi)) \quad \forall\, \varphi \in C^\infty(\bar{\Omega}) \ .$$

The lemma is proved. □

Consider now inequality (6.14), where $K = H_0^1(\Omega) \times H_0^1(\Omega) \times K_1$, and

$$K_1 = \{w \in H^{2,0}(\Omega) \mid w_\nu \geq 0 \quad \text{on} \quad \Gamma\}$$

and define

$$\Phi(w) = -\int_\Omega f w dx \ , \quad f \in L^2(\Omega) \ .$$

From (6.14) it follows that for the set K and the functional Φ, we can derive equations (6.3) and the inequality (we put $k_{12} = k_{21} = 0$)

$$B(w, \tilde{w} - w) + \langle k_{ij}N_{ij} - N_{ij}w_{x_ix_j}, \tilde{w} - w \rangle \geq \langle f, \tilde{w} - w \rangle \quad \forall\, \tilde{w} \in K_1 \ . \tag{6.16}$$

The space $C_0^\infty(\Omega)$ is contained in K_1, so we can take in (6.16) the functions $w \pm \varphi$, $\varphi \in C_0^\infty(\Omega)$, as test functions. This implies that the equation

$$\Delta^2 w = f_0 \tag{6.17}$$

is satisfied in the sense of distributions, with $f_0 = f + N_{ij}w_{x_ix_j} - k_{ij}N_{ij}$. The function f_0 belongs to the space $L^1(\Omega)$. Hence, $w \in H^{2,0}(\Omega)$, $\Delta^2 w \in L^1(\Omega)$. Taking into account generalized Green's formula (6.15) and the boundary conditions $w = \tilde{w} = 0$ on Γ we get

$$B(w, \tilde{w} - w) + \langle k_{ij}N_{ij} - N_{ij}w_{x_ix_j}, \tilde{w} - w \rangle$$
$$= \langle M(w), \tilde{w}_\nu - w_\nu \rangle_{\frac{1}{2}} + \langle f, \tilde{w} - w \rangle \quad \forall\, \tilde{w} \in K_1 \ .$$

So, for every $w \in H^{2,0}(\Omega)$, one can solve problem (6.22), (6.24) and find $u, v \in H_0^1(\Omega)$. Then we can substitute u, v into inequality (6.20), considering u, v as the image of the operator defined by (6.22) at w. The inequality obtained in this way has a solution w^g. Now, the existence of u^g, v^g follows from (6.22), (6.24) for $w = w^g$.

When proving Lemma 6.4 below we shall consider the transformation of the space $H^{2,0}(\Omega)$

$$\pi \ : \ w \mapsto Rw \ , \quad R > 0 \ . \tag{6.25}$$

It is worth noting that components of the vector $F(w)$ can be decomposed into two parts, corresponding to the first- and the second-order homogeneity with respect to R. Namely,

$$F^i(w) = F_1^i(w) + F_2^i(w) \ ,$$
$$F_1^i(Rw) = RF_1^i(w) \ , \quad F_2^i(Rw) = R^2 F_2^i(w) \ .$$

According to Lemma 6.3, one can find u, v for the above parts. For example, the problem

$$-\frac{\partial \sigma_{ij}}{\partial x_j} = F_2^i(w) \ , \quad i = 1, 2 \ , \tag{6.26}$$

with the boundary conditions (6.24) possesses a unique solution (u_2, v_2) corresponding to $\big(F_2^1(w), F_2^2(w)\big)$.

Let us formulate a convergence result for solutions ω^g.

Given $g \to 0$, we denote by $\varepsilon \to 0$ a sequence $g_\ell \to 0$ with $\ell \in \mathbb{N}$, $\ell \to \infty$, and we shall say that $\varepsilon \to 0$ is a subsequence of $g \to 0$. This terminology is used throughtout the book.

Theorem 6.1. *There exists a subsequence ω^ε of ω^g such that $\omega^\varepsilon \to \omega$ weakly in $H(\Omega)$ as $\varepsilon \to 0$. Moreover, ω satisfies equations (6.3) and*

$$B(w, \tilde{w}) + \langle\, k_{ij} N_{ij} - N_{ij} w_{x_i x_j}, \tilde{w} \,\rangle = \langle\, f, \tilde{w} \,\rangle \quad \forall\, \tilde{w} \in H^{2,0}(\Omega) \ . \tag{6.27}$$

Proof. We multiply equations (6.3) by u^g, v^g, respectively and sum up with (6.20) when $\tilde{w} = 0$. The resulting relation has the form

$$\begin{aligned} B(w^g, w^g) + 2 \int_\Omega \big(\varepsilon_{11}^2 + \varepsilon_{22}^2 + 2\sigma \varepsilon_{11} \varepsilon_{22} + \frac{1}{2}(1-\sigma)\varepsilon_{12}^2\big)\, dx \\ - \langle\, k_{ij} N_{ij}, w^g \,\rangle + g \int_\Gamma \Big|\frac{\partial w^g}{\partial \nu}\Big| d\Gamma \le \langle\, f, w^g \,\rangle \ . \end{aligned} \tag{6.28}$$

The functions ε_{ij}, N_{ij} are related to u^g, v^g, w^g by formulae (6.2), (6.5). We have to stress that equations (6.3) are valid as integral identities with the derivatives removed from N_{ij}. The multiplication performed at the beginning of the proof reduces to the substitution of u^g, v^g into these identities as test functions.

By Lemma 6.4 below, the functions w^g are bounded as $g \to 0$. We choose a subsequence w^ε such that $w^\varepsilon \to w$ weakly in $H^{2,0}(\Omega)$. Then from Lemma 6.3 it follows that $u^\varepsilon, v^\varepsilon$ are bounded in $H_0^1(\Omega)$. Hence, we can assume that $u^\varepsilon, v^\varepsilon \to u, v$ weakly in $H_0^1(\Omega)$. By imbedding theorems, $w_{x_i}^\varepsilon \to w_{x_i}$ strongly in $L^4(\Omega)$. Let $\omega = (u, v, w)$. This convergence allows us to pass to the limit in (6.3), (6.20) as $\varepsilon \to 0$. In fact, since $\dfrac{\partial w^\varepsilon}{\partial \nu}$ are bounded in $L^1(\Gamma)$, the term of (6.20) containing $\dfrac{\partial w^\varepsilon}{\partial \nu}$ converges to zero. Consider the nonlinear term in the left-hand side of (6.20). To emphasize the dependence of N_{ij} on ω^ε we shall write $N_{ij}(\omega^\varepsilon)$. We show that

$$\langle\, N_{ij}(\omega^\varepsilon) w_{x_i x_j}^\varepsilon, \tilde{w} - w^\varepsilon \,\rangle = -\langle\, N_{ij}(\omega^\varepsilon) w_{x_j}^\varepsilon, \tilde{w}_{x_i} - w_{x_i}^\varepsilon \,\rangle \ .$$

To get this we use the fact that $N_{ij}(\omega^\varepsilon)$ satisfy equations (6.3). Since ω^ε converges to ω, we get $N_{ij}(\omega^\varepsilon) \to N_{ij}(\omega)$ weakly in $L^2(\Omega)$. Taking into account the strong convergence of $w_{x_i}^\varepsilon$ in $L^4(\Omega)$ we can pass to the limit as $\varepsilon \to 0$ in the right-hand side of the above equality.

Convergence of the remaining terms is evident. As a result, from (6.20) it follows that

$$B(w, \tilde{w} - w) + \langle\, k_{ij} N_{ij} - N_{ij} w_{x_i x_j}, \tilde{w} - w \,\rangle \geq \langle\, f, \tilde{w} - w \,\rangle$$
$$\forall\, \tilde{w} \in H^{2,0}(\Omega) \ .$$

Since $\tilde{w}$ is arbitrary, relation (6.27) results. By Green's formula (6.15), we can assume that equation (6.27) contains the boundary condition $M(w) = 0$ on Γ. A passage to the limit in the equation (6.3) as $\varepsilon \to 0$ can be done in a similar way. The theorem is proved. □

Boundedness of w^g is established by the following lemma.

Lemma 6.4. *There exists a constant $c > 0$ independent of g such that*

$$\|w^g\|_2 \leq c \ . \tag{6.29}$$

Proof. Consider the functional

$$E(w) = B(w, w) + 2 \int_\Omega (\varepsilon_{11}^2 + \varepsilon_{22}^2 + 2\sigma \varepsilon_{11}\varepsilon_{22} + \frac{1}{2}(1 - \sigma)\varepsilon_{11}^2) dx - \langle\, k_{ij} N_{ij}, w \,\rangle \ ,$$

defined on $H^{2,0}(\Omega)$, where u, v satisfy (6.22), (6.24) for a given w, the functions ε_{ij}, N_{ij} are related to u, v, w by formulae (6.2), (6.5). By (6.5), there exists a constant c_1 independent of w such that

$$|k_{ij} N_{ij} w| \leq c_1 w^2 \sum_{i,j=1}^{2} k_{ij}^2 + \varepsilon_{11}^2 + \varepsilon_{22}^2 + 2\sigma\varepsilon_{11}\varepsilon_{22} + \frac{1}{2}(1 - \sigma)\varepsilon_{12}^2 \ . \tag{6.30}$$

Let S be the unit sphere in the space $H^{2,0}(\Omega)$ and let π be the transformation (6.25). Consider the set

$$S_0 = \left\{ w \in S \mid c_1 \max_{\bar{\Omega}} w^2 \int_\Omega \sum_{i,j=1}^{2} k_{ij}^2 dx \le \frac{1}{2} \right\} .$$

From (6.30) and the formula for $E(w)$ it follows that

$$E(w) \ge \frac{1}{2} R^2 \tag{6.31}$$

on $\pi(S_0)$. The functional $E(w)$ can be represented in the form $E(w) = \sum_{i=2}^{4} E_i(w)$, where $E_i(w)$ are homogeneous functionals of order i bounded on S by some constants.

Let us now consider the set $S_1 = S \backslash S_0$ such that

$$c_1 \max_{\bar{\Omega}} w^2 \int_\Omega \sum_{i,j=1}^{2} k_{ij}^2 dx > \frac{1}{2} , \quad w \in S_1 .$$

Obviously, it is not possible to construct a sequence w_k on the set S_1, which weakly converges to zero in $H^{2,0}(\Omega)$. If it were possible, we could choose a subsequence w_i strongly converging in the space $C(\bar{\Omega})$, contradictory to the definition of S_1.

Let us analyse more closely $E_4(w)$. This functional can be written as

$$E_4(w) = 2 \int_\Omega (\bar{\varepsilon}_{11}^2 + \bar{\varepsilon}_{22}^2 + 2\sigma \bar{\varepsilon}_{11} \bar{\varepsilon}_{22} + \frac{1}{2}(1-\sigma)\bar{\varepsilon}_{12}^2) dx ,$$

$$\bar{\varepsilon}_{11} = u_{2x_1} + \frac{1}{2} w_{x_1}^2 , \quad \bar{\varepsilon}_{22} = v_{2x_2} + \frac{1}{2} w_{x_2}^2 ,$$

$$\bar{\varepsilon}_{12} = u_{2x_2} + v_{2x_1} + w_{x_1} w_{x_2} ,$$

where u_2, v_2 satisfy (6.26), (6.24) and the functional is homogeneous of order two with respect to u_2, v_2.

It is easy to prove that $E_4(w) \ge c_2$, $c_2 > 0$ on S_1. Otherwise one can choose a sequence w_i such that $E_4(w_i) \to 0$. Assume $w_i \to w_0$ weakly in $H^{2,0}(\Omega)$. Then, by the formula for E_4,

$$\int_\Omega (\bar{\varepsilon}_{11} + \bar{\varepsilon}_{22}) dx \to 0 , \quad \bar{\varepsilon}_{11} = \varepsilon_{11}(w_i) , \quad \bar{\varepsilon}_{22} = \varepsilon_{22}(w_i) ,$$

or, in other words,

$$\int_\Omega \left(\frac{\partial u_{2i}}{\partial x_1} + \frac{\partial v_{2i}}{\partial x_2} \right) dx + \frac{1}{2} \int_\Omega (w_{ix_1}^2 + w_{ix_2}^2) dx \to 0 .$$

The first integral is equal to zero in virtue of the boundary conditions for u_2, v_2. By the compactness of the imbedding of $H^{2,0}(\Omega)$ in $H_0^1(\Omega)$, $w_0 = 0$, contradictory to the fact that on S_1 there is no sequence weakly converging to zero.

Thus, $E(w) \geq R^2 + c_2 R^4 - c_3 R^3 - c_4 R^2$ on $\pi(S_1)$. According to (6.31), for sufficiently large R, it follows that

$$E(w) \geq \frac{1}{2} R^2 .$$

Moreover, the estimate (6.29) results from (6.28), which proves the lemma. □

In proving the coercivity of E one can use similar arguments to that of (Vorovich, Lebedev, 1972), where the solvability of nonlinear boundary problems for shells was proved.

Theorem 6.2. *There exists a subsequence ω^ε of the sequence ω^g such that $\omega^\varepsilon \to \omega$ weakly in $H(\Omega)$ as $\varepsilon \to \infty$. Moreover, the function $\omega = (u, v, w)$ satisfies equations (6.3) and*

$$B(w, \tilde{w}) + \langle k_{ij} N_{ij} - N_{ij} w_{x_i x_j}, \tilde{w} \rangle = \langle f, \tilde{w} \rangle \quad \forall \, \tilde{w} \in H_0^2(\Omega) , \qquad (6.32)$$

and $w \in H_0^2(\Omega)$.

We give a sketch of the proof. As above, estimate (6.29) holds. From (6.28) it follows that

$$g \int_\Gamma \left| \frac{\partial w^g}{\partial \nu} \right| d\Gamma \leq c$$

with a constant independent of g. This means that

$$g \int_\Gamma \left| \frac{\partial w^g}{\partial \nu} \right| d\Gamma \to 0 \quad \text{when} \quad g \to \infty .$$

Choosing a subsequence w^ε weakly converging to w in $H^{2,0}(\Omega)$ and taking into account the compactness of the imbedding of $H^1(\Omega)$ in $L^1(\Gamma)$ we get that $w_\nu = 0$ on Γ. Thus, $w \in H_0^2(\Omega)$.

Let us now justify (6.32). As in the previous theorem, there exist functions $u^\varepsilon, v^\varepsilon$ weakly converging to u, v in $H_0^1(\Omega)$. We substitute $w^\varepsilon + \varphi$, where $\varphi \in C_0^\infty(\Omega)$, into inequality (6.20) as a test function. Then, assuming $w^\varepsilon_{x_i} \to w_{x_i}$ strongly in $L^4(\Omega)$ one can pass to the limit in (6.20) as $\varepsilon \to \infty$. The resulting relation coincides with (6.32).

6.2.3. Let us assume that the boundary Γ consists of two curves Γ_0 and Γ_1, and the length of Γ_1 is positive. Denote

$$H^1_{\Gamma_1}(\Omega) = \{ u \in H^1(\Omega) \mid u = 0 \quad \text{on} \quad \Gamma_1 \} ,$$

$$H_{\Gamma_1}(\Omega) = H^1_{\Gamma_1}(\Omega) \times H^1_{\Gamma_1}(\Omega) \times H_0^2(\Omega) .$$

Consider minimization problems (6.11) and (6.12), where K is a closed and convex set in the space $H^1_{\Gamma_1}(\Omega)$. In this case the existence result formulated in Section 6.1 applies.

For instance, take $f \in L^2(\Omega)$, $U_\nu \equiv U\nu$, $U = (u, v)$,

$$K = \left\{\omega = (u, v, w) \in H_{\Gamma_1}(\Omega) \mid U_\nu \le 0 \quad \text{a.e. on} \quad \Gamma_0\right\} .$$

A solution to problem (6.12) for $\Phi(\tilde{w}) = -\int_\Omega f\tilde{w}dx$ satisfies the inequality

$$\omega \in K \;:\; \langle \Pi'(\omega), \tilde{\omega} - \omega\rangle \ge \langle f, \tilde{w} - w\rangle \qquad \forall\, \tilde{\omega} \in K . \tag{6.33}$$

Defining

$$K_2 = \left\{(u, v) \in H^1_{\Gamma_1}(\Omega) \times H^1_{\Gamma_1}(\Omega) \mid U_\nu \le 0 \quad \text{a.e. on} \quad \Gamma_0\right\} ,$$

we can rewrite inequality (6.33) as

$$U \in K_2 \;:\; A(U, V - U) \ge \langle\, F(w), V - U\,\rangle \quad \forall\, V \in K_2 , \tag{6.34}$$

$$\begin{aligned} w \in H^2_0(\Omega) \;:\; & B(w, \psi) + \langle\, N_{ij} w_{x_j}, \psi_{x_i}\,\rangle + \\ & + \langle\, k_{ij} N_{ij}, \psi\,\rangle = \langle\, f, \psi\,\rangle \quad \forall\, \psi \in H^2_0(\Omega) , \end{aligned} \tag{6.35}$$

where the bilinear form A is the same as in formulae (6.7), $F(w)$ is the vector whose first component is defined by (6.23) and the second component can be obtained by replacing k_{11}, x_1 with k_{22}, x_2.

It follows from (6.35) that equation (6.17) is satisfied.

We show that $f_0 \in H^{-1-\varepsilon}(\Omega)$, $\varepsilon > 0$, where $H^{-1-\varepsilon}(\Omega)$ is the dual space to $H^{1+\varepsilon}_0(\Omega)$. In fact, $f_0 \in L^1(\Omega)$, and hence, by the continuity of the imbedding of $H^{1+\varepsilon}_0(\Omega)$ in $L^\infty(\Omega)$,

$$\langle\, f_0, \varphi\,\rangle \le \|f_0\|_{L^1(\Omega)} \|\varphi\|_{L^\infty(\Omega)} \le c \|f_0\|_{L^1(\Omega)} \|\varphi\|_{H^{1+\varepsilon}_0(\Omega)} ,$$

for any function $\varphi \in H^{1+\varepsilon}_0(\Omega)$. Consequently, $f_0 \in H^{-1-\varepsilon}(\Omega)$. Then, $w \in H^{3-\varepsilon}(\Omega)$, by the well known regularity results for the Dirichlet problem for biharmonic operators. By imbedding theorems, for $\varepsilon < 1$, $w_{x_i} \in L^\infty(\Omega)$. Hence, the components $F^1(w)$, $F^2(w)$ of $F(w)$ belong to the space $L^2(\Omega)$. In this case problem (6.34) can be viewed as a two-dimensional Signorini problem of finding U with the known right-hand side $F(w)$. In view of the smoothness of the boundary Γ from the results of (Fichera, 1974) it follows that $U \in H^2(\tilde{\Omega})$, $\tilde{\Omega} = \Omega \setminus \mathcal{O}$, where $\mathcal{O}$ is a neighbourhood of $\Gamma_0 \cap \Gamma_1$. Consequently, the equations

$$\frac{\partial N_{ij}}{\partial x_j} = 0 , \quad i = 1, 2 , \tag{6.36}$$

as well as the boundary conditions

$$U_\nu \le 0 , \quad N_\nu \le 0 , \quad U_\nu N_\nu = 0 , \quad N_\tau = 0 \;\text{ on }\; \Gamma_0 \tag{6.37}$$

are satisfied in the sense of L^2. Here $N_\nu = N_{ij}\nu_j\nu_i$, N_τ has components $(N_\tau)_i = N_{ij}\nu_j - \nu_i N_\nu$. This can be obtained by integrating by parts (6.34). Problem (6.35) - (6.37) with zero boundary conditions for U on Γ_1 describes a contact between a shell and a rigid punch. Conditions (6.37) say that $U_\nu < 0$, $N_\nu = 0$, $N_\tau = 0$ when the shell moves away from the rigid punch. In the converse case $U_\nu = 0$, $N_\nu \leq 0$, $N_\tau = 0$.

7 Boundary value problems for linear shells

In this section a boundary value problem for a linear model of a shallow shell is considered. Boundary conditions are of inequality type. They correspond to the unilateral contact between the shell and the rigid punch and allow the shell points to move in all directions.

The formulation of the problem is as follows. Let $\Omega \subset \mathbb{R}^2$ be a bounded domain with smooth boundary Γ. The boundary is represented as the union $\Gamma = \Gamma_0 \cup \Gamma_1$. For the sake of simplicity, we assume that Γ_0, Γ_1 are curves such that the length of Γ_1 is positive.

Denote by $H^1_{\Gamma_1}(\Omega)$ the Sobolev space obtained as the completion in $H^1(\Omega)$ of smooth functions vanishing near Γ_1. The space $H^2_{\Gamma_1}(\Omega)$ is obtained in a similar way. We use the notation $H(\Omega) = H^1_{\Gamma_1}(\Omega) \times H^1_{\Gamma_1}(\Omega) \times H^2_{\Gamma_1}(\Omega)$, and $\|\cdot\|_p$ is the norm in $H^p(\Omega)$. The space $H^1_{\Gamma_1}(\Omega)$ has been already introduced in Section 6.2.3. We shall also use some other notations introduced in Section 6. In particular, the bilinear form $B(\cdot\,,\,\cdot)$, the bending moment $M(w)$ and the transverse force $T(w)$ are defined in the same way as in Section 6.2.1. Moreover, $U = (u, v)$, $U_\nu \equiv u\nu_1 + v\nu_2$, $\nu = (\nu_1, \nu_2)$ is the outward unit normal to the boundary Γ.

Let us consider the energy functional on the space $H(\Omega)$

$$\Pi(\omega) = \Pi_1(\omega) - 2\int_\Omega F\omega dx \ , \quad \omega = (u, v, w) \ .$$

$$\Pi_1(\omega) = B(w, w) + \int_\Omega \left\{ \varepsilon_{11}^2 + \varepsilon_{22}^2 + 2\sigma\varepsilon_{11}\varepsilon_{22} + \frac{1}{2}(1-\sigma)\varepsilon_{12}^2 \right\} dx \ , \tag{7.1}$$

$$\varepsilon_{11} = u_{x_1} + k_{11}w \ , \quad \varepsilon_{22} = v_{x_2} + k_{22}w \ , \quad \varepsilon_{12} = u_{x_2} + v_{x_1} \ .$$

Here $F = (f_1, f_2, f) \in L^2(\Omega)$, $k_{11}, k_{22} \in C^1(\bar{\Omega})$, $0 < \sigma < \dfrac{1}{2}$.

Let K be a closed convex set in $H(\Omega)$ defined as

$$K = \left\{ \omega = (U, w) \in H(\Omega) \;\middle|\; w \geq 0, \ -U_\nu \geq 0 \quad \text{on} \quad \Gamma_0 \right\} .$$

Consider the problem of minimizing of $\Pi(\omega)$ over the set K. This problem is equivalent to the variational inequality

$$\omega \in K \ : \ \langle\, \Pi'(\omega), \tilde{\omega} - \omega \,\rangle \geq 0 \quad \forall\, \tilde{\omega} \in K \ . \tag{7.2}$$

A solution to problem (7.2) exists and is unique. The proof of this fact is similar to that of Lemma 6.1. Verification of the coercivity of the functional Π is simpler now since in absence of nonlinear components in the representation of ε_{ij} the corresponding formulae are considerably simplified. Uniqueness follows immediately.

From (7.2) it follows that the equations

$$\Delta^2 w + k_{11}N_{11} + k_{22}N_{22} = f \ , \tag{7.3}$$

$$-\frac{\partial N_{ij}}{\partial x_j} = f_i \ , \quad i = 1,2 \ , \tag{7.4}$$

is satisfied in Ω in the sense of distributions, where

$$N_{11} = \varepsilon_{11} + \sigma\varepsilon_{22} \ , \quad N_{22} = \varepsilon_{22} + \sigma\varepsilon_{11} \ , \quad N_{12} = \frac{1}{2}(1-\sigma)\varepsilon_{12} \ . \tag{7.5}$$

To see this, one has to substitute $\omega + \omega_0$ into (7.2) as a test function $\tilde{\omega}$, where $\omega_0 \in C_0^\infty(\Omega)$ is an arbitrary element.

Moreover, the following conditions

$$w \geq 0 \ , \quad T(w) \geq 0 \ , \quad wT(w) = 0 \ , \quad M(w) = 0 \ , \tag{7.6}$$

$$-U_\nu \geq 0 \ , \quad -N_\nu \geq 0 \ , \quad U_\nu N_\nu = 0 \ , \quad N_\tau = 0 \ , \tag{7.7}$$

are satisfied in Γ_0. Here $N_\nu = N_{ij}\nu_j\nu_i$, N_τ is a vector with components $(N_\tau)_i = N_{ij}\nu_j - N_\nu\nu_i$, $i = 1,2$. Under sufficient regularity of solutions the validity of (7.6), (7.7) can be verified as follows.

For any function $\psi \in H^2_{\Gamma_1}(\Omega)$, $\psi \geq 0$ on Γ_0, by the Green formula

$$B(w,\psi) = \langle \, \Delta^2 w, \psi \, \rangle + \int_\Gamma \psi T(w) d\Gamma + \int_\Gamma \frac{\partial \psi}{\partial \nu} M(w) d\Gamma \ , \tag{7.8}$$

where, as previously, $\langle \, \cdot \, , \, \cdot \, \rangle$ denote the integral over Ω. By (7.2), for any ψ,

$$B(w,\psi) + \langle \, k_{11}N_{11} + k_{22}N_{22} - f, \psi \, \rangle \geq 0 \ . \tag{7.9}$$

To see this, it is enough to substitute $\omega + \omega_1$ as a test function $\tilde{\omega}$, $\omega_1 = (0,0,\psi)$. One can also see that relations (7.6) follow from (7.3), (7.8), (7.9) and the boundary conditions $w = \dfrac{\partial w}{\partial \nu} = 0$ imposed on Γ_1. These conditions follow from the fact that $w \in H(\Omega)$. Relations (7.7) can be obtained in a similar way.

Lemma 6.2 allows us to make some additional observation concerning (7.6). Namely, it follows from (7.3) that $\Delta^2 w \in L^2(\Omega)$, and hence $M(w) \in H^{-1/2}(\Gamma)$, $T(w) \in H^{-3/2}(\Gamma)$. Consequently, the integral $\int_\Gamma wT(w)d\Gamma$ is well-defined, since $w \in H^{3/2}(\Gamma)$, $\dfrac{\partial w}{\partial \nu} \in H^{1/2}(\Gamma)$.

An analogous observation applies to (7.7). In fact, from equations (7.4) it follows that $\dfrac{\partial N_{ij}}{\partial x_j} \in L^2(\Omega)$. According to Lemma 1.5 of Chapter 3, $\varphi_i\nu_i$ can be defined in $H^{-1/2}(\Gamma)$ for every function $\varphi = (\varphi_1, \varphi_2)$ such that φ, div $\varphi \in L^2(\Omega)$. Hence, $N_{ij}\nu_j \in H^{-1/2}(\Gamma)$, $i = 1, 2$. Consequently, $N_\nu \in H^{-1/2}(\Gamma)$. Taking into account that $U_\nu \in H^{1/2}(\Gamma)$, we see that $\int_\Gamma U_\nu N_\nu d\Gamma$ is correctly defined.

It is worth noting here that in general it is not possible to accept the products $wT(w)$ and $U_\nu N_\nu$ as duality pairings between $H^{3/2}(\Gamma_0)$, $H^{-3/2}(\Gamma_0)$ and $H^{1/2}(\Gamma_0)$, $H^{-1/2}(\Gamma_0)$, respectively.

Now we construct on the subsets of $\Gamma_0 \backslash \partial\Gamma_0$ some nonnegative measures μ_1, μ_2 characterizing the interaction between the shell and the punch. More precisely, μ_1 corresponds to the interaction in a direction normal to the boundary Γ, while μ_2 describes the interaction in a direction normal to the middle surface.

The space $C_0(\Gamma_0)$ contains all functions continuous on $\Gamma_0 \backslash \partial\Gamma_0$ whose supports are compact with the following convergence: $\varphi_n \to \varphi$ in $C_0(\Gamma_0)$ if φ_n converges uniformly to φ and supports of all φ_n belong to a fixed compact $B \subset \Gamma_0 \backslash \partial\Gamma_0$.

Theorem 7.1. *There exist nonnegative measures μ_1, μ_2 defined on the $\mathfrak{S}$-algebra of Borel subsets of $\Gamma_0 \backslash \partial\Gamma_0$ such that the representation*

$$\langle\, \Pi'(\omega), \tilde{\omega} \,\rangle = \int_{\Gamma_0} \tilde{w} d\mu_1 - \int_{\Gamma_0} \tilde{U}_\nu d\mu_2 \tag{7.10}$$

is valid for all $\tilde{\omega} = (\tilde{U}, \tilde{w}) \in H(\Omega) \cap C_0(\Gamma_0)$.

Proof. First note that for $\chi = (0, 0, \tilde{w}) \in H(\Omega)$ and $\tilde{w} \geq 0$ on Γ_0,

$$\langle\, \Pi'(\omega), \chi \,\rangle \geq 0 \ . \tag{7.11}$$

To see this, it is enough to substitute $\omega + \chi \in K$ into (7.2) as a test function. Further, let $\tilde{w} \in H^2_{\Gamma_1}(\Omega) \cap C_0(\Gamma_0)$ and let $\tilde{w}^*$ be the trace on Γ_0 of this function. The linear variety of all such functions $\tilde{w}^*$ on Γ_0 will be denoted by $\mathcal{V}$.

We define a linear functional on $\mathcal{V}$ by the formula

$$L(\tilde{w}^*) = \langle\, \Pi'(\omega), \tilde{\omega} \,\rangle \ , \quad \tilde{\omega} = (0, 0, \tilde{w}) \ .$$

The functional L is well-defined because if $\tilde{w}_1^* = \tilde{w}_2^*$, then by (7.11) we have $L(\tilde{w}_1^*) = L(\tilde{w}_2^*)$.

Choose any element $\tilde{w}^* \in C_0^2(\Gamma_0)$, where $C_0^2(\Gamma_0)$ is the space of functions defined on $\Gamma_0 \backslash \partial\Gamma_0$ with first and second order continuous derivatives and compact supports. The function $\tilde{w}^*$ can be extended to the whole boundary Γ by putting its value equal zero. Next this function can be extended to the domain Ω so as to be a function of class $H^2_{\Gamma_1}(\Omega)$.

This means that the variety $\mathcal{V}$ contains all functions from $C_0^2(\Gamma_0)$. The functional L can be extended to $C_0(\Gamma_0)$, for it is positive. Since any linear positive functional on $C_0(\Gamma_0)$ is determined by a nonnegative measure,

$$L(\tilde{w}^*) = \int_{\Gamma_0} \tilde{w}^* d\mu_1 \ ,$$

we get

$$\langle\, \Pi'(\omega), \tilde{\omega}\,\rangle = \int_{\Gamma_0} \tilde{w}\, d\mu_1 \tag{7.12}$$

for all $\tilde{\omega} \in H(\Omega) \cap C_0(\Gamma_0)$ of the form $(0, 0, \tilde{w})$.

Note that equations (7.4) with boundary conditions (7.7) form actually a two-dimensional Signorini problem. It is not essential that functions N_{ij} depend on the displacement w and the curvatures k_{11}, k_{22}. The measure μ_2 can be constructed in the same way as in Section 1.2. Hence,

$$\langle\, \Pi'(\omega), \tilde{\omega}\,\rangle = -\int_{\Gamma_0} \tilde{U}_\nu d\mu_2 \ , \quad \tilde{U} = (\tilde{u}, \tilde{v}) \tag{7.13}$$

for all functions $\tilde{\omega} \in H(\Omega) \cap C_0(\Gamma_0)$ of the type $\tilde{\omega} = (\tilde{u}, \tilde{v}, 0)$. Now the required representation (7.10) follows from (7.12), (7.13).

The above constructed measures take finite values for all compacts $B \subset \Gamma_0 \backslash \partial\Gamma_0$.

In general, properties of the measure μ_2 depend on the regularity of the function U. In particular, known regularity results for the Signorini problem enable us to prove the absolute continuity of the measure μ_2 with respect to the Lebesgue measure on $\Gamma_0 \backslash \partial\Gamma_0$.

Namely, for any $x^0 \in \Gamma_0 \backslash \partial\Gamma_0$, there exists a neighbourhood Ω_0 such that $U \in H^2(\Omega_0 \cap \Omega)$. The density of the measure μ_2 is equal to $-N_\nu$, where $N_\nu \in H^{1/2}_{\text{loc}}(\Gamma_0 \backslash \partial\Gamma_0)$.

As far as the measure μ_1 is concerned, its properties depend on the regularity of the function w. The following result is valid.

Theorem 7.2. *For any $x^0 \in \Gamma_0 \backslash \partial\Gamma_0$ there exists a neighbourhood Ω_0 such that $w \in H^3(\Omega_0 \cap \Omega)$.*

Proof. Without loss of generality we can assume that x^0 coincides with the origin and the direction of the axis x_2 coincides with the outer normal to the boundary Γ.

We start by considering the case where a part of Γ_0 near x^0 is rectilinear. Put

$$d_\tau h(x) = \tau^{-1}\big[h(x + \tau e_1) - h(x)\big] \ , \quad \Delta_\tau = -d_{-\tau} d_\tau \ ,$$

where $|\tau| > 0$, e_1 is the unit vector of the axis x_1. Denote by R_δ the ball of radius δ and centre x^0. Suppose that $\varphi \in C_0^\infty(R_\delta)$, $\varphi \equiv 1$ on $R_{\delta/2}$, $0 \le \varphi \le 1$, $|\tau| < \dfrac{\delta}{2}$.

The function $w_\tau = w + \dfrac{\tau^2}{2}\varphi^2 \Delta_\tau w$ satisfies the inequality $w_\tau \ge 0$ on Γ_0 since for δ sufficiently small and $x \in \Gamma_0$ we have

$$w_\tau(x) = \big(1 - \varphi^2(x)\big)w(x) + \frac{\varphi^2(x)}{2}\big(w(x + \tau e_1) + w(x - \tau e_1)\big) \ge 0 \ .$$

This means that $(u, v, w_\tau) \in K$. Substituting the function (u, v, w_τ) into (7.2) as a test function we get

$$B(w, \varphi^2 \Delta_\tau w) - \langle\, f - k_{11}N_{11} - k_{22}N_{22},\ \varphi^2 \Delta_\tau w \,\rangle \geq 0 \ . \tag{7.14}$$

The following chain of formulae has the property that the difference of any two neighbouring terms can be bounded from above by the right-hand side of inequality (7.15)

$$B(w, \varphi^2 \Delta_\tau w) \to B(\varphi w, \Delta_\tau \varphi w) \to$$
$$\to -B(\varphi w, d_{-\tau} d_\tau \varphi w) \to -B(d_\tau \varphi w, d_\tau \varphi w) \ .$$

An estimation of the second term in (7.14) is simpler. Thus, from (7.14) it follows that

$$\|d_\tau(\varphi w)\|_2^2 \leq c\{\|f\|_0^2 + \|u\|_1^2 + \|v\|_1^2 + \|w\|_2^2 + \|d_\tau(\varphi w)\|_2\ \|w\|_2\} \tag{7.15}$$

with a constant c independent of τ. Consequently, $\|d_\tau(\varphi w)\|_2$ are bounded uniformly with respect to τ. This implies that the third order derivatives of w, except $D^3_{x_2} w$, belong to $L^2(R_{\frac{\delta}{2}} \cap \Omega)$. We can write equation (7.3) in the form

$$D^4_{x_2} w = g \ . \tag{7.16}$$

It follows that $g \in H^{-1}(R_{\frac{\delta}{2}} \cap \Omega)$. But if $w \in H^2(\Omega)$, then $D^3_{x_2} w \in H^{-1}(R_{\frac{\delta}{2}} \cap \Omega)$. This gives, by equation (7.16), that $D^3_{x_2} w \in L^2(R_{\frac{\delta}{2}} \cap \Omega)$, which proves the theorem in this case.

Note that we have also used the fact that if $\varphi, \varphi_{x_i} \in H^{-1}(\tilde{\Omega})$, $i = 1, 2$, then $\varphi \in L^2(\tilde{\Omega})$ (see Duvaut, Lions, 1972).

If Γ_0 has no rectilinear part near x^0 the transformation of independent variables can be performed

$$y_1 = x_1 \ , \quad y_2 = x_2 - \alpha(x_1) \ .$$

Here $x_2 = \alpha(x_1)$ is the equation of the boundary near x^0. The reasoning in this case is similar. □

8 Dynamic problems

8.1 Variational inequality for a beam

Let $\Omega = (0, 1)$, $Q = \Omega \times (0, T)$, $T > 0$ and let K be a closed convex set in $L^2(\Omega) \times L^2(\Omega)$. We assume that K contains zero. Derivatives with respect to t will be denoted by primes.

The operator of nonlinear vibration of a beam has the form

$$\begin{pmatrix} u \\ w \end{pmatrix} \mapsto \begin{pmatrix} u'' - U_x \\ w'' + w_{xxxx} - (w_x U)_x \end{pmatrix} , \quad U = u_x + \frac{1}{2}\, w_x^2 \ .$$

The symbols $\langle \cdot, \cdot \rangle$ denote duality pairings between $H^{-1}(\Omega)$, $H_0^1(\Omega)$ and between $L^p(\Omega)$, $L^q(\Omega)$, $p^{-1} + q^{-1} = 1$, $\| \cdot \|_s$ is a norm in $H^s(\Omega)$.

Theorem 8.1. *Let $u_0 \in H_0^1(\Omega)$, $w_0 \in H_0^2(\Omega)$, $u_1, w_1 \in L^2(\Omega)$, $f, g \in L^2(Q)$, $(u_0, w_0) \in K$. Then there exists a pair of functions (u, w) such that*

$$\begin{gathered} u \in L^\infty\big(0, T; H_0^1(\Omega)\big) , \quad w \in L^\infty\big(0, T; H_0^2(\Omega)\big) , \\ u', w' \in L^\infty\big(0, T; L^2(\Omega)\big) , \end{gathered} \tag{8.1}$$

$$\begin{aligned} &\int_0^T \langle u' - \int_0^t U_x d\tau, \varphi - u \rangle dt \\ &+ \int_0^T \langle w' + \int_0^t \Big[w_{xxxx} - (w_x U)_x \Big] d\tau, \psi - w \rangle dt \\ &\geq \int_0^T \langle G, \varphi - u \rangle dt + \int_0^T \langle F, \psi - w \rangle \, dt , \end{aligned} \tag{8.2}$$

for all functions $(\varphi, \psi) \in L^2\big(0, T; H_0^1(\Omega) \times H_0^2(\Omega)\big)$ such that

$$\big(\varphi(t), \psi(t)\big) \in K \quad \text{a.e. on} \quad (0, T) ,$$

$$u(0) = u_0 , \quad w(0) = w_0 , \tag{8.3}$$

$$\big(u(t), w(t)\big) \in K \quad \text{a.e. on} \quad (0, T) , \tag{8.4}$$

$$G = u_1 + \int_0^t g(\tau) d\tau , \quad F = w_1 + \int_0^t f(\tau) d\tau .$$

Proof. Putting

$$\bar{u}(t) = \int_0^t u(\tau) d\tau , \quad \bar{w}(t) = \int_0^t w(\tau) d\tau . \tag{8.5}$$

and replacing $\bar{u}, \bar{w}$ with u, w, respectively, we can rewrite relations (8.1)–(8.4) as

$$\int_0^T \langle L_1 - G, \varphi - u' \rangle dt + \int_0^T \langle L_2 - F, \psi - w' \rangle dt \geq 0 \tag{8.6}$$

for all φ, ψ satisfying the above conditions and

$$\begin{gathered} u, u' \in L^\infty\big(0, T; H_0^1(\Omega)\big) , \quad w, w' \in L^\infty\big(0, T; H_0^2(\Omega)\big) , \\ u'', w'' \in L^\infty\big(0, T; L^2(\Omega)\big) , \end{gathered} \tag{8.7}$$

$$u(0) = 0 , \quad u'(0) = u_0 , \quad w(0) = 0 , \quad w'(0) = w_0 , \tag{8.8}$$

$$\big(u'(t), w'(t)\big) \in K \quad \text{a.e. on} \quad (0, T) . \tag{8.9}$$

Here

$$L_1 \equiv u'' - \int_0^t V_x d\tau \ , \quad V \equiv u'_x + \frac{1}{2} w'^2_x \ ,$$

$$L_2 \equiv w'' + w_{xxxx} - \int_0^t (w'_x V)_x dt \ .$$

We shall prove the solvability of problem (8.6)–(8.9).

Let p_K be the operator of orthogonal projection of the space $L^2(\Omega) \times L^2(\Omega)$ onto the set K and let $p(v) = v - p_K v$ be the penalty operator for $v \in L^2(\Omega) \times L^2(\Omega)$. The penalty operator depends upon the set K and possesses continuity and monotonicity properties (see Lions, 1969).

Denoting by p_1, p_2 the components of the operator p we consider for $\varepsilon > 0$ the problem

$$L_1^\varepsilon + \frac{1}{\varepsilon} p_1 (u'_\varepsilon, w'_\varepsilon) = G \ , \tag{8.10}$$

$$L_2^\varepsilon + \frac{1}{\varepsilon} p_2 (u'_\varepsilon, w'_\varepsilon) = F \ , \tag{8.11}$$

$$u_\varepsilon = w_\varepsilon = w_{\varepsilon x} = 0 \quad \text{when} \quad x = 0, 1 \ , \tag{8.12}$$

$$u_\varepsilon(0) = 0 \ , \quad u'_\varepsilon(0) = u_0 \ , \quad w_\varepsilon(0) = 0 \ , \quad w'_\varepsilon(0) = w_0 \ . \tag{8.13}$$

The symbols $L_1^\varepsilon, L_2^\varepsilon$ denote the values of the operators L_1, L_2 at $u_\varepsilon, w_\varepsilon$, respectively.

To prove the solvability of problem (8.10)–(8.13) we use the Galerkin method.

We choose basis functions $\{\psi_j\}$, $j = 1, 2, \ldots$, in the space $H_0^2(\Omega)$ and assume that these functions are orthonormal in $L^2(\Omega)$. By omitting the symbol ε, we get the approximate solution to the problem in the form

$$u_m(t) = \sum_{i=1}^m b_{im}(t) \psi_i \ , \quad w_m(t) = \sum_{i=1}^m c_{im}(t) \psi_i \ .$$

where b_{im}, c_{im} satisfy the following system of ordinary differential equations

$$\langle L_1^m + \frac{1}{\varepsilon} p_1 (u'_m, w'_m), \psi_j \rangle = \langle G, \psi_j \rangle \ , \tag{8.14}$$

$$\langle L_2^m + \frac{1}{\varepsilon} p_2 (u'_m, w'_m), \psi_j \rangle = \langle F, \psi_j \rangle \ , \qquad 1 \le j < m \ . \tag{8.15}$$

with the initial data

$$u_m(0) = 0 \ , \quad u'_m(0) \to u_0 \quad \text{in the norm} \quad H_0^1(\Omega) \ ,$$

$$w_m(0) = 0 \ , \quad w'_m(0) \to w_0 \quad \text{in the norm} \quad H_0^2(\Omega) \ .$$

Let us consider equations (8.14), (8.15) for $t = 0$. Multiplying the j-th equation (8.14) by $b''_{jm}(0)$ and the j-th equation (8.15) by $c''_{jm}(0)$ and summing up the resulting formulae over all j from 1 to m, one gets

$$\begin{aligned} \|u''_m(0)\|^2 + \|w''_m(0)\|^2 = {} & \langle u_1, u''_m(0) \rangle + \langle w_1, w''_m(0) \rangle \\ & - \frac{1}{\varepsilon} \langle p_1(u'_m(0), w'_m(0)), u''_m(0) \rangle \\ & - \frac{1}{\varepsilon} \langle p_2(u'_m(0), w'_m(0)), w''_m(0) \rangle . \end{aligned} \tag{8.16}$$

Hence

$$\|u''_m(0)\| + \|w''_m(0)\| \le c , \tag{8.17}$$

where c is a constant independent of m.

Let us differentiate equations (8.14), (8.15) with respect to t, then multiply them by $b''_{jm}(t)$, $c''_{jm}(t)$ respectively and sum up over all j. The resulting relation has the form

$$\begin{aligned} & \frac{1}{2}\frac{d}{dt}\{\|u''_m\|^2 + \|w''_m\|^2 + \|w'_{mxx}\|^2 + \|V_m\|^2\} + \frac{1}{\varepsilon} \langle p_1(u'_m, w'_m)', u''_m \rangle \\ & + \frac{1}{\varepsilon} \langle p_2(u'_m, w'_m)', w''_m \rangle = \langle g, u''_m \rangle + \langle f, w''_m \rangle , \quad V_m = u'_{mx} + \frac{1}{2} w'^2_{mx} . \end{aligned}$$

The penalty term is nonnegative for almost all t. Neglecting this term, the following inequality is a consequence of the Gronwall lemma

$$\max_{0 \le t \le T} \{\|u''_m(t)\|^2 + \|w''_m(t)\|^2 + \|w'_{mxx}(t)\|^2 + \|V_m(t)\|^2\} \le c \tag{8.18}$$

with a constant c independent of m. Bearing in mind the boundedness of w'^2_{mx} in $L^\infty(0,T; L^2(\Omega))$ from (8.18) it follows that

$$\max_{0 \le t \le T} \|u'_{mx}(t)\| \le c , \tag{8.19}$$

where c does not depend on m. Since

$$u_m(t) = \int_0^t u'_m(\tau) d\tau , \quad w_m(t) = \int_0^t w'_m(\tau) d\tau ,$$

from (8.18), (8.19) it follows that

$$\max_{0 \le t \le T} \{\|u_{mx}(t)\| + \|w_{mxx}(t)\|\} \le c . \tag{8.20}$$

The system of equations (8.14), (8.15) can be written in the normal form. Estimates (8.18)–(8.20) give, in particular, the boundedness of $b_{im}, b'_{im}, c_{im}, c'_{im}$ on $(0,T)$. Thus, the Galerkin system is solvable on the interval $(0,T)$. Moreover, taking into account the boundary conditions, a subsequence u_m, w_m exists such that

$$\begin{aligned} u_m, u_m' &\to u_\varepsilon, u_\varepsilon' \quad \text{weakly-}(*) \text{ in} \quad L^\infty(0,T;H_0^1(\Omega)) \ , \\ w_m, w_m' &\to w_\varepsilon, w_\varepsilon' \quad \text{weakly-}(*) \text{ in} \quad L^\infty(0,T;H_0^2(\Omega)) \ , \\ u_m'', w_m'' &\to u_\varepsilon'', w_\varepsilon'' \quad \text{weakly-}(*) \text{ in} \quad L^\infty(0,T;L^2(\Omega)) \ , \\ u_m', w_m' &\to u_\varepsilon', w_\varepsilon' \quad \text{strongly} \quad \text{in} \quad L^2(Q) \ . \end{aligned} \tag{8.21}$$

The last formula is a consequence of compact imbedding of $H^1(Q)$ in $L^2(Q)$. By the well-known theorem (Lions, 1969) the space with the norm

$$\|v\|_{L^2(0,T;H_0^1(\Omega))} + \|v'\|_{L^2(0,T;H^{-1}(\Omega))}$$

has a compact injection in $L^2(0,T;L^2(\Omega))$. By the above estimates, w_{mx}', w_{mx}'' are bounded in $L^2(0,T;H_0^1(\Omega))$, $L^2(0,T;H^{-1}(\Omega))$ respectively. By the above-mentioned theorem, the subsequence w_{mx}' possesses the property that

$$w_{mx}' \to w_{\varepsilon x}' \quad \text{strongly in} \quad L^2(Q) \ . \tag{8.22}$$

The convergence (8.21), (8.22) enables us to pass to the limit in equations (8.14), (8.15) as $m \to \infty$ for fixed j.

In view of the density of the basis functions we get the existence of solutions to (8.10), (8.11). We omit the proof of this fact here. The convergence will be demonstrated later on (see (8.24) and below). The structure of nonlinear terms in both cases is the same.

Now we shall use the following result (Lions, Magenes, 1968). Any Banach space X with the norm

$$\|v\|_{L^p(0,T;X)} + \|v'\|_{L^p(0,T;X)}$$

has a continuous injection in $C(0,T;X)$. In particular,

$$\|v(0)\|_X \le c\big(\|v\|_{L^p(0,T;X)} + \|v'\|_{L^p(0,T;X)}\big)$$

with a constant c independent of v.

Applying this result to the sequences u_m and u_m' one gets

$$u_m(0) \to u_\varepsilon(0) \quad \text{weakly in} \quad H_0^1(\Omega) \ ,$$

$$u_m'(0) \to u_\varepsilon'(0) \quad \text{weakly in} \quad L^2(\Omega) \ .$$

Since $u_m'(0) \to u_0$ strongly in $H_0^1(\Omega)$ and $u_m(0) = 0$ we see that u_ε satisfies condition (8.13). The same reasoning applies to w_m and consequently w_ε satisfies (8.13).

Let us discuss now the passage to the limit as $\varepsilon \to 0$.

First we note that a priori estimates (8.18)–(8.20) are actually independent of ε. To verify this we turn back to the arguments used to derive estimate (8.17). Namely, we consider (8.16). It is clear that

$$\begin{aligned}&\frac{1}{\varepsilon}\left\langle p_1\left(u'_m(0), w'_m(0)\right), u'_m(0)\right\rangle + \frac{1}{\varepsilon}\left\langle p_2\left(u'_m(0), w'_m(0)\right), w'_m(0)\right\rangle \\ &\quad \leq \frac{1}{2\varepsilon^2}\left\|p(u'_m(0), w'_m(0))\right\|^2 + \frac{1}{2}\left(\|u'_m(0)\|^2 + \|w'_m(0)\|^2\right) .\end{aligned}$$

Since $(u_0, w_0) \in K$ and $u'_m(0), w'_m(0)$ tend to u_0, w_0 respectively, from (8.16) it follows that

$$\|u''_m(0)\|^2 + \|w''_m(0)\|^2 \leq \frac{1}{\varepsilon^2}\|p(u'_m(0), w)_m(0))\|^2 + c ,$$

where c is independent of ε. Hence, by

$$\frac{1}{2\varepsilon^2}\left\|p(u'_m(0), w'_m(0))\right\|^2 \leq \frac{1}{2}$$

estimate (8.17) is uniform with respect to ε in the sense that for every fixed $\varepsilon > 0$ there exists a number $m_0 = m(\varepsilon)$ such that for all $m \geq m_0$ estimate (8.17) holds.

The estimates which follow (8.17) will also be uniform with respect to ε. After passing to the limit as $m \to \infty$ we obtain the estimate for the solutions u_ε, w_ε the same as for u_m, w_m. This follows from the weak-(*) closedness of closed balls in the space $L^\infty(0,T;X)$, where X is a reflexive Banach space. Next we can pass to the limit as $\varepsilon \to 0$. Thus,

$$\begin{aligned}\max_{0\leq t\leq T}\big\{&\|u''_\varepsilon(t)\| + \|u'_\varepsilon(t)\|_1 + \|u_\varepsilon(t)\|_1 + \|w''_\varepsilon(t)\| \\ &+ \|w'_\varepsilon(t)\|_2 + \|w_\varepsilon(t)\|_2\big\} \leq c\end{aligned} \tag{8.23}$$

where c is a constant independent of ε.

In view of (8.23) we claim that subsequences $u_\varepsilon, w_\varepsilon$ converge in the sense of (8.21), (8.22), where m has to be changed by ε.

Now we shall show that $\big(u'(t), w'(t)\big) \in K$.

From (8.10), (8.11) it follows that $p(u'_\varepsilon, w'_\varepsilon) \to 0$ weakly-(*) in $L^\infty\big(0,T; H^{-1}(\Omega) \times H^{-2}(\Omega)\big)$. The operator P_K satisfies the Lipschitz condition, and so does the operator p. Consequently, $p(u'_\varepsilon, w'_\varepsilon) \to p(u', w')$ strongly in $L^2(Q) \times L^2(Q)$ since in view of (8.23), $u'_\varepsilon, w'_\varepsilon \to u', w'$ strongly in $L^2(Q)$. Therefore, $p(u', w') = 0$, and the proof of (8.9) is completed.

Let us show that the limiting functions u, w satisfy inequality (8.6). We take $(\varphi, \psi) \in L^2\big(0,T; H^1_0(\Omega) \times H^2_0(\Omega)\big)$, $\big(\varphi(t), \psi(t)\big) \in K$ almost everywhere. By the monotonicity of p, from (8.10), (8.11) it follows that

$$\langle L_1^\varepsilon - G, \varphi - u'_\varepsilon\rangle + \langle L_2^\varepsilon - F, \psi - w'_\varepsilon\rangle \geq 0 \quad \text{for almost all} \quad t .$$

Hence

$$
\begin{aligned}
&\int_0^T \langle L_1^\varepsilon, \varphi\rangle \, dt - \int_0^T \langle G, \varphi - u'_\varepsilon\rangle \, dt + \int_0^T \langle L_2^\varepsilon, \psi\rangle \, dt \\
&- \int_0^T \langle F, \psi - w'_\varepsilon\rangle \, dt \geq \int_0^T \langle L_1^\varepsilon, u'_\varepsilon\rangle \, dt + \int_0^T \langle L_2^\varepsilon, w'_\varepsilon\rangle \, dt \\
&= \frac{1}{2}(\|u'_\varepsilon(T)\|^2 + \|u_{\varepsilon x}(T)\|^2 - \|u_0\|^2 + \|w'_\varepsilon(T)\|^2 + \|w_{\varepsilon xx}(T)\|^2 \\
&- \|w_0\|^2) - \int_0^T \langle \int_0^t \frac{1}{2} w'^2_{\varepsilon x} d\tau, u'_{\varepsilon x}\rangle \, dt - \int_0^T \langle \int_0^t w'_{\varepsilon x} V_\varepsilon d\tau, w'_{\varepsilon x}\rangle \, dt \ .
\end{aligned}
\tag{8.24}
$$

We shall justify the passage to the limit in the expression $\int_0^T \langle \int_0^t w'_{\varepsilon x} u'_{\varepsilon x} d\tau, w'_{\varepsilon x}\rangle \, dt$.

By imbedding theorems, $w'_{\varepsilon x}$ are bounded in $L^\infty(Q)$. This implies the boundedness of $w'_{\varepsilon x} u'_{\varepsilon x}$ in $L^2(Q)$.

Suppose that $w'_{\varepsilon x} u'_{\varepsilon x} \to w'_x u'_x$ weakly in Q. Then

$$
\int_0^t w'_{\varepsilon x} u'_{\varepsilon x} d\tau \to \int_0^t w'_x u'_x d\tau \quad \text{weakly in} \quad L^2(Q) \ .
$$

Hence, in view of the strong convergence $w'_{\varepsilon x} \to w'_\varepsilon$ in $L^2(Q)$, we get the required conclusion.

For the remaining nonlinear terms, the arguments are simpler, and as $\varepsilon \to 0$ one can assume that

$$
u_\varepsilon(T) \to u(T) \quad \text{weakly in} \quad H_0^1(\Omega), \quad w_\varepsilon(T) \to w(T) \quad \text{weakly in} \quad H_0^2(\Omega) \ ,
$$

$$
u'_\varepsilon(T) \to u'(T) \ , \ w'_\varepsilon(T) \to w'(T) \quad \text{weakly in} \quad L^2(\Omega) \ .
$$

This means that the lower limit of the right-hand side of (8.24) is greater than or equal to

$$
\begin{aligned}
&\frac{1}{2}(\|u'(T)\|^2 + \|u_x(T)\|^2 - \|u_0\|^2 + \|w'(T)\|^2 + \|w_{xx}(T)\|^2 - \|w_0\|^2) \\
&- \int_0^T \langle \int_0^t \frac{1}{2} w'^2_x d\tau, u'_x\rangle \, dt - \int_0^T \langle \int_0^t w'_x V d\tau, w'_x\rangle \, dt \\
&= \int_0^T \langle L_1, u'\rangle \, dt + \int_0^T \langle L_2, w'\rangle \, dt \ .
\end{aligned}
$$

Thus, inequality (8.6) holds which proves the theorem.

In particular, if K_1 is a closed convex set containing zero in $L^2(\Omega)$ and $K = L^2(\Omega) \times K_1$, inequality (8.2) takes the form

$$u' - \frac{\partial}{\partial x} \int_0^t U d\tau = G ,$$

$$\int_0^T \langle w' + \int_0^t \{w_{xxxx} - (w_x U)_x\} d\tau, \psi - w \rangle \, dt \geq \int_0^T \langle F, \psi - w \rangle \, dt .$$

The latter holds for every function $\psi \in L^2(0,T; H_0^2(\Omega))$, $\psi(t) \in K_1$ almost everywhere.

Now consider the case where K is a subset of a smaller space, namely, $H_0^1(\Omega) \times H_0^2(\Omega)$. Let K be a closed convex set in $H_0^1(\Omega) \times H_0^2(\Omega)$ with nonempty interior $\operatorname{int} K$ and zero belong to K.

Theorem 8.2. *Let $(u_0, w_0) \in \operatorname{int} K$, $u_1, w_1 \in L^2(\Omega)$. Then there exist functions u, w satisfying the conditions*

$$u \in L^\infty(0,T; H_0^1(\Omega)), \quad w \in L^\infty(0,T; H_0^2(\Omega)), \quad u', w' \in L^\infty(0,T; L^2(\Omega)) ,$$

$$\begin{aligned} &\int_0^T \langle u' - \int_0^t U_x d\tau, \varphi - u \rangle dt \\ &+ \int_0^T \langle w' + \int_0^t \{w_{xxxx} - (w_x U)_x\} d\tau, \psi - w \rangle \, dt \\ &\geq \int_0^T \langle G, \varphi - u \rangle \, dt + \int_0^T \langle F, \psi - w \rangle \, dt \end{aligned}$$

which are valid for all functions (φ, ψ) such that

$$(\varphi, \psi) \in L^2(0,T; H_0^1(\Omega) \times H_0^2(\Omega)), \quad (\varphi(t), \psi(t)) \in K \quad \text{a.e.} ,$$

$$u(0) = u_0 , \quad w(0) = w_0 ,$$

$$(u(t), w(t)) \in K \quad \text{a.e. on} \quad (0,T) . \tag{8.25}$$

The proof of this theorem is analogous to that of the previous theorem. We sketch it only briefly.

Assume that the operator p acts from $H_0^1(\Omega) \times H_0^2(\Omega)$ into the dual space and depends on the set K. After the substitution of type (8.5) of unknown functions an approximate problem will coincide with (8.10)–(8.13). Its solvability can be proved by the Galerkin approach. The basis functions can be chosen in the space $H_0^2(\Omega)$. By (8.16), the equalities $p_i(u_n'(0), w_m'(0)) = 0$, $i = 1, 2$, hold for sufficiently large m. Hence, (8.17) and other estimates can be assumed to be uniform with respect to ε, m. This allows us to prove the solvability of problem (8.10)–(8.13) as well as passing to the limit as $\varepsilon \to 0$.

When passing to the limit as $m \to \infty$ in (8.14), (8.15) (for fixed ε) we need the convergence

$$p(u'_m, w'_m) \to p(u'_\varepsilon, w'_\varepsilon) \quad \text{weakly in} \quad L^2(0,T; H^{-1}(\Omega) \times H^{-2}(\Omega)) \ . \tag{8.26}$$

This convergence can be achieved by exploiting the monotonicity of the operator p.

The passage to the limit as $\varepsilon \to 0$ is based on similar arguments to those used in the proof of Theorem 8.1.

It is worth explaining the proof of (8.25) in a more detailed way. So, let a subsequence $u_\varepsilon, w_\varepsilon$ converge to u, w in the sense of (8.21), (8.22). From (8.10), (8.11) it follows that

$$p(u'_\varepsilon, w'_\varepsilon) \to 0 \quad \text{weakly-}(*) \text{ in} \quad L^\infty(0,T; H^{-1}(\Omega) \times H^{-2}(\Omega)) \ . \tag{8.27}$$

On the other hand, multiplying (8.10), (8.11) by $u'_\varepsilon, w'_\varepsilon$, respectively, we get

$$\int_0^T [\langle p_1(u'_\varepsilon, w'_\varepsilon), u'_\varepsilon\rangle + \langle p_2(u'_\varepsilon, w'_\varepsilon), w'_\varepsilon\rangle] dt \le c\varepsilon \ , \tag{8.28}$$

where c does not depend on ε. Since

$$\int_0^T [\langle p_1(u'_\varepsilon, w'_\varepsilon) - p_1(\varphi,\psi), u'_\varepsilon - \varphi\rangle] dt +$$
$$+ \int_0^T [\langle p_2(u'_\varepsilon, w'_\varepsilon) - p_2(\varphi,\psi), w'_\varepsilon - \psi\rangle] dt \ge 0 \ .$$

From (8.27), (8.28) as $\varepsilon \to 0$ it follows that

$$\int_0^T [\langle p_1(\varphi,\psi), u' - \varphi\rangle + \langle p_2(\varphi,\psi), w' - \psi\rangle] dt \le 0 \ .$$

Let us take $(\varphi,\psi) = (u' + \lambda\hat{\varphi}, w' + \lambda\hat{\psi})$, $\lambda > 0$, $(\hat{\varphi},\hat{\psi}) \in L^2(0,T; H_0^1(\Omega) \times H_0^2(\Omega))$. Dividing this inequality by λ and passing to the limit as $\lambda \to 0$ we get $p(u', w') = 0$ by the semicontinuity of the operator p. Thus, we obtain (8.25).

In the sequel we make frequent use of semicontinuity. Let us recall that an operator $A : V \mapsto V^*$ is semicontinuous if the function $t \mapsto \langle A(u + tv), w\rangle$ is a continuous function from $\mathbb{R}$ into $\mathbb{R}$ for every fixed $u, v, w \in V$.

8.2 Variational inequality for a shell

Let $\Omega \subset \mathbb{R}^2$ be a bounded domain with smooth boundary, $Q = \Omega \times (0,T)$ and K be a closed convex set in $L^2(\Omega)$, containing zero. The nonlinear operator for a shallow shell has the form

$$\begin{pmatrix} u \\ v \\ w \end{pmatrix} \mapsto \begin{pmatrix} u'' - \dfrac{\partial N_{1j}}{\partial x_j} \\ v'' - \dfrac{\partial N_{2j}}{\partial x_j} \\ w'' + \Delta^2 w - (N_{ij} w_{x_j})_{x_i} + k_{11} N_{11} + k_{22} N_{22} \end{pmatrix} . \tag{8.29}$$

The meaning of the functions u, v, w, N_{ij} is the same as in Sections 3, and 6, ie.,

$$N_{11} = \varepsilon_{11} + \sigma\varepsilon_{22}, \quad N_{22} = \varepsilon_{22} + \sigma\varepsilon_{11}, \quad N_{12} = \frac{1}{2}(1-\sigma)\varepsilon_{12} , \tag{8.30}$$

$$\begin{gathered} \varepsilon_{11} = u_{x_1} + k_{11} w + \frac{1}{2} w_{x_1}^2, \quad \varepsilon_{22} = v_{x_2} + k_{22} w + \frac{1}{2} w_{x_2}^2 , \\ \varepsilon_{12} = u_{x_2} + v_{x_1} + w_{x_1} w_{x_2} . \end{gathered} \tag{8.31}$$

Theorem 8.3. *Let* $u_0, v_0 \in H_0^1(\Omega)$, $w_0 \in H_0^2(\Omega)$, $u_1, v_1, w_1 \in L^2(\Omega)$, $w_0 \in K$, $f_i, f \in L^2(Q)$, $i = 1, 2$; $k_{11}, k_{22} \in L^\infty(\Omega)$. *Then there exist functions* u, v, w *satisfying the relations*

$$\begin{gathered} u, v \in L^\infty(0, T; H_0^1(\Omega)) , \\ w \in L^\infty(0, T; H_0^2(\Omega)), \quad u', v', w' \in L^\infty(0, T; L^2(\Omega)) , \end{gathered} \tag{8.32}$$

$$u' - \int_0^t \frac{\partial N_{1j}}{\partial x_j} \, d\tau = F_1 , \tag{8.33}$$

$$v' - \int_0^t \frac{\partial N_{2j}}{\partial x_j} \, d\tau = F_2 , \tag{8.34}$$

$$\begin{aligned} &\int_0^T \langle w' + \int_0^t \{\Delta^2 w - \left(N_{ij} w_{x_j}\right)_{x_i} + k_{11} N_{11} + k_{22} N_{22}\} d\tau, \varphi - w \rangle dt \\ &\quad \ge \int_0^T \langle F, \varphi - w \rangle dt \quad \forall\, \varphi \in L^2(0, T; H_0^2(\Omega)), \ \varphi(t) \in K , \end{aligned} \tag{8.35}$$

$$u = u_0 , \quad v = v_0 , \quad w = w_0 \quad \text{when} \quad t = 0 , \tag{8.36}$$

$$w(t) \in K \quad \text{a.e. on} \quad (0, T) , \tag{8.37}$$

$$F_1 = u_1 + \int_0^t f_1(\tau) d\tau , \quad F_2 = v_1 + \int_0^t f_2(\tau) d\tau , \quad F = w_1 + \int_0^t f(\tau) d\tau .$$

Proof. Let $l_1(u, v, w) = \dfrac{\partial N_{1j}}{\partial x_j}$, where N_{1j} depend on u, v, w through the formulae (8.30), (8.31). Respectively, we put $l_2(u, v, w) = \dfrac{\partial N_{2j}}{\partial x_j}$, $l(u, v, w) = (N_{ij} w_{x_j})_{x_i} - k_{11} N_{11} - k_{22} N_{22}$.

Introduce the notations

$$\tilde{u}(t) = \int_0^t u(\tau)d\tau \ , \quad \tilde{v}(t) = \int_0^t v(\tau)d\tau \ , \quad \tilde{w}(t) = \int_0^t w(\tau)d\tau \ .$$

In so doing the relations (8.32)–(8.37) may be rewritten as follows (omitting the tilda sign)

$$L_1 \equiv u'' - \int_0^t l_1(u', v', w')d\tau = F_1 \ , \tag{8.38}$$

$$L_2 \equiv v'' - \int_0^t l_2(u', v', w')d\tau = F_2 \ , \tag{8.39}$$

$$\int_0^T \langle L - F, \varphi - w' \rangle \, dt \geq 0 \tag{8.40}$$
$$\forall \, \varphi \in L^2(0, T; H_0^2(\Omega)) \ , \quad \varphi(t) \in K \quad \text{a.e.} \ ,$$

$$u, u', v, v' \in L^\infty(0, T; H_0^1(\Omega)) \ , \quad w, w' \in L^\infty(0, T; H_0^2(\Omega)) \ ,$$
$$u'', v'', w'' \in L^\infty(0, T; L^2(\Omega)) \ , \tag{8.41}$$

$$u = v = w = 0, \quad u' = u_0, \quad v' = v_0, \quad w' = w_0 \quad \text{when} \quad t = 0 \ , \tag{8.42}$$

$$w'(t) \in K \quad \text{a. e. on} \quad (0, T) \ , \tag{8.43}$$

$$L \equiv w'' + \Delta^2 w - \int_0^t l(u', v', w')d\tau \ .$$

To prove the solvability of problem (8.38)–(8.43) we introduce the penalty operator $p : L^2(\Omega) \mapsto L^2(\Omega)$, related to the set K. If P_K is the operator of orthogonal projection of $L^2(\Omega)$ onto K, then $p(w) = w - P_K w$. The operator p satisfies the Lipschitz condition.

Let $\varepsilon > 0$ be a fixed parameter. Now we prove the solvability of the following problem (the symbol ε is omitted)

$$L_1 = F_1 \ , \tag{8.44}$$

$$L_2 = F_2 \ , \tag{8.45}$$

$$L + \frac{1}{\varepsilon} p(w') = F \ , \tag{8.46}$$

$$u = v = w = \frac{\partial w}{\partial \nu} = 0 \quad \text{on} \quad \partial\Omega \times (0, T) \ , \tag{8.47}$$

$$u = v = w = 0, \quad u' = u_0, \quad v' = v_0, \quad w' = w_0 \quad \text{when} \quad t = 0 \ . \tag{8.48}$$

To this end we use the Galerkin method.

Let $\{\psi_j\}$ be basis functions in the space $H_0^2(\Omega)$. For the approximate solutions

$$\big(u_m(t), v_m(t), w_m(t)\big) = \sum_{i=1}^m \big(a_{im}(t), b_{im}(t), c_{im}(t)\big)\psi_i$$

we consider the problem

$$\langle L_1^m, \psi_j \rangle = \langle F_1, \psi_j \rangle \,, \tag{8.49}$$

$$\langle L_2^m, \psi_j \rangle = \langle F_2, \psi_j \rangle \,, \tag{8.50}$$

$$\langle L^m + \frac{1}{\varepsilon}\, p(w_m'), \psi_j \rangle = \langle F, \psi_j \rangle \,, \tag{8.51}$$

$$\big(u_m(0), v_m(0), w_m(0)\big) = (0,0,0) \;, \tag{8.52}$$

$$\big(u_m'(0), v_m'(0), w_m'(0)\big) = (u_{m1}, v_{m1}, w_{m1}) \;, \tag{8.53}$$

where $u_{m1}, v_{m1} \to u_0, v_0$ strongly in $H_0^1(\Omega)$, $w_{m1} \to w_0$ strongly in $H_0^2(\Omega)$. From (8.49)–(8.51) it follows that

$$\begin{aligned} \|u_m''(0)\|^2 + \|v_m''(0)\|^2 + \|w_m''(0)\|^2 &= \langle w_1, w_m''(0)\rangle \\ + \langle u_1, u_m''(0)\rangle + \langle v_1, v_m''(0)\rangle &- \frac{1}{\varepsilon}\langle p(w_{m1}), w_m''(0)\rangle \,. \end{aligned} \tag{8.54}$$

At the same time $w_{m1} \to w_0$ strongly in $H_0^2(\Omega)$ and $w_0 \in K$. Hence, by the Lipschitz condition

$$\|p(w_{m1})\| = \|p(w_{m1}) - p(w_0)\| \le c\, \|w_{m1} - w_0\| \to 0 \;.$$

This means that for every $\varepsilon > 0$ there exists $m_0 = m(\varepsilon)$ such that $\frac{1}{2\varepsilon^2}\, \|p(w_{m1})\| \le \frac{1}{2}$ for all $m \ge m_0$. Applying the Cauchy inequality to the right-hand side of (8.54) when $m \ge m_0$ we find that

$$\|u_m''(0)\| + \|v_m''(0)\| + \|w_m''(0)\| \le c \tag{8.55}$$

with a constant c independent of m and ε. Differentiating with respect to t and multiplying (8.49)–(8.51) by u_m'', v_m'', w_m'', respectively , we obtain

$$\begin{aligned} &\frac{1}{2}\frac{d}{dt}\Big\{\|u_m''\|^2 + \|v_m''\|^2 + \|w_m''\|^2 + \|\Delta w_m'\|^2 \\ &+ \sum_{(i,j)\neq(2,1)} \int_\Omega N_{ij}(u_m', v_m', w_m')\varepsilon_{ij}(u_m', v_m', w_m')dx\Big\} \\ &+ \frac{1}{\varepsilon}\langle p(w_m')', w_m''\rangle = \langle f_1, u_m''\rangle + \langle f_2, v_m''\rangle + \langle f, w_m''\rangle \,. \end{aligned} \tag{8.56}$$

Here $N_{ij}(u_m', v_m', w_m')$, $\varepsilon_{ij}(u_m', v_m', w_m')$ depend on u_m', v_m', w_m' in the same way as N_{ij}, ε_{ij} depend on u, v, w in the formulae (8.30), (8.31). By using the relation $\langle\, p'(w_m'(t)),\, w_m''(t)\rangle \ge 0$ and neglecting the penalty term in (8.56) the inequalities result

$$\max_{0\le t\le T} \big\{\|u_m''(t)\|^2 + \|v_m''(t)\|^2 + \|w_m''(t)\|^2 + \|\Delta w_m'(t)\|^2\big\} \le c \;, \tag{8.57}$$

$$\max_{0\le t\le T} \sum_{i,j=1}^{2} \|\varepsilon_{ij}(u_m', v_m', w_m')\|^2 \le c \;. \tag{8.58}$$

In these relations constants do not depend on m, ε $(m \geq m_0)$. In deriving (8.58) we exploited the positive definiteness of $\sum_{(i,j)\neq(2,1)} N_{ij}\varepsilon_{ij}$ with respect to ε_{ij}^2. By imbedding theorems, from (8.57) it follows that $w_{mx_i}'^2$ are bounded in $L^\infty(0,T;L^2(\Omega))$. It is also clear that by inequality (8.58)

$$\max_{0\leq t\leq T} \{\|u_m'(t)\|_1 + \|v_m'(t)\|_1\} \leq c \; . \tag{8.59}$$

Thus, by the formulae $u = \int_0^t u'(\tau)d\tau$ valid for u_m, v_m, w_m from (8.57), (8.59) we obtain that

$$\max_{0\leq t\leq T} \{\|u_m(t)\|_1 + \|v_m(t)\|_1 + \|w_m(t)\|_2\} \leq c \tag{8.60}$$

with a constant independent of ε, m $(m \geq m_0)$. Therefore, the solvability of the Galerkin equations (8.49)–(8.53) on the interval $(0,T)$ follows from the above estimates. Moreover, we can pass to the limit as $m \to \infty$ for every fixed $\varepsilon > 0$.

The next step consists in passing to the limit as $\varepsilon \to 0$ in equations (8.44)–(8.46). The structure of nonlinear terms is the same as $m \to \infty, \varepsilon \to 0$, hence it is enough to analyse carefully the second step concerning the case $\varepsilon \to 0$. According to what has been said before, a priori estimates (8.57), (8.59), (8.60) for $u_\varepsilon, v_\varepsilon, w_\varepsilon$ in problem (8.44)–(8.48) can be regarded as uniform with respect to ε. Using the same notation as before, we choose a subsequence $u_\varepsilon, v_\varepsilon, w_\varepsilon$, and as $\varepsilon \to 0$ we have

$$\begin{aligned}
&u_\varepsilon, u_\varepsilon', v_\varepsilon, v_\varepsilon' \to u, u', v, v' \quad \text{weakly-}(*) \text{ in } \quad L^\infty(0,T;H_0^1(\Omega)) \; ,\\
&w_\varepsilon, w_\varepsilon' \to w, w' \quad \text{weakly-}(*) \text{ in } \quad L^\infty(0,T;H_0^2(\Omega)) \; ,\\
&u_\varepsilon'', v_\varepsilon'', w_\varepsilon'' \to u'', v'', w'' \quad \text{weakly-}(*) \text{ in } \quad L^\infty(0,T;L^2(\Omega)) \; ,\\
&u_\varepsilon', v_\varepsilon' \to u', v' \quad \text{strongly in } \quad L^2(Q) \; ,\\
&w_\varepsilon' \to w' \quad \text{strongly in } \quad L^2(0,T;H^{2-\delta}(\Omega)), \quad \delta > 0 \; .
\end{aligned} \tag{8.61}$$

Here the last convergence is a consequence of the compactness theorems (Lions, 1969), since w_ε'' are bounded in $L^2(0,T;L^2(\Omega))$ and w_ε' are bounded in $L^2(0,T;H_0^2(\Omega))$.

First we prove that $w'(t) \in K$. From (8.46) it follows that $w_\varepsilon' \to w'$ weakly-$(*)$ in $L^\infty(0,T;H^{-2}(\Omega))$. Since $w_\varepsilon' \to w'$ strongly in $L^2(Q)$, we have $p(w_\varepsilon') \to p(w')$ strongly in $L^2(Q)$. This means that $p(w'(t)) = 0$ almost everywhere, which proves property (8.43).

Take $\varphi \in L^2(0,T;H_0^2(\Omega))$, $\varphi(t) \in K$ almost everywhere. By the monotonicity of p, from (8.46) it follows that

$$\langle L^\varepsilon - f,\; \varphi - w_\varepsilon' \rangle \geq 0$$

This inequality is valid almost everywhere on $(0,T)$ and after integrating it from 0 to T we obtain

$$\begin{aligned}\int_0^T \langle L^\varepsilon, \varphi\rangle\, dt - \int_0^T \langle F, \varphi - w'_\varepsilon\rangle\, dt \ge \int_0^T \langle L^\varepsilon, w'_\varepsilon\rangle\, dt \\ = \frac{1}{2}\left(\|w'_\varepsilon(T)\|^2 + \|\Delta w_\varepsilon(T)\|^2 - \|\Delta w_0\|^2\right) \\ - \int_0^T \langle \int_0^t l(u'_\varepsilon, v'_\varepsilon, w'_\varepsilon) d\tau, w'_\varepsilon\rangle\, dt\ .\end{aligned} \tag{8.62}$$

We discuss the passage to the limit in the last nonlinear term of (8.62). The typical nonlinear expression has the form

$$\int_0^T \langle \int_0^t \left(u'_{\varepsilon x_1} + \frac{1}{2}\, w'^2_{\varepsilon x_1}\right) w'_{\varepsilon x_1} d\tau, w'_{\varepsilon x_1}\rangle\, dt\ . \tag{8.63}$$

By imbedding theorems, from (8.61) it follows that, for small δ, $w'_{\varepsilon x_1} \to w'_{x_1}$ strongly in $L^2(0,T;L^6(\Omega))$. Hence, $u'_{\varepsilon x_1} w'_{\varepsilon x_1} \to u'_{x_1} w'_{x_1}$ weakly in $L^2(0,T;L^{6/5}(\Omega))$ and therefore

$$\int_0^t u'_{\varepsilon x_1} w'_{\varepsilon x_1} d\tau \to \int_0^t u'_{x_1} w'_{x_1} d\tau \quad \text{weakly in} \quad L^2(0,T;L^{6/5}(\Omega))\ .$$

Once again, taking into account the strong convergence of $w'_{\varepsilon x_1}$ in $L^2(0,T;\, L^6(\Omega))$, we conclude that

$$\int_0^T \langle \int_0^t u'_{\varepsilon x_1} w'_{\varepsilon x_1} d\tau, w'_{\varepsilon x_1}\rangle\, dt \to \int_0^T \langle \int_0^t u'_{x_1} w'_{x_1} d\tau, w'_{x_1}\rangle\, dt\ .$$

To justify the passage to the limit in the remaining term of (8.63) one can use the strong convergence of $w'_{\varepsilon x_1}$ in $L^2(0,T;L^6(\Omega))$. Due to the boundedness of $w'_{\varepsilon x_1}$ in $L^\infty(0,T;L^6(\Omega))$ we obtain strong convergence $w'^3_{\varepsilon x_1} \to w'^3_{x_1}$ in $L^2(Q)$, ie.,

$$\int_0^t w'^3_{\varepsilon x_1} d\tau \to \int_0^t w'^3_{x_1} d\tau \quad \text{strongly in} \quad L^2(Q)\ .$$

The passage to the limit in the remaining nonlinear terms of $l(u'_\varepsilon, v'_\varepsilon, w'_\varepsilon)$ is simpler. From (8.61) it follows that

$$w_\varepsilon(T) \to w(T) \text{ weakly in } H_0^2(\Omega), \quad w'_\varepsilon(T) \to w'(T) \text{ weakly in } L^2(\Omega)\ .$$

Hence

$$\begin{aligned}\underline{\lim}\left(\|w'_\varepsilon(T)\|^2 + \|\Delta w_\varepsilon(T)\|^2\right) \ge \|w'(T)\|^2 + \|\Delta w(T)\|^2 \\ = 2\int_0^T \langle w'' + \Delta^2 w, w'\rangle\, dt\ .\end{aligned}$$

Taking into account the arguments used previously we conclude that the lower limit of the right-hand side of (8.62) is greater than or equal to

$$\int_0^T \langle L, w' \rangle \, dt \ .$$

The passage to the limit in the left-hand side of (8.62) can be accomplished more easily since φ is a fixed function. As a result inequality (8.40) follows from (8.62). In conclusion we have to pass to the limit in (8.44), (8.45) as $\varepsilon \to 0$. Equation (8.44) is satisfied as identity, ie.,

$$\int_0^T \langle L_1^\varepsilon, \varphi \rangle \, dt = \int_0^T \langle F_1, \varphi \rangle \, dt \tag{8.64}$$

for every $\varphi \in L^2(0,T; H_0^1(\Omega))$. Since $w'^2_{\varepsilon x_1} \to w'^2_{x_1}$ strongly in $L^2(Q)$, we have

$$\int_0^t \left(u'_{\varepsilon x_1} + k_{11} w'_\varepsilon + \frac{1}{2} w'^2_{\varepsilon x_1}\right) d\tau \to \int_0^t \left(u'_{x_1} + k_{11} w' + \frac{1}{2} w'^2_{x_1}\right) d\tau$$

weakly in $L^2(Q)$. In particular, this implies that we can pass to the limit in the term

$$\int_0^T \langle \int_0^t \left(u'_{\varepsilon x_1} + k_{11} w'_\varepsilon + \frac{1}{2} w'^2_{\varepsilon x_1}\right) d\tau, \varphi_{x_1} \rangle \, dt \ .$$

A similar reasoning applies to the remaining terms of L_1^ε. Thus, (8.44) follows from (8.64) and the theorem is proved. □

In conclusion we consider the case where K is a closed convex set in $H_0^2(\Omega)$, containing zero. Moreover, it is assumed that the interior int$K$ is nonempty. The corresponding result can be formulated as follows.

Theorem 8.4. *Let* $u_0, v_0 \in H_0^1(\Omega)$, $w_0 \in \text{int}K$, $u_1, v_1, w_1 \in L^2(\Omega)$, $f_i, f \in L^2(Q)$, $i = 1,2$, $k_{11}, k_{22} \in L^\infty(\Omega)$. *Then there exist functions* u, v, w *satisfying the relations*

$$u, v \in L^\infty(0,T; H_0^1(\Omega)), \quad w \in L^\infty(0,T; H_0^2(\Omega)) \ ,$$

$$u', v', w' \in L^\infty(0,T; L^2(\Omega)) \ ,$$

$$u' - \int_0^t \frac{\partial N_{1j}}{\partial x_j} \, d\tau = F_1 \ ,$$

$$v' - \int_0^t \frac{\partial N_{2j}}{\partial x_j} \, d\tau = F_2 \ ,$$

$$\int_0^T \langle w' + \int_0^t \{\Delta^2 w - (N_{ij} w_{x_j})_{x_i} + k_{11} N_{11} + k_{22} N_{22}\} d\tau \ ,$$

$$\varphi - w \rangle \, dt \geq \int_0^T \langle F, \varphi - w \rangle \, dt$$

$$\forall \varphi \in L^2(0,T;H_0^2(\Omega)), \quad \varphi(t) \in K \quad \text{a.e. on} \quad (0,T) \ ,$$

$$u = u_0, \quad v = v_0, \quad w = w_0 \quad \text{when} \quad t = 0 \ ,$$

$$w(t) \in K \quad \text{a.e. on} \quad (0,T) \ .$$

The proof of this theorem is similar to that of the previous theorem and will be omitted here. The penalty operator acts from $H_0^2(\Omega)$ into $H^{-2}(\Omega)$. The condition $w_0 \in \operatorname{int} K$ allows us to use the inclusion $w_{m1} \in K$ for large m in deriving the estimates for $u_m''(0), v_m''(0), w_m''(0)$ in (8.54). Hence, $p(w_{m1}) = 0$. Consequently, the constant c appearing in inequality (8.55) is independent of ε, m for sufficiently large m.

Chapter 3
Variational Inequalities in Plasticity

In plasticity models which are widely used in mechanics the constitutive laws generalize ones corresponding to elasticity. An important feature of these models are inequality type constraints imposed upon stresses. Without such constraints they reduce to elasticity models.

In applications, the most widespread models of plasticity are the flow model and the Hencky model. What makes them essentially different are forms of the constitutive laws they use. The formulation of the constitutive law in the flow model is based on velocities while in the Hencky model it is based on displacements.

Although at the first glance these models seem to be equivalent, they are not equivalent from the standpoint of mechanics. For instance, it may happen that the Hencky model is inconsistent, since it is impossible to describe the unloading process in this model. The flow model is always consistent and therefore preferable.

On the other hand, the Hencky model often arises in studying the flow model. This motivates the interest in the Hencky model. Moreover, in many cases the Hencky model is easier to handle. Both types of models will be examined in this chapter.

1 Preliminaries

Let $\Omega \subset \mathbb{R}^3$ be a bounded domain and let $\mathbb{R}^6$ be the space of symmetric vectors $\sigma = \{\sigma_{ij}\}$, $i, j = 1, 2, 3$. We consider the set

$$\tilde{K} = \left\{\sigma \in \mathbb{R}^6 \mid \Phi(\sigma) \le 0\right\} ,$$

where $\Phi : \mathbb{R}^6 \mapsto \mathbb{R}$ is a convex and continuous function. Denote by π the operator of orthogonal projection of the space $\mathbb{R}^6$ onto the set $\tilde{K}$.

On the space of functions $\{\sigma_{ij}(x)\}$ belonging to $L^2(\Omega)$ we introduce the functional

$$P(\sigma) = \frac{1}{2} \|\sigma - \pi\sigma\|^2 .$$

As before, $\|\cdot\|$ is a norm in $L^2(\Omega)$.

Lemma 1.1. *The functional P is convex.*

Proof. The proof is a straightforward consequence of properties of the operator π. In fact, for any $\lambda \in (0,1)$,

$$\begin{aligned}&\big|\lambda\sigma_1(x)+(1-\lambda)\sigma_2(x)-\pi\big(\lambda\sigma_1(x)+(1-\lambda)\sigma_2(x)\big)\big|\\ &\qquad\le \big|\lambda\sigma_1(x)+(1-\lambda)\sigma_2(x)-\sigma(x)\big| \quad \forall\, \sigma \in K\end{aligned}$$

a.e. in Ω, where

$$K=\left\{\sigma \in L^2(\Omega) \;\middle|\; \sigma(x)\in \tilde{K} \quad \text{a.e. in} \quad \Omega\right\} . \tag{1.1}$$

Substituting in the above formula $\lambda\pi\sigma_1+(1-\lambda)\pi\sigma_2$ as a test function for σ, next squaring and integrating it over Ω we immediately get the convexity of P. □

Lemma 1.2. *The derivative of the functional P is given by the formula*

$$P'(\sigma)=\sigma-\pi\sigma \ .$$

Proof. For any $\sigma_0 \in L^2(\Omega)$, we have

$$\begin{aligned}\lim_{\lambda\to 0}\ &\frac{P(\sigma+\lambda\sigma_0)-P(\sigma)}{\lambda}=\langle\, \sigma-\pi\sigma,\sigma_0\,\rangle\\ &+\lim_{\lambda\to 0}\ \langle\, \sigma-\pi\sigma\ ,\ \frac{\pi\sigma-\pi(\sigma+\lambda\sigma_0)}{\lambda}\,\rangle \ .\end{aligned} \tag{1.2}$$

We prove that the second term in the right-hand side is equal to zero.

Since the operator π satisfies the Lipschitz condition, the function

$$\frac{\pi\sigma-\pi(\sigma+\lambda\sigma_0)}{\lambda}$$

is bounded in $L^2(\Omega)$. Hence, letting $\lambda_m \to 0$ we can choose a subsequence weakly converging to a certain $\sigma^* \in L^2(\Omega)$.

We claim that

$$\langle\, \sigma-\pi\sigma,\sigma^*\,\rangle=0 \ . \tag{1.3}$$

In fact, in view of the well-known property of the operator π

$$\langle\, \sigma-\pi\sigma\ ,\ \frac{\pi\sigma-\pi(\sigma+\lambda_m\sigma_0)}{\lambda_m}\,\rangle \ge 0 \ .$$

On the other hand,

$$\langle\, \sigma+\lambda_m\sigma_0-\pi(\sigma+\lambda_m\sigma_0)\ ,\ \frac{\pi(\sigma+\lambda_m\sigma_0)-\pi\sigma}{\lambda_m}\,\rangle\ge 0$$

and formula (1.3) follows.

Letting $\lambda = \lambda_m \to 0$, from (1.2) we get

$$\frac{P(\sigma + \lambda\sigma_0) - P(\sigma)}{\lambda} \to \langle\, \sigma - \pi\sigma, \sigma_0 \,\rangle \ .$$

For convex functionals the quotient $\dfrac{P(\sigma + \lambda\sigma_0) - P(\sigma)}{\lambda}$ is a nondecreasing function of λ. Therefore, when λ converges to zero, the limit is unique. This concludes the proof. □

Remark. 1.1. The fact that the quotient $\dfrac{P(\sigma + \lambda\sigma_0) - P(\sigma)}{\lambda}$ is a nondecreasing function of λ was proved in Chapter 1, Section 1.3, under the assumption that the functional P is differentiable. This fact, however, remains true if we drop the differentiability assumption.

As before, we put $\varepsilon_{ij}(u) = \dfrac{1}{2}\,(u_{i,j} + u_{j,i})$. In the sequel we need the following lemma.

Lemma 1.3. *There exists a constant $c > 0$ such that for any $u = (u_1, u_2) \in H_0^1(\Omega)$*

$$\|u\| \le c \sum_{i,j=1}^{2} \|\varepsilon_{ij}(u)\|_{L^1(\Omega)} \ . \tag{1.4}$$

Proof. We extend the function u beyond Ω by assigning it the value zero and introduce new coordinates η_1, η_2 related to the old ones by the formulae $x_1 = \dfrac{1}{\sqrt{2}}\,(\eta_1 + \eta_2)$, $x_2 = \dfrac{1}{\sqrt{2}}\,(\eta_1 - \eta_2)$.

Let $v = u_1 + u_2$. Then,

$$\frac{\partial v}{\partial \eta_1} = \frac{1}{\sqrt{2}}\left(\frac{\partial u_1}{\partial x_1} + \frac{\partial u_1}{\partial x_2} + \frac{\partial u_2}{\partial x_1} + \frac{\partial u_2}{\partial x_2}\right) \ .$$

Hence,

$$\left\|\frac{\partial v}{\partial \eta_1}\right\|_{L^1(\Omega)} \le \frac{1}{\sqrt{2}} \sum_{i,j=1}^{2} \|\varepsilon_{ij}(u)\|_{L^1(\Omega)} \ .$$

Since

$$|u_1(x)| \le \int_{-\infty}^{\infty} \left|\frac{\partial u_1}{\partial x_1}\,(y, x_2)\right| dy \equiv L_1(x_2) \ ,$$

$$|u_2(x)| \le \int_{-\infty}^{\infty} \left|\frac{\partial u_2}{\partial x_2}\,(x_1, y)\right| dy \equiv L_2(x_1) \ ,$$

we have $|v(x)| \le L_1(x_2) + L_2(x_1)$ and

$$|v(x)| \le \int_{-\infty}^{\infty} \left|\frac{\partial v}{\partial \eta_1}(x_1 + \frac{y}{\sqrt{2}}, x_2 + \frac{y}{\sqrt{2}})\right| dy \equiv L(x) \ .$$

Consequently,

$$\int_{\mathbb{R}^2} |v(x)|^2 dx \leq \int_{\mathbb{R}^2} L(x)L_1(x_2)dx + \int_{\mathbb{R}^2} L(x)L_2(x_1)dx \ ,$$

and $L(x)$ is independent of η_1. Changing the variables in the first term of the right-hand side, we obtain

$$\int_{\mathbb{R}^2} L(x)L_1(x_2)dx = \int_{-\infty}^{\infty} L\left[\int_{-\infty}^{\infty} L_1(x_2)d\eta_1\right] d\eta_2$$

$$= \sqrt{2}\left\|\frac{\partial v}{\partial \eta_1}\right\|_{L^1(\Omega)} \left\|\frac{\partial u_1}{\partial x_1}\right\|_{L^1(\Omega)} \leq c \sum_{i,j=1}^{2} \|\varepsilon_{ij}(u)\|^2_{L^1(\Omega)} \ .$$

Thus,

$$\|u_1 + u_2\|^2 \leq c \sum_{i,j=1}^{2} \|\varepsilon_{ij}(u)\|^2_{L^1(\Omega)} \ .$$

In a similar way we get

$$\|u_1 - u_2\|^2 \leq c \sum_{i,j=1}^{2} \|\varepsilon_{ij}(u)\|^2_{L^1(\Omega)} \ .$$

The assertion of the lemma follows from the last two inequalities. □

Assume that the domain $\Omega \subset \mathbb{R}^n$ is bounded and its boundary Γ is smooth. Define the space

$$V_0(\Omega) = \{u = (u_1, u_2, \ldots, u_n) \mid u \in L^2(\Omega), \ \text{div} \ u \in L^2(\Omega)\}$$

with the norm

$$\|u\|^2_{V_0(\Omega)} = \|u\|^2 + \|\text{div} \ u\|^2 \ . \tag{1.5}$$

We are interested in the case $n = 2, 3$. The elements $u \in V_0(\Omega)$ have the traces $u \cdot \nu$ on the boundary Γ in the sense of the space $H^{-1/2}(\Gamma)$. This statement is proved in the following lemma.

Lemma 1.4. *There exists a linear and continuous operator*

$$\gamma \ : \ V_0(\Omega) \mapsto H^{-1/2}(\Gamma)$$

such that

$$\gamma(u) = u \cdot \nu \quad \text{for all} \quad u \in C^\infty(\bar{\Omega}) \ , \tag{1.6}$$

and, for any $w \in H^1(\Omega)$, the following generalized formula for integration by parts holds

$$\int_\Omega u \nabla w dx + \int_\Omega w \ \text{div} \ u \ dx = \langle \gamma(u), w \rangle_{1/2} \ . \tag{1.7}$$

Proof. Choose functions $\varphi \in H^{1/2}(\Gamma)$, $w \in H^1(\Omega)$ such that $\gamma_0 w = \varphi$, where γ_0 is the trace operator on Γ. For $u \in V_0(\Omega)$, define

$$L_u(\varphi) = \int_\Omega (w \text{ div } u + u\nabla w)dx \ .$$

We show that $L_u(\varphi)$ is independent of w. Take $w_1, w_2 \in H^1(\Omega)$, $\gamma_0 w_1 = \gamma_0 w_2 = \varphi$. Set $w = w_1 - w_2$. It is enough to verify that

$$\int_\Omega w \text{ div } u \, dx + \int_\Omega u\nabla w dx = 0 \ . \tag{1.8}$$

In fact, since $w \in H_0^1(\Omega)$, one can assume that $w = \lim w_m$, $w_m \in C_0^\infty(\Omega)$ and

$$\int_\Omega w_m \text{ div } u \, dx + \int_\Omega u\nabla w_m dx = 0 \ . \tag{1.9}$$

By passing to the limit in (1.9) we get (1.8). This proves that $L_u(\varphi)$ is independent of w.

Take $w = l\varphi$, where l is the operator which maps $H^{1/2}(\Gamma)$ in $H^1(\Omega)$. By

$$\left|L_u(\varphi)\right| \le 2\|u\|_{V_0(\Omega)} \, \|w\|_1$$

and in view of the continuity of l, we have

$$\left|L_u(\varphi)\right| \le c_0\|u\|_{V_0(\Omega)} \, \|\varphi\|_{H^{1/2}(\Gamma)} \ . \tag{1.10}$$

This implies that $\varphi \mapsto L_u(\varphi)$ is a linear and continuous function from $H^{1/2}(\Gamma)$ into R. Therefore, there exists an element $g = g(u) \in H^{-1/2}(\Gamma)$ such that

$$L_u(\varphi) = \langle \, g, \varphi \, \rangle_{1/2} \ . \tag{1.11}$$

The function $u \mapsto g(u)$ is linear and, by (1.10), bounded. This proves that the function $u \mapsto g(u) = \gamma(u)$ from $V_0(\Omega)$ into $H^{-1/2}(\Gamma)$ is continuous.

We now verify (1.6). Choose any element $u = (u_1, u_2, \ldots, u_n) \in C^\infty(\bar{\Omega})$ and $w \in C^\infty(\bar{\Omega})$. From what has already been shown

$$L_u(\gamma_0 w) = \int_\Omega \text{div } (uw)dx = \int_\Gamma \gamma_0 w \cdot u\nu d\Gamma = \langle \, u\nu, \gamma_0 w \, \rangle_{1/2} \ .$$

Since the traces $\gamma_0 w$ of w are dense in $H^{1/2}(\Gamma)$, the formula

$$L_u(\varphi) = \langle \, u\nu, \varphi \, \rangle_{1/2}$$

follows by continuity for any $\varphi \in H^{1/2}(\Gamma)$. Comparing with (1.11), we get $\gamma(u) = u\nu$. This finishes the proof. □

Below we formulate an useful lemma which follows from the lemma just established.

Let the boundary Γ of the domain $\Omega \subset \mathbb{R}^n$ be smooth and

$$V(\Omega) = \{\sigma = \{\sigma_{ij}\} \in L^2(\Omega) \mid \sigma_{ij,j} \in L^2(\Omega)\ ,\ i = 1, 2, \dots, n\}\ , \tag{1.12}$$

where

$$\sigma_{ij,j} = \sum_{j=1}^{n} \frac{\partial \sigma_{ji}}{\partial x_j}\ .$$

Lemma 1.5. *For $\sigma \in V(\Omega)$, we can define the traces $\sigma_{ij}\nu_j \in H^{-1/2}(\Gamma)$ such that*

$$\|\sigma_{ij}\nu_j\|_{H^{-1/2}(\Gamma)} \le c \left(\sum_{j=1}^{n} \|\sigma_{ij}\| + \sum_{j=1}^{n} \|\sigma_{ij,j}\| \right)\ , \quad i = 1, 2, \dots, n\ . \tag{1.13}$$

Now we prove the Green formula for integration by parts. This formula will be needed in studying elastoplastic plates.

Let $M = \{M_{ij}\}$, $i, j = 1, 2$, $M_{ij} = M_{ji}$ and

$$T(M) = M_{ij,j}\nu_i - \frac{\partial}{\partial s}\left[M_{11}\nu_1\nu_2 - M_{22}\nu_1\nu_2 + M_{12}(\nu_2^2 - \nu_1^2)\right]\ , \tag{1.14}$$

where $s = (-\nu_2, \nu_1)$ is the unit tangent along Γ. Hence, $w_\nu \equiv \dfrac{\partial w}{\partial \nu}$.

Lemma 1.6. *For all sufficiently smooth functions w and M_{ij},*

$$\int_\Omega M_{ij,ij} w dx = \int_\Omega M_{ij} w_{,ij} dx - \int_\Gamma M_{ij}\nu_j\nu_i w_\nu d\Gamma + \int_\Gamma T(M) w d\Gamma\ .$$

Proof. Denote

$$\begin{gathered} M_i = M_{ij}\nu_j\ ,\quad i = 1, 2\ , \\ M_\nu = M_1\nu_1 + M_2\nu_2\ ,\quad M_s = M_1\nu_2 - M_2\nu_1\ . \end{gathered} \tag{1.15}$$

Then

$$\begin{gathered} M_\nu = M_{ij}\nu_j\nu_i\ , \\ M_s = M_{11}\nu_1\nu_2 + M_{12}\nu_2^2 - M_{21}\nu_1^2 - M_{22}\nu_2\nu_1\ . \end{gathered}$$

Since the corresponding determinant in formulae (1.15) is different from zero we can write

$$M_1 = M_\nu\nu_1 + M_s\nu_2\ ,\quad M_2 = M_\nu\nu_2 - M_s\nu_1\ . \tag{1.16}$$

Integrating twice by parts we get

$$\int_\Omega M_{ij,ij} w dx = \int_\Omega M_{ij} w_{,ij} dx + \int_\Gamma M_{ij,j}\nu_i w d\Gamma - \int_\Gamma M_{ij}\nu_j w_{,i} d\Gamma \tag{1.17}$$

and, moreover,

$$\int_\Gamma M_{ij}\nu_j w_{,i} d\Gamma = \int_\Gamma (M_1 w_{,1} + M_2 w_{,2}) d\Gamma \ .$$

With formulae (1.16) we obtain

$$\int_\Gamma M_{ij}\nu_j w_{,i} d\Gamma = \int_\Gamma (M_\nu \nu_1 + M_s \nu_2) w_{,1} d\Gamma + \int_\Gamma (M_\nu \nu_2 - M_s \nu_1) w_{,2} d\Gamma$$

$$= \int_\Gamma M_\nu (\nu_1 w_{,1} + \nu_2 w_{,2}) d\Gamma + \int_\Gamma M_s (\nu_2 w_{,1} - \nu_1 w_{,2}) d\Gamma$$

$$= \int_\Gamma M_\nu \frac{\partial w}{\partial \nu} d\Gamma - \int_\Gamma M_s \frac{\partial w}{\partial s} d\Gamma$$

$$= \int_\Gamma M_\nu \frac{\partial w}{\partial \nu} d\Gamma + \int_\Gamma w \frac{\partial M_s}{\partial s} d\Gamma \ .$$

In getting the last equality we exploit the fact that Γ is closed, and therefore integration by parts is possible. Now the proof is completed by taking into account the integral from (1.17). □

Lemma 1.6 enables us to establish one more result. We introduce the space

$$V_1(\Omega) = \{M = \{M_{ij}\} \in L^2(\Omega) \mid M_{ij,ij} \in L^2(\Omega)\}$$

with the norm

$$\|M\|^2_{V_1(\Omega)} = \|M\|^2 + \|M_{ij,ij}\|^2 \ .$$

Lemma 1.7. *There exists a linear and continuous operator*

$$\Lambda \ : \ V_1(\Omega) \to H^{-3/2}(\Gamma) \times H^{-1/2}(\Gamma) \ ,$$

$$\Lambda(M) = \big(\Lambda_0(M),\ \Lambda_1(M)\big) \ ,$$

such that

$$\Lambda(M) = \big(T(M), M_{ij}\nu_j\nu_i\big) \quad \forall\, M \in C^\infty(\bar{\Omega}) \ , \tag{1.18}$$

and, for all $M \in V_1(\Omega)$, $w \in H^2(\Omega)$, the generalized Green formula

$$\int_\Omega (w M_{ij,ij} - M_{ij} w_{,ij}) dx = \langle\, \Lambda_0(M), w \,\rangle_{3/2} - \langle\, \Lambda_1(M), w_\nu \,\rangle_{1/2} \tag{1.19}$$

holds true.

Proof. Take $w \in H^2(\Omega)$ and define $\gamma_0 w \in H^{3/2}(\Gamma)$, $\gamma_1 w \in H^{1/2}(\Gamma)$, where, as before, γ_0 is the trace operator on Γ, and γ_1 is the trace operator for the derivative along the exterior normal to Γ.

There exists a linear and continuous operator

$$l \ : \ H^{3/2}(\Gamma) \times H^{1/2}(\Gamma) \mapsto H^2(\Omega)$$

such that $\gamma_0\big(l(\varphi_1, \varphi_2)\big) = \varphi_1$, $\gamma_1\big(l(\varphi_1, \varphi_2)\big) = \varphi_2$.

Let $\varphi_1 \in H^{3/2}(\Gamma)$, $\varphi_2 \in H^{1/2}(\Gamma)$ be functions such that $\varphi_1 = \gamma_0 w$, $\varphi_2 = \gamma_1 w$ for $w \in H^2(\Omega)$.

For any element $M \in V_1(\Omega)$, define

$$L_M(\varphi_1, \varphi_2) = \int_\Omega (w M_{ij,ij} - M_{ij} w_{,ij}) dx \ . \tag{1.20}$$

We show that $L_M(\varphi_1, \varphi_2)$ is independent of w. For $w_1, w_2 \in H^2(\Omega)$, $\gamma_0 w_i = \varphi_1$, $\gamma_1 w_i = \varphi_2$, $i = 1, 2$, we put $w = w_1 - w_2$. Obviously, $\gamma_0 w = 0$, $\gamma_1 w = 0$, and thus, $w \in H_0^2(\Omega)$.

Formula (1.20) is clearly valid for w_1 and w_2. It is enough to prove that

$$\int_\Omega (w M_{ij,ij} - M_{ij} w_{,ij}) dx = 0 \ . \tag{1.21}$$

Formula (1.21) holds for smooth functions w with compact supports defined on Ω. These functions are dense in $H_0^2(\Omega)$, hence $L_M(\varphi_1, \varphi_2)$ is independent of w.

Choose $w = l(\varphi_1, \varphi_2)$, where $\varphi_1 \in H^{3/2}(\Gamma)$, $\varphi_2 \in H^{1/2}(\Gamma)$. By the definition of L_M, and the continuity of the operator l, we readily obtain

$$\begin{aligned} \left| L_M(\varphi_1, \varphi_2) \right| &\le 2 \, \|M\|_{V_1(\Omega)} \, \|w\|_2 \\ &\le c \, \|M\|_{V_1(\Omega)} \left(\|\varphi_1\|_{H^{3/2}(\Gamma)} + \|\varphi_2\|_{H^{1/2}(\Gamma)} \right) \ . \end{aligned} \tag{1.22}$$

Thus, the linear function $(\varphi_1, \varphi_2) \mapsto L_M(\varphi_1, \varphi_2)$ is continuous on $H^{3/2}(\Gamma) \times H^{1/2}(\Gamma)$, and one can find an element $\Lambda(M) = \big(\Lambda_0(M), \Lambda_1(M)\big) \in H^{-3/2}(\Gamma) \times H^{-1/2}(\Gamma)$ such that

$$L_M(\varphi_1, \varphi_2) = \langle \Lambda_0(M), \varphi_1 \rangle_{3/2} - \langle \Lambda_1(M), \varphi_2 \rangle_{1/2} \ .$$

Moreover, by (1.22),

$$\|\Lambda(M)\|_{H^{-3/2}(\Gamma) \times H^{-1/2}(\Gamma)} \le c \, \|M\|_{V_1(\Omega)} \ .$$

We show that $\Lambda_0(M) = T(M)$, $\Lambda_1(M) = M_{ij}\nu_j\nu_i$ for all $M \in C^\infty(\bar{\Omega})$. Let w be a function from $C^2(\bar{\Omega})$. By Lemma 1.6,

$$\begin{aligned} L_M(\gamma_0 w, \gamma_1 w) &= \int_\Gamma T(M) w \, d\Gamma - \int_\Gamma M_{ij}\nu_j\nu_i w_\nu \, d\Gamma \\ &= \langle T(M), w \rangle_{3/2} - \langle M_{ij}\nu_j\nu_i, w_\nu \rangle_{1/2} \ . \end{aligned}$$

Since $C^2(\bar{\Omega})$ is dense in $H^2(\Omega)$, this equality holds for $w \in H^2(\Omega)$. Moreover, for any $(\varphi_1, \varphi_2) \in H^{3/2}(\Gamma) \times H^{1/2}(\Gamma)$, we have

$$L_M(\varphi_1, \varphi_2) = \langle T(M), \varphi_1 \rangle_{3/2} - \langle M_{ij}\nu_j\nu_i, \varphi_2 \rangle_{1/2} \ .$$

Comparing this formula with the previous formula for $L_M(\varphi_1, \varphi_2)$ we get (1.18), which proves the lemma. □

We conclude this section with a result on the solvability of nonlinear equations. The proof can be found in (Lions, 1969).

Lemma 1.8. *Let V be a reflexive separable Banach space, and let V^* be its dual. Assume that $A : V \mapsto V^*$ is a bounded semicontinuous and monotone operator, and, for a certain $u^0 \in V$,*

$$\frac{\langle A(u), u - u^0 \rangle}{\|u\|_V} \to \infty \quad \text{when} \quad \|u\|_V \to \infty \ .$$

Then, for any $f \in V^*$, there exists $u \in V$ such that

$$A(u) = f \ .$$

2 The Hencky model

2.1 The three-dimensional elastoplastic body

Our aim here is to find functions $u = (u_1, u_2, u_3)$, $\sigma = \{\sigma_{ij}\}$, ξ_{ij}, which satisfy, in a bounded domain $\Omega \subset \mathbb{R}^3$ with smooth boundary Γ, the following system of equations and inequalities

$$-\sigma_{ij,j} = f_i \ , \quad i = 1, 2, 3 \ , \tag{2.1}$$

$$\varepsilon_{ij}(u) = c_{ijkl}\sigma_{kl} + \xi_{ij} \ , \quad i, j = 1, 2, 3 \ , \tag{2.2}$$

$$\Phi(\sigma) \le 0 \ , \tag{2.3}$$

$$\xi_{ij}(\bar{\sigma}_{ij} - \sigma_{ij}) \le 0 \quad \forall\, \bar{\sigma} \ , \ \Phi(\bar{\sigma}) \le 0 \ , \tag{2.4}$$

and the boundary condition

$$u = 0 \quad \text{on} \quad \Gamma \ . \tag{2.5}$$

The functions f_i are given, $\Phi : \mathbb{R}^6 \mapsto \mathbb{R}$ describes the plastic yield condition. We assume that Φ is convex and continuous, and zero belongs to the interior of the set

$$\tilde{K} = \{\sigma = \{\sigma_{ij}\} \in \mathbb{R}^6 \ | \ \Phi(\sigma) \le 0\} \ .$$

We assume that $\varepsilon_{ij}(u) = \dfrac{1}{2}\,(u_{i,j} + u_{j,i})$.

In physical terms, u is the displacement vector, σ_{ij} is the stress tensor, $\varepsilon_{ij}(u)$ and ξ_{ij} are the tensors of full and plastic deformations, respectively.

Moreover, (2.1) is an equilibrium equation, (2.2) give a representation of the tensor $\varepsilon_{ij}(u)$ as the sum of elastic $c_{ijkl}\sigma_{kl}$ and plastic ξ_{ij} parts. Inequality (2.3) is typical for elastoplastic models. Its meaning is that stresses in the model cannot be "too large", that is, they cannot exceed the surface $\Phi(\sigma) = 0$ in the space $\{\sigma_{ij}\}$.

Inequality (2.4) defines a direction of the plastic deformation tensor up to the yield surface.

Observe that the condition $\Phi \equiv 0$ implies $\xi_{ij} \equiv 0$. In such a case equations (2.1), (2.2) coincide with the linear elastic model. Moreover, if in the general case $\Phi(\sigma) < 0$, then from (2.4) it follows that the corresponding values ξ_{ij} vanish. This means that we obtain an elastic equilibrium state at the point in question.

Some particular cases of model (2.1)–(2.4) are well known. For instance, under the assumptions (Duvaut, Lions, 1972) of dependence of the functions considered upon coordinates we get a simple model of a cylindrical rod formulated as follows.

In a domain $\tilde{\Omega} \subset \mathbb{R}^2$ find a solution to variational inequality

$$v \in K_1 \ : \ \int_{\tilde{\Omega}} \nabla v \cdot (\nabla \bar{v} - \nabla v) dx \geq \int_{\tilde{\Omega}} \tilde{f}(\bar{v} - v) dx \quad \forall\, \bar{v} \in K_1 \ ,$$

where

$$K_1 = \left\{ v \in H_0^1(\tilde{\Omega}) \ \middle| \ |\nabla v| \leq 1 \quad \text{a.e. in} \quad \tilde{\Omega} \right\} .$$

The domain $\tilde{\Omega}$ corresponds to a cross section of the cylindrical rod. The sought function v is called a stress function. Components of the stress tensor can be represented in terms of the first derivatives of v.

It is worth noting that the above variational inequality can be derived provided that the function Φ corresponds to the Huber-von Mises plastic yield condition and the coefficients c_{ijkl} correspond to the isotropic case.

Let us return to problem (2.1)–(2.5). Assume that $c_{ijkl} = c_{ijlk} = c_{klij}$, $c_{ijkl}\alpha_{kl}\alpha_{ij} \geq c\,|\alpha|^2$, $c > 0$, $c_{ijkl} \in L^\infty(\Omega)$ and denote

$$K = \left\{ \sigma \in L^2(\Omega) \ \middle| \ \sigma(x) \in \tilde{K} \quad \text{a.e. in} \quad \Omega \right\} ,$$

$$C(\sigma, \tau) = \int_\Omega c_{ijkl}\sigma_{kl}\tau_{ij} dx \ ,$$

$$V(\Omega) = \left\{ \sigma \in L^2(\Omega) \ \middle| \ \sigma_{ij,j} \in L^3(\Omega), \ i = 1,2,3 \right\} .$$

In view of (2.2) the functions ξ_{ij} can be substituted into (2.4). Integrating over Ω and using boundary condition (2.5), we easily get the variational inequality.

A result on the existence of solutions to this inequality will be proved in this section.

Theorem 2.1. *Let $f_i \in L^3(\Omega)$, $i = 1,2,3$. Assume that there exists a solution σ^0 to (2.1) such that $(1+\kappa)\sigma^0 \in K$, where $\kappa > 0$ is a constant. Then there exist functions u, σ satisfying (2.1) such that*

$$\begin{gathered} u \in BD(\Omega) \ , \quad \sigma \in K \ , \\ C(\sigma, \bar{\sigma} - \sigma) + \int_\Omega u_i(\bar{\sigma}_{ij,j} - \sigma_{ij,j}) dx \geq 0 \quad \forall\, \bar{\sigma} \in K \cap V(\Omega) \ . \end{gathered} \tag{2.6}$$

Here $BD(\Omega)$ is the Banach space of bounded deformations with the norm

$$\|u\|_{BD(\Omega)} = \|u\|_{L^1(\Omega)} + \sum_{i,j=1}^{3} \|\varepsilon_{ij}(u)\|_{\mathcal{M}^1(\Omega)} .$$

By $\mathcal{M}^1(\Omega)$ we denote bounded measures on Ω. The space $\mathcal{M}^1(\Omega)$ is dual to the normed space $C_0(\Omega)$ of continuous functions with compact supports, equipped with the uniform convergence topology, so that

$$\|v\|_{\mathcal{M}^1(\Omega)} = \sup_{\varphi \in C_0(\Omega)} \frac{(v,\varphi)}{\|\varphi\|_{C_0(\Omega)}} .$$

We should note that functions from $BD(\Omega)$ belong to $L^{3/2}(\Omega)$, and, moreover, for such functions, the trace on Γ in the sense of $L^1(\Omega)$ can be defined (see Temam, 1983). As far as the boundary values for a solution u to problem (2.6) are concerned one can refer to Remark 2.1 at the end of this section.

Proof of Theorem 2.1. Let π be the operator of orthogonal projection of $\mathbb{R}^6$ onto the set $\tilde{K}$ and let δ, λ be positive parameters.

Consider the regularized problem (in the notation we ignore the dependence of the solution on δ, λ)

$$-\delta \Delta u_i - \sigma_{ij,j} = f_i , \quad i = 1,2,3 , \tag{2.7}$$

$$c_{ijkl}\sigma_{kl} - \varepsilon_{ij}(u) + \frac{1}{\lambda}(\sigma - \pi\sigma)_{ij} = 0 , \quad i,j = 1,2,3 , \tag{2.8}$$

$$u = 0 \quad \text{on} \quad \Gamma . \tag{2.9}$$

To obtain a priori estimates we multiply (2.7), (2.8) by u_i, $\sigma_{ij} - \sigma^0_{ij}$, respectively, and integrate them over Ω. By the monotonicity of the operator $\sigma \mapsto \sigma - \pi\sigma$, the nonnegative term $\frac{1}{\lambda}\langle \sigma - \pi\sigma, \sigma - \sigma^0 \rangle$ can be neglected. Therefore,

$$\delta \sum_{i=1}^{3} \|\nabla u_i\|^2 + C(\sigma,\sigma) \le C(\sigma,\sigma^0) + \langle f_i, u_i \rangle + \langle u_i, \sigma^0_{ij,j} \rangle . \tag{2.10}$$

The function σ^0 solves (2.1), thus, by (2.10),

$$\delta \|u\|_1^2 + \|\sigma\|^2 \le c \tag{2.11}$$

with a constant c independent of δ, λ. The operator of problem (2.7)–(2.9) satisfies all the conditions of Lemma 1.8, thus the problem is solvable and $u \in H^1_0(\Omega)$, $\sigma \in L^2(\Omega)$.

For the neglected penalty term we have the estimate

$$\frac{1}{\lambda}\langle \sigma - \pi\sigma, \sigma - \sigma^0 \rangle \le c \tag{2.12}$$

which is uniform with respect to δ, λ.

Consider the functional $P(\sigma) = \frac{1}{2\lambda} \|\sigma - \pi\sigma\|^2$. According to Lemma 1.2, the derivative of this functional is given by the formula $P'(\sigma) = \frac{1}{\lambda} (\sigma - \pi\sigma)$. By the requirements imposed on σ^0, we get $\sigma^0 + \tau \in K$ provided $\|\tau\|_{L^\infty(\Omega)} \leq \alpha$ and α is sufficiently small.

By the convexity of P,

$$\langle P'(\sigma), \tau \rangle \leq \langle P'(\sigma), \sigma - \sigma^0 \rangle + P(\sigma^0 + \tau) - P(\sigma) \ .$$

Since $P(\sigma^0 + \tau) = 0$, by (2.12), the right-hand side is bounded from above and thus, for all τ as above,

$$\langle P'(\sigma), \tau \rangle \leq c \ .$$

Consequently, we get the estimate

$$\frac{1}{\lambda} \|\sigma - \pi\sigma\|_{L^1(\Omega)} \leq c \tag{2.13}$$

which holds uniformly with respect to δ, λ. From (2.8) it follows that $\varepsilon_{ij}(u)$ are bounded in $L^1(\Omega)$. Since $L^1(\Omega) \subset \mathcal{M}^1(\Omega)$, $\varepsilon_{ij}(u)$ are bounded in $\mathcal{M}^1(\Omega)$. It is well known (Temam, 1983) that in the three-dimensional case for $u = (u_1, u_2, u_3) \in H_0^1(\Omega)$, $\varepsilon_{ij}(u) \in \mathcal{M}^1(\Omega)$, we have

$$\|u\|_{L^{3/2}(\Omega)} \leq c \sum_{i,j=1}^{3} \|\varepsilon_{ij}(u)\|_{\mathcal{M}^1(\Omega)} \ .$$

Thus, the estimate

$$\|u\|_{L^{3/2}(\Omega)} + \sum_{i,j=1}^{3} \|\varepsilon_{ij}(u)\|_{\mathcal{M}^1(\Omega)} \leq c \tag{2.14}$$

is uniform with respect to δ, λ.

Now we analyse the passage to the limit by letting first $\delta \to 0$, and next $\lambda \to 0$. At each step solutions will be indexed by the respective parameter.

Assume that $u^\delta \in H_0^1(\Omega)$, $\sigma^\delta \in L^2(\Omega)$ solve problem (2.7)–(2.9). By a priori estimates (2.11), (2.14), one can choose a subsequence still denoted by u^δ, σ^δ,

$$u^\delta \to u^\lambda \quad \text{weakly in} \quad L^{3/2}(\Omega) \ ,$$

$$\delta u^\delta \to 0 \quad \text{strongly in} \quad H_0^1(\Omega) \ ,$$

$$\sigma^\varepsilon \to \sigma^\lambda \quad \text{weakly in} \quad L^2(\Omega) \ ,$$

$$\varepsilon_{ij}(u^\delta) \to \varepsilon_{ij}(u^\lambda) \quad \text{weakly-}(*) \text{ in} \quad \mathcal{M}^1(\Omega) \ , \quad i,j = 1,2,3 \ ,$$

when $\delta \to 0$. By weak-$(*)$ convergence in the space $\mathcal{M}^1(\Omega)$ we mean that for any continuous function φ with the compact support we have $(\varepsilon_{ij}(u^\delta), \varphi) \to$

$(\varepsilon_{ij}(u^\lambda),\varphi)$. Note that the limiting functions $\varepsilon_{ij}(u^\lambda)$ belong to the same ball in the space $\mathcal{M}^1(\Omega)$ as $\varepsilon_{ij}(u^\delta)$. In the limit equations (2.7), (2.8) have the form

$$-\sigma^\lambda_{ij,j} = f_i \ , \quad i = 1,2,3 \ , \tag{2.15}$$

$$c_{ijkl}\sigma^\lambda_{kl} - \varepsilon_{ij}(u^\lambda) + \frac{1}{\lambda}\,(\sigma^\lambda - \pi\sigma^\lambda)_{ij} = 0 \ , \quad i,j = 1,2,3 \ . \tag{2.16}$$

The weak convergence $P'(\sigma^\delta) \to P'(\sigma^\lambda)$ in $L^2(\Omega)$ follows from the monotonicity of the operator P'. Now, from the sequence $u^\lambda, \sigma^\lambda$ one can choose a subsequence such that

$$\begin{aligned} u^\lambda \to u \quad &\text{weakly in} \quad L^{3/2}(\Omega) \ , \\ \sigma^\lambda \to \sigma \quad &\text{weakly in} \quad L^2(\Omega) \ , \\ \varepsilon_{ij}(u^\lambda) \to \varepsilon_{ij}(u) \quad &\text{weakly-}(*) \text{ in} \quad \mathcal{M}^1(\Omega) \ , \quad i,j = 1,2,3 \end{aligned} \tag{2.17}$$

when $\lambda \to 0$. From (2.15) it clearly follows that after passing to the limit when $\lambda \to 0$ we get

$$-\sigma_{ij,j} = f_i \ , \quad i = 1,2,3 \ .$$

We multiply (2.16) by $\bar\sigma_{ij} - \sigma^\lambda_{ij}$ as test functions and integrate over Ω, where $\bar\sigma \in K \cap V(\Omega)$. The integral identity obtained from (2.16), leads to the inequality

$$C(\sigma^\lambda, \bar\sigma - \sigma^\lambda) + \int_\Omega u^\lambda_i(\bar\sigma_{ij,j} - \sigma^\lambda_{ij,j})dx \geq 0 \ .$$

Replacing $\sigma^\lambda_{ij,j}$ by f_i and using the property that $\underline{\lim}\ C(\sigma^\lambda,\sigma^\lambda) \geq C(\sigma,\sigma)$ we can pass to the limit in view of (2.17).

By standard arguments, $\sigma \in K$. □

Remark. 2.1. Inequality (2.6) has been derived from (2.1)–(2.5). However, one cannot expect that the solution u to problem (2.6) satisfies boundary condition (2.5) since, in general, smoothness of u does not ensure this boundary condition.

2.2 The perfect plastic body

2.2.1. The three-dimensional perfect plastic body. The equations of the three-dimensional perfect plastic body can be recovered from (2.1)–(2.4) by putting $c_{ijkl} = 0$.

We have to find functions u, σ such that in a domain $\Omega \subset \mathbb{R}^3$ with smooth boundary Γ we have

$$-\sigma_{ij,j} = f_i \ , \quad i = 1,2,3 \ , \tag{2.18}$$

$$\varepsilon_{ij}(u)(\bar\sigma_{ij} - \sigma_{ij}) \leq 0 \quad \forall\, \bar\sigma, \Phi(\bar\sigma) \leq 0 \ , \tag{2.19}$$

$$\Phi(\sigma) \leq 0 \ , \tag{2.20}$$

$$u = u^0 \quad \text{on} \quad \Gamma \ . \tag{2.21}$$

To prove the existence of solution to the variational inequality resulting from (2.18)–(2.21) we can apply the approach developed in Section 2.1.

We need the following assumption. We assume that the set $\tilde{K} = \{\sigma \in \mathbb{R}^6 \mid \Phi(\sigma) \leq 0\}$ is bounded in $\mathbb{R}^6$ and $u^0 \in H^{1/2}(\Gamma)$.

In the sequel by $\langle \cdot, \cdot \rangle_{1/2}$ we denote the duality pairing between $H^{-1/2}(\Gamma)$ and $H^{1/2}(\Gamma)$.

Theorem 2.2. *Under the hypotheses of Theorem 2.1 and the assumption stated above there exist functions u, σ satisfying equations (2.18) and the inequality*

$$\int_\Omega u_i(\bar{\sigma}_{ij,j} - \sigma_{ij,j})dx - \langle u_i^0, \bar{\sigma}_{ij}\nu_j - \sigma_{ij}\nu_j \rangle_{1/2} \geq 0 \qquad \forall \bar{\sigma} \in K \cap V(\Omega) \,, \tag{2.22}$$

and, moreover,

$$u \in BD(\Omega) \ , \quad \sigma \in K \ .$$

Proof. The regularized boundary value problem has the form (cf. (2.7)–(2.9))

$$-\delta \Delta u_i - \sigma_{ij,j} = f_i \ , \quad i = 1,2,3 \ , \tag{2.23}$$

$$-\varepsilon_{ij}(u) + \frac{1}{\lambda}(\sigma - \pi\sigma)_{ij} = 0 \ , \quad i,j = 1,2,3 \ , \tag{2.24}$$

$$u = u^0 \quad \text{on} \quad \Gamma \ . \tag{2.25}$$

An a priori estimate for this problem can be written as follows

$$\delta \, \|u\|_1^2 + \|\sigma\|^2 + \|u\|^2_{L^{3/2}(\Omega)} + \sum_{i,j=1}^{3} \|\varepsilon_{ij}(u)\|^2_{\mathcal{M}^1(\Omega)} \leq c \tag{2.26}$$

with a constant c independent of δ, λ. Despite of the assumption $c_{ijkl} = 0$, the estimate for σ in the space $L^2(\Omega)$ is true since the functions $\pi\sigma$ in (2.24) are bounded in $L^2(\Omega)$.

Estimate (2.26) follows in much the same way as those in Section 2.1, and therefore, we skip the details here.

It is convenient to reduce the problem to a homogeneous one. It can be done by putting $w = u - U$, where $U \in H^1(\Omega)$ is an extension of u^0 to the domain Ω.

By passing to the limit when $\delta \to 0$, $\lambda \to 0$, we get (2.22). Note that $\sigma_{ij}\nu_j \in H^{-1/2}(\Gamma)$ for $\sigma \in V(\Omega)$, and by Lemma 1.5

$$\|\sigma_{ij}\nu_j\|_{H^{-1/2}(\Gamma)} \leq c \left(\sum_{j=1}^{3} \|\sigma_{ij}\| + \|\sigma_{ij,j}\| \right) \ , \qquad i = 1,2,3 \ ,$$

with a constant c independent of σ. One should only note that the notation $V(\Omega)$ is used in Section 1 and Section 2 to denote different spaces. Thus, after passing to the limit when $\delta \to 0$ we get

$$\int_\Omega u_i^\lambda(\bar\sigma_{ij,j} - \sigma_{ij,j}^\lambda)dx - \langle\, u_i^0, \bar\sigma_{ij}\nu_j - \sigma_{ij}^\lambda \nu_j \,\rangle_{1/2} \geq 0 \qquad \forall\, \bar\sigma \in K \cap V(\Omega) \ . \tag{2.27}$$

When $\lambda \to 0$, by the a priori estimates one can assume that

$$\sigma^\lambda \to \sigma \quad \text{weakly in} \quad L^2(\Omega) \ ,$$
$$u^\lambda \to u \quad \text{weakly in} \quad L^{3/2}(\Omega) \ ,$$
$$\varepsilon_{ij}(u^\lambda) \to \varepsilon_{ij}(u) \quad \text{weakly-}(*) \text{ in} \quad \mathcal{M}^1(\Omega) \ , \ i,j = 1,2,3 \ .$$

Moreover,

$$\sigma_{ij}^\lambda \nu_j \to \sigma_{ij}\nu_j \quad \text{weakly in} \quad H^{-1/2}(\Gamma) \ , \quad i = 1,2,3 \ .$$

We can pass to the limit in (2.27) as in Theorem 2.1.

Remark. 2.2. Variational inequality (2.22) is a consequence of (2.18)–(2.21). But, as in Section 2.1, it is not guaranteed that the solution $u \in BD(\Omega)$ satisfies the boundary condition.

2.2.2. The perfect plastic plate. We seek functions $w, M = \{M_{ij}\}$, $i,j = 1,2$, such that in a bounded domain $\Omega \subset \mathbb{R}^2$ with smooth boundary Γ we have

$$-M_{ij,ij} = f \ , \tag{2.28}$$

$$-w_{,ij}(\bar M_{ij} - M_{ij}) \leq 0 \quad \forall\, \bar M, \Phi(\bar M) \leq 0 \ , \tag{2.29}$$

$$\Phi(M) \leq 0 \ , \tag{2.30}$$

$$w = w^0 \ , \ w_\nu = w^1 \quad \text{on} \quad \Gamma \ . \tag{2.31}$$

Here $\Phi : \mathbb{R}^3 \mapsto \mathbb{R}$ is a convex and continuous function describing the plastic yield condition. We assume that the set

$$\tilde K = \{M \in \mathbb{R}^3 \mid \Phi(M) \leq 0\}$$

is bounded in $\mathbb{R}^3$ and zero is its interior point.

Denote

$$K = \{M \in L^2(\Omega) \mid \Phi(M(x)) \leq 0 \quad \text{a.e. in} \quad \Omega\} \ ,$$

$$V_1(\Omega) = \{M \in L^2(\Omega) \mid M_{ij,ij} \in L^2(\Omega)\} \ , \ R(M) = M_{ij}\nu_j\nu_i \ ,$$

$$T(M) = M_{ij,j}\nu_i - \frac{\partial}{\partial s}\left[M_{11}\nu_1\nu_2 - M_{22}\nu_1\nu_2 + M_{12}(\nu_2^2 - \nu_1^2)\right] \ ,$$

where $s = (-\nu_2, \nu_1)$ is the unit tangent vector to Γ. From Lemma 1.7 it follows that for $M \in V_1(\Omega)$ the values $R(M)$ and $T(M)$ on the boundary can be defined in the sense of $H^{-1/2}(\Gamma)$ and $H^{-3/2}(\Gamma)$, respectively, and, moreover,

$$\|R(M)\|_{H^{-1/2}(\Gamma)} + \|T(M)\|_{H^{-3/2}(\Gamma)} \le c \, \|M\|_{V_1(\Omega)} \, . \tag{2.32}$$

We assume that $w^0 \in H^{3/2}(\Gamma)$, $w^1 \in H^{1/2}(\Gamma)$. Under this hypothesis a function $U \in H^2(\Omega)$ such that $U = w^0$, $U_\nu = w^1$ on Γ, obviously exists. Assume also that there exists a solution M^0 to equation (2.28) such that $(1+\kappa)M^0 \in K$, where $\kappa > 0$ is a constant.

Theorem 2.3. *Let $f \in L^2(\Omega)$. Then functions w, M_{ij} can be found which satisfy (2.28), the first boundary condition (2.31), and the inequality*

$$\begin{aligned} &\int_\Omega w(\bar{M}_{ij,ij} - M_{ij,ij})dx - \langle\, w^0, T(\bar{M} - M)\,\rangle_{3/2} \\ &\quad + \langle\, w^1, R(\bar{M} - M)\,\rangle_{1/2} \ge 0 \quad \forall\, \bar{M} \in K \cap V_1(\Omega) \, , \end{aligned} \tag{2.33}$$

and, moreover,

$$M \in K \ , \quad w \in H^1(\Omega) \ , \quad \nabla w \in BD(\Omega) \ . \tag{2.34}$$

Proof. Let π be the operator of orthogonal projection of $\mathbb{R}^3$ onto the set $\tilde{K}$ and let δ, λ be positive parameters.

Consider the regularized boundary value problem

$$\delta \Delta^2 w - M_{ij,ij} = f \ , \tag{2.35}$$

$$w_{,ij} + \frac{1}{\lambda}\,(M - \pi M)_{ij} = 0 \ , \quad i,j = 1,2 \ , \tag{2.36}$$

$$w = w^0 \ , \ w_\nu = w^1 \quad \text{on} \quad \Gamma \ . \tag{2.37}$$

This problem can be reduced to a homogeneous one by the substitution $u = w - U$. The resulting problem takes the form

$$\delta \Delta^2 u - M_{ij,ij} = f - \delta \Delta^2 U \ , \tag{2.38}$$

$$u_{,ij} + \frac{1}{\lambda}\,(M - \pi M)_{ij} = -U_{,ij} \ , \quad i,j = 1,2 \ , \tag{2.39}$$

$$u = 0 \ , \ u_\nu = 0 \quad \text{on} \quad \Gamma \ . \tag{2.40}$$

It should be noted that the components of πM are bounded in $L^2(\Omega)$. We multiply (2.38) and (2.39) by u, and $M_{ij} - M^0_{ij}$, respectively, and next integrate them over Ω. The following inequalities

$$\delta\, \|\Delta u\|^2 + \|M\|^2 \le c \ , \tag{2.41}$$

$$\frac{1}{\lambda}\,\langle\, M - \pi M, M - M^0 \,\rangle \le c \ . \tag{2.42}$$

are satisfied uniformly with respect to $\delta \leq \delta_0$, λ. As in Section 2.1, inequality (2.42) yields the boundedness of $\frac{1}{\lambda}(M - \pi M)$ in $L^1(\Omega)$. Hence, from (2.39) it follows that $u_{,ij}$ are bounded in $\mathcal{M}^1(\Omega)$. By Lemma 1.3, equations (2.39) imply that $u_{,i}$ are bounded in $L^2(\Omega)$, and, therefore, u is bounded in $H_0^1(\Omega)$.

Thus,

$$\|u\|_1^2 + \sum_{i,j=1}^{2} \|u_{,ij}\|^2_{\mathcal{M}^1(\Omega)} \leq c \tag{2.43}$$

holds uniformly with respect to δ, λ. Clearly, estimate (2.41) allows us to apply Lemma 1.8 in proving solvability of problem (2.38)–(2.40) for fixed δ, λ. The operator of this problem maps $H_0^2(\Omega) \times L^2(\Omega)$ into the dual space and satisfies all the conditions of Lemma 1.8. Hence, there exists a solution $u \in H_0^2(\Omega)$, $M \in L^2(\Omega)$.

Now we analyse the passage to the limit by letting first $\delta \to 0$, and next $\lambda \to 0$. Let us denote the solution to (2.38)–(2.40) by u^δ, M^δ. By estimates (2.41), (2.43) a subsequence u^δ, M^δ can be chosen such that

$$\begin{aligned} u^\delta &\to u^\lambda \quad \text{weakly in} \quad H_0^1(\Omega) \ , \\ \delta u^\delta &\to 0 \quad \text{strongly in} \quad H_0^2(\Omega) \ , \\ u^\delta_{,ij} &\to u^\lambda_{,ij} \quad \text{weakly-}(*) \text{ in} \quad \mathcal{M}^1(\Omega) \ , \ i,j = 1,2 \ , \\ M^\delta &\to M^\lambda \quad \text{weakly in} \quad L^2(\Omega) \end{aligned} \tag{2.44}$$

when $\delta \to 0$. Passing to the limit in (2.38)–(2.39) we easily obtain

$$-M^\lambda_{ij,ij} = f \ , \tag{2.45}$$

$$u^\lambda_{,ij} + \frac{1}{\lambda}(M^\lambda - \pi M^\lambda)_{ij} = -U_{,ij} \ . \tag{2.46}$$

The weak convergence $M^\delta - \pi M^\delta \to M^\lambda - \pi M^\lambda$ in $L^2(\Omega)$ follows from monotonicity.

Let $\bar{M} \in K \cap V_1(\Omega)$. From (2.46) it follows that

$$\begin{aligned} \int_\Omega w^\lambda(\bar{M}_{ij,ij} - M^\lambda_{ij,ij})dx &- \langle\, w^0, T(\bar{M} - M^\lambda)\,\rangle_{3/2} \\ &+ \langle\, w^1, R(\bar{M} - M^\lambda)\,\rangle_{1/2} \geq 0 \ . \end{aligned} \tag{2.47}$$

Here we use again the notation $w^\lambda = u^\lambda + U$. Note that the boundedness of $u^\delta_{,ij}$ results from (2.39) (generally speaking, nonuniformly in λ). Thus, in addition to (2.44), for any fixed λ

$$u^\delta_{,ij} \to u^\lambda_{,ij} \quad \text{weakly in} \quad L^2(\Omega) \ , \quad i,j = 1,2 \ .$$

In particular, this implies that the limiting functions w^λ belong to $H^2(\Omega)$, and, moreover, $w^\lambda = w^0$, $w^\lambda_\nu = w^1$ on Γ. According to Lemma 1.7, the boundedness of M^λ and $M^\lambda_{ij,ij}$ in $L^2(\Omega)$ implies the boundedness of $T(M^\lambda)$ and $R(M^\lambda)$ in $H^{-3/2}(\Gamma)$ and $H^{-1/2}(\Gamma)$, respectively. Hence, it can be assumed that

$$\begin{aligned} w^\lambda \to w \quad &\text{weakly in} \quad H^1(\Omega) \ , \\ w^\lambda_{,ij} \to w_{,ij} \quad &\text{weakly-}(*) \text{ in} \quad \mathcal{M}^1(\Omega) \ , \ i,j = 1,2 \ , \\ M^\lambda \to M \quad &\text{weakly in} \quad L^2(\Omega) \ , \\ T(M^\lambda) \to T(M) \quad &\text{weakly in} \quad H^{-3/2}(\Gamma) \ , \\ R(M^\lambda) \to R(M) \quad &\text{weakly in} \quad H^{-1/2}(\Gamma) \end{aligned} \tag{2.48}$$

when $\lambda \to 0$. In view of (2.45) the quantities $M^\lambda_{ij,ij}$ can be replaced by f in (2.47). Now, taking into account (2.48), one can easily pass to the limit. Therefore, we get (2.33).

By standard arguments we obtain that $M \in K$. □

Remark. 2.3. We cannot guarantee that the limiting function w satisfies the second boundary condition $w_\nu = w^1$ on Γ.

Remark. 2.4. Note that the trace of a function from $BD(\Omega)$ on a smooth variety is defined through a given orientation. This means, in particular, that if a smooth variety Γ_0 is contained in Ω and divides Ω into two open sets Ω_1, Ω_2, then, for any function $u \in BD(\Omega)$, one can define on Γ_0 two traces from the space $L^1(\Gamma_0)$.

In general, these traces are different. The first trace is called an interior trace and the second one is called an exterior trace (for a chosen normal ν on Γ_0). For instance, a smooth function on Ω having a jump of the first kind on Γ_0 belongs to $BD(\Omega)$ and has two different traces on Γ_0.

3 Dynamic problem for generalized equations of the flow model

In this section the so-called generalized equations of the flow model will be treated. In this model the constitutive law is formulated in terms of the velocities. Moreover, the velocity strain tensor is represented as the sum of three parts: elastic, plastic, and creep. As mentioned at the beginning of the present chapter, from the standpoint of mechanics, the flow models have some advantages over the Hencky type models.

Our main concern in this section is to prove the existence theorem for the corresponding initial-boundary value problem.

Let $\Omega \subset \mathbb{R}^3$ be a bounded domain with smooth boundary Γ and let $\Phi : \mathbb{R}^6 \mapsto \mathbb{R}$ be a convex continuous function. Let

$$K = \{\sigma = \{\sigma_{ij}\} \in L^2(\Omega) \mid \Phi(\sigma(x)) \le 0 \quad \text{a.e. in} \quad \Omega\} \ .$$

The deviator components of the tensor σ_{ij} will be denoted by s_{ij}, ie. $s_{ij} = \sigma_{ij} - \frac{1}{3}\delta_{ij}\sigma_{kk}$, where δ_{ij} is the Kronecker symbol.

The problem can be formulated as follows. We want to find functions $v = (v_1, v_2, v_3)$, $\sigma = \{\sigma_{ij}\}$, $\lambda = \{\lambda_{ij}\}$ satisfying in the domain $Q = \Omega \times (0,T)$ the relations

$$\dot{v}_i - \sigma_{ij,j} = f_i \ , \quad i = 1,2,3 \ , \tag{3.1}$$

$$\varepsilon_{ij}(v) = c_{ijkl}\dot{\sigma}_{kl} + B(t)I(s)^{n-1}s_{ij} + \lambda_{ij} \ , \quad i,j = 1,2,3 \ , \tag{3.2}$$

$$\Phi(\sigma) \le 0 \ , \tag{3.3}$$

$$\lambda_{ij}(\tau_{ij} - \sigma_{ij}) \le 0 \quad \forall\, \tau, \Phi(\tau) \le 0 \ , \tag{3.4}$$

and the boundary and initial conditions

$$\sigma_{ij}\nu_j = 0 \quad \text{on} \quad \Gamma \times (0,T) \ , \quad i = 1,2,3 \ , \tag{3.5}$$

$$v = v_0 \ , \quad \sigma = \sigma_0 \quad \text{when} \quad t = 0 \ . \tag{3.6}$$

Here $B(t)$ is a given function, $I(s) = \left(\frac{1}{2}s_{ij}s_{ij}\right)^{1/2}$, $n \ge 1$ is an integer.

Let $\overset{\circ}{V}(\Omega)$ be the space of tensors $\sigma \in L^2(\Omega)$ such that $\sigma_{ij,j} \in L^2(\Omega)$, $i = 1,2,3$, $s_{ij} \in L^{n+1}(\Omega)$ and $\sigma_{ij}\nu_j = 0$ on Γ, $i = 1,2,3$. According to Lemma 1.5, the quantities $\sigma_{ij}\nu_j$ are defined on the boundary as elements of $H^{-1/2}(\Gamma)$. The stress deviator of the stress tensor σ_0 is denoted by s_0.

The main result of this section is an existence theorem for (3.1)–(3.6). In what follows we eliminate λ_{ij} according to (3.2), (3.4). The derivative with respect to t is denoted by a dot.

Theorem 3.1. *Let $v_0 = (v_{10}, v_{20}, v_{30}) \in H^1(\Omega)$, $\sigma_0 = \{\sigma_{0ij}\} \in K \cap \overset{\circ}{V}(\Omega)$, $s_0 = \{s_{0ij}\} \in L^{2n}(\Omega)$, $0 \in K$, $B \in C^2[0,T]$, $B(t) > 0$, $t \in [0,T]$. $f_i, \dot{f}_i \in L^2(Q)$, $i = 1,2,3$. There exist unique functions $v = (v_1, v_2, v_3)$, $\sigma = \{\sigma_{ij}\}$ satisfying the relations*

$$v, \sigma \in L^\infty(0,T; L^2(\Omega)) \ , \quad \dot{v}, \dot{\sigma} \in L^2(Q) \ , \quad s \in L^{n+1}(Q) \ ,$$

$$\int_0^T \langle \dot{v}_i, \varphi\rangle dt + \int_0^T \langle \sigma_{ij}, \varphi_{,j}\rangle dt = \int_0^T \langle f_i, \varphi\rangle dt \quad \forall\, \varphi \in L^2(0,T; H^1(\Omega)) \ , \tag{3.7}$$

$$\begin{aligned} &\int_0^T C(\dot{\sigma}, \tau - \sigma) dt + \int_0^T \langle\, v_i, \tau_{ij,j} - \sigma_{ij,j}\,\rangle dt \\ &+ \int_0^T \langle B(t)I(s)^{n-1}s_{ij}, \tau_{ij} - \sigma_{ij}\rangle dt \ge 0 \ \forall\, \tau \in L^{n+1}(0,T; K \cap \overset{\circ}{V}(\Omega)) \ , \end{aligned} \tag{3.8}$$

$$\sigma(t) \in K \quad \text{a.e. on} \quad (0,T)$$

and initial conditions (3.6).

Proof. Let π be the operator of orthogonal projection of the space $L^2(\Omega)$ onto the set K and let $\varepsilon > 0$ be a parameter.

Consider the approximate problem obtained from (3.1)–(3.6) by replacing (3.2)–(3.4) with the following system of equations

$$\varepsilon_{ij}(v) = c_{ijkl}\dot{\sigma}_{kl} + B(t)I(s)^{n-1}s_{ij} + \frac{1}{\varepsilon}(\sigma - \pi\sigma)_{ij} \ , \quad i,j = 1,2,3 \ . \tag{3.9}$$

We shall prove that problem (3.1), (3.5), (3.6), (3.9) has a solution. A priori estimates for a solution are of the form

$$\max_{0\le t\le T}\{\|v(t)\| + \|\sigma(t)\|\} \le c \ , \quad \|s\|_{L^{n+1}(Q)} \le c \ , \tag{3.10}$$

$$\|\dot{v}\|_{L^2(Q)} + \|\dot{\sigma}\|_{L^2(Q)} \le c \tag{3.11}$$

where c depends on v_0, σ_0, f_i, T and is independent of ε.

Inequalities (3.10) follow from (3.1), and (3.9) after multiplying them, respectively, by v_i, and σ_{ij}, and next integrating over Ω. We use the equality $\langle v_i, \sigma_{ij,j}\rangle + \langle \sigma_{ij}, \varepsilon_{ij}(v)\rangle = 0$ which is valid for all smooth functions v, σ satisfying the boundary conditions $\sigma_{ij}\nu_j = 0$ on Γ. In particular, this equality is satisfied for $v \in H^1(\Omega)$, $\sigma \in \overset{\circ}{V}(\Omega)$. Moreover, we use the relations

$$B(t)\int_\Omega I(s)^{n-1}s_{ij}\sigma_{ij}dx = 2B(t)\int_\Omega I(s)^{n+1}dx \ ,$$

$$B(t) \ge c > 0 \ , \quad t \in [0,T] \ ,$$

which holds under the assumptions of the present theorem.

The estimate (3.11) follows from (3.1), (3.9). We differentiate these equations with respect to t, and next multiply them by $\dot{v}_i, \dot{\sigma}_{ij}$ respectively. Integrating over Ω and summing up the resulting equations we get the equality containing the penalty term which is nonnegative, and hence negligible. When dealing with equation (3.9) we have to obtain an estimate for the following expression

$$\int_\Omega \frac{d}{dt}\left(B(t)I(s)^{n-1}s_{ij}\right)\dot{\sigma}_{ij}dx$$

$$\equiv \int_\Omega \dot{B}(t)I(s)^{n-1}s_{ij}\dot{\sigma}_{ij}dx + \int_\Omega B(t)\frac{d}{dt}\left(I(s)^{n-1}s_{ij}\right)\dot{\sigma}_{ij}dx \ .$$

It is clearly seen that the second term of the right-hand side is nonnegative since it can be rewritten as

$$B(t)\int_\Omega \{2(n-1)I(s)^{n-1}\dot{I}(s)^2 + I(s)^{n-1}\dot{s}_{ij}\dot{s}_{ij}\}dx \ .$$

The first term is equal to

$$\frac{2}{n+1}\int_\Omega \dot{B}(t)\frac{d}{dt}I(s)^{n+1}dx \ .$$

Thus, after multiplication by $\dot{v}_i$, $\dot{\sigma}_{ij}$ and integration over Ω and t, we obtain the following inequality

$$\dot{R}(t) + \frac{2}{n+1}\int_\Omega \dot{B}(\eta)I(s)^{n+1}dx \Big|_{\eta=0}^{\eta=t} - \frac{2}{n+1}\int_0^t\int_\Omega \ddot{B}(\eta)I(s)^{n+1}dxd\eta$$

$$\leq R(t) + \frac{1}{2}\int_0^t \|f(\eta)\|^2 d\eta + \dot{R}(0) \ , \quad f = (f_1, f_2, f_3) \ ,$$

$$R(t) = \frac{1}{2}\int_0^t \{\|\dot{v}(\eta)\|^2 + C(\dot{\sigma}(\eta), \dot{\sigma}(\eta))\}d\eta \ .$$

By the assumptions $\dot{R}(0)$ is bounded. This follows from (3.1), (3.9) by putting $t = 0$. Thus, with this inequality, and by (3.10) we get (3.11).

The existence of a solution can be shown by the Galerkin method. We take the basis functions $\{\varphi^k\}$, $k = 1, 2, \ldots$, and $\{\psi^k\}$, $k = 1, 2, \ldots$, in the spaces $[H^1(\Omega)]^3$ and $\overset{\circ}{V}(\Omega) \cap L^{2n}(\Omega)$, $\varphi^k = (\varphi_1^k, \varphi_2^k, \varphi_3^k)$, $\psi^k = \{\psi_{ij}^k\}$, $i, j = 1, 2, 3$. To avoid ambiguity we use subscripts to denote vector components.

We seek the Galerkin approximation v^m, σ^m in the form

$$v^m(t) = \sum_{i=1}^m a_i^m(t)\varphi^i \ , \quad \sigma^m(t) = \sum_{i=1}^m b_i^m(t)\psi^i \ .$$

The functions a_i^m, b_i^m are defined by a system of ordinary differential equations resulting from (3.1), (3.9). The initial data for v^m, σ^m are found from the decompositions of v_0, σ_0. A priori estimates (3.10), (3.11) are valid for v^m, σ^m with constants independent of n. In particular, this yields the solvability of the Galerkin system on $(0, T)$.

If so, there exists a subsequence v^m, σ^m (with the previous notation) such that

$$v^m, \sigma^m \to v^\varepsilon, \sigma^\varepsilon \quad \text{weakly-}(*) \text{ in} \quad L^\infty(0, T; L^2(\Omega)) \ ,$$

$$\dot{v}^m, \dot{\sigma}^m \to \dot{v}^\varepsilon, \dot{\sigma}^\varepsilon \quad \text{weakly in} \quad L^2(Q) \ ,$$

$$s^m \to s^\varepsilon \quad \text{weakly in} \quad L^{n+1}(Q)$$

when $m \to \infty$. This convergence allows us to justify the passage to limit in the Galerkin equations as $m \to \infty$. The convergence

$$B(t)I(s^m)^{n-1}s_{ij}^m \to B(t)I(s^\varepsilon)^{n-1}s_{ij}^\varepsilon \quad \text{weakly in} \quad L^{\frac{n+1}{n}}(Q)$$

follows from the monotonicity of the operator

$$\sigma \mapsto \{I(s)^{n-1}s_{ij}\} \ . \tag{3.12}$$

Moreover, the monotonicity of the operator $\sigma \mapsto \frac{1}{\varepsilon}(\sigma - \pi\sigma)$ ensures the convergence (ε is fixed)

$$\frac{1}{\varepsilon}(\sigma^m - \pi\sigma^m) \to \frac{1}{\varepsilon}(\sigma^\varepsilon - \pi\sigma^\varepsilon) \quad \text{weakly in} \quad L^2(Q) \ .$$

Thus, after passing to the limit in the Galerkin equations, we can easily see that a solution to (3.1), (3.5), (3.6), (3.9) satisfies the following relations

$$\begin{gathered}\int_0^T \langle \dot{v}_i^\varepsilon, \varphi \rangle dt + \int_0^T \langle \sigma_{ij}^\varepsilon, \varphi_{,j} \rangle dt = \int_0^T \langle f_i, \varphi \rangle dt \\ \forall \varphi \in L^2(0,T; H^1(\Omega)) \ ,\end{gathered} \tag{3.13}$$

$$\begin{gathered}\int_0^T C(\dot{\sigma}^\varepsilon, \psi) dt + \int_0^T \langle v_i^\varepsilon, \psi_{ij,j} \rangle dt + \int_0^T B(t) \langle I(s^\varepsilon)^{n-1} s_{ij}^\varepsilon, \psi_{ij} \rangle dt \\ + \frac{1}{\varepsilon} \int_0^T \langle \sigma^\varepsilon - \pi\sigma^\varepsilon, \psi \rangle dt = 0 \quad \forall \psi \in L^{n+1}(0,T; \overset{\circ}{V}(\Omega)) \ .\end{gathered} \tag{3.14}$$

Boundary condition (3.5) is fulfilled in the weak sense, ie. it is included in identity (3.13). Initial conditions (3.6) are fulfilled in the strong sense.

Now we analyse the passage to the limit when $\varepsilon \to 0$. As has been already mentioned, a priori estimates (3.10), (3.11) are independent of ε. Therefore, these estimates will be valid for $v^\varepsilon, \sigma^\varepsilon$.

Let us choose a subsequence $v^\varepsilon, \sigma^\varepsilon$ such that

$$v^\varepsilon, \sigma^\varepsilon \to v, \sigma \quad \text{weakly-}(*) \text{ in} \quad L^\infty(0,T; L^2(\Omega)) \ ,$$

$$\dot{v}^\varepsilon, \dot{\sigma}^\varepsilon \to \dot{v}, \dot{\sigma} \quad \text{weakly in} \quad L^2(Q) \ ,$$

$$s^\varepsilon \to s \quad \text{weakly in} \quad L^{n+1}(Q)$$

when $\varepsilon \to 0$. By taking the function $\psi = \tau - \sigma^\varepsilon$ in (3.14), where $\tau \in L^{n+1}(0,T; K \cap \overset{\circ}{V}(\Omega))$ and substituting it into (3.14) we get the inequality

$$\begin{gathered}\int_0^T C(\dot{\sigma}^\varepsilon, \tau - \sigma^\varepsilon) dt + \int_0^T \langle v_i^\varepsilon, \tau_{ij,j} - \sigma_{ij,j}^\varepsilon \rangle dt \\ + \int_0^T B(t) \langle I(s^\varepsilon)^{n-1} s_{ij}^\varepsilon, \tau_{ij} - \sigma_{ij}^\varepsilon \rangle dt \geq 0 \ .\end{gathered}$$

Here, by the same arguments as before, we can pass to the limit when $\varepsilon \to 0$. Indeed, the quantities $\sigma_{ij,j}^\varepsilon$ can be replaced by $\dot{v}_i^\varepsilon - f_i$. We also use the fact that the limiting function σ belongs to $L^{n+1}(0,T; K \cap \overset{\circ}{V}(\Omega))$ which readily follows from (3.13), (3.14). Therefore, inequality (3.8) is valid, and by standard arguments we can pass to the limit in (3.13).

Now we prove the uniqueness of solution. Assume that there exist two different solutions v^1, σ^1 and v^2, σ^2 satisfying inequality (3.8). For $\sigma = \sigma^1 - \sigma^2$ we easily get

$$\int_0^{t_1} C(\dot{\sigma}, \sigma)dt + \int_0^{t_1} B(t)\langle\, I(s^1)^{n-1} s^1_{ij} - I(s^2)^{n-1} s^2_{ij}, \sigma^1_{ij} - \sigma^2_{ij}\,\rangle dt \leq 0 \ , \quad t_1 \in (0,T) \ .$$

Since the second term is nonnegative, $\sigma \equiv 0$. From equations (3.1) it follows that $v^1 - v^2 = 0$ which concludes the proof. □

4 The Kirchhoff-Love shell. Existence of solutions to the dynamic problem

The main purpose of this section is to study the dynamic elastoplastic problem for the Kirchhoff-Love shell. As in the preceding chapter, we consider a shallow shell with the middle surface close to a plane. This means, in particular, that distances defined on the middle surface coincide with those on the plane.

4.1 Problem formulation

Let $\Omega \subset \mathbb{R}^2$ be a bounded domain with smooth boundary Γ, $Q = \Omega \times (0,T)$, $T > 0$. We consider the space of vectors $(N_{11}, N_{12}, N_{22}, M_{11}, M_{12}, M_{22})$ denoted by (N, M) for symmetric 2×2 matrices N, M,

$$\mathbb{R}^6 = \{E = (N,M) \mid N = \{N_{ij}\} \ , \ M = \{M_{ij}\} \ , \ i,j = 1,2\} \ .$$

Let $\Phi : \mathbb{R}^6 \mapsto \mathbb{R}$ be a convex continuous function.

The formulation of the dynamic elastoplastic problem for a shallow shell is as follows. We want to find functions $v = (v_1, v_2)$, w, $E = (N, M)$, ξ_{ij}, λ_{ij}, $i, j = 1, 2$, satisfying in the domain Q the following system of equations and inequalities

$$\dot{v}_i - N_{ij,j} = f_i \ , \ i = 1,2 \ , \tag{4.1}$$

$$\dot{w} - M_{ij,ij} + k_{ij} N_{ij} = f \ , \tag{4.2}$$

$$\varepsilon_{ij}(v) + k_{ij} w = a_{ijkl} \dot{N}_{kl} + \xi_{ij} \ , \ i,j = 1,2 \ , \tag{4.3}$$

$$-w_{,ij} = b_{ijkl} \dot{M}_{kl} + \lambda_{ij} \ , \ i,j = 1,2 \ , \tag{4.4}$$

$$\Phi(E) \leq 0 \ , \tag{4.5}$$

$$\xi_{ij}(\bar{N}_{ij} - N_{ij}) + \lambda_{ij}(\bar{M}_{ij} - M_{ij}) \leq 0 \quad \forall\, (\bar{N}, \bar{M}),\ \Phi(\bar{N}, \bar{M}) \leq 0 \ , \tag{4.6}$$

and the boundary and initial conditions

$$N_{ij} \nu_j = 0 \ , \ M_{ij} \nu_j \nu_i = T(M) = 0 \quad \text{on} \quad \Gamma \times (0,T) \ , \tag{4.7}$$

$$v = v_0 \ , \ w = w_0 \ , \ E = E_0 \quad \text{when} \quad t = 0 \ . \tag{4.8}$$

Here $k_{ij} = k_{ji}$ are given functions. The coefficients a_{ijkl}, b_{ijkl} belong to the space $L^\infty(\Omega)$ and are symmetric and positive definite, $T(M)$ was introduced in (1.14), Section 1. Equations (4.1), (4.2) are motion equations. Relations (4.3), (4.4) give representations of the velocity strain tensor $\varepsilon_{ij}(v) + k_{ij}w$ and the velocity curvature tensor $-w_{,ij}$ as the sums of elastic and plastic parts. Inequality (4.6) is the maximum principle for the dissipation capacity.

It is clear that the condition $\Phi \equiv 0$ implies $\xi_{ij} = \lambda_{ij} = 0$. In this case (4.1)–(4.4) coincide with the dynamic equations of an elastic shell.

Let us integrate inequality (4.6) over Q after eliminating ξ_{ij}, λ_{ij} according to (4.3), (4.4). Under sufficient smoothness of the solution we can integrate by parts. In this way we obtain the variational inequality which is the object of our consideration.

To begin we introduce some notations,

$$W(\Omega) = \{E = (N, M) \in L^2(\Omega) \mid N_{ij} = N_{ij} \ , \ M_{ij} = M_{ij} \ , \ i,j = 1,2,$$
$$N_{ij,j} \in L^2(\Omega), \ i = 1,2, \ M_{ij,ij} \in L^2(\Omega)\} \ ,$$

$$K = \{E \in L^2(\Omega) \mid \Phi(E(x)) \le 0 \quad \text{a.e. in} \quad \Omega\} \ .$$

Here the notation, eg. $E \in L^2(\Omega)$ means that all components of $E = (N, M)$ are elements of $L^2(\Omega)$.

Let $v_0 = (v_{10}, v_{20})$. According to Lemma 1.5 and Lemma 1.7, $N_{ij}\nu_j$, $M_{ij}\nu_j\nu_i$, $T(M)$ are well-defined in the sense of $H^{-1/2}(\Gamma)$, $H^{-1/2}(\Gamma)$ and $H^{-3/2}(\Gamma)$, respectively, for the elements of $W(\Omega)$. The set of functions from $W(\Omega)$ satisfying the conditions $N_{ij}\nu_j = 0$, $M_{ij}\nu_j\nu_i = 0$, $T(M) = 0$ on Γ will be denoted by $\overset{\circ}{W}(\Omega)$, $C(E, \alpha) = \langle \, a_{ijkl}N_{kl}, \varphi_{ij} \, \rangle + \langle \, b_{ijkl}M_{kl}, \tau_{ij} \, \rangle$, $\alpha = (\varphi, \tau)$.

4.2 The main result

Theorem 4.1. *Let $0 \in K$, $f, \dot{f}, f_i, \dot{f}_i \in Q$, $i = 1,2$; $k_{ij} \in L^\infty(\Omega)$, $i,j = 1,2$; $v_{i0} \in H^1(\Omega)$, $i = 1,2$, $w_0 \in H^2(\Omega)$; $E_0 \in K \cap \overset{\circ}{W}(\Omega)$. Then there exist unique functions $v = (v_1, v_2)$, w, $E = (N, M)$ satisfying (4.1)–(4.2), boundary and initial conditions (4.7), (4.8), and the relations*

$$v, \dot{v}, w, \dot{w} \in L^\infty(0,T; L^2(\Omega)), \quad N_{ij,j}, M_{ij,ij} \in L^\infty(0,T; L^2(\Omega)) \ , \tag{4.9}$$

$$\int_0^T C(\dot{E}, \alpha - E)dt + \int_0^T \langle \, v_i, \varphi_{ij,j} - N_{ij,j} \, \rangle dt$$
$$+ \int_0^T \langle \, w, \tau_{ij,ij} - M_{ij,ij} \, \rangle dt - \int_0^T \langle \, k_{ij}w, \varphi_{ij} - N_{ij} \, \rangle dt \ge 0 \tag{4.10}$$
$$\forall \, \alpha = (\varphi, \tau) \in L^2(0,T; K \cap \overset{\circ}{W}(\Omega)) \ ,$$

$$E(t) \in K \quad \text{a.e. on} \quad (0,T) \ . \tag{4.11}$$

Proof. Let p be the penalty operator related to the set K. This operator acts from $L^2(\Omega)$ into $L^2(\Omega)$ and satisfies the Lipschitz condition.

For fixed $\varepsilon > 0$, consider the regularized system of equations

$$l_i \equiv \dot{v}_i - N_{ij,j} - f_i = 0 \ , \quad i = 1, 2 \ , \tag{4.12}$$

$$l_3 \equiv \dot{w} - M_{ij,ij} + k_{ij} N_{ij} - f = 0 \ , \tag{4.13}$$

$$a_{ijkl} \dot{N}_{kl} - \varepsilon_{ij}(v) - k_{ij} w + \frac{1}{\varepsilon}\, p(E)^1_{ij} = 0 \ , \quad i, j = 1, 2 \ , \tag{4.14}$$

$$b_{ijkl} \dot{M}_{kl} + w_{,ij} + \frac{1}{\varepsilon}\, p(E)^2_{ij} = 0 \ , \quad i, j = 1, 2 \ , \tag{4.15}$$

with boundary and initial conditions (4.7)–(4.8). We assume that $p(E) = \big(p(E)^1, p(E)^2\big)$, $p(E)^k = \{p(E)^k_{ij}\}$, $i, j, k = 1, 2$, $p(E)^k_{ij} = p(E)^k_{ji}$. To simplify the notation we ignore the symbol ε.

We may number equations (4.14), (4.15) in a natural way. The system obtained can be written as

$$L \equiv (l_4, l_5, \ldots, l_{11}) = 0 \ .$$

Namely, let us put equation (4.14) for $i = 1$, $j = 1$ in the form $l_4 = 0$; next we put $l_5 = 0$ for $i = 1$, $j = 2$; further we write $l_6 = 0$ for $i = 2$, $j = 1$, and so on.

Let $\{u_j\}$, $j = 1, 2, \ldots$, be basis functions in the space $H^2(\Omega)$ and let $\{\beta_j\}$, $j = 1, 2, \ldots$, be basis functions in $\overset{\circ}{W}(\Omega)$. For any component of the element $\{\beta_j\}$ we use the same enumeration as used above to define $L = (l_4, l_5, \ldots, l_{11})$. The Galerkin solutions to problem (4.12)–(4.15), (4.7), (4.8) are sought in the form

$$\big(v^n(t), w^n(t)\big) = \sum_{i=1}^{n} \big(a_i^n(t), b_i^n(t)\big) u_i \ , \quad E^n(t) = \sum_{i=1}^{n} c_i^n(t) \beta_i \ ,$$

where the coefficients a_i^n, b_i^n, c_i^n are defined by the following system of ordinary differential equations (note that a_i^n has two components)

$$\langle\, l_i^n, u_j \,\rangle = 0 \ , \ i = 1, 2, 3 \ ; \ \langle\, L^n, \beta_j \,\rangle = 0 \ ; \ j = 1, 2, \ldots, n \ . \tag{4.16}$$

The symbols l_i^n, L^n denote the values of the operators l_i, L at v^n, w^n, E^n. The initial data v_0^n, w_0^n, E_0^n for the solutions $v^n(t)$, $w^n(t)$, $E^n(t)$ can be found by using the expansions of v_0, w_0 with respect to basis functions $\{u_j\}$, and the expansion of E_0 with respect to basis functions $\{\beta_j\}$.

To obtain a priori estimates for problem (4.12)–(4.15), (4.7), (4.8) we observe that for sufficiently smooth functions v, w, N, M, satisfying boundary conditions (4.7), we have the following equalities

$$\langle\, v_i, N_{ij,j} \,\rangle + \langle\, N_{ij}, \varepsilon_{ij}(v) \,\rangle = 0 \ ,$$

$$\langle\, w, M_{ij,ij} \,\rangle - \langle\, M_{ij}, w_{,ij} \,\rangle = 0 \ .$$

In view of Lemma 1.7 and Lemma 1.5 the Galerkin solutions are sufficiently regular to use the latter equalities. After multiplying (4.12)–(4.15) by v_i, w, N_{ij}, M_{ij}, respectively, integrating over Ω, and summing up the relations obtained we get

$$\frac{1}{2}\frac{d}{dt}\left\{\int_\Omega (v^2+w^2)dx + C(E,E)\right\} + \frac{1}{\varepsilon}\langle\, p(E), E\,\rangle$$
$$= \langle\, f_i, v_i\,\rangle + \langle\, f, w\,\rangle\ .$$

Since the penalty term is nonnegative, the inequality

$$\max_{0\le t\le T}\left\{\|v(t)\|^2 + \|w(t)\|^2 + \|E(t)\|^2\right\} \le c \tag{4.17}$$

holds with a constant c which depends on v_0, w_0, E_0, T, f_i, f and is independent of ε. From (4.12)–(4.15) it follows that if $t=0$, then $\dot v(0)$, $\dot w(0)$, $\dot E(0)$ are bounded in $L^2(\Omega)$ (uniformly with respect to ε since $E_0 \in K \cap \overset{\circ}{W}(\Omega)$, and therefore $p(E_0)=0$). Next we differentiate equations (4.12)–(4.15) with respect to t and multiply by $\dot v_i, \dot w, \dot N_{ij}, \dot M_{ij}$, respectively. In view of the inequality $\langle\, \frac{dp(E(t))}{dt}, \dot E(t)\,\rangle \ge 0$ which is valid almost everywhere we easily deduce that

$$\max_{0\le t\le T}\left\{\|\dot v(t)\|^2 + \|\dot w(t)\|^2 + \|\dot E(t)\|^2\right\} \le c\ . \tag{4.18}$$

The dependence of the constant c is the same as in (4.17). The estimates (4.17), (4.18) in the continuous case can be obtained for v^n, w^n, E^n in a standard way. For instance, the multiplication of (4.13) by w and integration over Ω is equivalent in the discrete case to multiplication of the equation $\langle\, l_3^n, \psi_j\,\rangle = 0$ by b_j^n and summation over j from 1 to n. Thus, taking into account the strong convergence of the initial data v_0^n, w_0^n, E_0^n we deduce that estimates (4.17), (4.18) are valid for v^n, w^n, E^n with constants independent of n. Since the matrices a_{ijkl}, b_{ijkl} are positive definite, Galerkin equations (4.16) can be written in the normal form. The estimates obtained above yield the solvability of these equations on the interval $(0,T)$. Therefore, by passing to a subsequence, if necessary, we obtain that

$$\begin{aligned} &v^n, \dot v^n, w^n, \dot w^n \to v, \dot v, w, \dot w \quad \text{weakly-}(*) \text{ in } \quad L^\infty(0,T;L^2(\Omega))\ ,\\ &E^n, \dot E^n \to E, \dot E \quad \text{weakly-}(*) \text{ in } \quad L^\infty(0,T;L^2(\Omega)) \end{aligned} \tag{4.19}$$

when $n\to\infty$. The derivatives with respect to spatial variables are removed from l_i^n, L^n in (4.16) through the integration by parts for the basis functions are smooth enough. By (4.19), we can pass to the limit in (4.16) when $n\to\infty$. Therefore, we get the identities

$$[\dot v_i, u] + [N_{ij}, u_{,j}] = [f_i, u] \quad \forall\, u \in L^2(0,T;H^1(\Omega))\ , \tag{4.20}$$

$$[\dot{w}, z] - [M_{ij}, z_{,ij}] + [k_{ij}N_{ij}, z] = [f, z] \quad \forall\, z \in L^2(0,T; H^2(\Omega)) \ , \tag{4.21}$$

$$\int_0^T C(\dot{E}, \alpha)dt + [v_i, \varphi_{ij,j}] + [w, \tau_{ij,ij}] - [k_{ij}w, \varphi_{ij}] + \frac{1}{\varepsilon}\,[\eta, \alpha] = 0 \quad \forall\, \alpha = (\varphi, \tau) \in L^2(0,T; \overset{\circ}{W}(\Omega)) \ . \tag{4.22}$$

Here η denotes the weak limit in $L^2(Q)$ of the sequence $p(E^n)$. We also use the notation $[\cdot,\cdot] = \int_0^T \langle \cdot, \cdot \rangle \, dt$.

We show that the equality $\eta = p(E)$ holds. Denote

$$d_n = [p(E^n) - p(\alpha), E^n - \alpha] \geq 0 \quad \forall\, \alpha \in L^2(0,T; \overset{\circ}{W}(\Omega)) \ .$$

From equations (4.16) (for simplicity we put $\varepsilon = 1$) it follows that

$$[p(E^n), E^n] = -\int_0^T C(\dot{E}^n, E^n)dt + [k_{ij}w^n, N^n_{ij}] - [v^n_i, N^n_{ij,j}] - [w^n, M^n_{ij,ij}] \ .$$

Hence,

$$d_n = -[p(\alpha), E^n - \alpha] - [p(E^n), \alpha] - \frac{1}{2}\, C(E^n(T), E^n(T)) + \frac{1}{2}\, C(E^n(0), E^n(0)) + [k_{ij}w^n, N^n_{ij}] - [v^n_i, N^n_{ij,j}] - [w^n, M^n_{ij,ij}] \ .$$

Equations (4.16) imply

$$\langle \dot{v}^n_i, v^n_i \rangle - \langle v^n_i, N^n_{ij,j} \rangle = \langle f_i, v^n_i \rangle \ .$$

We integrate this relation with respect to t over $(0,T)$ and obtain

$$\frac{1}{2}(\|v^n(T)\|^2 - \|v^n_0\|^2) - [v^n_i, N^n_{ij,j}] = [f_i, v^n_i] \ .$$

In a similar way we get

$$\frac{1}{2}(\|w^n(T)\|^2 - \|w^n_0\|^2) - [w^n, M^n_{ij,ij}] + [k_{ij}N^n_{ij}, w^n] = [f, w^n] \ .$$

Therefore, the last three terms in the formula for d_n can be written as

$$\gamma^n = -\frac{1}{2}\,(\|v^n(T)\|^2 + \|w^n(T)\|^2 - \|v^n_0\|^2 - \|w^n_0\|^2) + [f_i, v^n_i] + [f, w^n] \ .$$

By (4.19), we can assume that $v^n(T)$, $w^n(T)$, $E^n(T)$ converge weakly in $L^2(\Omega)$ to $v(T)$, $w(T)$, $E(T)$, respectively. Hence,

$$\underline{\lim}\, \|v^n(T)\| \geq \|v(T)\| \ , \quad \underline{\lim}\, \|w^n(T)\| \geq \|w(T)\| \ ,$$
$$\underline{\lim}\, C(E^n(T), E^n(T)) \geq C(E(T), E(T)) \ .$$

Here we make use of the weak lower semicontinuity of $C(\cdot,\cdot)$. Replacing the last three terms in the formula for d_n by γ^n defined above and passing to the upper limit we get

$$0 \leq \overline{\lim}\ d_n \leq -[p(\alpha), E-\alpha] - [\eta, \alpha] - \int_0^T C(\dot{E}, E)dt \\ - [v_i, N_{ij,j}] - [w, M_{ij,ij}] + [k_{ij}N_{ij}, w] \ . \tag{4.23}$$

Continuing the proof, one can sum up identity (4.22) corresponding to $\alpha = E$, $\varepsilon = 1$ with inequality (4.23). This yields $[\eta - p(\alpha), E-\alpha] \geq 0$, $\forall\ \alpha \in L^2(0,T;\overset{\circ}{W}(\Omega))$. Thus, by the semicontinuity of the operator p, $p(E) = \eta$.

In this way we have proved the existence of a solution to (4.7)–(4.8), (4.12)–(4.15) such that $v^\varepsilon, \dot{v}^\varepsilon, w^\varepsilon, \dot{w}^\varepsilon, E^\varepsilon, \dot{E}^\varepsilon \in L^\infty(0,T;L^2(\Omega))$, $N^\varepsilon_{ij,j}$, $M^\varepsilon_{ij,ij} \in L^\infty(0,T;L^2(\Omega))$. The last three inclusions follow from equations (4.12), (4.13), since under the requirements imposed upon $f, f_i, \dot{f}, \dot{f}_i$ we have $f, f_i \in L^\infty(0,T;$ $L^2(\Omega))$. The derivatives of solutions with respect to t exist in the sense of $L^\infty(0,T;L^2(\Omega))$, hence the initial data are fulfilled in the sense of L^2; boundary conditions (4.7) are included in (4.20)–(4.21).

To complete the proof we have to pass to the limit when $\varepsilon \to 0$. The counterpart of estimate (4.18) for v^n, w^n, E^n does not provide the ε-independence, since estimating $\dot{v}^n(0)$, $\dot{w}^n(0)$, $\dot{E}^n(0)$ we cannot guarantee that the penalty term is equal to zero.

Let us examine this problem more closely and consider equations (4.16) for $t = 0$. Multiplying the first equality

$$\langle\ l_1^n(0), u_j\ \rangle = 0$$

by the first component of $\dot{a}_j^n(0)$, the second equality by the second component of $\dot{a}_j^n(0)$, and so on, and summing up over j, we get

$$\|\dot{v}^n(0)\|^2 - \langle\ N^n_{ij,j}(0), \dot{v}^n_i(0)\ \rangle = \langle\ f_i(0), \dot{v}^n_i(0)\ \rangle \ ,$$

$$\|\dot{w}^n(0)\|^2 - \langle\ M^n_{ij,ij}(0), \dot{w}^n(0)\ \rangle + \langle\ k_{ij}N^n_{ij}(0), \dot{w}^n(0)\ \rangle = \langle\ f(0), \dot{w}^n(0)\ \rangle \ ,$$

$$C(\dot{E}^n(0), \dot{E}^n(0)) - \langle\ \varepsilon_{ij}(\dot{v}^n(0)), \dot{N}^n_{ij}(0)\ \rangle - \langle\ k_{ij}\dot{w}^n(0), \dot{N}^n_{ij}(0)\ \rangle \\ + \langle\ \dot{M}^n_{ij}(0), \dot{w}^n_{,ij}(0)\ \rangle + \frac{1}{\varepsilon}\langle\ p(E^n(0)), \dot{E}^n(0)\ \rangle = 0 \ .$$

Therefore, $\|\dot{v}^n(0)\|^2 = \langle\ N^n_{ij,j}(0), \dot{v}^n_i(0)\ \rangle + \langle\ f_i(0), \dot{v}^n_i(0)\ \rangle$ and in the same way we can obtain equalities for $\|\dot{w}^n(0)\|^2$ and $C(\dot{E}^n(0), \dot{E}^n(0))$. This results in the following estimate by an application of the Cauchy inequality

$$\|\dot{v}^n(0)\| + \|\dot{w}^n(0)\| + c_0\|\dot{E}^n(0)\| \leq \sum_{i=1}^{2} \|f_i(0)\| + \|f(0)\|$$
$$+ \sum_{i=1}^{2} \|N^n_{ij,j}(0)\| + \|M^n_{ij,ij}(0)\| + \sum_{i,j=1}^{2} \|\varepsilon_{ij}(v^n(0))\| \tag{4.24}$$
$$+ \|k_{ij}N^n_{ij}(0)\| + \sum_{i,j=1}^{2} \|k_{ij}w^n(0)\| + \sum_{i,j=1}^{2} \|w^n_{,ij}(0)\| + \frac{1}{\varepsilon}\|p(E^n(0))\| \ ,$$

where $c_0 > 0$ is a constant depending on a_{ijkl}, b_{ijkl}. Here we use the equivalence of the norms $C(\cdot,\cdot)^{1/2}$ and $\|\cdot\|$. Because of the term $\frac{1}{\varepsilon}\|p(E^n(0))\|$, from (4.24) it follows that the derivatives with respect to t of v^n, w^n, E^n can be estimated at $t = 0$ by a constant depending on ε. All the remaining terms of the right-hand side of (4.24) are bounded when $n \to \infty$. Therefore, we obtain for the approximate solutions the a priori estimate of the same form as (4.18),

$$\max_{0 \leq t \leq T} \{\|\dot{v}^n(t)\|^2 + \|\dot{w}^n(t)\|^2 + \|\dot{E}^n(t)\|^2\}$$
$$\leq c + \frac{c}{\varepsilon^2} \|p(E^n(0)\|^2 \ ,$$

where constants are independent of ε, n. Since $p(E_0) = 0$ and the operator p is Lipschitz, the right-hand side is bounded from above by the expression

$$c + \frac{c}{\varepsilon^2} \|E^n(0) - E_0\|^2 \tag{4.25}$$

with the same constants as before. Thus, for any fixed $\varepsilon > 0$, (4.25) is bounded uniformly with respect to ε, for sufficiently large n, which means that for all $\varepsilon > 0$

$$\max_{0 \leq t \leq T} \{\|\dot{v}^\varepsilon(t)\| + \|\dot{w}^\varepsilon(t)\| + \|\dot{E}^\varepsilon(t)\|\} \leq c_1$$

where c_1 is a constant independent of ε. The first a priori estimate (4.17) for v^n, w^n, E^n can be obtained with a constant c independent of ε.

To conclude we can assume that estimates (4.17), (4.18) hold for $v^\varepsilon, w^\varepsilon, E^\varepsilon$ with constants independent of ε. Hence, one can choose a subsequence of the sequence $\{v^\varepsilon, \dot{v}^\varepsilon, w^\varepsilon, \dot{w}^\varepsilon\}$, still denoted by $\{v^\varepsilon, \dot{v}^\varepsilon, w^\varepsilon, \dot{w}^\varepsilon\}$, such that

$$v^\varepsilon, \dot{v}^\varepsilon, w^\varepsilon, \dot{w}^\varepsilon \to v, \dot{v}, w, \dot{w} \quad \text{weakly-}(*) \text{ in} \quad L^\infty(0,T; L^2(\Omega)) \ ,$$

$$E^\varepsilon, \dot{E}^\varepsilon \to E, \dot{E} \quad \text{weakly-}(*) \text{ in} \quad L^\infty(0,T; L^2(\Omega))$$

as $\varepsilon \to 0$. Since the operator p is monotone, from (4.22) it follows that for any $\alpha = (\varphi, \tau) \in L^2(0,T; K \cap \overset{\circ}{W}(\Omega))$ we have

$$\int_0^T C(\dot{E}^\varepsilon, \alpha - E^\varepsilon)dt + [v_i^\varepsilon, \varphi_{ij,j} - N_{ij,j}^\varepsilon] + [w^\varepsilon, \tau_{ij,ij} - M_{ij,ij}^\varepsilon] - [k_{ij}w^\varepsilon, \varphi_{ij} - N_{ij}^\varepsilon] \geq 0 \ .$$

Substituting the functions N_{ij}^ε, M_{ij}^ε according to (4.12), (4.13) we get

$$\int_0^T C(\dot{E}^\varepsilon, \alpha)dt + [v_i^\varepsilon, \varphi_{ij,j}] + [w^\varepsilon, \tau_{ij,ij}] - [k_{ij}w^\varepsilon, \varphi_{ij}] + [f_i, v_i^\varepsilon] + [f, w^\varepsilon] \geq \frac{1}{2}\, C\big(E^\varepsilon(T), E^\varepsilon(T)\big) - \frac{1}{2}\, C(E_0, E_0) + \frac{1}{2}\, \|v^\varepsilon(T)\|^2 - \frac{1}{2}\, \|v_0\|^2 + \frac{1}{2}\, \|w^\varepsilon(T)\|^2 - \frac{1}{2}\, \|w_0\|^2 \ .$$

In view of the weak convergence of $v^\varepsilon(T)$, $w^\varepsilon(T)$, $E^\varepsilon(T)$, using the weak lower semicontinuity of the norm $\|\cdot\|$ and bilinear form $C(\cdot,\cdot)$, we can pass to the lower limit as $\varepsilon \to 0$. Moreover, the right-hand side of the limiting inequality can be written as an integral over t. With (4.12), (4.13), we obtain the inequality

$$\int_0^T C(\dot{E}, \alpha - E)dt + [v_i, \varphi_{ij,j} - N_{ij,j}] + [w, \tau_{ij,ij} - M_{ij,ij}] - [k_{ij}w, \varphi_{ij} - N_{ij}] \geq 0 \quad \forall\, \alpha \in L^2(0,T; K \cap \overset{\circ}{W}(\Omega)).$$

We show that $E(t) \in K$ almost everywhere on $(0,T)$. From (4.14), (4.15) it follows that

$$[p(E^\varepsilon), E^\varepsilon] = -\varepsilon \int_0^T C(\dot{E}^\varepsilon, E^\varepsilon)dt - \varepsilon[v_i^\varepsilon, N_{ij,j}^\varepsilon] - \varepsilon[w^\varepsilon, M_{ij,ij}^\varepsilon] + \varepsilon[k_{ij}w^\varepsilon, N_{ij}^\varepsilon] \to 0 \ , \quad \varepsilon \to 0 \ .$$

For every $\alpha \in L^2(0,T; \overset{\circ}{W}(\Omega))$, if $[p(E^\varepsilon) - p(\alpha), E^\varepsilon - \alpha] \geq 0$, then $[p(\alpha), E - \alpha] \leq 0$. Put $\alpha = E + \lambda\psi$, where $\psi \in L^2(0,T; \overset{\circ}{W}(\Omega))$ is an arbitrary function, $\lambda > 0$ is a number. Dividing this inequality by λ and letting $\lambda \to 0$ we find, in view of the semicontinuity of the operator p, that $p(E) = 0$, and consequently, $E(t) \in K$ almost everywhere on $(0,T)$.

To prove uniqueness we consider the difference of two possible solutions $v = v^1 - v^2$, $w = w^1 - w^2$, $E = E^1 - E^2$. Clearly,

$$\int_0^{t_1} \big(C(\dot{E}(t), E(t)) + \langle\, \dot{v}_i(t), v_i(t)\,\rangle + \langle\, \dot{w}(t), w(t)\,\rangle\big)dt \leq 0 \ , \quad t_1 \in (0,T) \ .$$

Since the initial data for v, w, E are zero, we get $v = w = E = 0$. □

5 Existence of solutions to one-dimensional problems

In this section we prove existence theorems for one-dimensional elastoplastic problems. Some of them follow from the corresponding results for the two-dimensional model of shallow shells proved in the preceding section.

5.1 Elastoplastic problems for a beam and cylindrical shell

5.1.1. Kirchhoff beam. This model of a beam can be recovered from (4.1)–(4.6) when we lower the dimension by one. Let $Q = (a, b) \times (0, T)$ and let

$$\Phi \ : \ \mathbb{R}^2 \mapsto \mathbb{R} \tag{5.1}$$

be a convex continuous function. The problem is to find functions u, w, n, m, ξ_1, ξ_2 satisfying in the domain Q the following system of equations and inequalities

$$u_t - n_x = f_1 \ , \quad w_t - m_{xx} = f_2 \ ,$$

$$u_x = n_t + \xi_1 \ , \quad -w_{xx} = m_t + \xi_2 \ ,$$

$$\Phi(n, m) \le 0 \ ,$$

$$\xi_1(\bar{n} - n) + \xi_2(\bar{m} - m) \le 0 \quad \forall\, (\bar{n}, \bar{m}), \Phi(\bar{n}\, \bar{m}) \le 0 \ .$$

and the initial-boundary conditions

$$n = m = m_x = 0 \quad \text{when} \quad x = a, b \ , \tag{5.2}$$

$$u = u_0, \ \ w = w_0, \ \ n = n_0, \ \ m = m_0 \quad \text{when} \quad t = 0 \ . \tag{5.3}$$

In this section derivatives are indicated by subscripts x and t. Define

$$K = \big\{(n, m) \ \big| \ n, m \in L^2(a, b), \ \ \Phi\big(n(x), m(x)\big) \le 0 \ \text{a.e. in} \ (a, b)\big\} \ . \tag{5.4}$$

Set $f = (f_1, f_2)$. The following result on solvability of our problem is valid.

Theorem 5.1. *Let* $0 \in K$ *and*

$$f, f_t \in L^2(Q), \ \ u_0 \in H^1(a, b), \ \ w_0 \in H^2(a, b) \ ,$$

$$n_0 \in H_0^1(a, b), \ \ m_0 \in H_0^2(a, b), \ \ (n_0, m_0) \in K \ .$$

There exist functions u, w, n, m *such that*

$$u, u_t, w, w_t, n_t, m_t \in L^\infty(0, T; L^2(a, b)) \ , \tag{5.5}$$

$$n \in L^\infty(0, T; H_0^1(a, b)) \ , \quad m \in L^\infty(0, T; H_0^2(a, b)) \ ,$$

$$[u_t, g] + [n, g_x] = [f_1, g] \quad \forall\, g \in L^2(0, T; H^1(a, b)) \ , \tag{5.6}$$

$$[w_t, h] - [m, h_{xx}] = [f_2, h] \quad \forall\, h \in L^2(0, T; H^2(a, b)) \ , \tag{5.7}$$

$$[n_t, \bar{n} - n] + [m_t, \bar{m} - m] + [w, \bar{m}_{xx} - m_{xx}] \\ + [u, \bar{n}_x - n_x] \geq 0 \quad \forall\, (\bar{n}, \bar{m}) \in L^2(0, T; K \cap \overset{\circ}{V}) \ . \tag{5.8}$$

Moreover,

$$\big(n(t), m(t)\big) \in K \quad \text{a.e. in} \quad (0, T)$$

and initial conditions (5.3) are fulfilled.

Here $[\cdot,\cdot]$ denotes the scalar product in $L^2(Q)$ and

$$\overset{\circ}{V} = \left\{(n, m) \;\middle|\; n \in H_0^1(a, b),\;\; m \in H_0^2(a, b)\right\} \ .$$

5.1.2. Timoshenko beam. From the standpoint of mechanics the Timoshenko model of a beam is more accurate than the previous one. For instance, in this model we can describe transverse forces and turning angles of the normal to the middle axis. The plastic yield condition is formulated in terms of integrated stresses n, moments m and transverse forces q.

Let $\Phi \;:\; \mathbb{R}^3 \mapsto \mathbb{R}$ be a convex and continuous function. The problem is formulated as follows. We want to find functions $E = (n, m, q)$, $F = (u, v, w)$, ξ_1, ξ_2, ξ_3 satisfying in the domain Q the following system of equations and inequalities

$$u_t - n_x = f_1, \;\; v_t - m_x + q = 0, \;\; w_t - q_x = f_2 \ , \tag{5.9}$$

$$u_x = n_t + \xi_1, \;\; v_x = m_t + \xi_2, \;\; w_x + v = q_t + \xi_3 \ , \tag{5.10}$$

$$\Phi(E) \leq 0 \ , \tag{5.11}$$

$$\xi_1(\bar{n} - n) + \xi_2(\bar{m} - m) + \xi_3(\bar{q} - q) \leq 0 \\ \forall\, (\bar{n}, \bar{m}, \bar{q}) \ , \;\; \Phi(\bar{n}, \bar{m}, \bar{q}) \leq 0 \ , \tag{5.12}$$

and the initial-boundary conditions

$$E = 0 \quad \text{when} \quad x = a, b \ , \tag{5.13}$$

$$E = E_0, \;\; F = F_0 \quad \text{when} \quad t = 0 \ . \tag{5.14}$$

Let $f = (f_1, 0, f_2)$ and

$$K = \left\{(n, m, q) \;\middle|\; n, m, q \in L^2(a, b), \Phi\big(n(x), m(x), q(x)\big) \leq 0 \text{ a.e. in } \;\; (a, b)\right\} \ .$$

For problem (5.9)–(5.14) the following solvability result holds true.

Theorem 5.2. *Assume that* $0 \in K$, $f, f_t \in L^2(Q)$,

$$E_0 = (n_0, m_0, q_0) \in K \cap H_0^1(a, b), \;\; F_0 = (u_0, v_0, w_0) \in H^1(a, b) \ .$$

Then there exist unique functions $E = (n, m, q)$, $F = (u, v, w)$ satisfying equations (5.9) and the inequality

$$\begin{gathered}[E_t, \bar{E} - E] + [F, \bar{E}_x - E_x] - [B, \bar{E} - E] \geq 0 \\ \forall\, \bar{E} \in L^2(0,T; K \cap H_0^1(a,b)) \ .\end{gathered} \tag{5.15}$$

Moreover,

$$E \in L^\infty(0,T; H_0^1(a,b)), \quad F, F_t, E_t \in L^\infty(0,T; L^2(a,b)) \ , \tag{5.16}$$

initial conditions (5.14) are fulfilled and

$$E(t) \in K \quad \text{a.e. on} \quad (0,T) \ .$$

Here B has components $(0, 0, v)$. Inequality (5.15) follows from (5.10), (5.12), (5.13) by eliminating ξ_1, ξ_2, ξ_3 and integrating over Q.

5.1.3. Cylindrical shell with axial symmetry. Let the function Φ and the set K be defined by (5.1), (5.4). The formulation of the elastoplastic dynamic problem for a cylindrical shell with the axial symmetry is as follows.

Find functions w, n, m, ξ_1, ξ_2 such that in the domain Q the following conditions are satisfied

$$w_t - m_{xx} - n = f \ , \tag{5.17}$$

$$-w = n_t + \xi_1, \quad -w_{xx} = m_t + \xi_2 \ , \tag{5.18}$$

$$\Phi(n,m) \leq 0 \ , \tag{5.19}$$

$$\xi_1(\bar{n} - n) + \xi_2(\bar{m} - m) \leq 0 \quad \forall\, (\bar{n}, \bar{m}), \ \Phi(\bar{n}, \bar{m}) \leq 0 \ , \tag{5.20}$$

$$m = m_x = 0 \quad \text{when} \quad x = a, b \ , \tag{5.21}$$

$$w = w_0, \ n = n_0, \ m = m_0 \quad \text{when} \quad t = 0 \ . \tag{5.22}$$

Let

$$V = \{(n,m) \mid n, m, m_{xx} \in L^2(a,b), \ m = m_x = 0 \text{ when } x = a, b\} \ .$$

Now we formulate a solvability result for problem (5.17)–(5.22).

Theorem 5.3. *Assume that $0 \in K$ and $f, f_t \in L^2(Q)$,*

$$w_0 \in H^2(a,b), \ n_0 \in L^2(a,b), \ m_0 \in H_0^2(a,b), \ (n_0, m_0) \in K \ .$$

There exist unique functions w, n, m satisfying the relations

$$w, w_t, n, n_t, m, m_t \in L^\infty(0,T; L^2(a,b)) \ , \tag{5.23}$$

$$[w_t, g] - [m, g_{xx}] - [n, g] = [f, g] \quad \forall\, g \in L^2(0,T; H^2(a,b)) \ , \tag{5.24}$$

$$[m_t, \bar{m} - m] + [n_t, \bar{n} - n] + [w, \bar{m}_{xx} - m_{xx}] + [w, \bar{n} - n] \geq 0 \quad \forall\, (\bar{n}, \bar{m}) \in L^2(0,T; K \cap V) \ , \tag{5.25}$$

$$(n(t), m(t)) \in K \quad \text{for almost all} \quad t \in (0,T) \ , \tag{5.26}$$

and initial conditions (5.22).

Proofs of the theorems formulated in this section follow the lines of the proofs of the corresponding results given in Section 3, Section 4 and therefore are not given here. Observe only that relations (5.21)–(5.26) of Theorem 5.3 have to be approximated by the following relations (with the parameter δ tending to zero)

$$w_t^\delta - m_{xx}^\delta - n^\delta = f \ ,$$

$$m_t^\delta + w_{xx}^\delta + \delta^{-1} p(n^\delta, m^\delta)_2 = 0 \ ,$$

$$n_t^\delta + w^\delta + \delta^{-1} p(n^\delta, m^\delta)_1 = 0 \ ,$$

$$m^\delta = m_x^\delta = 0 \quad \text{when} \quad x = a, \ x = b \ ,$$

$$w^\delta = w_0, \ n^\delta = n_0, \ m^\delta = m_0 \quad \text{when} \quad t = 0 \ .$$

Here $p(n^\delta, m^\delta)_i$, $i = 1, 2$, denote the components of the penalty operator related to the set K. The operator acts from $L^2(a,b)$ into $L^2(a,b)$. The estimates of solutions to the above regularized problem take the form

$$\max_{0 \leq t \leq T} \{\|w^\delta(t)\| + \|n^\delta(t)\| + \|m^\delta(t)\|\} \leq c \ ,$$

$$\max_{0 \leq t \leq T} \{\|w_t^\delta(t)\| + \|n_t^\delta(t)\| + \|m_t^\delta(t)\|\} \leq c$$

with constants depending on w_0, n_0, m_0, f, T and being uniform with respect to δ. These estimates enable us to prove the existence of solutions to the problem and to pass to the limit when $\delta \to 0$.

As mentioned before, the details of the proofs of the results in Section 5.1 are omitted. The perfectly plastic problem will be considered in the next section, where we shall actually prove Theorem 5.2.

Observe that the square integrability over Ω of the derivatives with respect to x follows from the motion equations.

5.2 The perfectly plastic problem for a beam

In this section we show that for the Timoshenko model of a beam the solvability of the perfectly plastic problem can be achieved by passing to the limit in the elastoplastic problem.

The formulation of the perfectly plastic problem follows from (5.9)–(5.14) by putting $n_t = m_t = q_t = 0$, which, in turn, entails $u_x = \xi_1$, $v_x = \xi_2$, $w_x + v = \xi_3$.

Let $\Phi : \mathbb{R}^3 \mapsto \mathbb{R}$ be a convex and continuous function. We seek functions $E = (n, m, q)$, $F = (u, v, w)$ which satisfy in the domain Q the relations

$$u_t - n_x = f_1, \quad v_t - m_x + q = 0, \quad w_t - q_x = f_2 \ , \tag{5.27}$$

$$\Phi(E) \leq 0 \ , \tag{5.28}$$

$$\begin{gathered} u_x(\bar{n} - n) + v_x(\bar{m} - m) + (w_x + v)(\bar{q} - q) \leq 0 \\ \forall\, \bar{E} = (\bar{n}, \bar{m}, \bar{q}) \ , \ \Phi(\bar{E}) \leq 0 \ , \end{gathered} \tag{5.29}$$

and the initial-boundary conditions

$$E = 0 \quad \text{when} \quad x = a, b \ , \tag{5.30}$$

$$F = 0 \quad \text{when} \quad t = 0 \ . \tag{5.31}$$

We shall prove the solvability of problem (5.27)–(5.31). Let

$$K = \{E = (n, m, q) \mid E \in L^2(a, b), \ \Phi(E(x)) \leq 0 \text{ a.e. in } (a, b)\} \ .$$

Theorem 5.4. *Assume that* $0 \in K$ *and* $f_i, f_{it} \in L^2(Q)$. *There exist unique functions* $E = (n, m, q)$, $F = (u, v, w)$ *satisfying the relations*

$$E \in L^\infty(0, T; H_0^1(a, b)), \quad F, F_t \in L^\infty(0, T; L^2(a, b)) \ . \tag{5.32}$$

$$[u_t, g] + [n, g_x] = [f_1, g] \quad \forall\, g \in L^2(0, T; H^1(a, b)) \ , \tag{5.33}$$

$$[v_t, g] + [m, g_x] + [q, g] = 0 \quad \forall\, g \in L^2(0, T; H^1(a, b)) \ , \tag{5.34}$$

$$[w_t, g] + [q, g_x] = [f_2, g] \quad \forall\, g \in L^2(0, T; H^1(a, b)) \ . \tag{5.35}$$

$$\begin{gathered} [u, \bar{n}_x - n_x] + [v, \bar{m}_x - m_x] + [w, \bar{q}_x - q_x] \\ - [v, \bar{q} - q] \geq 0 \quad \forall\, (\bar{n}, \bar{m}, \bar{q}) \in L^2(0, T; K \cap H_0^1(a, b)) \ , \end{gathered} \tag{5.36}$$

and, moreover,

$$E(t) \in K \quad \text{a.e. on} \quad (0, T) \ ,$$

$$F = 0 \quad \text{when} \quad t = 0 \ .$$

Proof. Let $p : L^2(a, b) \mapsto L^2(a, b)$ be the penalty operator related to the set K and let $\varepsilon, \delta > 0$ be parameters.

Consider the regularized problem (for simplicity, the dependence of functions on ε, δ is not indicated)

$$l_1 \equiv u_t - n_x - f_1 = 0 \ , \tag{5.37}$$

$$l_2 \equiv v_t - m_x + q = 0 \ , \tag{5.38}$$

$$l_3 \equiv w_t - q_x - f_2 = 0 \ , \tag{5.39}$$

$$l_4 \equiv \varepsilon n_t - u_x + \delta^{-1} p(E)_1 = 0 \ , \tag{5.40}$$

$$l_5 \equiv \varepsilon m_t - v_x + \delta^{-1} p(E)_2 = 0 \ , \tag{5.41}$$

$$l_6 \equiv \varepsilon q_t - w_x - v + \delta^{-1} p(E)_3 = 0 \ , \tag{5.42}$$

$$E = 0 \quad \text{when} \quad x = a, b \ , \tag{5.43}$$

$$E = 0, \ F = 0 \quad \text{when} \quad t = 0 \ . \tag{5.44}$$

Here $p(E)_i$ are components of the penalty operator. The fact that we introduce the parameter ε is equivalent to replacing the perfectly plastic problem by the elastoplastic one (with the Young modulus ε^{-1}). The latter problem is, in turn, approximated by the penalty problem (5.37)–(5.44).

We begin by proving the existence of functions E, F which depend on ε, δ and satisfy (5.37)–(5.44). We use the Galerkin procedure.

Let $\{\varphi_j\}$, $\{\psi_j\}$, $j = 1, 2, \ldots$, be basis functions in the spaces $H^1(a,b)$ and $H_0^1(a,b)$, respectively. The approximate solutions E^s and F^s are sought in the form

$$E^s(t) = \sum_{i=1}^{s} a_i^s(t)\psi_i \ , \quad F^s(t) = \sum_{i=1}^{s} b_i^s(t)\varphi_i \ ,$$

where the three-component vectors a_i^s, b_i^s are defined by the following system of ordinary differential equations

$$\langle \, l_i^s, \varphi_j \, \rangle = 0, \ \ i = 1, 2, 3, \ \ \langle \, l_i^s, \psi_j \, \rangle = 0, \ \ i = 4, 5, 6; \ \ j = 1, 2, \ldots, s \ .$$

By l_i^s we denote the values of the operator l_i at E^s, F^s. In these equations, the basis functions are differentiated with respect to x only. The initial data for E^s, F^s are zero. A priori estimate for problem (5.37)–(5.44) has the form

$$\max_{0 \le t \le T} \{\|F(t)\| + \|F_t(t)\| + \varepsilon^{\frac{1}{2}} \, \|E(t)\| + \varepsilon^{\frac{1}{2}} \, \|E_t(t)\|\} \le c \ . \tag{5.45}$$

The constant c depends only on f_1, f_2, T. The estimate (5.45) can be obtained in two steps. First, by the multiplication of equations (5.37)–(5.42) by u, v, w, n, m, q, respectively, next by the differentiation of equations (5.37)–(5.42) with respect to t, and the multiplication of the resulting equations by $u_t, v_t, w_t, n_t, m_t, q_t$, respectively.

In fact, since $0 \in K$, and $E = 0$ when $t = 0$ we have $p(E) = 0$ for $t = 0$. Moreover, in view of (5.40)–(5.42), we have $E_t = 0$ when $t = 0$.

From (5.37)–(5.42) it follows that

$$\frac{1}{2} \frac{d}{dt} \{\|F_t(t)\|^2 + \varepsilon \, \|E_t(t)\|^2\} \le \langle \, f_t(t), F_t(t) \, \rangle \ , \tag{5.46}$$

where f has components $(f_1, 0, f_2)$. Since $E_t = 0$ when $t = 0$, by integrating (5.46), we get that the second and fourth terms in (5.45) are independent of ε, δ. The first and third terms in (5.45) are bounded uniformly with respect to ε, δ.

Estimate (5.45) provides the solvability of the Galerkin equations on the interval $(0, T)$. Moreover, we can choose a subsequence E^s, F^s such that

$$E^s, E_t^s \to E, E_t \quad \text{weakly-}(*) \text{ in} \quad L^\infty(0, T; L^2(a,b)) \ ,$$

$$F^s, F_t^s \to F, F_t \quad \text{weakly-}(*) \text{ in} \quad L^\infty(0, T; L^2(a,b))$$

when $s \to \infty$. After passing to the limit in the Galerkin equations as $s \to \infty$ the following identities result

$$[F_t, \varphi] + [E, \varphi_x] + [A, \varphi] = [f, \varphi] \quad \forall\, \varphi \in L^2(0,T; H^1(a,b)) \ , \tag{5.47}$$

$$\begin{aligned} \varepsilon[E_t, \psi] + [F, \psi_x] - [B, \psi] + \delta^{-1}[p(E), \psi] = 0 \\ \forall\, \psi \in L^2(0,T; H_0^1(a,b)) \ , \end{aligned} \tag{5.48}$$

where the vectors A and B have the components $(0, q, 0)$ and $(0, 0, v)$, respectively. The weak convergence of $p(E^s)$ to $p(E)$ in $L^2(Q)$ follows from the monotonicity of the operator p.

In particular, it follows from the above identities that equations (5.37)–(5.42) are fulfilled in the sense of distributions in Q and thus

$$E \in L^\infty(0,T; H_0^1(a,b)), \quad F \in L^\infty(0,T; H^1(a,b)) \ .$$

Actually, the above arguments show that the functions E and F depend on the parameters ε, δ. Note that the function F is estimated in $L^\infty(0,T; H^1(a,b))$ nonuniformly with respect to ε, δ.

Now we analyse passing to the limit in problem (5.37)–(5.44) when $\varepsilon \to 0$. We tag the solutions with the symbols ε, δ. The estimates

$$\max_{0 \le t \le T} \{\|F^{\varepsilon\delta}(t)\| + \|F_t^{\varepsilon\delta}(t)\|\} \le c \ , \tag{5.49}$$

$$\max_{0 \le t \le T} \|E^{\varepsilon\delta}(t)\|_1 \le c \tag{5.50}$$

are valid with constants independent of ε, δ. The first estimate is a direct consequence of the uniform boundedness (with respect to ε, δ) of estimates (5.45).

The second estimate results from (5.37)–(5.39) and the first estimate. Indeed, from (5.37), (5.39) we readily conclude that $n_x^{\varepsilon\delta}$, $q_x^{\varepsilon\delta}$ are bounded in $L^\infty(0,T; L^2(a,b))$. Hence, taking into account the boundary conditions for $n^{\varepsilon\delta}, q^{\varepsilon\delta}$ we obtain boundedness of $n^{\varepsilon\delta}, q^{\varepsilon\delta}$ in $L^\infty(0,T; H_0^1(a,b))$. Then, from (5.38) it follows that $m^{\varepsilon\delta}$ are bounded in $L^\infty(0,T; H_0^1(a,b))$.

By (5.49), (5.50), we can choose a subsequence such that

$$F^{\varepsilon\delta}, F_t^{\varepsilon\delta} \to F^\delta, F_t^\delta \quad \text{weakly-}(*) \text{ in} \quad L^\infty(0,T; L^2(a,b)) \ ,$$

$$E^{\varepsilon\delta} \to E^\delta \quad \text{weakly-}(*) \text{ in} \quad L^\infty(0,T; H_0^1(a,b)) \ ,$$

$$\varepsilon E_t^{\varepsilon\delta} \to 0 \quad \text{weakly in} \quad L^2(Q)$$

as $\varepsilon \to 0$, and $\delta > 0$ is fixed. The last convergence follows from the uniform boundedness with respect to ε, δ of the functions $\varepsilon^{\frac{1}{2}} E_t^{\varepsilon\delta}$ in the space $L^\infty(0,T; L^2(a,b))$. Passing to the limit in (5.47), and (5.48) as $\varepsilon \to 0$ we readily obtain

$$[F_t^\delta, \varphi] + [E^\delta, \varphi_x] + [A^\delta, \varphi] = [f, \varphi] \quad \forall\, \varphi \in L^2(0,T; H^1(a,b)) \ , \tag{5.51}$$

$$[F^\delta, \psi_x] - [B^\delta, \psi] + \delta^{-1}[p(E^\delta), \psi] = 0 \quad \forall\, \psi \in L^2(0,T; H_0^1(a,b)) \ , \tag{5.52}$$

where $A^\delta = (0, q^\delta, 0)$, $B^\delta = (0, 0, v^\delta)$. The boundary condition $E^\delta = 0$ for $x = a, b$ is fulfilled in the sense of L^2.The initial condition $F^\delta = 0$ for $t = 0$ is fulfilled in the same sense.

To conclude we have to show that we can pass to the limit when $\delta \to 0$. Since estimates (5.49), and (5.50) are uniform with respect to δ, F^δ, F_t^δ are bounded in $L^\infty(0,T; L^2(\Omega))$ and E^δ are bounded in $L^\infty(0,T; H_0^1(a,b))$.

Let a subsequence of F^δ, E^δ, still denoted by F^δ, E^δ, possess the following property as $\delta \to 0$

$$F^\delta, F_t^\delta \to F, F_t \quad \text{weakly-}(*) \text{ in} \quad L^\infty(0,T; L^2(a,b)) \ ,$$

$$E^\delta \to E \quad \text{weakly-}(*) \text{ in} \quad L^\infty(0,T; H_0^1(a,b)) \ .$$

Basing on the same arguments as previously we can pass to the limit in (5.51).

Consider now (5.52). Take a function $\bar{E} \in L^2(0,T; K \cap H_0^1(a,b))$. It follows that

$$[F^\delta, \bar{E}_x - E_x^\delta] - [B^\delta, \bar{E} - E^\delta] \geq 0 \ .$$

Substituting the components of E_x^δ according to equations (5.51) we obtain

$$[F^\delta, \bar{E}_x] - [B^\delta, \bar{E}] + [f, F^\delta] \geq \frac{1}{2} \, \|F^\delta(T)\|^2 \ .$$

By the convergence of F^δ, F_t^δ and the weak lower semicontinuity of the norm $\|\cdot\|$, we can pass to the limit. Thus

$$[F, \bar{E}_x - E_x] - [B, \bar{E} - E] \geq 0 \quad \forall\, \bar{E} \in L^2(0,T; K \cap H_0^1(a,b)) \ .$$

This inequality coincides with (5.36). As in the preceding section, from the monotonicity of the operator p it follows that $E(t) \in K$.

Let us prove the uniqueness of solution. The difference of two possible solutions $E = E_1 - E_2$, $F = F_1 - F_2$ satisfies the equations

$$u_t - n_x = 0, \ \ v_t - m_x + q = 0, \ \ w_t - q_x = 0 \ , \tag{5.53}$$

where $E = (n, m, q)$, $F = (u, v, w)$. Substitute into (5.36) the test function $\bar{E}(t) = E_2(t)$, $t \in (0, t_1)$, $\bar{E}(t) = E_1(t)$, $t \in (t_1, T)$. Next, substitute the test function $\bar{E}(t) = E_1(t)$, $t \in (0, t_1)$, $\bar{E}(t) = E_2(t)$, $t \in (t_1, T)$. Summing up the relations obtained, we easily find that

$$\|F(t_1)\|^2 \leq 0 \ , \ \ t_1 \in (0, T) \ .$$

Hence $F \equiv 0$, and, by the boundary conditions, (5.53) implies that $E \equiv 0$. □

6 Existence of solutions for a quasistatic shell

6.1 Formulation of the problem

Let, as in Section 4, $\Omega \subset \mathbb{R}^2$ be a bounded domain with smooth boundary Γ, $Q = \Omega \times (0,T)$. We seek functions $v = (v_1, v_2)$, w, $E = (N, M)$, ξ_{ij}, λ_{ij} satisfying in the domain Q the relations

$$-N_{ij,j} = f_i \ , \quad i = 1,2 \ ,$$

$$-M_{ij,ij} + k_{ij} N_{ij} = f \ ,$$

$$\varepsilon_{ij}(v) + k_{ij} w = a_{ijkl} \dot{N}_{kl} + \xi_{ij} \ , \quad i,j = 1,2 \ ,$$

$$-w_{,ij} = b_{ijkl} \dot{M}_{kl} + \lambda_{ij} \ , \quad i,j = 1,2 \ ,$$

$$\Phi(E) \leq 0 \ ,$$

$$\xi_{ij}(\bar{N}_{ij} - N_{ij}) + \lambda_{ij}(\bar{M}_{ij} - M_{ij}) \leq 0 \quad \forall\, \bar{E} = (\bar{N}, \bar{M}),\ \Phi(\bar{E}) \leq 0 \ ,$$

$$v = w = w_\nu = 0 \quad \text{on} \quad \Gamma \times (0,T) \ ,$$

$$E = 0 \quad \text{when} \quad t = 0 \ .$$

The notation used in this section follows the notation used in Section 4. In particular, the space $W(\Omega)$ and the set K as introduced at the end of Section 4.1 will be used in Theorem 6.1.

Throughout this section we assume that $f, f_i \in C\big(0,T; L^2(\Omega)\big)$, and $f(t) = f_i(t) = 0$ for $t \in (0,\delta)$, where $\delta > 0$ is a constant, $k_{ij} \in L^\infty(\Omega)$. By $B = (N^0, M^0) \in C^2(\bar{Q})$ we denote a solution to equations (6.1), (6.2) (see below) such that $B(t) \equiv 0$ for $t \in (0,\delta)$. The operator π is the operator of orthogonal projection of the space of vectors $(N, M) \in \mathbb{R}^6$ for symmetric 2×2 matrices $N = \{N_{ij}\}$, $M = \{M_{ij}\}$, onto the set

$$\tilde{K} = \big\{(N,M) \in \mathbb{R}^6 \ \big|\ \Phi(N,M) \leq 0\big\} \ .$$

6.2 Theorem of existence

Our aim now is to prove the following theorem.

Theorem 6.1. *Let the assumptions formulated above hold. If there exists a constant $\kappa > 0$ such that $(1+\kappa)B(t) \in K$, $t \in (0,\delta)$, then there exist functions $v = (v_1, v_2)$, w, $E = (N, M)$ such that*

$$-N_{ij,j} = f_i \ , \quad i = 1,2 \ , \tag{6.1}$$

$$-M_{ij,ij} + k_{ij} N_{ij} = f \ , \tag{6.2}$$

$$\int_0^T C(\dot{E},\alpha - E)dt + \int_0^T \langle\, v_i, \varphi_{ij,j} - N_{ij,j}\,\rangle dt$$
$$+ \int_0^T \langle\, w, \tau_{ij,ij} - M_{ij,ij}\,\rangle dt - \int_0^T \langle\, k_{ij}w, \varphi_{ij} - N_{ij}\,\rangle dt \geq 0 \tag{6.3}$$
$$\forall\, \alpha = (\varphi, \tau) \in L^2(0,T; K \cap W(\Omega))\ ,$$

$$E \in L^\infty(0,T; L^2(\Omega)),\ \ \dot{E}, v \in L^2(Q),\ \ w \in L^2(0,T; H_0^1(\Omega))\ , \tag{6.4}$$

$$E(t) \in K \quad \text{for almost all} \quad t \in (0,T)\ , \tag{6.5}$$

$$E = 0 \quad \text{when} \quad t = 0. \tag{6.6}$$

Proof. Let us first give the idea of the proof. We begin by proving the existence of solutions for every time-level of some problems approximating (6.1)–(6.6). The approximation is characterized by a parameter $\varepsilon > 0$. Next we establish a priori estimates which are uniform both with respect to ε and with respect to the discretization parameter. Finally, we pass to the limit.

Let s be an integer. Denote $h = \dfrac{T}{s}$ and put $dE^n = (E^n - E^{n-1})h^{-1}$ for every $n = 1, 2, \ldots, s$. Consider the following closed and convex sets in $L^2(\Omega)$

$$A(hn) = \{E \in L^2(\Omega)\ \big|\ -N_{ij,j} = f_i(hn),\ \ -M_{ij,ij} + k_{ij}N_{ij} = f(hn)\}\ ,$$

$$A_0 = \{E \in L^2(\Omega)\ \big|\ -N_{ij,j} = 0,\ \ -M_{ij,ij} + k_{ij}N_{ij} = 0\}$$

and the functionals

$$H(E) = \frac{1}{2h}\, C(E,E) - \frac{1}{h}\, C(E, E^{n-1}) + P(E)\ , \quad n = 1, 2, \ldots, s\ ,$$

where $P(E) = \frac{1}{2\varepsilon}\,\|E - \pi E\|^2$, $E^0 = 0$. It is easily seen that under our assumptions the sets $A(hn)$ and A_0 are nonempty.

Below we shall prove the existence of solutions to a problem which is discrete in t and approximates (6.1)–(6.6). By u we denote vectors such that $u = (u^1, u^2)$, $u^m = \{u_{ij}^m\}$, $i, j = 1, 2$, $m = 1, 2$.

We define the functional

$$G(u) = \sup_{E \in L^2(\Omega)} \{\langle\, E, u\,\rangle - H(E)\}\ .$$

As before, by $\langle\,\cdot,\cdot\,\rangle$ we denote the scalar product in $L^2(\Omega)$, and Y is a subspace of $L^2(\Omega)$ containing vectors $u = (u^1, u^2)$ such that

$$u_{ij}^1 = \varepsilon_{ij}(v) + k_{ij}w,\ \ u_{ij}^2 = -w_{,ij},\ \ v = (v_1, v_2) \in H_0^1(\Omega),\ \ w \in H_0^2(\Omega)\ .$$

Taking into account the zero boundary conditions for w and the first Korn inequality it is easy to prove that Y is closed in the space $L^2(\Omega)$.

On the space $L^2(\Omega)$ we introduce another functional

$$F(u) = \begin{cases} -\langle f_i(hn), v_i \rangle - \langle f(hn), w \rangle, & \text{if} \quad u \in Y \\ +\infty & \text{otherwise} . \end{cases}$$

As a pointwise supremum of affine functions the functional G is weakly lower semicontinuous and so is the functional F for the subspace Y is closed.

Consider the problem

$$\inf_{u \in L^2(\Omega)} \{G(u) + F(u)\}.$$

By the definitions of G and F the problem is equivalent to

$$\begin{aligned} \inf_{\substack{v \in H_0^1(\Omega) \\ w \in H_0^2(\Omega)}} \Big\{ \sup_{E \in L^2(\Omega)} &\big[\langle N_{ij}, \varepsilon_{ij}(v) + k_{ij} w \rangle - \langle M_{ij}, w_{,ij} \rangle \\ &- H(E)\big] - \langle f_i(hn), v_i \rangle - \langle f(hn), w \rangle\Big\} . \end{aligned} \tag{6.7}$$

We first construct the dual problem to (6.7), and next prove the solvability of problem (6.7). By $*$ we denote the operation of conjugation.

According to (Ekeland, Temam, 1976), the dual problem to (6.7) with respect to the perturbation $\Lambda(u, q) = G(u) + F(u - q)$ is of the form

$$\inf_{E \in L^2(\Omega)} \{G^*(E) + F^*(-E)\} .$$

By the convexity of H, we have $G^* = H$, and

$$\begin{aligned} F^*(-E) &= \sup_{u \in L^2(\Omega)} \{ -\langle u, E \rangle - F(u)\} \\ &= \sup_{\substack{v \in H_0^1(\Omega) \\ w \in H_0^2(\Omega)}} \{ -\langle N_{ij}, \varepsilon_{ij}(v) + k_{ij} w \rangle + \langle M_{ij}, w_{,ij} \rangle \\ &\quad + \langle f_i(hn), v_i \rangle + \langle f(hn), w \rangle\} = \Psi_{A(hn)}(E) , \end{aligned}$$

where $\Psi_{A(hn)}$ is the indicator function of the set $A(hn)$, i.e., the function which is equal to zero on $A(hn)$ and $+\infty$ beyond $A(hn)$. Therefore, the dual problem to (6.7) can be rewritten in the form

$$\inf_{E \in A(hn)} H(E) . \tag{6.8}$$

By the coercivity and the weak lower semicontinuity of H, problem (6.8) has a solution $E = E^n \in A(hn)$. This solution is characterized by the inequality

$$\langle H'(E), \ \alpha - E \rangle \geq 0 \quad \forall \, \alpha \in A(hn) \ ,$$

where $H'(E)$ is the derivative of H at E.

Let $\alpha = E \pm E_0$, $E_0 \in A_0$. With this substitution

$$C(dE, E_0) + \langle P'(E), E_0 \rangle = 0 \quad \forall\, E_0 \in A_0 \ . \tag{6.9}$$

Here $P'(E) = \dfrac{1}{\varepsilon}\,(E - \pi E)$ is the derivative of the functional P. We assume that $P'(E) = (P'(E)^1, P'(E)^2)$, $P'(E)^m = \{P'(E)^m_{ij}\}$, $m, i, j = 1, 2$. We shall use identity (6.9) in deriving a priori estimates.

We now prove the solvability of problem (6.7). At the point $\bar{E} = (\bar{N}, \bar{M})$, where the supremum is attained, we have

$$C(d\bar{E}, \alpha) - \langle \varphi_{ij}, \varepsilon_{ij}(v) + k_{ij}w \rangle + \langle \tau_{ij}, w_{,ij} \rangle$$
$$+ \langle P'(\bar{E}), \alpha \rangle = 0 \quad \forall\, \alpha = (\varphi, \tau) \in L^2(\Omega);\ \ d\bar{E} = (\bar{E} - E^{n-1})h^{-1} \ .$$

Since α is arbitrary

$$\begin{aligned} \varepsilon_{ij}(v) + k_{ij}w &= a_{ijkl}d\bar{N}_{kl} + P'(\bar{E})^1_{ij}, \quad i, j = 1, 2 \ , \\ -\, w_{,ij} &= b_{ijkl}d\bar{M}_{kl} + P'(\bar{E})^2_{ij}, \quad i, j = 1, 2 \ . \end{aligned} \tag{6.10}$$

Substituting this expressions into (6.7) when $E = \bar{E}$, we find that the bracketed term equals

$$\frac{1}{2h}\, C(\bar{E}, \bar{E}) - P(\bar{E}) + \langle P'(\bar{E}), \bar{E} \rangle \ .$$

Consequently, for $u^1_{ij} = \varepsilon_{ij}(v) + k_{ij}w$, $u^2_{ij} = -w_{,ij}$, the functional $G(u) + F(u)$ can be bounded from below by

$$\begin{aligned} &\frac{1}{2h}\, C(\bar{E}, \bar{E}) - \frac{1}{2\varepsilon}\, \|\bar{E} - \pi\bar{E}\|^2 + \frac{1}{\varepsilon}\, \langle\, \bar{E} - \pi\bar{E}, \bar{E}\, \rangle \\ &\qquad - \|f_i(hn)\|\ \|v_i\|_1 - \|f(hn)\|\ \|w\|_2 \ . \end{aligned} \tag{6.11}$$

With the estimates of the norms $\|v\|_1$ and $\|w\|_2$ which follow from (6.10) we deduce that (6.11) tends to $+\infty$ whenever $\|v\|_1 + \|w\|_2 \to \infty$ and h is sufficiently small. Thus, the functional in braces in (6.7) is coercive. Its lower semicontinuity guarantees that problem (6.7) has a solution $v = v^n \in H^1_0(\Omega)$, $w = w^n \in H^2_0(\Omega)$. The vector $u = u^n$ corresponds to this solution.

Solutions to problem (6.7) and its dual (6.8) satisfy the relation

$$G(u^n) + G^*(E^n) = \langle\, u^n, E^n \,\rangle \ .$$

Substituting $G(u^n)$ and $G^*(E^n)$ we get

$$\langle\, \alpha, u^n \,\rangle - H(\alpha) \le \langle\, E^n, u^n \,\rangle - H(E^n) \quad \forall\, \alpha \in L^2(\Omega) \ .$$

This means that the functional $H(\cdot) - \langle\, \cdot, u^n \,\rangle$ attains its minimum at E^n, hence its derivative is equal to zero at this point. Thus, in view of the equalities $u^1_{ij} = \varepsilon_{ij}(v^n) + k_{ij}w^n$, $u^2_{ij} = -w^n_{,ij}$, we obtain

$$\begin{aligned} \varepsilon_{ij}(v^n) + k_{ij}w^n &= a_{ijkl}dN^n_{kl} + P'(E^n)^1_{ij}, \quad i, j = 1, 2 \ , \\ -\, w^n_{,ij} &= b_{ijkl}dM^n_{kl} + P'(E^n)^2_{ij}, \quad i, j = 1, 2 \ . \end{aligned} \tag{6.12}$$

Basing ourselves on (6.9) we derive a priori estimates for the solution. To this end we substitute the value $\bar{E}^n = E^n - B^n$ as a test function into (6.9) with $E = E^n$. Here B solves equations (6.1), (6.2), so that $B^n = B(hn) \in K \cap A(hn)$. Multiplying the resulting relations by h and summing them up from 1 to s we obtain

$$\sum_{n=1}^{s} C(d\bar{E}^n, \bar{E}^n)h + \sum_{n=1}^{s} \langle P'(E^n), \bar{E}^n \rangle h = -\sum_{n=1}^{s} C(dB^n, \bar{E}^n)h \ . \tag{6.13}$$

The operator P' is monotone for P is convex. Since $P'(B^n) = 0$, the second sum in the left-hand side of (6.13) is nonnegative. Moreover,

$$\sum_{n=1}^{m} C(d\bar{E}^n, \bar{E}^n)h = \frac{1}{2}\|\bar{E}^m\|_C^2 - \frac{1}{2}\|\bar{E}^0\|_C^2 + \frac{1}{2}\sum_{n=1}^{m}\|\bar{E}^n - \bar{E}^{n-1}\|_C^2 \ ,$$

$$\sum_{n=1}^{m} C(dB^n, \bar{E}^n)h = -\sum_{n=2}^{m} C(d\bar{E}^n, B^{n-1})h - C(B^0, \bar{E}^1) + C(B^m, \bar{E}^m) \ ,$$

where $\|\cdot\|_C^2 = C(\cdot,\cdot)$. Therefore, from (6.13) it follows that

$$\max_n \|\bar{E}^n\|_C \le c \ , \quad \sum_{n=1}^{s} \langle P'(E^n), \bar{E}^n \rangle h \le c \tag{6.14}$$

with constants independent of s, ε.

By our assumptions, $(1+\kappa)B^n \in K$. Moreover, the set $\tilde{K}$ is convex and contains zero in its interior. Thus, there exists a constant $\bar{\kappa}$ depending on κ and $\tilde{K}$, such that $\operatorname{dist}\big(B^n(x), C\tilde{K}\big) \ge \bar{\kappa}$ for almost all $x \in \Omega$. Here $C\tilde{K}$ denotes the complement of the set $\tilde{K}$.

Assume that for some $x \in \Omega$ the vector $E^n(x) = B^n(x) + \bar{E}^n(x)$ belongs to $C\tilde{K}$. Then, the hyperplane $\{\xi \in \mathbb{R}^6 \mid [\xi, l(x)] = a\}$ separates $\tilde{K}$ and $E^n(x)$, where $[\cdot,\cdot]$ denotes the scalar product corresponding to the Euclidean norm, a is a number, and

$$l(x) = P'(E^n)(x)\ \big|P'(E^n)(x)\big|^{-1} \ . \tag{6.15}$$

Since $E^n(x) \in C\tilde{K}$, $B^n(x) + \bar{\kappa}l(x) \in \tilde{K}$, we get

$$\big[B^n(x) + \bar{E}^n(x), l(x)\big] \ge a, \quad \big[B^n(x) + \bar{\kappa}l(x), l(x)\big] \le a \ .$$

This yields the inequality $\big[\bar{E}^n(x), l(x)\big] \ge \bar{\kappa}$. Thus, multiplying (6.15) by $\bar{E}^n(x)$, we get

$$\big|P'(E^n)(x)\big| \le \bar{\kappa}^{-1}\big[P'(E^n)(x), \bar{E}^n(x)\big] \ .$$

This inequality also remains true when $E^n(x) \in \tilde{K}$. After integrating it over Ω, with (6.14) we obtain

$$\|P'(E^n)\|_{L^1(\Omega)} \le c\,\langle P'(E^n), \bar{E}^n \rangle, \quad \sum_{n=1}^{s} \|P'(E^n)\|_{L^1(\Omega)}h \le c \ . \tag{6.16}$$

By taking the difference quotient in (6.9) and substituting $d\bar{E}^n \in A_0$ as a test function in the resulting integral identity we get

$$C(d^2\bar{E}^n, d\bar{E}^n) + \langle\, dP'(E^n), d\bar{E}^n \,\rangle = -C(d^2B^n, d\bar{E}^n)\ , \quad n = 2, 3, \ldots, s\ .$$

Multiplying these equalities by h, summing them up over n from 2 to $m \le s$, and integrating by parts we obtain

$$\sum_{n=2}^{m} \langle\, dP'(E^n), dB^n \,\rangle h = -\sum_{n=2}^{m-1} \langle\, P'(E^n), d^2B^{n+1} \,\rangle h$$
$$+ \langle\, P'(E^m), dB^m \,\rangle - \langle\, P'(E^1), dB^2 \,\rangle\ .$$

Note that

$$C(d^2\bar{E}^n, d\bar{E}^n)h \ge \frac{1}{2}\,\|d\bar{E}^n\|_C^2 - \frac{1}{2}\,\|d\bar{E}^{n-1}\|_C^2\ ,$$
$$\langle\, dP'(E^n), dE^n \,\rangle \ge 0\ .$$

In view of the second inequality of (6.16) and since $E^1 = B^2 = 0$ for sufficiently small h we obtain

$$\frac{1}{2}\,\|d\bar{E}^m\|_C^2 \le c + \frac{1}{2}\sum_{n=2}^{m} \|d\bar{E}^n\|_C^2 h + \langle\, P'(E^m), dB^m \,\rangle\ ,$$

with a constant c which does not depend on m, ε. By the Gronwall inequality and (6.16) we finally get the estimate

$$\sum_{n=2}^{s} \|d\bar{E}^n\|_C^2\, h \le c\ . \tag{6.17}$$

For $P'(E^n)$ an estimate stronger than (6.16) can be obtained. To see this let us substitute $\bar{E}^n$ as a test function into (6.9). By using (6.17) and the first inequality of (6.16) we obtain

$$\sum_{n=1}^{s} \|P'(E^n)\|_{L^1(\Omega)}^2\, h \le c \sum_{n=1}^{s} \left[C(dE^n, \bar{E}^n)\right]^2 h \le c \tag{6.18}$$

with the constants independent of s, ε. We shall use inequality (6.18) in deriving estimates for v^n, w^n below.

To complete the proof we should pass to the limit as $h \to 0$, $\varepsilon \to 0$.

Let $\tilde{E}_h$ be a piece-wise linear function of t such that $\tilde{E}_h(hn) = E^n$ and let v_h, w_h, E_h be step functions taking the values v^n, w^n, E^n at the points hn. By (6.14), (6.17), (6.18),

$$\|E_h\|_{L^\infty(0,T;L^2(\Omega))} + \left\|\frac{d\tilde{E}_h}{dt}\right\|_{L^2(Q)} + \|P'(E_h)\|_{L^2(0,T;L^1(\Omega))} \le c \tag{6.19}$$

with a constant independent of h, ε. From (6.12), and (6.19) it follows that

$$\|v_h\|_{L^2(0,T;H_0^1(\Omega))} + \|w_h\|_{L^2(0,T;H_0^2(\Omega))} + \|P'(E_h)\|_{L^2(Q)} \le c(\varepsilon) \ . \tag{6.20}$$

The right-hand side of (6.20) depends on ε. Thus, by (6.19) and (6.20), there exists a subsequence of $v_h, w_h, E_h, \tilde{E}_h$, still denoted by $v_h, w_h, E_h, \tilde{E}_h$, such that

$$E_h \to E^\varepsilon \quad \text{weakly-}(*) \text{ in } \quad L^\infty(0,T;L^2(\Omega)) \ ,$$

$$\frac{d\tilde{E}_h}{dt} \to \dot{E}^\varepsilon, \ P'(E_h) \to I^\varepsilon \quad \text{weakly in} \quad L^2(Q) \ ,$$

$$w_h \to w^\varepsilon \text{ weakly in } L^2(0,T;H_0^2(\Omega)), \ v_h \to v^\varepsilon \text{ weakly in } L^2(0,T;H_0^1(\Omega))$$

as $h \to 0$ (ε is fixed). By these formulae we can pass to the limit in (6.12).

To prove the equality $I^\varepsilon = P'(E^\varepsilon)$ we exploit the monotonicity of the operator P'. The limiting relations take the form

$$\begin{aligned} \varepsilon_{ij}(v^\varepsilon) + k_{ij}w^\varepsilon &= a_{ijkl}\dot{N}^\varepsilon_{kl} + P'(E^\varepsilon)^1_{ij}, \ i,j=1,2 \ , \\ -w^\varepsilon_{,ij} &= b_{ijkl}\dot{M}^\varepsilon_{kl} + P'(E^\varepsilon)^2_{ij} \ , \ i,j=1,2 \ . \end{aligned} \tag{6.21}$$

Estimate (6.19) is uniform with respect to ε. Hence, the limiting functions E^ε satisfy the inequality

$$\|E^\varepsilon\|_{L^\infty(0,T;L^2(\Omega))} + \|\dot{E}^\varepsilon\|_{L^2(Q)} + \|P'(E^\varepsilon)\|_{L^2(0,T;L^1(\Omega))} \le c$$

with the constant c independent of ε.

Therefore,

$$\sum_{i,j=1}^{2}\left(\|w^\varepsilon_{,ij}\|_{L^2(0,T;L^1(\Omega))} + \|\varepsilon_{ij}(v^\varepsilon)\|_{L^2(0,T;L^1(\Omega))}\right) \le c \ . \tag{6.22}$$

Since the space $L^1(\Omega)$ is not reflexive, (6.22) does not imply that one can choose a convergent subsequence from $w^\varepsilon, v^\varepsilon$ as $\varepsilon \to 0$.

On the other hand, for any $\varphi \in H_0^1(\Omega)$, the inequality

$$\|\varphi\| \le c\sum_{i=1}^{2}\|\varphi_{,i}\|_{L^1(\Omega)} \tag{6.23}$$

holds with a constant independent of φ. To see this, let us extend φ by putting its values equal to zero beyond Ω, which gives us

$$|\varphi(x_1,x_2)| = \left|\int_{-\infty}^{x_1}\frac{\partial\varphi}{\partial x_1}(y,x_2)dy\right| \le \int_{-\infty}^{\infty}\left|\frac{\partial\varphi}{\partial x_1}(y,x_2)\right|dy \equiv L_1(x_2) \ .$$

By the same arguments, $|\varphi(x_1, x_2)| \le L_2(x_1)$, where the meaning of $L_2(x_1)$ is clear. Thus,

$$\int_\Omega \varphi^2 dx \le \int_{-\infty}^{\infty}\int_{-\infty}^{\infty} L_1(x_2)L_2(x_1)dx_1dx_2 \le \frac{1}{2}\sum_{i=1}^{2} \|\varphi_{,i}\|^2_{L^1(\Omega)} ,$$

which establishes estimate (6.23).

According to Lemma 1.3, for $\varphi = (\varphi_1, \varphi_2) \in H_0^1(\Omega)$, we have

$$\|\varphi\| \le c \sum_{i,j=1}^{2} \|\varepsilon_{ij}(\varphi)\|_{L^1(\Omega)} . \tag{6.24}$$

In view of (6.23) and (6.24), from (6.22) it follows that

$$\|w^\varepsilon\|_{L^2(0,T;H_0^1(\Omega))} + \|v^\varepsilon\|_{L^2(Q)} \le c .$$

Having established these estimates we can pass to the limit as $\varepsilon \to 0$.

Assume that a subsequence of $E^\varepsilon, w^\varepsilon, v^\varepsilon$, still denoted by $E^\varepsilon, w^\varepsilon, v^\varepsilon$, possesses the property that

$$E^\varepsilon \to E \text{ weakly-}(*) \text{ in } L^\infty(0,T;L^2(\Omega)) , \quad \dot{E}^\varepsilon \to \dot{E} \text{ weakly in } L^2(Q) ,$$

$$w^\varepsilon \to w \text{ weakly in } L^2(0,T;H_0^1(\Omega)) , \quad v^\varepsilon \to v \text{ weakly in } L^2(Q) .$$

Multiplying (6.21) by $N_{ij}^\varepsilon - \varphi_{ij}$, $M_{ij}^\varepsilon - \tau_{ij}$, respectively, where $\alpha = (\varphi, \tau) \in L^2(0,T;K \cap W(\Omega))$ we obtain

$$\begin{aligned} &C(\dot{E}^\varepsilon, E^\varepsilon - \alpha) + \langle v_i^\varepsilon, N_{ij,j}^\varepsilon - \varphi_{ij,j} \rangle \\ &+ \langle w^\varepsilon, M_{ij,ij}^\varepsilon - \tau_{ij,ij} \rangle - \langle k_{ij}w^\varepsilon, N_{ij}^\varepsilon - \varphi_{ij} \rangle \le 0 . \end{aligned} \tag{6.25}$$

The functions $E^n = (N^n, M^n)$ belong to $A(hn)$. After passing to the limit when $h \to 0$, for $E^\varepsilon = (N^\varepsilon, M^\varepsilon)$, we have

$$-N_{ij,j}^\varepsilon = f_i, \ i = 1, 2; \ -M_{ij,ij}^\varepsilon + k_{ij}N_{ij}^\varepsilon = f . \tag{6.26}$$

Consequently, taking into account (6.26) and integrating (6.25) over t from 0 to T we get

$$\begin{aligned} &\frac{1}{2}\|E^\varepsilon(T)\|_C^2 - \int_0^T C(\dot{E}^\varepsilon, \alpha)dt - \int_0^T \langle v_i^\varepsilon, f_i + \varphi_{ij,j} \rangle dt \\ &- \int_0^T \langle w^\varepsilon, f + \tau_{ij,ij} \rangle dt + \int_0^T \langle k_{ij}w^\varepsilon, \varphi_{ij} \rangle dt \le 0 . \end{aligned} \tag{6.27}$$

In view of the convergence of E^ε to E, $\underline{\lim}\ \|E^\varepsilon(T)\|_C \geq \|E(T)\|_C$. This, after passing in (6.27) to the lower limit as $\varepsilon \to 0$, gives the inequality

$$\int_0^T C(\dot{E}, \alpha - E)dt + \int_0^T \langle\, v_i, \varphi_{ij,j} - N_{ij,j}\,\rangle dt + \int_0^T \langle\, w, \tau_{ij,ij} - M_{ij,ij}\,\rangle dt$$
$$- \int_0^T \langle\, k_{ij}w, \varphi_{ij} - N_{ij}\,\rangle dt \geq 0 \quad \forall\, \alpha = (\varphi, \tau) \in L^2\big(0, T; K \cap W(\Omega)\big)$$

Here we use the fact that the limiting functions N, M satisfy equations (6.26).

Finally, it is easy to verify that $E(t) \in K$ for almost all $t \in (0,T)$. In fact, in view of the convexity of P, $P(E^n) \leq \langle\, P'(E^n), \bar{E}^n\,\rangle$. Thus, by the second inequality of (6.14), we obtain

$$\frac{1}{\varepsilon}\int_0^T \|E_h - \pi E_h\|^2 dt \leq c$$

with a constant c uniform with respect to ε. By the weak convergence $P'(E_h) \to P'(E^\varepsilon)$ in $L^2(Q)$,

$$\int_0^T \|E^\varepsilon - \pi E^\varepsilon\|^2 dt \leq c\varepsilon\ .$$

With this result, the standard argument implies the required conclusion, which completes the proof. □

Remark. 6.1. Variational inequality (6.3) is derived from the exact formulation of the problem by eliminating ξ_{ij}, λ_{ij} and with the help of the boundary conditions $v = w = w_\nu = 0$ on $\Gamma \times (0,T)$. Nevertheless, it is not automatically guaranteed that the solutions to the variational inequality satisfy the boundary condition $w_\nu = 0$ on $\Gamma \times (0,T)$. Since $L^1(\Omega) \subset \mathcal{M}^1(\Omega)$, it is clear that $w_{,ij} \in L^2\big(0,T;\mathcal{M}^1(\Omega)\big)$, which means that $\nabla w \in L^2\big(0,T;BD(\Omega)\big)$, and the boundary values of w_ν on $\Gamma \times (0,T)$ make sense. In particular, the function $w_\nu(t)$ is defined on Γ for almost all $t \in (0,T)$. The same is true for the function v, since it has been actually proved that $v \in L^2\big(0,T;BD(\Omega)\big)$.

Remark. 6.2. It is quite obvious that the set of functions $\{f, f_i, B\}$ providing the existence of solutions to (6.1)–(6.6) is nonempty. In fact, if the functions $\{f, f_i, B\}$ satisfy equations (6.1), (6.2), then $\{\varepsilon f, \varepsilon f_i, \varepsilon B\}$ satisfy them too, and, moreover, $(1+\kappa)(\varepsilon B)(t) \in K$, $t \in (0,T)$, for small ε.

On the other hand, suppose we are given functions f, f_i which are sufficiently regular and vanish for $t \in (0,\delta)$. Then we can choose arbitrary functions $N^0_{12} = N^0_{21}$ vanishing on $(0,\delta)$ and find N^0_{11}, N^0_{22} with the help of (6.1). Moreover, by taking arbitrary functions $M^0_{11}, M^0_{12} = M^0_{21}$ such that $M^0_{11}(t) = M^0_{12}(t) = 0$, $t \in (0,\delta)$, M^0_{22} can be derived from (6.2). Thus, it is evident that the functions $\{\varepsilon f, \varepsilon f_i, \varepsilon B\}$ satisfy all the required conditions for small ε.

Note also that the condition $B \in C^2(\bar{Q})$ can be weakened. Actually, to state the results the continuity of derivatives of B with respect to t is required up to order two.

Remark. 6.3. To get a priori estimate for w^ε, we can use another approach. Namely, it is enough to make use of (6.24) for the vector $(w^\varepsilon_{,1}, w^\varepsilon_{,2})$.

We claim that the function E is unique. Assume that v^1, w^1, E^1 and v^2, w^2, E^2 are two different solutions. For the first solution v^1, w^1, E^1, substitute into (6.3) the following function

$$\alpha(t) = \begin{cases} E^2(t) \ , & t \in (0, t_1) \\ E^1(t) \ , & t \in (t_1, T) \ . \end{cases}$$

Next, for the second solution v^2, w^2, E^2, substitute into (6.3) the following test function

$$\alpha(t) = \begin{cases} E^1(t) \ , & t \in (0, t_1) \\ E^2(t) \ , & t \in (t_1, T) \ . \end{cases}$$

It is clear that $E^1 = (N^1, M^1)$ and $E^2 = (N^2, M^2)$ satisfy equations (6.1), (6.2). Thus, summing up the resulting relations, for $E = E^1 - E^2$, we easily get

$$\int_0^{t_1} C(\dot{E}, E)dt \le 0 \ , \quad t_1 \in (0, T) \ .$$

Hence $E \equiv 0$.

7 Contact problem for the Kirchhoff plate

In this section the contact elastoplastic problem for a plate is examined within the framework of the Hencky model. We shall prove the solvability of a variational inequality which follows from the exact (original) formulation of the problem. The same will be proved for a perfectly plastic plate. Finally, we shall study the relationship between the two problems.

7.1 Elastoplastic problem

We seek functions w, $M = \{M_{ij}\}$, ξ_{ij} satisfying in a bounded domain $\Omega \subset \mathbb{R}^2$ with smooth boundary Γ the following system of equations and inequalitites

$$w - \psi \ge 0 \ , \quad (M_{ij,ij} + f)(\bar{w} - w) \le 0 \quad \forall\, \bar{w}, \bar{w} \ge \psi \ , \tag{7.1}$$

$$-w_{,ij} = c_{ijkl}M_{kl} + \xi_{ij} \ , \quad i, j = 1, 2 \ , \tag{7.2}$$

$$\Phi(M) \le 0 \ , \quad \xi_{ij}(\bar{M}_{ij} - M_{ij}) \le 0 \quad \forall\, \bar{M}, \Phi(\bar{M}) \le 0 \ , \tag{7.3}$$

and the boundary conditions

$$w = w_\nu = 0 \quad \text{on} \quad \Gamma \ . \tag{7.4}$$

The function $\Phi : \mathbb{R}^3 \mapsto \mathbb{R}$ describes the plastic yield condition. We assume that this function is convex and continuous. The given coefficients $c_{ijkl} \in L^\infty(\Omega)$ are symmetric and positive definite. The function ψ describes a punch shape, $\psi \in H^2(\Omega)$, $\psi\big|_\Gamma < 0$, $f \in L^2(\Omega)$ is given.

Relations (7.1)–(7.4) represent the unilateral boundary problem with the geometrical restriction $w - \psi \geq 0$ and the physical restriction $\Phi(M) \leq 0$.

These relations describe, as a particular case, the contact problem for an elastic plate. In fact, let $\Phi \equiv 0$, ie., there is no restriction on M. Then, $\xi_{ij} \equiv 0$ and we can calculate $M_{ij} = a_{ijkl}w_{,kl}$ from (7.2) and substitute it into (7.1). In this way, we derive the variational inequality of order four which describes the contact problem for an elastic plate. In particular, if the plate is isotropic we get the inequality for the operator

$$w \mapsto a\Delta^2 w - f \ ,$$

which has been already considered in Chapter 2. On the other hand, if there is no geometrical restriction, ie. $\psi \equiv -\infty$, relations (7.1) reduce to

$$-M_{ij,ij} = f \ , \tag{7.5}$$

and (7.1)–(7.4) correspond to the equilibrium problem for an elastoplastic plate (with no contact).

We now introduce some notation and formulate the main assumptions ensuring the existence of a solution. We assume that the set $\tilde{K} = \{M \in \mathbb{R}^3 \ \big| \ \Phi(M) \leq 0\}$ contains zero in its interior and there exists a solution M^0 to equation (7.5) such that $(1+\kappa)M^0 \in K$, where $\kappa > 0$ is a constant and

$$K = \left\{ M \in L^2(\Omega) \ \big| \ M(x) \in \tilde{K} \quad \text{a.e. in} \quad \Omega \right\} \ .$$

As before, the space of bounded measures defined on Ω is denoted by $\mathcal{M}^1(\Omega)$. For convenience we introduce the space of functions

$$V_1(\Omega) = \left\{ M \in L^2(\Omega) \ \big| \ M_{ij,ij} \in L^2(\Omega) \right\} \ .$$

Observe that in absence of appropriate estimates the existence of smooth solutions to (7.1)–(7.4) is difficult to prove. This is, in particular, a motivation for introducing the space $\mathcal{M}^1(\Omega)$.

Moreover, it is difficult to give any reasonable interpretation to the term $M_{ij,ij}w$ of (7.1). This term is eliminated in the following way. We substitute $\xi_{ij} =$

$-c_{ijkl}M_{kl} - w_{,ij}$ into (7.3), then sum it up with the second inequality of (7.1) and integrate over Ω. This gives

$$\begin{gathered} w - \psi \geq 0 \ , \ \ \Phi(M) \leq 0 \ , \\ \int_\Omega \big(\bar{M}_{ij,ij}w - M_{ij}\bar{w}_{,ij} + c_{ijkl}M_{kl}\bar{M}_{ij} - c_{ijkl}M_{kl}M_{ij} \\ + f(w - \bar{w})\big)dx \geq 0 \\ \forall\, \bar{w}, \bar{w} \geq \psi \ , \quad \forall\, \bar{M}, \Phi(\bar{M}) \leq 0 \ . \end{gathered} \tag{7.6}$$

We assume here that test functions $\bar{w}$ satisfy boundary conditions (7.4).

Variational inequality (7.6) is equivalent to the relations resulting from (7.1)–(7.4) after eliminating ξ_{ij},

$$\begin{gathered} w - \psi \geq 0 \ , \ \ \Phi(M) \leq 0 \ , \\ \int_\Omega (M_{ij,ij} + f)(\bar{w} - w)dx \leq 0 \quad \forall\, \bar{w}, \bar{w} \geq \psi \ , \\ \int_\Omega c_{ijkl}M_{kl}(\bar{M}_{ij} - M_{ij})dx + \int_\Omega w\, (\bar{M}_{ij,ij} - M_{ij,ij})dx \geq 0 \\ \forall\, \bar{M}, \Phi(\bar{M}) \leq 0 \ . \end{gathered}$$

In order to verify these relations, we add and subtract $M_{ij,ij}w$ from (7.6) and next use the following simple fact.

If the variational inequality

$$\begin{gathered} u \in K_1 \ , \ \ v \in K_2 \ : \\ (L_1, \bar{u} - u) + (L_2, \bar{v} - v) \geq 0 \quad \forall\, \bar{u} \in K_1 \ , \forall\, \bar{v} \in K_2 \end{gathered}$$

holds, then the variational inequalities

$$(L_1, \bar{u} - u) \geq 0 \ \forall\, \bar{u} \in K_1 \ ; \quad (L_2, \bar{v} - v) \geq 0 \ \forall\, \bar{v} \in K_2$$

also hold. The converse is also valid.

Theorem 7.1. *Let the assumptions formulated above hold. Then there exists a solution of variational inequality (7.6) such that*

$$M_{ij} \in L^2(\Omega) \ , \ \ w \in H_0^1(\Omega) \ , \ \ w_{,ij} \in \mathcal{M}^1(\Omega) \ .$$

Proof. Put $p(w) = -(w - \psi)^-$. Let π be the operator of the orthogonal projection of $\mathbb{R}^3$ onto the set $\tilde{K}$ and $q(M) = M - \pi M$.

Consider the regularized boundary value problem involving three positive parameters $\varepsilon, \delta, \lambda$. The problem is to find functions w, M_{ij} such that

$$\delta \Delta^2 w - M_{ij,ij} + \lambda^{-1} p(w) = f \ , \tag{7.7}$$

$$c_{ijkl}M_{kl} + w_{,ij} + \varepsilon^{-1}q(M)_{ij} = 0 \ , \quad i,j = 1,2 \ , \tag{7.8}$$

$$w = w_\nu = 0 \quad \text{on} \quad \Gamma \ . \tag{7.9}$$

For notational simplicity the dependence of w, M_{ij} on the parameters is omitted here. It is an immediate consequence of Lemma 1.8 that problem (7.7)–(7.9) has a solution such that $w \in H_0^2(\Omega)$, $M \in L^2(\Omega)$.

It is more convenient for us to verify the coercivity and monotonicity of the operator after having established a priori estimate. Thus, we skip that over at this point.

We now derive estimates for solutions to problem (7.7)–(7.9) which explain the character of dependence on the parameters.

We choose a function $w^0 \in H_0^2(\Omega)$ such that $w^0 \geq \psi$ in Ω. We multiply (7.7) and (7.8) by $w - w^0$ and $M_{ij} - M_{ij}^0$, respectively, integrate over Ω and sum up. Taking into account the equation for M^0, we easily derive

$$\begin{aligned} &\delta\|\Delta w\|^2 + C(M,M) + \lambda^{-1}\langle p(w), w - w^0\rangle + \varepsilon^{-1}\langle q(M), M - M^0\rangle \\ &\quad = -\langle\, f, w^0 \,\rangle - \langle\, M_{ij}, w^0_{,ij} \,\rangle + C(M, M^0) + \delta\langle\, \Delta w, \Delta w^0 \,\rangle \ . \end{aligned} \tag{7.10}$$

Here $C(M, M^0) = \langle\, c_{ijkl}M_{kl}, M_{ij}^0 \,\rangle$. The terms containing the parameters λ, ε are nonnegative, hence from (7.10) it follows that

$$\delta \ \|\Delta w\|^2 + \|M\|^2 \leq c \ . \tag{7.11}$$

Here the constant in the right-hand side does not depend on $\lambda, \varepsilon, \delta \leq \delta_0$. Besides, from (7.10) it follows that

$$\varepsilon^{-1}\langle\, q(M), M - M^0 \,\rangle \leq c \tag{7.12}$$

uniformly with respect to $\lambda, \varepsilon, \delta \leq \delta_0$.

From (7.12) the boundedness of $\varepsilon^{-1}q(M)$ in the space $L^1(\Omega)$ follows. In fact, let $P(M) = \dfrac{1}{2\varepsilon} \ \|M - \pi M\|^2$. By Lemma 1.2, $P'(M) = \dfrac{1}{\varepsilon} \ q(M)$, where $P'(M)$ is the derivative of the functional P at M. By requirements imposed on M^0, we have $M^0 + N \in K$ provided $\|N\|_{L^\infty(\Omega)} \leq \alpha$ and α is small enough. In particular, $P(M^0 + N) = 0$. The convexity of the functional P implies that $\langle\, P'(M), N \,\rangle \leq \langle\, P'(M), M - M^0 \,\rangle + P(M^0 + N) - P(M)$, where the right-hand side is bounded and nonnegative. Thus, $\langle\, P'(M), N \,\rangle \leq c$ and

$$\varepsilon^{-1} \ \|q(M)\|_{L^1(\Omega)} \leq c \ . \tag{7.13}$$

Estimate (7.13) is, as the previous one, uniform with respect to $\lambda, \varepsilon, \delta \leq \delta_0$.

Equations (7.7), (7.8) are satisfied as identities,

$$\begin{aligned} &\delta\langle\, \Delta w, \Delta\tilde{w} \,\rangle - \langle\, M_{ij}, \tilde{w}_{,ij} \,\rangle + \lambda^{-1}\langle\, p(w), \tilde{w} \,\rangle = \langle\, f, \tilde{w} \,\rangle \quad \forall\, \tilde{w} \in H_0^2(\Omega) \ , \\ &C(M, \tilde{M}) + \langle\, w_{,ij}, \tilde{M}_{ij} \,\rangle + \varepsilon^{-1}\langle\, q(M), \tilde{M} \,\rangle = 0 \quad \forall\, \tilde{M} \in L^2(\Omega) \ . \end{aligned}$$

Above we have multiplied (7.7), (7.8) by $w - w^0$ and $M_{ij} - M_{ij}^0$, respectively, and integrated over Ω. This means that the functions $w - w^0$, $M_{ij} - M_{ij}^0$ are substituted into these identities as trial functions. From (7.8) and (7.13) it follows that $w_{,ij}$ are bounded in $L^1(\Omega)$ uniformly with respect to λ, ε, $\delta \le \delta_0$.

The inequality

$$\|\varphi\| \le c \sum_{i=1}^{2} \|\varphi_{,i}\|_{L^1(\Omega)}$$

holds for an arbitrary smooth function φ vanishing on Γ (compare with (6.23)). Hence, $w_{,i}$ are bounded in $L^2(\Omega)$. Consequently, in view of boundary conditions (7.9), we readily obtain

$$\|w\|_1 \le c \tag{7.14}$$

with a constant uniform with respect to λ, ε, $\delta \le \delta_0$. Moreover, since the space $L^1(\Omega)$ is continuously imbedded in $\mathcal{M}^1(\Omega)$, (7.8) gives

$$\sum_{i,j=1}^{2} \|w_{,ij}\|_{\mathcal{M}^1(\Omega)} \le c \ . \tag{7.15}$$

We now pass to the limit by letting the parameters tend to zero. At each step we indicate the dependence of the solution on the corresponding parameter and we omit other parameters.

We first analyse the passage to the limit as $\delta \to 0$. By estimates (7.11), (7.14), (7.15), there exists a sequence w^δ, M^δ such that

$$M^\delta \to M^\varepsilon \quad \text{weakly in} \quad L^2(\Omega) \ ,$$

$$\delta w^\delta \to 0 \quad \text{strongly in} \quad H_0^2(\Omega) \ ,$$

$$w^\delta \to w^\varepsilon \quad \text{weakly in } H_0^1(\Omega) \ , \quad \text{strongly in} \quad L^2(\Omega) \ ,$$

$$w_{,ij}^\delta \to w_{,ij}^\varepsilon \quad \text{weakly-}(*) \text{ in } \mathcal{M}^1(\Omega), \quad i,j = 1,2 \ .$$

By the continuity of the operator p from $L^2(\Omega)$ into $L^2(\Omega)$,

$$p(w^\delta) \to p(w^\varepsilon) \quad \text{strongly in} \quad L^2(\Omega) \ . \tag{7.16}$$

This allows us to pass to the limit in (7.7) as $\delta \to 0$. In fact, the equation

$$-M_{ij,ij}^\varepsilon + \lambda^{-1} p(w^\varepsilon) = f \tag{7.17}$$

is obviously satisfied. Equations (7.8) can be rewritten in the form

$$c_{ijkl} M_{kl}^\varepsilon + w_{,ij}^\varepsilon + \varepsilon^{-1} q(M^\varepsilon)_{ij} = 0 \ , \quad i,j = 1,2 \tag{7.18}$$

provided $q(M^\delta) \to q(M^\varepsilon)$ weakly in $L^2(\Omega)$.

We now prove this convergence, by using the monotonicity of the operator q. First of all, it is clear that $q(M^\delta)$ are bounded in the space $L^2(\Omega)$. Hence, choosing a subsequence, if necessary, we can assume that

$$q(M^\delta) \to \chi \quad \text{weakly in} \quad L^2(\Omega) \ .$$

as $\delta \to 0$.

We now show that $q(M^\varepsilon) = \chi$. A formal passage to the limit in (7.8) as $\delta \to 0$ implies that

$$c_{ijkl} M^\varepsilon_{kl} + w^\varepsilon_{,ij} + \varepsilon^{-1} \chi_{ij} = 0 \ , \quad i,j = 1,2 \ . \tag{7.19}$$

From equations (7.8) it follows that

$$\varepsilon^{-1} \langle \, q(M^\delta), M^\delta \, \rangle = -C(M^\delta, M^\delta) - \langle \, w^\delta_{,ij}, M^\delta_{ij} \, \rangle \ . \tag{7.20}$$

In view of (7.19), in a similar way we get

$$C(M^\varepsilon, M^\varepsilon) + \langle \, w^\varepsilon_{,ij}, M^\varepsilon_{ij} \, \rangle + \varepsilon^{-1} \langle \, \chi_{ij}, M^\varepsilon_{ij} \, \rangle = 0 \ . \tag{7.21}$$

Multiplying (7.7) by w^δ, and next integrating it we calculate the last term in the right-hand side of (7.20), namely we have

$$\begin{aligned} \langle \, w^\delta_{,ij}, M^\delta_{ij} \, \rangle =& \lambda^{-1} \, \langle \, p(w^\delta), w^\delta \, \rangle - \langle \, f, w^\delta \, \rangle \\ &+ \delta \, \langle \, \Delta w^\delta, \Delta w^\delta \, \rangle \ . \end{aligned} \tag{7.22}$$

Substituting this term into (7.20) and passing to the upper limit we obtain

$$\overline{\lim_{\delta \to 0}} \, \langle \, q(M^\delta), M^\delta \, \rangle \le \langle \, \chi, M^\varepsilon \, \rangle \ . \tag{7.23}$$

This is a consequence of (7.21), the weak lower semicontinuity of the norm $\| \cdot \|$, and equation (7.17).

It follows from the monotonicity of q that $\langle \, q(M^\delta) - q(M). M^\delta - M \, \rangle \ge 0$ for any $M \in L^2(\Omega)$. Taking into account (7.23) and passing to the upper limit, we obtain $\langle \, \chi - q(M), M^\varepsilon - M \, \rangle \ge 0$.

Put $M = M^\varepsilon + \alpha N$, $N \in L^2(\Omega)$, $\alpha > 0$. Dividing it by α and passing to the limit as $\alpha \to 0$ we get $\langle \, \chi - q(M^\varepsilon), N \, \rangle \ge 0 \ \forall \ N \in L^2(\Omega)$. Since N is arbitrary, the required equality $\chi = q(M^\varepsilon)$ follows.

We proceed to show that we can pass to the limit when $\varepsilon \to 0$. Assume that a subsequence of $w^\varepsilon, M^\varepsilon$, still denoted by $w^\varepsilon, M^\varepsilon$, possesses the property

$$\begin{gathered} w^\varepsilon \to w^\lambda \quad \text{weakly in} \quad H^1_0(\Omega) \ , \ \text{strongly in} \quad L^2(\Omega) \ , \\ w^\varepsilon_{,ij} \to w^\lambda_{,ij} \quad \text{weakly-}(*) \text{ in} \quad \mathcal{M}^1(\Omega) \ , \ i,j = 1,2 \ , \\ M^\varepsilon \to M^\lambda \quad \text{weakly in} \quad L^2(\Omega) \ . \end{gathered} \tag{7.24}$$

We multiply equations (7.17), (7.18) by $\varepsilon w^\varepsilon$, M^ε_{ij}, respectively, integrate them over Ω and sum them up. This implies $\langle \, q(M^\varepsilon), M^\varepsilon \, \rangle \to 0$.

With this fact, it is evident that $q(M^\lambda) = 0$, ie., $\Phi(M^\lambda) \le 0$ almost everywhere in Ω. In fact, from (7.18) it follows that $q(M^\varepsilon) \to 0$ weakly in $H^{-1}(\Omega)$. On the other hand, in view of the boundedness of $q(M^\varepsilon)$ in $L^2(\Omega)$, $q(M^\varepsilon) \to \chi$ weakly in $L^2(\Omega)$. Thus, $\chi = 0$.

Passing to the limit when $\varepsilon \to 0$ the inequality $\langle\, q(M^\varepsilon) - q(M), M^\varepsilon - M \,\rangle \ge 0$ implies that $\langle\, q(M), M^\lambda - M \,\rangle \le 0$, for all $M \in L^2(\Omega)$.

Set $M = M^\lambda + \alpha N$, $N \in L^2(\Omega)$, divide it by α and pass to the limit when $\alpha \to 0$. This yields $q(M^\lambda) = 0$ as required.

Multiply (7.18) by $\bar{M}_{ij} - M^\varepsilon_{ij}$, sum it up over i, j and integrate over Ω. The term containing the parameter ε is nonpositive for $\bar{M} \in K$. Thus,

$$C(M^\varepsilon, \bar{M} - M^\varepsilon) + \langle\, w^\varepsilon, \bar{M}_{ij,ij} - M^\varepsilon_{ij,ij} \,\rangle \ge 0 \quad \forall\, \bar{M} \in K \cap V_1(\Omega) \ . \tag{7.25}$$

From equation (7.17) it follows that $M^\varepsilon_{ij,ij}$ are bounded in $L^2(\Omega)$ (in general, nonuniformly with respect to λ), hence,

$$M^\varepsilon_{ij,ij} \to M^\lambda_{ij,ij} \quad \text{weakly in} \quad L^2(\Omega) \ . \tag{7.26}$$

Letting $\varepsilon \to 0$, in view of (7.24), (7.26), we can pass to the limit in (7.25). As a result, we obtain

$$\begin{gathered} C(M^\lambda, \bar{M} - M^\lambda) + \langle\, w^\lambda, \bar{M}_{ij,ij} - M^\lambda_{ij,ij} \,\rangle \ge 0 \\ \forall\, \bar{M} \in K \cap V_1(\Omega) \ . \end{gathered} \tag{7.27}$$

We can also pass to the limit in (7.17) when $\varepsilon \to 0$. Hence

$$-M^\lambda_{ij,ij} + \lambda^{-1} p(w^\lambda) = f \ . \tag{7.28}$$

Finally, we analyse a passage to the limit when $\lambda \to 0$. We choose a subsequence of w^λ, M^λ, still denoted by w^λ, M^λ, such that

$$\begin{gathered} w^\lambda \to w \quad \text{weakly in} \quad H^1_0(\Omega) \ , \quad \text{strongly in} \quad L^2(\Omega) \ , \\ M^\lambda \to M \quad \text{weakly in} \quad L^2(\Omega) \ , \\ w^\lambda_{,ij} \to w_{,ij} \quad \text{weakly-}(*) \text{ in} \quad \mathcal{M}^1(\Omega) \ , \quad i,j = 1,2 \end{gathered}$$

as $\lambda \to 0$. Then $p(w^\lambda) \to p(w)$ strongly in $L^2(\Omega)$. Moreover, the estimate of the functions M^λ implies the boundedness of $M^\lambda_{ij,ij}$ in $H^{-2}(\Omega)$. Hence, as can be seen from (7.28), $p(w^\lambda) \to 0$ strongly in $H^{-2}(\Omega)$. Thus, $p(w) = 0$, and consequently $w \ge \psi$ in Ω.

Let us now multiply (7.28) by $\bar{w} - w^\lambda$, integrate it over Ω and sum up with (7.27). The following inequality results

$$C(M^\lambda, \bar{M} - M^\lambda) + \langle\, w^\lambda, \bar{M}_{ij,ij} \,\rangle - \langle\, M^\lambda_{ij}, \bar{w}_{,ij} \,\rangle - \langle\, f, \bar{w} - w^\lambda \,\rangle \ge 0 \ ,$$

where $\bar{w} \in H^2_0(\Omega)$, $\bar{w} \ge \psi$, $\bar{M} \in K \cap V_1(\Omega)$. As noted above, we can pass to the upper limit here and, therefore, inequality (7.6) follows. It was shown above that $\Phi(M^\lambda) \le 0$. The set K is closed and convex in $L^2(\Omega)$, hence, weakly closed. Consequently, the limiting element M satisfies the inequality $\Phi(M) \le 0$. □

7.2 The perfectly plastic problem

The formulation of the perfectly plastic problem results from (7.2) by putting the coefficients c_{ijkl} equal to zero. The problem is to find functions satisfying the following equations and inequalities

$$w - \psi \geq 0 \ , \quad (M_{ij,ij} + f)(\bar{w} - w) \leq 0 \quad \forall\, \bar{w}, \bar{w} \geq \psi \ .$$

$$\Phi(M) \leq 0 \ , \quad -w_{,ij}(\bar{M}_{ij} - M_{ij}) \leq 0 \quad \forall\, \bar{M}, \Phi(\bar{M}) \leq 0 \ ,$$

$$w = w_\nu = 0 \quad \text{on} \quad \Gamma \ .$$

Note that the terms $c_{ijkl}M_{kl}$ have been essentially used in Theorem 7.1 to derive a priori estimates.

In the case where $c_{ijkl} = 0$ an estimate of M can be obtained for some particular functions Φ. In showing this we need the assumption that the set

$$\tilde{K} = \{M \mid \Phi(M) \leq 0\} \tag{7.29}$$

is bounded in $\mathbb{R}^3$. This assumption is not particularily restrictive in applications, since it is always fulfilled for practical problems [see Erkhov, 1978].

Theorem 7.2. *Let the assumptions of Theorem 7.1 hold and the set $\tilde{K}$ be bounded in $\mathbb{R}^3$. Then there exists a solution to inequality (7.6) with $c_{ijkl} = 0$.*

Proof. Consider the regularized problem (7.7)–(7.9) for $c_{ijkl} = 0$. A priori estimates of solutions to this problem have the form

$$\delta\, \|\Delta w\|^2 \leq c \ , \tag{7.30}$$

$$\|M\|^2 \leq c \ . \tag{7.31}$$

These estimates can be obtained from (7.7)–(7.9) similarily as in Theorem 7.1. Note that now πM are bounded in $L^2(\Omega)$. The boundedness of M in the space $L^2(\Omega)$ is guaranteed by the presence of the terms $\dfrac{1}{\varepsilon} M_{ij}$ in equations (7.8).

The rest of the proof runs as in Theorem 7.1. First, in addition to (7.30), (7.31) we obtain for the function w estimates of the type (7.14), (7.15) which are uniform with respect to the parameters.

Second, we pass to the limit in the same way as before. Finally, we obtain the inequality

$$\begin{aligned} & w - \psi \geq 0 \ , \quad \Phi(M) \leq 0 \ , \\ & \langle\, w, \bar{M}_{ij,ij} \,\rangle - \langle\, M_{ij}, \bar{w}_{,ij} \,\rangle - \langle\, f, \bar{w} - w \,\rangle \geq 0 \ , \\ & \forall\, \bar{w} \in H_0^2(\Omega) \ , \quad \bar{w} \geq \psi \ , \quad \forall\, \bar{M} \in K \cap V_1(\Omega) \end{aligned} \tag{7.32}$$

where

$$M_{ij} \in L^2(\Omega) \ , \quad w \in H_0^1(\Omega) \ , \quad w_{,ij} \in \mathcal{M}^1(\Omega) \ .$$

This completes the proof. □

Observe that a solution to the perfectly plastic problem can be regarded as the limit of solutions to some elastoplastic problems provided the set $\tilde{K}$ is bounded in $\mathbb{R}^3$.

To see this, consider problem (7.7)–(7.9), where $\delta_1 c_{ijkl}$ are substituted instead of c_{ijkl}, $\delta_1 > 0$. By Theorem 7.1, for any fixed δ_1, a solution $w^{\delta_1}, M^{\delta_1}$ can be found. Moreover, the following estimate

$$\delta \, \|\Delta w\|^2 + \|M\|^2 + \|w\|_1^2 + \sum_{i,j=1}^{2} \|w_{,ij}\|_{\mathcal{M}^1(\Omega)} \leq c \tag{7.33}$$

holds uniformly with respect to $\delta_1, \lambda, \varepsilon$, $\delta \leq \delta_0$. Passing to the limit when $\delta \to 0$, $\varepsilon \to 0$, $\lambda \to 0$ we maintain estimate (7.33) for M. Thus, after passing to the limit as $\lambda \to 0$ we get

$$\delta_1 C(M^{\delta_1}, \bar{M} - M^{\delta_1}) + \langle\, w^{\delta_1}, \bar{M}_{ij,ij} \,\rangle - \langle\, M_{ij}^{\delta_1}, \bar{w}_{,ij} \,\rangle$$
$$- \langle\, f, \bar{w} - w^{\delta_1} \,\rangle \geq 0 \quad \forall\, \bar{w} \in H_0^2(\Omega), \;\; \bar{w} \geq \psi, \;\; \forall\, \bar{M} \in K \cap V_1(\Omega) \; .$$

It is easily seen, that we can here pass to the limit as $\delta_1 \to 0$. This gives (7.32) which is our assertion.

Remark. 7.1. In the general case, solutions to inequalities (7.6), (7.32) do not satisfy boundary condition $w_\nu = 0$ on Γ. However, $\nabla w \in BD(\Omega)$ and, therefore, the function w_ν is defined on Γ in the sense of $L^1(\Gamma)$.

8 Contact problem for the Timoshenko beam

Let us consider the elastoplastic contact problem for the Timoshenko model of a beam. This model has some advantages over the Kirchhoff model. For instance, in this model bending moments are smooth functions. This enables us to formulate the existence theorem for the exact formulation of the problem. In particular, the boundary conditions both for displacements and bending moments will be fulfilled.

The problem is to find functions w, m, ξ such that on the interval $(0,1)$ the following conditions are satisfied

$$\psi_1 \leq w \leq \psi_2 \; , \tag{8.1}$$

$$(m_{xx} + f)(\bar{w} - w) \leq 0 \quad \forall\, \bar{w}, \;\; \psi_1 \leq \bar{w} \leq \psi_2 \; , \tag{8.2}$$

$$-w_{xx} = m - m_{xx} + \xi \; , \tag{8.3}$$

$$|m| \leq c_0 \; , \tag{8.4}$$

$$\xi(\bar{m} - m) \leq 0 \quad \forall\, \bar{m}, \;\; |\bar{m}| \leq c_0 \; , \tag{8.5}$$

$$w = 0, \;\; m = 0 \quad \text{when} \quad x = 0, 1 \; . \tag{8.6}$$

Here $\psi_1(x), \psi_2(x)$ are given functions describing the punch shape, $w(x)$ is the beam displacement, $m(x)$ is the bending moment. The problem contains geometrical restrictions (8.1) imposed upon the beam displacement and physical restriction (8.4) imposed upon the moment; $c_0 > 0$ is a constant, $f \in L^2(0,1)$.

By dropping restriction (8.1) we obtain the problem of the beam equilibrium without contact condition, and by dropping restriction (8.4) we obviously obtain the contact problem for an elastic beam.

We now formulate basic assumptions under which the existence theorem will be proved. Let

$$K = \{m \in H_0^1(0,1) \mid |m(x)| \le c_0 \quad \text{on} \quad (0,1)\} .$$

We assume that there exists a function m^0 such that

$$-m^0_{xx} = f \tag{8.7}$$

and $(1+\kappa)m^0 \in K$, where $\kappa > 0$ is a constant.

Let $\psi_1, \psi_2 \in H^1(0,1)$. The pair (ψ_1, ψ_2) will be denoted by ψ. We assume that there exists a function $w^0 \in H_0^1(0,1)$ satisfying the inequalities

$$\psi_1(x) \le w^0(x) \le \psi_2(x) \quad \text{on} \quad (0,1) . \tag{8.8}$$

Put

$$B_\psi = \{w \in H_0^1(0,1) \mid \psi_1(x) \le w(x) \le \psi_2(x) \text{ on } (0,1)\} .$$

The next theorem establishes the solvability of problem (8.1)–(8.6).

Theorem 8.1. *Let the assumptions formulated above hold. Then there exist functions w, m, ξ which satisfy the relations*

$$w \in B_\psi, \quad m \in K, \quad \xi \in H^{-1}(0,1) , \tag{8.9}$$

$$\langle m_{xx} + f, \bar{w} - w \rangle \le 0 \quad \forall \, \bar{w} \in B_\psi , \tag{8.10}$$

$$m - m_{xx} + w_{xx} + \xi = 0 , \tag{8.11}$$

$$\langle \xi, \bar{m} - m \rangle \le 0 \quad \forall \, \bar{m} \in K . \tag{8.12}$$

Proof. Let π be the operator of orthogonal projection of $L^2(0,1)$ onto the set $\{m \in L^2(0,1) \mid |m| \le c_0\}$.

Denote $p(m) = m - \pi m$ and put

$$q(w)(x) = \begin{cases} w(x) - \psi_2(x), & \text{if} \quad w(x) > \psi_2(x) \\ 0, & \text{if} \quad \psi_1(x) \le w(x) \le \psi_2(x) \\ w(x) - \psi_1(x), & \text{if} \quad w(x) < \psi_1(x) . \end{cases}$$

Let $\varepsilon, \delta, \lambda$ be positive parameters. We consider the regularized problem (without indicating explicitely the dependence of solutions on these parameters)

$$-\varepsilon w_{xx} - m_{xx} + \frac{1}{\delta}\, q(w) = f \ , \tag{8.13}$$

$$m - m_{xx} + w_{xx} + \frac{1}{\lambda}\, p(m) = 0 \ , \tag{8.14}$$

$$w = 0, \ \ m = 0 \quad \text{when} \quad x = 0, 1 \ . \tag{8.15}$$

To obtain a priori estimates for solutions to (8.13)–(8.15), we multiply (8.13) by $w - w^0$ and (8.14) by $m - m^0$, where w^0, m^0 satisfy (8.7) and (8.8). Then we integrate the resulting relations over $(0,1)$ and sum them up. By the monotonicity of the operators p, q,

$$\varepsilon \, \|w_x\|^2 + \|m\|^2 + \|m_x\|^2 + \frac{1}{\lambda} \, \langle \, p(m), m - m^0 \, \rangle \le c \ . \tag{8.16}$$

The constant in the right-hand side of (8.16) is uniform with respect to $\varepsilon \le \varepsilon_0$, δ, λ. Thus,

$$\frac{1}{\lambda} \, \langle \, p(m), m - m^0 \, \rangle \le c \ , \quad \frac{1}{\lambda} \, \langle \, p(m), \bar{m} - m \, \rangle \le 0 \ .$$

The second inequality follows from the monotonicity of the operator p and holds for every $\bar{m} \in K$. Summing up these inequalities, we easily obtain $\frac{1}{\lambda} \, \langle \, p(m), \bar{m} - m^0 \, \rangle \le c$. By our assumptions, the element $\bar{m} = m^0 + m^1$ belongs to the set K provided $m^1 \in H_0^1(0,1)$, $|m^1| \le \dfrac{c_0 \kappa}{1+\kappa}$. Therefore,

$$\frac{1}{\lambda} \, \langle \, p(m), m^1 \, \rangle \le c \quad \forall \, m^1 \in H_0^1(0,1), \ \ |m^1| \le \frac{c_0 \kappa}{1+\kappa} \ .$$

Consequently, $\frac{1}{\lambda}\, p(m)$ are bounded in $H^{-1}(0,1)$ uniformly with respect to $\varepsilon \le \varepsilon_0$, $\delta > 0$, $\lambda > 0$. Taking into account the boundary conditions, from estimate (8.16) it follows that m are bounded in $H_0^1(0,1)$. In view of the boundedness of $\frac{1}{\lambda}\, p(m)$, the boundedness of w in $H_0^1(0,1)$ follows from (8.14). Thus, a priori estimate of a solution has the form

$$\|w\|_1 + \|m\|_1 + \frac{1}{\lambda} \, \|p(m)\|_{H^{-1}(0,1)} \le c \ . \tag{8.17}$$

A priori estimate (8.16) actually shows that for fixed $\varepsilon, \delta, \lambda$ the operator of (8.13)–(8.15) satisfies the coercivity conditions of Lemma 1.8. Accordingly, a solution $w, m \in H_0^1(0,1)$ to problem (8.13)–(8.15) exists for fixed parameters $\varepsilon, \delta, \lambda$.

Monotonicity and semicontinuity of the operator follow from the respective properties of the operators p, q. At the same time, a priori estimate (8.17) provides an additional information about solutions which will be exploited in analysing passages to the limit.

We pass to the limit by letting first $\varepsilon \to 0$, next $\lambda \to 0$, and finally $\delta \to 0$. Each time, the dependence of the solution upon the corresponding parameter is indicated.

We first pass to the limit as $\varepsilon \to 0$. By (8.16), (8.17), there exists a sequence denoted by $w^\varepsilon, m^\varepsilon$ possessing the properties

$$w^\varepsilon, m^\varepsilon \to w^\lambda, m^\lambda \text{ weakly in } H_0^1(0,1), \text{ strongly in } L^2(0,1) \ .$$

$$\varepsilon w^\varepsilon \to 0 \quad \text{weakly in} \quad H_0^1(0,1) \ .$$

Then, by the continuity of the operators p and q acting from $L^2(0,1)$ into $L^2(0,1)$, we get the convergence $p(m^\varepsilon) \to p(m^\lambda)$, $q(w^\varepsilon) \to q(w^\lambda)$ in the space $L^2(0,1)$.

The limiting system of equations (8.13), (8.14) has the form

$$-m_{xx}^\lambda + \frac{1}{\delta}\, q(w^\lambda) = f \ , \tag{8.18}$$

$$m^\lambda - m_{xx}^\lambda + w_{xx}^\lambda + \frac{1}{\lambda}\, p(m^\lambda) = 0 \ . \tag{8.19}$$

Now we pass to the limit as $\lambda \to 0$. The limiting functions w^λ, m^λ will satisfy the same estimates as $w^\varepsilon, m^\varepsilon$.

Let us choose from w^λ, m^λ a subsequence, still denoted by w^λ, m^λ, such that

$$\begin{gathered} m^\lambda \to m^\delta \quad \text{weakly in} \quad H_0^1(0,1) \ , \\ w^\lambda \to w^\delta \text{ weakly in } H_0^1(0,1) \ , \quad \text{strongly in } L^2(0,1) \ , \\ \frac{1}{\lambda}\, p(m^\lambda) \to \xi^\delta \quad \text{weakly in} \quad H^{-1}(0,1) \end{gathered} \tag{8.20}$$

as $\lambda \to 0$. Multiply (8.19) by $\bar{m} - m^\lambda$. Dropping the nonpositive term $\frac{1}{\lambda}\langle\, p(m^\lambda), \bar{m} - m^\lambda \,\rangle$, we get

$$\langle\, m^\lambda - m_{xx}^\lambda + w_{xx}^\lambda, \bar{m} - m^\lambda \,\rangle \geq 0 \quad \forall\, \bar{m} \in K \ .$$

Multiplying (8.18) by w^λ we find the value $\langle\, w_{xx}^\lambda, m^\lambda \,\rangle \equiv -\langle\, w_x^\lambda, m_x^\lambda \,\rangle$. Thus,

$$\langle\, m^\lambda - m_{xx}^\lambda, \bar{m} - m^\lambda \,\rangle + \langle\, w_{xx}^\lambda, \bar{m} \,\rangle + \langle\, f - \frac{1}{\delta}\, q(w^\lambda), w^\lambda \,\rangle \geq 0 \quad \forall\, \bar{m} \in K \ .$$

Taking into account that $\varliminf_{\lambda \to 0} \|m_x^\lambda\|^2 \geq \|m_x^\delta\|^2$, and passing to the limit we obtain

$$\begin{gathered} \langle\, m^\delta - m_{xx}^\delta, \bar{m} - m^\delta \,\rangle + \langle\, w_{xx}^\delta, \bar{m} \,\rangle \\ + \langle\, f - \frac{1}{\delta}\, q(w^\delta), w^\delta \,\rangle \geq 0 \quad \forall\, \bar{m} \in K \ . \end{gathered} \tag{8.21}$$

Passing to the limit in (8.18), we get

$$-m^\delta_{xx} + \frac{1}{\delta}\, q(w^\delta) = f \ , \tag{8.22}$$

and thus, (8.21) may be written in the form

$$\langle\, m^\delta - m^\delta_{xx} + w^\delta_{xx}, \bar{m} - m^\delta \,\rangle \geq 0 \quad \forall\, \bar{m} \in K \ . \tag{8.23}$$

From (8.19) it follows that

$$m^\delta - m^\delta_{xx} + w^\delta_{xx} + \xi^\delta = 0 \ ,$$

and finally,

$$\langle\, \xi^\delta, \bar{m} - m^\delta \,\rangle \leq 0 \quad \forall\, \bar{m} \in K \ .$$

It can be easily verified that, by (8.19), $m^\delta \in K$.

We now pass to the limit as $\delta \to 0$. As above, from $w^\delta, m^\delta, \xi^\delta$ we can choose a subsequence, still denoted by $w^\delta, m^\delta, \xi^\delta$, such that

$$w^\delta, m^\delta \to w, m \quad \text{weakly in} \quad H^1_0(0,1) \ ,$$

$$\xi^\delta \to \xi \quad \text{weakly in} \quad H^{-1}(0,1)$$

as $\delta \to 0$. In view of the monotonicity of the operator q, from (8.22) it follows that

$$-\langle\, m^\delta_{xx} + f, \bar{w} - w^\delta \,\rangle \geq 0 \quad \forall\, \bar{w} \in B_\psi \ .$$

Adding this inequality to (8.23) and passing to the limit in the resulting relation we get

$$\begin{aligned} -\langle\, m_{xx} + f, \bar{w} \,\rangle + \langle\, f, w \,\rangle + \langle\, m - m_{xx}, \bar{m} - m \,\rangle \\ + \langle\, w_{xx}, \bar{m} \,\rangle \geq 0 \quad \forall\, \bar{w} \in B_\psi, \ \ \forall\, \bar{m} \in K \ . \end{aligned} \tag{8.24}$$

If we add $\langle\, m_x, w_x \,\rangle$ to (8.24) and next subtract it, we obtain the inequality

$$\begin{gathered} w \in B_\psi, \ \ m \in K \ : \\ \langle\, m_{xx} + f, w - \bar{w} \,\rangle + \langle\, m - m_{xx} + w_{xx}, \bar{m} - m \,\rangle \geq 0 \\ \forall\, \bar{w} \in B_\psi \ , \quad \forall\, \bar{m} \in K \ , \end{gathered}$$

which is, in turn, equivalent to the following relations

$$w \in B_\psi \ : \ \langle\, m_{xx} + f, \bar{w} - w \,\rangle \leq 0 \quad \forall\, \bar{w} \in B_\psi \ , \tag{8.25}$$

$$m \in K \ : \ \langle\, m - m_{xx} + w_{xx}, \bar{m} - m \,\rangle \geq 0 \quad \forall\, \bar{m} \in K \ . \tag{8.26}$$

From (8.22) it follows that $w \in B_\psi$. As before, introducing the function ξ, we can rewrite inequality (8.26) as

$$m - m_{xx} + w_{xx} + \xi = 0 \ , \tag{8.27}$$

$$\langle\, \xi, \bar{m} - m \,\rangle \leq 0 \quad \forall\, \bar{m} \in K \ . \tag{8.28}$$

This completes the proof of Theorem 8.1. □

We now show that we can pass to the limit in another order, by letting first $\varepsilon \to 0$, next $\delta \to 0$, and finally $\lambda \to 0$.

After passing to the limit as $\varepsilon \to 0$, we obtain

$$-m^{\delta}_{xx} + \frac{1}{\delta}\, q(w^{\delta}) = f \ , \tag{8.29}$$

$$m^{\delta} - m^{\delta}_{xx} + w^{\delta}_{xx} + \frac{1}{\lambda}\, p(m^{\delta}) = 0 \ , \tag{8.30}$$

$$w^{\delta} \in H^1_0(0,1) \ , \ m^{\delta} \in H^1_0(0,1) \ .$$

Here w^{δ}, m^{δ} are the limiting functions. By our estimates, when $\delta \to 0$, we have

$$w^{\delta} \to w^{\lambda} \quad \text{weakly in} \quad H^1_0(0,1) \ ,$$

$$m^{\delta} \to m^{\lambda} \text{ weakly in } H^1_0(0,1) \ , \text{ strongly in } L^2(0,1) \ .$$

Multiplying (8.29) by $\bar{w} - w^{\delta}$, $\bar{w} \in B_{\psi}$, and taking into account the monotonicity of q we obtain

$$\langle\, m^{\delta}_{xx} + f, \bar{w} - w^{\delta} \,\rangle \le 0 \ .$$

Substituting here m^{δ}_{xx} according to (8.30), the inequality takes the form

$$\langle\, m^{\delta} + w^{\delta}_{xx} + \frac{1}{\lambda}\, p(m^{\delta}) + f, \bar{w} - w^{\delta} \,\rangle \le 0 \ .$$

Since $\underline{\lim}_{\delta \to 0} \ \|w^{\delta}_x\|^2 \ge \|w^{\lambda}_x\|^2$, for any fixed $\bar{w} \in B_{\psi}$, we can pass to the limit which gives

$$\langle\, m^{\lambda} + w^{\lambda}_{xx} + \frac{1}{\lambda}\, p(m^{\lambda}) + f, \bar{w} - w^{\lambda} \,\rangle \le 0 \ .$$

Using the limiting relation obtained from (8.30), namely

$$m^{\lambda} - m^{\lambda}_{xx} + w^{\lambda}_{xx} + \frac{1}{\lambda}\, p(m^{\lambda}) = 0 \ , \tag{8.31}$$

the above inequality can be rewritten as

$$w^{\lambda} \in B_{\psi} \ : \ \langle\, m^{\lambda}_{xx} + f, \bar{w} - w^{\lambda} \,\rangle \le 0 \quad \forall\, \bar{w} \in B_{\psi} \ . \tag{8.32}$$

Now we analyse a passage to the limit in (8.31), (8.32) when $\lambda \to 0$. Note that if in (8.32) $\bar{w}$ is replaced by w^0, and (8.31) is multiplied by $m^{\lambda} - m^0$, we easily get the estimate

$$\frac{1}{\lambda}\, \langle\, p(m^{\lambda}), m^{\lambda} - m^0 \,\rangle \le c \ ,$$

which is uniform with respect to λ. Hence, as in the proof of Theorem 8.1, $\frac{1}{\lambda}\, p(m^{\lambda})$ is bounded in $H^{-1}(0,1)$. Letting $\lambda \to 0$, and passing to a subsequence, if necessary, we get

$$w^{\lambda}, m^{\lambda} \to w, m \quad \text{weakly in} \quad H^1_0(0,1) \ ,$$

$$\frac{1}{\lambda}\, p(m^{\lambda}) \to \xi \quad \text{weakly in} \quad H^{-1}(0,1) \ .$$

As before, when $\lambda \to 0$, by (8.31), (8.32), and the monotonicity of the operator p, we obtain

$$w \in B_\psi \ , \quad m \in K \ , \quad \xi \in H^{-1}(0,1) \ ,$$
$$\langle\, m_{xx} + f, \bar{w} - w \,\rangle \leq 0 \quad \forall\, \bar{w} \in B_\psi \ ,$$
$$m - m_{xx} + w_{xx} + \xi = 0 \ ,$$
$$\langle\, \xi, \bar{m} - m \,\rangle \leq 0 \quad \forall\, \bar{m} \in K \ .$$

This proves that we can pass to the limit by letting first $\varepsilon \to 0$, next $\delta \to 0$, and finally $\lambda \to 0$.

The uniqueness of the function m from Theorem 8.1 follows by the same arguments. In general, the functions w, ξ are not unique.

9 The case of tangential displacements

In addition to the previous section we study the case where beam points have both normal and tangential displacements. The boundary conditions are also satisfied. The formulation of the problem is as follows.

We have to find functions v, w, m, n, ξ satisfying on the interval $(0, 1)$ the relations

$$w - v\varphi_x \geq \varphi \ , \tag{9.1}$$
$$(m_{xx} + f)(\bar{w} - w) + (n_x + g)(\bar{v} - v) \leq 0 \quad \forall\, (\bar{v}, \bar{w}) \ , \quad \bar{w} - \bar{v}\varphi_x \geq \varphi \ , \tag{9.2}$$
$$-w_{xx} = m - m_{xx} + \xi \ , \quad v_x = n \ , \tag{9.3}$$
$$|m| \leq c^* \ , \quad \xi(\bar{m} - m) \leq 0 \ , \quad \forall\, \bar{m}, \ |\bar{m}| \leq c^* \ , \tag{9.4}$$
$$v = w = m = 0 \ , \quad x = 0, 1 \ . \tag{9.5}$$

Here $f, g \in L^2(I)$ are given exterior forces, $I = (0, 1)$, c^* is a positive constant. Restriction (9.1) is a nonpenetration condition, the equation $y = \varphi(x)$ describes the punch shape, v, w are the tangential and normal displacements of the beam, respectively, m is the bending moment, n is the integrated stress, ξ is the plastic part of the beam curvature.

We make some assumptions which allow us to prove the solvability of problem (9.1)–(9.5).

Let $\varphi \in H(I)$,

$$B = \big\{ (v, w) \ \big| \ v, w \in H_0^1(I) \ , \ w - v\varphi_x \geq \varphi \ \text{ on } I \big\} \ ,$$
$$K = \big\{ m \in H_0^1(I) \ \big| \ |m| \leq c^* \ \text{ on } I \big\} \ .$$

We assume that there exists a solution m^0 to the equation

$$m_{xx}^0 + f = 0 \ , \tag{9.6}$$

which satisfies the condition $(1 + \kappa)m^0 \in K$, where $\kappa > 0$ is a constant.

We assume also that there exists at least one element $(v^0, w^0) \in B$. This evidently depends on the boundary conditions of φ.

For our problem the following solvability result is true.

Theorem 9.1. *Let the assumptions formulated above hold. Then there exist functions $v, w, m \in H_0^1(I)$, $\xi \in H^{-1}(I)$ such that*

$$(v, w) \in B \ , \quad m \in K \ , \tag{9.7}$$

$$\langle\, m_{xx} + f, \bar{w} - w \,\rangle + \langle\, n_x + g, \bar{v} - v \,\rangle \leq 0 \quad \forall\, (\bar{v}, \bar{w}) \in B \ , \tag{9.8}$$

$$-w_{xx} = m - m_{xx} + \xi \ , \quad v_x = n \ , \tag{9.9}$$

$$\langle\, \xi, \bar{m} - m \,\rangle \leq 0 \quad \forall\, \bar{m} \in K \ . \tag{9.10}$$

Proof. Let $\varepsilon, \delta, \lambda$ be positive parameters. We consider the regularized boundary value problem (without indicating the dependence of the solution on the parameters)

$$-\varepsilon w_{xx} - m_{xx} + \delta^{-1}\beta_1(v, w) - f = 0 \ , \tag{9.11}$$

$$-v_{xx} + \delta^{-1}\beta_2(v, w) - g = 0 \ , \tag{9.12}$$

$$w_{xx} + m - m_{xx} + \lambda^{-1}p(m) = 0 \ , \tag{9.13}$$

$$v = w = m = 0 \ , \quad x = 0, 1 \ . \tag{9.14}$$

Here $(\beta_1, \beta_2) \ : \ \left[L^2(I)\right]^2 \mapsto \left[L^2(I)\right]^2$ is the penalty operator related to the restriction $w - v\varphi_x \geq \varphi$; $p(m) = m - \pi m$, π is the operator of the orthogonal projection of $L^2(I)$ onto the set $\{m \in L^2(I) \mid |m| \leq c^*\}$.

To obtain a priori estimates for solutions, we multiply (9.11)–(9.13) by $w - w^0$, $v - v^0$, $m - m^0$, respectively. By the monotonicity of p, (β_1, β_2) and in view of equation (9.6),

$$\varepsilon\, \|w_x\|^2 + \|v_x\|^2 + \|m_x\|^2 + \|m\|^2 + \lambda^{-1}\, \langle\, p(m), m - m^0 \,\rangle \leq c \ , \tag{9.15}$$

where the constant is independent of $\varepsilon \leq \varepsilon_0$, δ, λ. By the monotonicity of p,

$$\lambda^{-1}\, \langle\, p(m), \bar{m} - m \,\rangle \leq 0 \quad \forall\, \bar{m} \in K \ .$$

Bearing in mind (9.15) the inequality

$$\lambda^{-1}\, \langle\, p(m), \bar{m} - m^0 \,\rangle \leq c \quad \forall\, \bar{m} \in K \tag{9.16}$$

follows. This condition on m^0 implies that $\bar{m} \in K$ if $\bar{m} = m^0 + m^1$, $m^1 \in H_0^1(I)$, $|m^1| \leq c^*\kappa(1 + \kappa)^{-1}$.

Substitute $\bar{m} = m^0 + m^1$ into (9.16) as a trial function. Since m^1 is arbitrary, the inequality

$$\lambda^{-1}\, \|p(m)\|_{H^{-1}(I)} \leq c \tag{9.17}$$

holds uniformly with respect to all the parameters. With this estimate from (9.13) it follows that w_{xx} are bounded in $H^{-1}(I)$ and hence

$$\|w_x\| \leq c \ . \tag{9.18}$$

Estimate (9.15) enables us to prove that the operator of problem (9.11)–(9.14) acts from the space $[H_0^1(I)]^3$ into its dual, and satisfy all the conditions of Lemma 1.8 for fixed parameters. Consequently a unique solution $v, w, m \in H_0^1(I)$ to problem (9.11)–(9.14) exists for fixed $\varepsilon, \delta, \lambda$.

To continue the proof we pass to the limit by letting first $\varepsilon \to 0$, next $\lambda \to 0$, and finally $\delta \to 0$. At each step we indicate the dependence of the solution on the parameter by superscript. So, let $v^\varepsilon, w^\varepsilon, m^\varepsilon$ be a solution to problem (9.11)–(9.14). By (9.15), (9.18), there exists a subsequence of $v^\varepsilon, w^\varepsilon, m^\varepsilon$, still denoted by $v^\varepsilon, w^\varepsilon, m^\varepsilon$, and possessing the property

$$v^\varepsilon, w^\varepsilon, m^\varepsilon \to v^\lambda, w^\lambda, m^\lambda \ \text{weakly in} \ H_0^1(I) \ , \ \text{strongly in} \ L^2(I) \ .$$

Passing to the limit in (9.11)–(9.14) when $\varepsilon \to 0$, we obtain

$$v^\lambda, w^\lambda, m^\lambda \in H_0^1(I) \ , \tag{9.19}$$

$$-m_{xx}^\lambda + \delta^{-1}\beta_1(v^\lambda, w^\lambda) - f = 0 \ , \tag{9.20}$$

$$-v_{xx}^\lambda + \delta^{-1}\beta_2(v^\lambda, w^\lambda) - g = 0 \ , \tag{9.21}$$

$$w_{xx}^\lambda + m^\lambda - m_{xx}^\lambda + \lambda^{-1}p(m^\lambda) = 0 \ . \tag{9.22}$$

The convergence of the nonlinear terms is a consequence of the strong convergence of $v^\varepsilon, w^\varepsilon, m^\varepsilon$. Since estimates (9.15), (9.17), (9.18) are uniform with respect to the parameters, we have

$$v^\lambda, w^\lambda, m^\lambda \to v^\delta, w^\delta, m^\delta \quad \text{weakly in} \quad H_0^1(I) \ ,$$

$$v^\lambda, w^\lambda \to v^\delta, w^\delta \quad \text{strongly in} \quad L^2(I) \ ,$$

$$\lambda^{-1}p(m^\lambda) \to \xi^\delta \quad \text{weakly in} \quad H^{-1}(I) \ .$$

Therefore, when $\lambda \to 0$, from (9.19)–(9.22) it follows that

$$v^\delta, w^\delta \in H_0^1(I) \ , \quad m^\delta \in K \ , \quad \xi^\delta \in H^{-1}(I) \ , \tag{9.23}$$

$$-m_{xx}^\delta + \delta^{-1}\beta_1(v^\delta, w^\delta) - f = 0 \ , \tag{9.24}$$

$$-v_{xx}^\delta + \delta^{-1}\beta_2(v^\delta, w^\delta) - g = 0 \ , \tag{9.25}$$

$$w_{xx}^\delta + m^\delta - m_{xx}^\delta + \xi^\delta = 0 \ , \tag{9.26}$$

$$\langle \ \xi^\delta, \bar{m} - m^\delta \ \rangle \leq 0 \quad \forall \, \bar{m} \in K \ . \tag{9.27}$$

By usual arguments, the monotonicity of p implies that $m^\delta \in K$.

We now pass to the limit as $\delta \to 0$. Assuming

$$v^\delta, w^\delta, m^\delta \to v, w, m \quad \text{weakly in} \quad H_0^1(I) \ ,$$
$$\xi^\delta \to \xi \quad \text{weakly in} \quad H^{-1}(I) \ ,$$

we multiply (9.24)–(9.25) by $\bar{w} - w^\delta$, $\bar{v} - v^\delta$, respectively, and integrate over I. In view of the monotonicity of (β_1, β_2) we obtain the inequality

$$\langle\, m_{xx}^\delta + f, \bar{w} - w^\delta \,\rangle + \langle\, v_{xx}^\delta + g, \bar{v} - v^\delta \,\rangle \le 0 \ . \tag{9.28}$$

From (9.26), (9.27) it follows that

$$\langle\, -w_{xx}^\delta - m^\delta + m_{xx}^\delta, \bar{m} - m^\delta \,\rangle \le 0 \ . \tag{9.29}$$

Let us sum up inequalities (9.28), (9.29) and pass to the limit in the obtained relation. By adding to and subtracting from the resulting inequality the term $\langle\, m_x, w_x \,\rangle$ we obtain two inequalities of the type (9.28), (9.29), respectively. Hence, the limiting relations are of the form

$$(v, w) \in B \ , \quad m \in K \ , \quad \xi \in H^{-1}(I) \ ,$$

$$\langle\, m_{xx} + f, \bar{w} - w \,\rangle + \langle\, n_x + g, \bar{v} - v \,\rangle \le 0 \quad \forall\, (\bar{v}, \bar{w}) \in B \ ,$$

$$w_{xx} + m - m_{xx} + \xi = 0 \ , \quad n = v_x \ ,$$

$$\langle\, \xi, \bar{m} - m \,\rangle \le 0 \quad \forall\, \bar{m} \in K \ .$$

This completes the proof. □

Assuming the existence of two solutions v^1, w^1, m^1, ξ^1, v^2, w^2, m^2, ξ^2, we can easily prove that $m^1 = m^2$, $v^1 = v^2$. Hence, the functions v, m are unique.

Let us construct a nonnegative measure ν_φ characterizing interaction forces. First of all, observe that if $v^0, w^0 \in H_0^1(I)$ and $w^0 - v^0\varphi_x \ge 0$, then

$$-\langle\, m_{xx} + f, w^0 \,\rangle - \langle\, n_x + g, v^0 \,\rangle \ge 0 \ . \tag{9.30}$$

Indeed, the element $(v + v^0, w + w^0)$ belongs to B, thus inequality (9.30) follows from (9.8) with $(v + v^0, w + w^0)$ substituted as a trial function.

By $C_0(I)$ we denote the space of continuous functions on I having compact supports with the topology of uniform convergence on compacts.

Theorem 9.2. *There exists a nonnegative measure ν_φ defined on the $\mathfrak{S}$-algebra of Borel subsets of I such that for all $\bar{v}, \bar{w} \in H_0^1(I) \cap C_0(I)$ the representation*

$$\langle\, m_{xx} + f, \bar{w} \,\rangle + \langle\, n_x + g, \bar{v} \,\rangle = -\int_0^1 \frac{\bar{w} - \bar{v}\varphi_x}{\sqrt{1 + \varphi_x^2}}\, d\nu_\varphi \tag{9.31}$$

holds.

Proof. Let $\mathcal{V}$ be the linear space of functions of the form

$$\chi = \frac{\bar{w} - \bar{v}\varphi_x}{\sqrt{1+\varphi_x^2}} \ , \tag{9.32}$$

where $\bar{v}, \bar{w} \in H_0^1(I) \cap C_0(I)$.

On $\mathcal{V}$ we define the following linear functional

$$\Pi(\chi) = -\langle\, m_{xx} + f, \bar{w} \,\rangle - \langle\, n_x + g, \bar{v} \,\rangle \ . \tag{9.33}$$

Here χ corresponds to the functions $\bar{v}, \bar{w}$. By (9.30) this functional is positive.

By formula (9.33), the functional Π is well defined. Indeed, if $\chi^1 = \chi^2$, then $\chi^1 - \chi^2 = 0$, where the functions χ^1, χ^2 correspond to $(\bar{v}^1, \bar{w}^1)$, $(\bar{v}^2, \bar{w}^2)$, respectively. Consequently, bearing in mind (9.30),

$$\langle\, m_{xx} + f, \bar{w}^1 - \bar{w}^2 \,\rangle + \langle\, n_x + g, \bar{v}^1 - \bar{v}^2 \,\rangle = 0 \ ,$$

which means that $\Pi(\chi^1) = \Pi(\chi^2)$.

This functional can be extended to the whole space $C_0(I)$. To see this, let us take an arbitrary function $h \in C_0(I)$ and a sequence $h_n \in C_0^\infty(I)$ such that $h_n \to h$ in $C_0(I)$. Denote by $S(h_n)$ the supports of the functions h_n and assume that those supports belong to a fixed compact $M \subset I$. Select a function $\psi \in C_0^\infty(I)$ such that $\psi \equiv 1$ on M, $0 \le \psi \le 1$ everywhere. Then

$$\left|h_m(x) - h_n(x)\right| \le \alpha_{mn}\psi(x) \ , \quad \alpha_{mn} = \max_I \left|h_m(x) - h_n(x)\right| \ .$$

Since the functional Π is positive,

$$\left|\Pi(h_m) - \Pi(h_n)\right| \le \alpha_{mn}\Pi(\psi) \ .$$

Since $\alpha_{mn} \to 0$, when $m, n \to \infty$, the limit of $\Pi(h_n)$ exists and we denote it by $\Pi(h)$. This limit does not depend on approximating sequences. Moreover, the extended functional is linear and positive on $C_0(I)$. This means that there exists a nonnegative measure ν_φ such that

$$\Pi(\chi) = \int_0^1 \chi d\nu_\varphi \quad \forall\, \chi \in C_0(I) \ .$$

Now representation (9.31) for the functions of the type (9.32) immediately follows which completes the proof. □

We have

$$\nu_\varphi = -\sqrt{1+\varphi_x^2}\,(m_{xx} + f) \ .$$

In fact, $m_{xx} + f \in H^{-1}(I)$, and, moreover, $\sqrt{1+\varphi_x^2}\,(m_{xx}+f) \in H^{-1}(I)$. Putting $\bar{v} = 0$ in (9.31) we find that

$$-\langle\, m_{xx} + f, \bar{w} \,\rangle = \int_0^1 \frac{\bar{w} d\nu_\varphi}{\sqrt{1+\varphi_x^2}} \ .$$

Consequently,

$$-\langle \sqrt{1+\varphi_x^2}\,(m_{xx}+f), \bar{w} \rangle = -\langle m_{xx}+f, \bar{w}\sqrt{1+\varphi_x^2} \rangle = \int_0^1 \bar{w}\, d\nu_\varphi \ .$$

Hence, the sought formula for ν_φ results.

The support of the measure ν_φ belongs to the contact set $\{w - v\varphi_x = \varphi\}$. In fact, from (9.8) it follows that the equations

$$m_{xx} + f = 0 \ , \quad v_{xx} + g = 0$$

are satisfied in the domain $\{w - v\varphi_x > \varphi\}$. In particular, $v, m \in H^2_{\text{loc}}$ in a vicinity of each point from the domain.

The equation

$$-\varphi_x(m_{xx}+f) = v_{xx} + g \tag{9.34}$$

is always fulfilled in I. In fact, it is easily seen that $(v+\psi, w+\psi\varphi_x) \in B$ for any $\psi \in C_0^\infty(I)$. Substitute $(v+\psi, w+\psi\varphi_x)$ into (9.8) as a test function. This clearly implies that

$$\langle (m_{xx}+f)\varphi_x, \psi \rangle + \langle v_{xx}+g, \psi \rangle \le 0 \ .$$

Since ψ is arbitrary, equation (9.34) results. If $\varphi_x = 0$ in I, then $v_{xx} + g = 0$ in I, and hence $v \in H^2(I) \cap H_0^1(I)$.

The inequalities $\varphi(0) < 0$, $\varphi(1) < 0$ guarantee that the relation $w - v\varphi_x > \varphi$ is satisfied in some neighbourhood of the points $x = 0$, $x = 1$. Then $\nu_\varphi(I) < +\infty$, since the measure of every compact from I is finite.

10 Beam under plasticity and creep conditions

To extend the results of the previous section we shall study here the case where the constitutive law describes the elasticity, plasticity and creep. The problem we shall analyse is nonstationary, its solutions depend on t.

Let $Q = I \times (0,T)$, $I = (0,1)$, $T > 0$. We want to find functions v, w, m, n, ξ which satisfy in the domain Q the following relations

$$\begin{gathered} w - v\varphi_x \ge \varphi \ , \quad (m_{xx}+f)(\bar{w}-w) + (n_x+g)(\bar{v}-v) \le 0 \\ \forall\, \bar{v}, \bar{w}, \quad \bar{w} - \bar{v}\varphi_x \ge \varphi \ , \end{gathered} \tag{10.1}$$

$$v_x - n - \int_0^t n\, d\tau = 0 \ , \tag{10.2}$$

$$-w_{xx} + m_{xx} + \int_0^t m_{xx}\, d\tau = \xi \ , \tag{10.3}$$

$$|m| \le c^* \ , \quad \xi(\bar{m} - m) \le 0 \quad \forall\, \bar{m}, \ |\bar{m}| \le c^* \ , \tag{10.4}$$

$$v = w = m = 0 \ , \quad x = 0, 1 \ . \tag{10.5}$$

Relations (10.2)–(10.4) represent the constitutive law.

We formulate basic hypotheses under which we prove the existence theorem.

Assume that $f, g \in L^2(I)$, and B, K are the convex sets introduced in the previous section. Suppose that the equation

$$m^0_{xx} + f = 0 \tag{10.6}$$

has a solution m^0 such that $(1+\kappa)m^0 \in K$, where $\kappa = \text{const} > 0$. For the sake of convenience we choose a function n^0 satisfying the equation

$$n^0_x + g = 0 \ . \tag{10.7}$$

We assume that $\varphi \in H^1(I)$; $\varphi(0), \varphi(1) \le 0$. In particular, this yields the existence of functions v^0, w^0 such that $(v^0, w^0) \in B$.

The main result of this section is formulated as follows.

Theorem 10.1. *Let the assumptions formulated above hold. There exist functions* v, w, m, n, ξ *such that*

$$\begin{gathered} v, w, m, m_t \in L^2\big(0, T; H^1_0(I)\big) \ , \\ n, n_t \in L^2(Q) \ , \quad \xi \in L^2\big(0, T; H^{-1}(I)\big) \ , \\ m(t) \in K \ , \quad \big(v(t), w(t)\big) \in B \quad \text{a.e. on } (0, T) \ , \end{gathered} \tag{10.8}$$

$$\begin{gathered} \int_0^T \langle\, m_{xx} + f, \bar{w} - w \,\rangle dt + \int_0^T \langle\, n_x + g, \bar{v} - v \,\rangle dt \le 0 \\ \forall\, \bar{v}, \bar{w} \in L^2\big(0, T; H^1_0(I)\big), \ \big(\bar{v}(t), \bar{w}(t)\big) \in B \ , \end{gathered} \tag{10.9}$$

$$v_x - n - \int_0^t n d\tau = 0 \ , \tag{10.10}$$

$$-w_{xx} + m_{xx} + \int_0^t m_{xx} d\tau = \xi \ , \tag{10.11}$$

$$\int_0^T \langle\, \xi, \bar{m} - m \,\rangle dt \le 0 \quad \forall\, \bar{m} \in L^2\big(0, T; H^1_0(I)\big), \ \bar{m}(t) \in K \ . \tag{10.12}$$

Proof. Let (β_1, β_2) and p be the operators introduced in the previous section.

Consider the regularized boundary value problem in Q with three positive parameters $\varepsilon, \delta, \lambda$ (the dependence of solutions on the parameters is not shown explicitely in the notation adopted)

$$-\varepsilon v_{xx} - n_x + \delta^{-1}\beta_1(v, w) = g \ , \tag{10.13}$$

$$-\varepsilon w_{xx} - m_{xx} + \delta^{-1}\beta_2(v, w) = f \ , \tag{10.14}$$

$$-v_x + n + \int_0^t n d\tau = 0 \ , \tag{10.15}$$

$$-m_{xx} - \int_0^t m_{xx} d\tau + w_{xx} + \lambda^{-1} p(m) = 0 \ , \tag{10.16}$$

$$v = w = m = 0 \ , \quad x = 0, 1 \ . \tag{10.17}$$

The idea of the proof is as follows. First we establish the solvability of (10.13)–(10.17) for fixed parameters and next we pass to the limit as $\varepsilon \to 0$, $\lambda \to 0$, $\delta \to 0$.

We begin by showing a priori estimates for solutions to (10.13)–(10.17). To this end we multiply (10.13)–(10.16) by $v - v^0$, $w - w^0$, $n - n^0$, $m - m^0$, respectively, and integrate them over Q. By the monotonicity of (β_1, β_2), and p, and in view of equations (10.6), (10.7)

$$\varepsilon \ \|v_x\|^2_{L^2(Q)} + \varepsilon \ \|w_x\|^2_{L^2(Q)} \le c \ , \tag{10.18}$$

$$\|m_x\|^2_{L^2(Q)} + \|n\|^2_{L^2(Q)} \le c \ , \tag{10.19}$$

$$\left\| \int_0^T m_x d\tau \right\|^2 + \left\| \int_0^T n d\tau \right\|^2 \le c \ . \tag{10.20}$$

For $t = 0$, equations (10.13)–(10.16) takes the form

$$-\varepsilon v_{xx} - v_{xx} + \delta^{-1} \beta_1(v, w) = g \ , \tag{10.21}$$

$$-\varepsilon w_{xx} - m_{xx} + \delta^{-1} \beta_2(v, w) = f \ , \tag{10.22}$$

$$-m_{xx} + w_{xx} + \lambda^{-1} p(m) = 0 \ . \tag{10.23}$$

We see that (10.21)–(10.23), (10.17) is an elliptic boundary value problem possessing at least one solution v_1, w_1, m_1.

The following a priori estimate

$$\|v_{1x}\|^2 + \|m_{1x}\|^2 + \varepsilon \ \|w_{1x}\|^2 \le c \tag{10.24}$$

holds with the constant c independent of $\varepsilon \le \varepsilon_0, \delta, \lambda$. The estimate follows when we multiply (10.21)–(10.23) by $v - v^0$, $w - w^0$, $m - m^0$, respectively.

Now we can differentiate the equations (10.13)–(10.16) with respect to t and multiply them by v_t, w_t, n_t, m_t, respectively. Since

$$\langle \ (\beta_{1t}, \beta_{2t}), (v_t, w_t) \ \rangle \ge 0 \ , \ \ \langle \ p(m)_t, m_t \ \rangle \ge 0$$

a.e. on $(0, T)$ (see Lions, 1969), we get

$$\varepsilon \ \|v_{tx}\|^2_{L^2(Q)} + \varepsilon \ \|w_{tx}\|^2_{L^2(Q)} \le c \ , \tag{10.25}$$

$$\|m_{tx}\|^2_{L^2(Q)} + \|n_t\|^2_{L^2(Q)} \le c \ , \tag{10.26}$$

$$\|m_x(T)\|^2 + \|n(T)\|^2 \le c \ . \tag{10.27}$$

The initial data used in getting (10.25)–(10.27) are the above functions $m_1, n_1 \equiv v_{1x}$. In general, the solution v_1, w_1, m_1 depends on $\varepsilon, \delta, \lambda$, but what really matters here is that a priori estimate (10.24) is uniform with respect to those parameters.

Inequalities (10.18)–(10.19) ensure the solvability of problem (10.13)–(10.17) for fixed parameters since the operator of problem (10.13)–(10.17) satisfies all the conditions of Lemma 1.8. Observe that estimate (10.27) follows also from (10.19), (10.26).

We now obtain an additional a priori estimate. Consider the functional $q(m) = \|m - \pi m\|^2$. The derivative of this functional is given by the formula $q'(m) = 2p(m)$. Take a function $\hat{m}$ from $L^\infty(Q)$ such that the element $\bar{m} \equiv m^0 + \hat{m}$ satisfies the inequality $|\bar{m}(t)| \le c^*$, $t \in (0,T)$. Since q is a convex functional

$$\lambda^{-1}\langle p(m), \hat{m}\rangle \le \lambda^{-1}\langle p(m), m - m^0\rangle + \frac{1}{2}\lambda^{-1}q(m^0 + \hat{m}) - \frac{1}{2}\lambda^{-1}q(m) \ . \tag{10.28}$$

In view of (10.18)–(10.20), (10.25)–(10.27), the multiplication of (10.13)–(10.16) by $v - v^0$, $w - w^0$, $n - n^0$, $m - m^0$ respectively, implies that

$$\lambda^{-1}\langle\, p(m), m - m^0 \,\rangle \quad \text{are bounded in} \quad L^2(0,T) \ .$$

Thus, from (10.28) it follows that $\lambda^{-1}p(m)$ are bounded in $L^2\big(0,T;L^1(I)\big)$. Consequently, in view of the inclusion $L^1(I) \subset H^{-1}(I)$, the estimate

$$\lambda^{-1}\|p(m)\|_{L^2\left(0,T;H^{-1}(I)\right)} \le c$$

holds uniformly with respect to $\varepsilon \le \varepsilon_0, \delta, \lambda$. Taking into account the boundary conditions for w we find that

$$w \quad \text{are bounded in} \quad L^2(0,T;H_0^1(I)) \ . \tag{10.29}$$

From (10.15) if follows that

$$v \quad \text{are bounded in} \quad L^2(0,T;H_0^1(I)) \ . \tag{10.30}$$

We now analyse passages to the limit. The solution to problem (10.13)–(10.17) will be denoted by $v^\varepsilon, w^\varepsilon, m^\varepsilon, n^\varepsilon$. Without loss of generality a subsequence, denoted by $v^\varepsilon, w^\varepsilon, m^\varepsilon, n^\varepsilon$, is assumed to possess the following properties as $\varepsilon \to 0$

$$\varepsilon v^\varepsilon, \varepsilon v_t^\varepsilon \to 0 \quad \text{weakly in} \quad L^2(0,T;H_0^1(I)) \ ,$$

$$\varepsilon w^\varepsilon, \varepsilon w_t^\varepsilon \to 0 \quad \text{weakly in} \quad L^2(0,T;H_0^1(I)) \ ,$$

$$n^\varepsilon, n_t^\varepsilon \to n^\lambda, n_t^\lambda \quad \text{weakly in} \quad L^2(Q) \ ,$$

$$m^\varepsilon, m_t^\varepsilon \to m^\lambda, m_t^\lambda \quad \text{weakly in} \quad L^2(0,T;H_0^1(I)) \ ,$$

$$m^\varepsilon \to m^\lambda \quad \text{strongly in} \quad L^2(Q) \ ,$$

$$\int_0^t m_x^\varepsilon d\tau, \int_0^t n^\varepsilon d\tau \to \int_0^t m_x^\lambda d\tau, \int_0^t n^\lambda d\tau \ \text{weakly in} \ L^2(Q) \ ,$$

$$v^\varepsilon, w^\varepsilon \to v^\lambda, w^\lambda \quad \text{weakly in} \quad L^2(0,T;H_0^1(I)) \ .$$

When $\varepsilon \to 0$, equations (10.13)–(10.16) take the form

$$-n_x^\lambda + \delta^{-1}\beta_1(v^\lambda, w^\lambda) = g \ , \tag{10.31}$$

$$-m_{xx}^\lambda + \delta^{-1}\beta_2(v^\lambda, w^\lambda) = f \ , \tag{10.32}$$

$$v_x^\lambda - n^\lambda - \int_0^t n^\lambda d\tau = 0 \ , \tag{10.33}$$

$$-m_{xx}^\lambda - \int_0^t m_{xx}^\lambda d\tau + w_{xx}^\lambda + \lambda^{-1} p(m^\lambda) = 0 \ . \tag{10.34}$$

The convergence of the nonlinear terms $\beta_i(v^\varepsilon, w^\varepsilon)$ follows from the monotonicity of the operator (β_1, β_2).

Let us now demonstrate that we can pass to the limit when $\lambda \to 0$. We choose a subsequence such that

$$n^\lambda, n_t^\lambda \to n^\delta, n_t^\delta \quad \text{weakly in} \quad L^2(Q) \ ,$$

$$m^\lambda, m_t^\lambda, v^\lambda, w^\lambda \to m^\delta, m_t^\delta, v^\delta, w^\delta \ \text{weakly in} \ L^2(0,T; H_0^1(I)) \ ,$$

$$\int_0^t m_x^\lambda d\tau, \int_0^t n^\lambda d\tau \to \int_0^t m_x^\delta d\tau, \int_0^t n^\delta d\tau \ \text{weakly in} \ L^2(Q) \ ,$$

$$\lambda^{-1} p(m^\lambda) \to \xi^\delta \quad \text{weakly in} \quad L^2(0,T; H^{-1}(I))$$

when $\lambda \to 0$. Since $\lambda \to 0$, from (10.33)–(10.34) it follows that

$$v_x^\delta - n^\delta - \int_0^t n^\delta d\tau = 0 \ , \tag{10.35}$$

$$-m_{xx}^\delta - \int_0^t m_{xx}^\delta d\tau + w_{xx}^\delta + \xi^\delta = 0 \ . \tag{10.36}$$

By the monotonicity of the operator p,

$$\begin{gathered} \int_0^T \langle\, w_{xx}^\lambda - m_{xx}^\lambda - \int_0^t m_{xx}^\lambda d\tau, \bar{m} - m^\lambda \,\rangle dt \geq 0 \\ \forall\, \bar{m} \in L^2(0,T; H_0^1(I)), \ \ \bar{m}(t) \in K \ . \end{gathered} \tag{10.37}$$

From (10.32) it is seen that when $\lambda \to 0$

$$m^\lambda \quad \text{are bounded in} \quad L^2(0,T; H^2(I) \cap H_0^1(I))$$

(in general, not uniformly with respect to δ). Consequently, in view of the above estimates we can assume that for every fixed δ

$$m^\lambda \to m^\delta \quad \text{strongly in} \quad L^2(0,T; H_0^1(I)) \ .$$

In particular,

$$\int_0^T \langle w_{xx}^\lambda, m^\lambda \rangle dt \to \int_0^T \langle w_{xx}^\delta, m^\delta \rangle dt$$

and we can pass to the limit in (10.37) as $\lambda \to 0$, which means that

$$\int_0^T \langle \xi^\delta, \bar{m} - m^\delta \rangle dt \leq 0 \tag{10.38}$$

for every $\bar{m}$.

The limiting functions $v^\delta, w^\delta, m^\delta, n^\delta$ satisfy (10.35), (10.36), (10.38) and

$$-n_x^\delta + \delta^{-1}\beta_1(v^\delta, w^\delta) = g \ ,$$

$$-m_{xx}^\delta + \delta^{-1}\beta_2(v^\delta, w^\delta) = f \ .$$

The condition $m^\delta(t) \in K$ follows immediately from (10.34).

By similar arguments we pass to the limit when $\delta \to 0$ and get (10.8)–(10.12), which completes the proof. □

11 The contact viscoelastoplastic problem for a beam

In this section we discuss the contact viscoelastoplastic problem for the Timoshenko model of a beam. In this model the constitutive law describes the elasticity property as well as the viscocity and plasticity ones. The main result of the section is the existence theorem for this problem.

The formulation of the problem is as follows. Let $I = (0,1)$. We have to find functions v, w, m, n, ξ satisfying in $Q = I \times (0,1)$ the following relations

$$w - v\varphi_x \geq \varphi \ , \tag{11.1}$$

$$\begin{gathered}(m_{xx} + f)(\bar{w} - w) + (n_x + g)(\bar{v} - v) \leq 0 \\ \forall\, \bar{v}, \bar{w}, \ \ \bar{w} - \bar{v}\varphi_x \geq \varphi \ ,\end{gathered} \tag{11.2}$$

$$n_t - v_x = 0 \ , \tag{11.3}$$

$$m_{txx} - w_{xx} = \eta \ , \tag{11.4}$$

$$|m| \leq c^*, \ \ \eta(\bar{m} - m) \geq 0 \ \ \ \forall\, \bar{m}, \ \ |\bar{m}| \leq c^* \ , \tag{11.5}$$

$$n = n^0 \ , \quad m = m^0 \ , \quad t = 0 \ , \tag{11.6}$$

$$v = w = m = 0 \ , \quad x = 0, 1 \ . \tag{11.7}$$

To simplify the formulae we have dropped the terms corresponding to the "elastic" part in (11.3), (11.4). These terms have no impact on the existence theorem.

Let $f, g \in L^2(I)$. The convex sets B, K are defined as in the two previous sections.

Assume that for the initial data m^0, n^0 the following equations

$$m^0_{xx} + f = 0 \ , \quad n^0_x + g = 0 \tag{11.8}$$

hold in I, and moreover $(1+\kappa)m^0 \in K$, where $\kappa = \text{const} > 0$.

We assume also that $\varphi \in H^2(I)$; $\varphi(0), \varphi(1) \le 0$. Under this assumption there exist functions v^0, w^0 such that $(v^0, w^0) \in B$ and $w^0_{xx} \in L^2(I)$. From (11.3), (11.6) it is seen that $n(t) = n^0 + \int_0^t v_x d\tau$. In particular, taking into account (11.8), one has $n_x = -g + \int_0^t v_{xx} d\tau$.

The main result of this section is as follows.

Theorem 11.1. *Let the above assumptions hold. Then there exist functions v, w, m, n, ξ such that*

$$m, m_t, v \in L^\infty\big(0,T; H^1_0(I)\big) \ , \quad n, n_t \in L^\infty\big(0,T; L^2(I)\big) \ ,$$

$$w \in L^2\big(0,T; H^1_0(I)\big) \ , \quad \eta \in L^2\big(0,T; H^{-1}(I)\big) \ ,$$

$$m(t) \in K \ , \quad \big(v(t), w(t)\big) \in B \quad \text{a.e. on } (0,T) \ ,$$

$$\int_0^T \langle\, m_{xx} + f, \bar{w} - w \,\rangle dt + \int_0^T \langle\, n_x + g, \bar{v} - v \,\rangle dt \le 0$$

$$\forall\, \bar{v}, \bar{w} \in L^2\big(0,T; H^1_0(I)\big) \ , \quad \big(\bar{v}(t), \bar{w}(t)\big) \in B \ ,$$

$$n_t - v_x = 0 \ ,$$

$$m_{txx} - w_{xx} = \eta \ ,$$

$$\int_0^T \langle\, \eta, \bar{m} - m \,\rangle dt \le 0 \quad \forall\, \bar{m} \in L^2\big(0,T; H^1_0(I)\big) \ , \quad \bar{m}(t) \in K \ ,$$

and boundary conditions (11.6) hold.

Proof. Let (β_1, β_2) and p be the operators defined as in the two previous sections. Consider the regularized boundary value problem with positive parameters $\varepsilon, \delta, \lambda$

$$\varepsilon v_t - \int_0^t v_{xx} d\tau + \delta^{-1}\beta_1(v,w) = 0 \ , \tag{11.9}$$

$$\varepsilon w_t - \varepsilon w_{xx} - m_{xx} + \delta^{-1}\beta_2(v,w) = f \ , \tag{11.10}$$

$$-m_{txx} + w_{xx} + \lambda^{-1}p(m) = 0 \ , \tag{11.11}$$

$$v = v^0 \ , \quad w = w^0 \ , \quad m = m^0 \ , \quad t = 0 \ , \tag{11.12}$$

$$v = w = m = 0 \ , \quad x = 0, 1 \ . \tag{11.13}$$

The idea of the proof is the same as in the previous results. First we establish the solvability of (11.9)–(11.13) for arbitrary but fixed parameter values and next we pass to the limit by letting $\varepsilon \to 0$, $\lambda \to 0$, $\delta \to 0$.

Now we find a priori estimates of the solution to (11.9)–(11.13). To do this we multiply (11.9)–(11.11) by $v - v^0$, $w - w^0$, $m - m^0$, respectively, and integrate them over I.

By the first equation (11.8) and the nonnegativity of the terms connected with the operators $(\beta_1, \beta_2), p$, we get the inequality

$$\frac{1}{2}\frac{d}{dt}\{\varepsilon\,\|v(t)\|^2 + \varepsilon\|w(t)\|^2 + \|m_x(t)\|^2 + \Big\|\int_0^t v_x d\tau\Big\|^2\}$$
$$+\frac{\varepsilon}{2}\,\|w_x(t)\|^2 \le \varepsilon\langle\, v_t(t), v^0\,\rangle + \varepsilon\langle\, w_t(t), w^0\,\rangle + \frac{1}{2}\,\|m_x(t)\|^2$$
$$+\frac{1}{2}\,\Big\|\int_0^t v_x d\tau\Big\|^2 + \langle\, m_{tx}(t), m_x^0\,\rangle + c\ .$$

The constant c is uniform with respect to $\varepsilon \le \varepsilon_0, \delta, \lambda$. The differential inequality can be integrated for the functions v^0, w^0, m^0 are independent of t. Hence, the inequality

$$\max_{0\le t\le T}\{\varepsilon\,\|v(t)\|^2 + \varepsilon\,\|w(t)\|^2\} \le c\ , \tag{11.14}$$

$$\|m_x\|^2_{L^2(Q)} + \Big\|\int_0^T v_x d\tau\Big\|^2 \le c\ , \quad \varepsilon\,\|w_x\|^2 \le c \tag{11.15}$$

holds uniformly with respect to $\varepsilon \le \varepsilon_0, \delta, \lambda$. From (11.9) for $t = 0$ it follows that $v_t(0) = 0$. Equations (11.10), (11.11) imply $w_t(0) = w^0_{xx} \in L^2(I)$, $m_{txx}(0) = w^0_{xx}$, ie. $m_t(0)$ is bounded in $H^2(I)\cap H^1_0(I)$.

Differentiating equations (11.9)–(11.11) with respect to t and multiplying them by v_t, w_t, m_t, respectively, in view of the inequalities

$$\langle\, (\beta_{1t}, \beta_{2t}), (v_t, w_t)\,\rangle \ge 0\ , \quad \langle\, p(m)_t, m_t\,\rangle \ge 0\ ,$$

which hold a.e. on $(0,T)$, we obtain

$$\frac{1}{2}\frac{d}{dt}\{\varepsilon\,\|v_t(t)\|^2 + \varepsilon\,\|w_t(t)\|^2 + \|v_x(t)\|^2 + \|m_{tx}(t)\|^2\}$$
$$+\varepsilon\,\|w_{tx}(t)\|^2 \le 0\ .$$

After integrating the above inequality we get

$$\max_{0\le t\le T}\{\varepsilon\,\|v_t(t)\|^2 + \varepsilon\,\|w_t(t)\|^2 + \|v_x(t)\|^2 + \|m_{tx}(t)\|^2\} \le c\ , \tag{11.16}$$

$$\varepsilon\,\|w_{tx}\|^2_{L^2(Q)} \le c\ . \tag{11.17}$$

Another estimate can also be established. The multiplication of (11.9)–(11.11) by $v-v^0$, $w-w^0$, $m-m^0$, respectively, implies that $\lambda^{-1}\langle\, p(m), m-m^0\,\rangle$ are bounded in $L^2(0,T)$ uniformly with respect to $\varepsilon \le \varepsilon^0, \delta, \lambda$.

Consider the functional $q(m) = \|m - \pi m\|^2$. Its derivative is equal to $q'(m) = 2p(m)$. Let us choose $\hat{m} \in L^\infty(Q)$ such that $|\hat{m}(t) + m^0| \leq c^*$ a.e. on $(0,T)$. Then, by the convexity of q,

$$\lambda^{-1}\langle\, p(m), \hat{m}\,\rangle \leq \lambda^{-1}\langle\, p(m), m - m^0\,\rangle + \frac{1}{2}\,\lambda^{-1}q(\hat{m} + m^0) - \frac{1}{2}\,\lambda^{-1}q(m)\ .$$

Taking into account the boundedness of $\lambda^{-1}\langle\, p(m), m - m^0\,\rangle$ we obtain

$$\lambda^{-1}p(m) \quad \text{are bounded in} \quad L^2\big(0,T;L^1(I)\big)\ .$$

Consequently, bearing in mind (11.11), the inclusion $L^1(I) \subset H^{-1}(I)$ implies that

$$w \quad \text{are bounded in} \quad L^2\big(0,T;H_0^1(I)\big)\ , \tag{11.18}$$

uniformly with respect to $\varepsilon \leq \varepsilon_0, \delta, \lambda$. When estimating w we have actually used the same arguments as in the previous section.

Now we are in a position to prove the solvability of problem (11.9)–(11.13) for arbitrary but fixed parameter values by using the Galerkin method. An essential feature is that estimates of the type (11.14)–(11.17) can be obtained for the Galerkin solutions. This enables us to prove the solvability of the Galerkin equations and to pass to the limit when the dimension of the Galerkin approximation tends to the infinity. In a similar way estimate (11.18) follows from (11.9)–(11.13) provided (11.9)–(11.13) is solvable.

Therefore, for fixed parameters, there exists a solution to problem (11.9)–(11.13)

$$v, w, v_t, w_t \in L^\infty\big(0,T;L^2(I)\big)\ ,$$

$$w, w_t \in L^2\big(0,T;H_0^1(I)\big)\ ,\quad v, m, m_t \in L^\infty\big(0,T;H_0^1(I)\big)\ .$$

We now analyse the passage to the limit. At each step the solution is indexed by the corresponding parameter without referring to the dependence on the remaining parameters.

First, we let $\varepsilon \to 0$. The solution to (11.9)–(11.13) is denoted by $v^\varepsilon, w^\varepsilon, m^\varepsilon$. We can choose a subsequence of $v^\varepsilon, w^\varepsilon, m^\varepsilon$, still denoted by $v^\varepsilon, w^\varepsilon, m^\varepsilon$, such that

$$\varepsilon v^\varepsilon, \varepsilon v_t^\varepsilon, \varepsilon w^\varepsilon, \varepsilon w_t^\varepsilon \to 0 \quad \text{weakly-}(*) \text{ in } L^\infty\big(0,T;L^2(I)\big)\ ,$$

$$\varepsilon w^\varepsilon \to 0 \quad \text{weakly in} \quad L^2\big(0,T;H_0^1(I)\big)\ ,$$

$$v^\varepsilon, m^\varepsilon, m_t^\varepsilon \to v^\lambda, m^\lambda, m_t^\lambda \ \text{weakly-}(*) \text{ in } L^\infty\big(0,T;H_0^1(I)\big)\ ,$$

$$w^\varepsilon \to w^\lambda \quad \text{weakly in} \quad L^2\big(0,T;H_0^1(I)\big)\ ,$$

$$m^\varepsilon \to m^\lambda \quad \text{strongly in} \quad L^2(Q)\ ,$$

$$\int_0^t m_x^\varepsilon d\tau \to \int_0^t m_x^\lambda d\tau \quad \text{weakly in} \quad L^2(Q)$$

when $\varepsilon \to 0$. From (11.9)–(11.11) after passing to the limit when $\varepsilon \to 0$ it follows that the limiting functions $v^\lambda, w^\lambda, m^\lambda$ satisfy the relations

$$-\int_0^t v^\lambda_{xx} d\tau + \delta^{-1}\beta_1(v^\lambda, w^\lambda) = 0 \ , \tag{11.19}$$

$$-m^\lambda_{xx} + \delta^{-1}\beta_2(v^\lambda, w^\lambda) = f \ , \tag{11.20}$$

$$-m^\lambda_{txx} + w^\lambda_{xx} + \lambda^{-1}p(m^\lambda) = 0 \ . \tag{11.21}$$

The convergence of nonlinear terms $\beta_i(v^\varepsilon, w^\varepsilon)$ follows from the monotonicity of the operator (β_1, β_2).

We can pass to the limit in (11.19)–(11.21) when $\lambda \to 0$. Indeed, there exixts a subsequence of $v^\lambda, w^\lambda, m^\lambda$, still denoted by $v^\lambda, w^\lambda, m^\lambda$, possessing the following properties

$$v^\lambda, m^\lambda, m^\lambda_t \to v^\delta, m^\delta, m^\delta_t \quad \text{weakly-}(*) \text{ in } \quad L^\infty(0,T; H^1_0(I)),$$

$$w^\lambda \to w^\delta \quad \text{weakly in} \quad L^2(0,T; H^1_0(I)) \ ,$$

$$\lambda^{-1}p(m^\lambda) \to \eta^\delta \quad \text{weakly in} \quad L^2(0,T; H^{-1}(I)) \ ,$$

$$\int_0^t v^\lambda_x d\tau \to \int_0^t v^\delta_x d\tau \quad \text{weakly in} \quad L^2(Q)$$

when $\lambda \to 0$. From (11.20) when $\lambda \to 0$ it evidently follows that

$$m^\lambda \quad \text{are bounded in} \quad L^2(0,T; H^2(I) \cap H^1_0(I)) \ .$$

Clearly, this boundedness is in general not uniform with respect to δ. Therefore, we may additionally assume that for every fixed δ

$$m^\lambda \to m^\delta \quad \text{strongly in} \quad L^2(0,T; H^1_0(I)) \ . \tag{11.22}$$

From (11.21) it follows that

$$-m^\delta_{txx} + w^\delta_{xx} + \eta^\delta = 0 \ . \tag{11.23}$$

By (11.21) and monotonicity,

$$\int_0^T \langle -m^\lambda_{txx} + w^\lambda_{xx}, \bar{m} - m^\lambda \rangle dt \geq 0 \ .$$

In view of what has been shown above we can pass to the limit and we get

$$\int_0^T \langle \eta^\delta, \bar{m} - m^\delta \rangle dt \leq 0 \quad \forall\, \bar{m} \in L^2(0,T; H^1_0(I)) \ , \quad \bar{m}(t) \in K \text{ a.e.} \tag{11.24}$$

Thus, the limiting functions $v^\delta, w^\delta, m^\delta, \eta^\delta$ satisfy (11.23), (11.24) and the equations

$$-\int_0^t v^\delta_{xx} d\tau + \delta^{-1}\beta_1(v^\delta, w^\delta) = 0 \ , \tag{11.25}$$

$$-m^\delta_{xx} + \delta^{-1}\beta_2(v^\delta, w^\delta) = f \ . \tag{11.26}$$

The inclusion $m^\delta(t) \in K$ follows from (11.21).

Finally, we demonstrate that we can pass to the limit when $\delta \to 0$. Choosing a subsequence with the same notation we assume that

$$v^\delta, m^\delta, m_t^\delta \to v, m, m_t \quad \text{weakly-}(*) \text{ in} \quad L^\infty(0,T;H_0^1(I)) \ ,$$

$$w^\delta \to w \quad \text{weakly in} \quad L^2(0,T;H_0^1(I)) \ ,$$

$$\eta^\delta \to \eta \quad \text{weakly in} \quad L^2(0,T;H^{-1}(I)) \ ,$$

$$\int_0^t v_x^\delta d\tau \to \int_0^t v_x d\tau \quad \text{weakly in} \quad L^2(Q) \ .$$

Let us multiply (11.25), (11.26) by $\bar{v} - v^\delta$, $\bar{w} - w^\delta$, respectively, and sum up with (11.24). The term $\int_0^T \langle \, m_{xx}^\delta, w^\delta \, \rangle \, dt$ vanishes in the resulting relation. Therefore, after passing to the limit we get the equation

$$-m_{txx} + w_{xx} + \eta = 0$$

and the inequality

$$\int_0^T \langle \, m_{xx} + f, \bar{w} - w \, \rangle dt + \int_0^T \langle \, \int_0^t v_{xx} d\tau, \bar{v} - v \, \rangle dt$$

$$+ \int_0^T \langle \, \eta, \bar{m} - m \, \rangle dt \le 0 \quad \forall \, \bar{v}, \bar{w}, \bar{m} \in L^2(0,T;H_0^1(I)) \ ,$$

$$\big(\bar{v}(t), w(t)\big) \in B \ , \quad \bar{m}(t) \in K \quad \text{a.e. on} \quad (0,T) \ .$$

Denote

$$n(t) = n^0 + \int_0^t v_x d\tau \ . \tag{11.27}$$

It follows from the above relations that all equations and inequalities appearing in the statement of the theorem are satisfied. Also, the initial condition $m = m^0$ is not affected by passing to the limit when $\varepsilon \to 0$, $\lambda \to 0$, $\delta \to 0$. The initial condition for n follows from (11.27). This completes the proof. □

Chapter 4
Optimal Control Problems

1 Optimal distribution of external forces for plates with obstacles

In the present section we study optimization problems for elastic plates with obstacles. An optimal distribution of external forces is attained via the minimization of a functional which depends on the plate displacement. Therefore, the right-hand side of the variational inequality describing the displacement of an elastic plate with an obstacle loaded by distributed external forces, serves as a control.

1.1 Cost functionals with measures

Let $\Omega \subset \mathbb{R}^2$ be a bounded domain with smooth boundary Γ and let ϕ, $\phi|_\Gamma < 0$, be a smooth function in $\bar{\Omega}$. As before, $\|\cdot\|_s$ denotes the norm in $H^s(\Omega)$.

Consider the convex and closed set

$$K = \{w \in H_0^2(\Omega) | w(x) \geq \phi(x), \quad x \in \Omega\} .$$

For a given $u \in L^2(\Omega)$, w is a solution to the variational inequality

$$w \in K : \quad \int_\Omega \Delta w \Delta(\varphi - w) dx \geq \int_\Omega u(\varphi - w) dx \quad \forall\, \varphi \in K . \tag{1.1}$$

This inequality is equivalent to the minimization over K of the energy functional of the plate

$$\mathcal{I}(\varphi) = \int_\Omega (\Delta\varphi)^2 dx - 2 \int_\Omega u\varphi dx .$$

This is a mathematical model of an elastic plate in the state of static equilibrium, interacting with an obstacle. The obstacle shape is defined by the equation $z = \phi(x)$.

From (1.1) it follows that

$$\mu_u = \Delta^2 w - u$$

is a positive distribution on Ω, and consequently, a nonnegative Radon measure in Ω. This measure describes the work of interaction forces between the plate and the obstacle.

Let $\mathcal{U}_{ad} \subset L^2(\Omega)$ be a closed and convex set. Consider the control problem

$$\inf_{\mathcal{U}_{ad}} \mathcal{J}(u) \equiv \inf_{\mathcal{U}_{ad}} \big(\mu_u(\Omega) + \|u\|_0^2\big). \tag{1.2}$$

Problem (1.2) is solvable. Indeed, let $\{u_k\} \subset \mathcal{U}_{ad}$ be a minimizing sequence, bounded in $L^2(\Omega)$. Thus, there exists a subsequence, without loss of generality denoted by $\{u_k\}$, and an element $u \in L^2(\Omega)$ such that

$$u_k \to u \quad \text{weakly in } L^2(\Omega) .$$

Obviously, $u \in \mathcal{U}_{ad}$, since $\mathcal{U}_{ad}$ is weakly closed in $L^2(\Omega)$. For a fixed $\varphi \in K$, by putting in (1.1) $u = u_k$ the a priori estimate

$$\|\Delta w_k\|_0 \le c$$

follows with c being independent of $k = 1, 2, \ldots$ Therefore, there exists a subsequence, without loss of generality denoted by $\{w_k\}$, such that

$$w_k \to w \quad \text{weakly in } H_0^2(\Omega) .$$

Put $\mu_k \equiv \mu_{u_k} = \Delta^2 w_k - u_k$. For every compact set $B \subset \Omega$, $\mu_k(B) \le c$ uniformly with respect to $k = 1, 2, \ldots$ Indeed, let $\eta \in C_0^\infty(\Omega)$, $\eta \equiv 1$ on B, $\eta \ge 0$. Then

$$\mu_k(B) \le \int_\Omega \eta d\mu_k = \int_\Omega (\Delta\eta \Delta w_k - \eta u_k) dx \le c .$$

Hence, $\mu_k \to \mu_u \equiv \Delta^2 w - u$ weakly-$(*)$, and for every function φ which is continuous in Ω and finite, ie. has a compact support in Ω,

$$\int \varphi d\mu_k \to \int \varphi d\mu_u .$$

The measure μ_u depends on the pair of functions (u, w), where w solves variational inequality (1.1) with u as the right-hand side (see also the explanation below). In particular, from the weak-$(*)$ convergence of the sequence of measures $\{\mu_k\}$, it follows that

$$\liminf_{k\to\infty} \mu_k(\Omega) \ge \mu_u(\Omega) . \tag{1.3}$$

For u_k , w_k , $k = 1, 2, \ldots,$ inequality (1.1) takes the form

$$\int_\Omega \Delta w_k \Delta\varphi dx \ge \|\Delta w_k\|_0^2 + \int_\Omega u_k(\varphi - w_k) dx . \tag{1.4}$$

By the Sobolev imbedding theorem,

$$w_k \to w \quad \text{strongly in } L^2(\Omega) ,$$

since the sequence $\{w_k\}$ converges weakly in $H_0^2(\Omega)$. Finally, passing to the limit in (1.4), we see that the limits w, u satisfy inequality (1.1). Let

$$j = \inf_{u \in \mathcal{U}_{ad}} \mathcal{J}(u) .$$

By the lower semicontinuity of the norm $\|\cdot\|_0$ and by inequality (1.3),

$$j = \liminf_{k \to \infty} \mathcal{J}(u_k) \geq \mathcal{J}(u) \geq j ,$$

which means that u solves (1.1)–(1.2).

For $\varepsilon > 0$, $u \in \mathcal{U}_{ad}$, consider the equation

$$w \in H_0^2(\Omega) \ : \quad \Delta^2 w + \frac{1}{\varepsilon} p(w) = u, \tag{1.5}$$

where

$$p(w)(x) = \begin{cases} w(x) - \phi(x), & \text{if } w(x) - \phi(x) \leq 0 \\ 0, & \text{if } w(x) - \phi(x) > 0 . \end{cases}$$

For every u, the solutions w_ε to nonlinear equation (1.5) converge to a solution of variational inequality (1.1) as $\varepsilon \downarrow 0$.

Let $\mu_{u,\varepsilon} \equiv \Delta^2 w - u$, where w is defined by (1.5). Consider the optimal control problem

$$\inf_{\mathcal{U}_{ad}} \mathcal{J}_\varepsilon(u) \equiv \inf_{\mathcal{U}_{ad}} \left(\mu_{u,\varepsilon}(\Omega) + \|u\|_0^2\right). \tag{1.6}$$

We first show that for every $\varepsilon > 0$ problem (1.5),(1.6) is solvable.

Solutions to problem (1.5) are regular in the sense that $w \in H^{4,0}(\Omega) \equiv H^4(\Omega) \cap H_0^2(\Omega)$. Therefore,

$$\mu_{u,\varepsilon}(\Omega) = \int_\Omega (\Delta^2 w - u) dx.$$

Any minimizing sequence $u_k \in \mathcal{U}_{ad}$ is bounded in $L^2(\Omega)$. The corresponding solutions w_k belong to $H^{4,0}(\Omega)$. By choosing any element $\varphi \in K$ and multiplying the equation

$$\Delta^2 w_k + \frac{1}{\varepsilon} p(w_k) = u_k \tag{1.7}$$

by $w_k - \varphi$ we get

$$\|\Delta w_k\|_0 \leq c$$

which holds uniformly with respect to k. From (1.7) it follows that

$$\|w_k\|_4 \leq c$$

holds uniformly with respect to k, and, since ε is fixed, the dependence of c on ε can be neglected.

Thus, without loss of generality we can assume that

$$\begin{aligned} u_k &\to u \quad \text{weakly in} \quad L^2(\Omega), \\ w_k &\to w \quad \text{weakly in} \quad H^{4,0}(\Omega), \\ w_k &\to w \quad \text{strongly in} \quad L^2(\Omega) \end{aligned}$$

as $k \to \infty$. Passing to the limit in equation (1.7), as $k \to \infty$, we get equation (1.5) which entails that $w = w(u)$.

Denote

$$j_\varepsilon = \inf_{u \in \mathcal{U}_{ad}} \mathcal{J}_\varepsilon(u) . \tag{1.8}$$

In view of the convergences of the sequences $\{u_k\}$, $\{w_k\}$, as $k \to \infty$,

$$j_\varepsilon = \liminf_{k \to \infty} \mathcal{J}_\varepsilon(u_k) \geq \mathcal{J}_\varepsilon(u) \geq j_\varepsilon .$$

Therefore, $u = u_\varepsilon$ solves problem (1.6). A solution to (1.5) for $u = u_\varepsilon$ is denoted by $w_\varepsilon = w_\varepsilon(u_\varepsilon)$.

Theorem 1.1. *There exist subsequences, denoted by $\{w_\varepsilon\}$, $\{u_\varepsilon\}$, such that*

$$u_\varepsilon \to u \quad \textit{weakly in } L^2(\Omega),$$
$$w_\varepsilon \to w \quad \textit{weakly in } H_0^2(\Omega) ,$$

as $\varepsilon \downarrow 0$. Moreover,

$$j_\varepsilon \to j,$$

u is a solution to (1.2) and $w = w(u)$ solves (1.1).

Proof. For $v \in \mathcal{U}_{ad}$, let μ_v be the measure defined by (1.1) with $u = v$. We begin by showing that

$$\lim_{\varepsilon \downarrow 0} \mathcal{J}_\varepsilon(v) = \mathcal{J}(v), \tag{1.9}$$

where $\mathcal{J}(v) = \mu_v(\Omega) + \|v\|_0^2$. Let $w_\varepsilon(v)$ be a solution to (1.5) for $u = v$. As before, the a priori estimate

$$\|w_\varepsilon(v)\|_2 \leq c \tag{1.10}$$

holds uniformly with respect to $\varepsilon > 0$. Hence, there exists a subsequence, without loss of generality denoted by $\{w_\varepsilon(v)\}$, such that

$$w_\varepsilon(v) \to \widetilde{w} \quad \text{weakly in } H_0^2(\Omega),$$

as $\varepsilon \downarrow 0$, and furthermore,

$$\mu_{v,\varepsilon} \equiv \Delta^2 w_\varepsilon(v) - v \to \mu_v \equiv \Delta^2 \widetilde{w} - v \quad \text{weakly-}(*) .$$

The measure μ_v is well-defined since the pair $(v, \widetilde{w})$ satisfies variational inequality (1.1) — all the details are given below.

From the imbedding theorems it follows that $w_\varepsilon(v) \to \widetilde{w}$ uniformly in $\bar{\Omega}$. There exists a compact $B \subset \Omega$ such that for $\varepsilon > 0$ small enough

$$\mu_{v,\varepsilon}(\Omega \setminus B) = 0 . \tag{1.11}$$

Indeed, $\widetilde{w}$ is continuous in $\bar{\Omega}$ and vanishes on the boundary Γ. Moreover, the function $p(w_\varepsilon(v))$ is equal to zero when $w_\varepsilon(v) > \phi$, and $\phi < 0$ in an open neighbourhood of Γ.

Hence, we can assume that $\mu_v(\partial B) = 0$ on the boundary ∂B of the compact set B. By this, $\mu_{v,\varepsilon}(B) \to \mu_v(B)$ (for details see Chapter 1, Section 1.10).

By (1.11),

$$\mu_{v,\varepsilon}(\Omega) = \mu_{v,\varepsilon}(B) \to \mu_v(B) = \mu_v(\Omega)$$

as $\varepsilon \downarrow 0$. Hence,

$$\mathcal{J}_\varepsilon(v) = \mu_{v,\varepsilon}(\Omega) + \|v\|_0^2 \to \mathcal{J}(v) = \mu_v(\Omega) + \|v\|_0^2 \,.$$

Now, to get (1.9) it remains to show that the pair $(v, \widetilde{w})$ satisfies the variational inequality

$$\widetilde{w} \in K : \quad \int_\Omega \Delta\widetilde{w}\Delta(\varphi - \widetilde{w})dx \geq \int_\Omega v(\varphi - \widetilde{w})dx \quad \forall\, \varphi \in K \,.$$

This, however, follows in a standard way from the nonlinear equation

$$\Delta^2 w_\varepsilon(v) + \frac{1}{\varepsilon}p(w_\varepsilon(v)) = v \,.$$

This establishes (1.9).

Let u be a solution to (1.1), (1.2). Then

$$j_\varepsilon \leq \mathcal{J}_\varepsilon(u) \to \mathcal{J}(u) \,.$$

Consequently,

$$\limsup_{\varepsilon\downarrow 0} j_\varepsilon \leq \mathcal{J}(u). \tag{1.12}$$

By the definition of a solution to (1.5)–(1.6),

$$\mu_{u_\varepsilon,\varepsilon}(\Omega) + \|u_\varepsilon\|_0^2 \leq \mu_{v,\varepsilon}(\Omega) + \|v\|_0^2$$

and, by the above arguments, the right-hand side of this inequality is bounded uniformly with respect to $\varepsilon > 0$. In particular,

$$\|u_\varepsilon\|_0 \leq c \,.$$

Then, from the equation

$$\Delta^2 w_\varepsilon + \frac{1}{\varepsilon}p(w_\varepsilon) = u_\varepsilon \tag{1.13}$$

we get the a priori estimate

$$\|w_\varepsilon\|_2 \leq c$$

which holds uniformly with respect to $\varepsilon > 0$. In view of these a priori estimates there exist elements $\widetilde{u} \in \mathcal{U}_{ad}$ and $\widetilde{w} \in H_0^2(\Omega)$ such that

$$\begin{aligned}
&u_\varepsilon \to \widetilde{u} \quad \text{weakly in } L^2(\Omega),\\
&w_\varepsilon \to \widetilde{w} \quad \text{weakly in } H_0^2(\Omega) \text{ and strongly in } L^2(\Omega),\\
&\mu_{u_\varepsilon,\varepsilon} \equiv \Delta^2 w_\varepsilon - u_\varepsilon \to \mu_{\widetilde{u}} \equiv \Delta^2\widetilde{w} - \widetilde{u} \quad \text{weakly-}(*) \,.
\end{aligned}$$

Passing to the limit in equation (1.13) as $\varepsilon \downarrow 0$ we can show that $\widetilde{w} = w(\widetilde{u})$, ie. $\widetilde{w}$ solves (1.1) for $u = \widetilde{u}$. On the other hand,

$$\mu_{u_\varepsilon,\varepsilon}(\Omega) \to \mu_{\widetilde{u}}(\Omega) \tag{1.14}$$

when $\varepsilon \downarrow 0$. Thus,

$$\liminf_{\varepsilon \downarrow 0} j_\varepsilon \geq \mu_{\widetilde{u}}(\Omega) + \|\widetilde{u}\|_0^2 \ . \tag{1.15}$$

Comparing (1.12) with (1.15), we conclude that $\widetilde{u}$ solves problem (1.2), and $j_\varepsilon \to j$ as $\varepsilon \downarrow 0$. Since we have already shown that $\widetilde{w} = w(\widetilde{u})$, the theorem is proved. □

1.2 Cost functionals with norms

Consider now an optimal control problem with a cost functional other than that in (1.2), namely,

$$\inf_{u \in \mathcal{U}_{ad}} \mathcal{J}(u) \equiv \inf_{u \in \mathcal{U}_{ad}} \{\|w(u) - w_0\|_0^2 + \|u\|_0^2\}, \tag{1.16}$$

where, for $u \in \mathcal{U}_{ad}$, $w(u)$ solves variational inequality (1.1) and $w_0 \in L^2(\Omega)$ is given. Using the same arguments as in Section 1.1 the existence of an optimal solution to problem (1.16) can be established.

As before, we consider a family of optimization problems depending on $\varepsilon > 0$. For a fixed $\varepsilon > 0$, an approximate problem is of the form

$$\inf_{u \in \mathcal{U}_{ad}} \mathcal{J}_\varepsilon(u) \equiv \inf_{u \in \mathcal{U}_{ad}} \{\|w_\varepsilon(u) - w_0\|_0^2 + \|u\|_0^2\} \ . \tag{1.17}$$

Here $w_\varepsilon(u) \in H_0^2(\Omega)$ is a solution to the nonlinear equation

$$\Delta^2 w_\varepsilon(u) + \frac{1}{\varepsilon} p(w_\varepsilon(u)) = u \ . \tag{1.18}$$

For any $\varepsilon > 0$, problem (1.17) is solvable. Introduce the notations

$$\begin{aligned} j &= \inf_{u \in \mathcal{U}_{ad}} \mathcal{J}(u), \\ j_\varepsilon &= \inf_{u \in \mathcal{U}_{ad}} \mathcal{J}_\varepsilon(u), \\ w_\varepsilon &= w_\varepsilon(u_\varepsilon) \ . \end{aligned}$$

Theorem 1.2. *There exist subsequences, without loss of generality denoted by $\{u_\varepsilon\}$ and $\{w_\varepsilon\}$, such that*

$$\begin{aligned} u_\varepsilon &\to u \quad \textit{weakly in } L^2(\Omega), \\ w_\varepsilon &\to w \quad \textit{weakly in } H_0^2(\Omega), \\ j_\varepsilon &\to j, \end{aligned}$$

as $\varepsilon \downarrow 0$, where u is a solution to problem (1.16), (1.1), and $w = w(u)$ solves variational inequality (1.1).

Proof. For any fixed $v \in \mathcal{U}_{ad}$,

$$\mathcal{J}_\varepsilon(u_\varepsilon) \le \mathcal{J}_\varepsilon(v).$$

Hence, the sequence $\{u_\varepsilon\}$ is bounded in $L^2(\Omega)$ for $\varepsilon_0 > \varepsilon > 0$. From the equation

$$\Delta^2 w_\varepsilon + \frac{1}{\varepsilon} p(w_\varepsilon) = u_\varepsilon \tag{1.19}$$

it follows that

$$\|w_\varepsilon\|_2 \le c$$

where c is independent of ε.

Therefore, there exist elements $\widetilde{u} \in \mathcal{U}_{ad}$, $\widetilde{w} \in H_0^2(\Omega)$ such that for subsequences, without loss of generality denoted by $\{u_\varepsilon\}$ and $\{w_\varepsilon\}$, respectively,

$$\begin{aligned} u_\varepsilon &\to \widetilde{u} \quad \text{weakly in } L^2(\Omega), \\ w_\varepsilon &\to \widetilde{w} \quad \text{weakly in } H_0^2(\Omega) \end{aligned} \tag{1.20}$$

as $\varepsilon \downarrow 0$. Multiplying (1.19) by $\varphi - w_\varepsilon$, $\varphi \in K$, and integrating over Ω we get

$$\int_\Omega \Delta w_\varepsilon \Delta(\varphi - w_\varepsilon) dx \ge \int_\Omega u_\varepsilon(\varphi - w_\varepsilon) dx \ . \tag{1.21}$$

In view of (1.20)

$$w_\varepsilon \to \widetilde{w} \quad \text{strongly in } L^2(\Omega),$$

and after passing to the upper limit in (1.21) as $\varepsilon \downarrow 0$ we obtain

$$\int_\Omega \Delta\widetilde{w}\Delta(\varphi - \widetilde{w}) dx \ge \int_\Omega \widetilde{u}(\varphi - \widetilde{w}) dx \quad \forall\, \varphi \in K \ ,$$

hence $\widetilde{w} = w(\widetilde{u})$. By (1.20),

$$\liminf_{\varepsilon\downarrow 0} j_\varepsilon \ge \mathcal{J}(\widetilde{u}) \ . \tag{1.22}$$

On the other hand,

$$w_\varepsilon(v) \to w(v) \quad \text{strongly in } L^2(\Omega),$$

for every $v \in L^2(\Omega)$, as $\varepsilon \downarrow 0$, and hence,

$$\mathcal{J}_\varepsilon(v) \equiv \|w_\varepsilon(v) - w_0\|_0^2 + \|v\|_0^2 \to \mathcal{J}(v) \equiv \|w(v) - w_0\|_0^2 + \|v\|_0^2 \ .$$

Here $w_\varepsilon(v)$ is a solution to equation (1.18), and $w(v)$ denotes a solution to (1.1) for $u = v$. Therefore, $\mathcal{J}_\varepsilon(v) \to \mathcal{J}(v)$, and

$$j_\varepsilon \le \mathcal{J}_\varepsilon(u) \to \mathcal{J}(u) \ .$$

Here u is a solution to problem (1.16), (1.1). Consequently,

$$\limsup_{\varepsilon\downarrow 0} j_\varepsilon \le \mathcal{J}(u) \ . \tag{1.23}$$

From (1.22) and (1.23) it follows that $\widetilde{u}$ is a solution to (1.16), (1.1), and

$$\lim_{\varepsilon\downarrow 0} j_\varepsilon = \mathcal{J}(\widetilde{u}) = j \ .$$

This concludes the proof. □

It follows from Theorem 1.2 that solutions to problem (1.16), (1.1) can be approximated by solutions to problem (1.17), (1.18). Furthermore, solutions to problem (1.17)–(1.18) can be approximated by solutions to the problem

$$\inf_{(v,\varphi)\in\mathcal{V}} \mathcal{J}_{\varepsilon,\delta}(v,\varphi), \tag{1.24}$$

where $\delta > 0$ is an auxiliary parameter, $\mathcal{V} = \mathcal{U}_{ad} \times H^{4,0}(\Omega)$, and

$$\mathcal{J}_{\varepsilon,\delta}(v,\varphi) = \|\varphi - w_0\|_0^2 + \|v\|_0^2 + \frac{1}{\delta}\|\Delta^2\varphi + \frac{1}{\varepsilon}p(\varphi) - v\|_0^2 .$$

The penalty term corresponds to nonlinear equation (1.18), and therefore the functions v, φ are independent. Denote

$$j_{\varepsilon,\delta} = \inf_{(v,\varphi)\in\mathcal{V}} \mathcal{J}_{\varepsilon,\delta}(v,\varphi) .$$

For any $\varepsilon > 0$, $\delta > 0$, there exists a solution to problem (1.24). To prove this fact we begin by showing that the functional $\mathcal{J}_{\varepsilon,\delta}(\cdot,\cdot)$ is coercive on $\mathcal{V}$.

We have

$$\|p(v)\|_0 \le c_1\|v\|_0 + c_2 \quad \forall\, v \in L^2(\Omega),$$
$$\|\varphi\|_4 \le c_3\|\Delta^2\varphi\|_0 \quad \forall\, \varphi \in H^{4,0}(\Omega),$$

where constants c_i, $i = 1,2,3$, are independent of v,φ.

Moreover, for every sequence $\{(v_m,\varphi_m)\}_{m=1}^{\infty}$, $(v_m,\varphi_m) \to (v,\varphi)$ weakly in $L^2(\Omega) \times H^{4,0}(\Omega)$,

$$\liminf_{m\to\infty} \mathcal{J}_{\varepsilon,\delta}(v_m,\varphi_m) \ge \mathcal{J}_{\varepsilon,\delta}(v,\varphi) .$$

Indeed, we can assume that $\varphi_m \to \varphi$ strongly in $L^2(\Omega)$, and hence $p(\varphi_m) \to p(\varphi)$ strongly in $L^2(\Omega)$.

Fix $\varepsilon > 0$. Let u_ε be a solution to problem (1.17)–(1.18), $w_\varepsilon = w_\varepsilon(u_\varepsilon)$. By $(u_{\varepsilon,\delta}, w_{\varepsilon,\delta})$ we denote a solution to (1.24).

Obviously, $\|\Delta^2 w_\varepsilon + \frac{1}{\varepsilon}p(w_\varepsilon) - u_\varepsilon\|_0 = 0$, and hence,

$$\mathcal{J}_{\varepsilon,\delta}(u_{\varepsilon,\delta}, w_{\varepsilon,\delta}) \le \mathcal{J}_{\varepsilon,\delta}(u_\varepsilon, w_\varepsilon) = \mathcal{J}_\varepsilon(u_\varepsilon) = j_\varepsilon . \tag{1.25}$$

Moreover, since j_ε does not depend on δ,

$$\|u_{\varepsilon,\delta}\|_0 \le c_1,$$
$$\|w_{\varepsilon,\delta}\|_0 \le c_2,$$
$$\|\Delta^2 w_{\varepsilon,\delta} + \frac{1}{\varepsilon}p(w_{\varepsilon,\delta}) - u_{\varepsilon,\delta}\|_0 \le c_3\delta^{\frac{1}{2}}$$

with constants c_i, $i = 1,2,3$, being independent of δ. Therefore, there exist elements $\widetilde{u}_\varepsilon, \widetilde{w}_\varepsilon$ and a subsequence, without loss of generality denoted by $\{(u_{\varepsilon,\delta}, w_{\varepsilon,\delta})\}$, such that

$$u_{\varepsilon,\delta} \to \widetilde{u}_\varepsilon \quad \text{weakly in } L^2(\Omega),$$
$$w_{\varepsilon,\delta} \to \widetilde{w}_\varepsilon \quad \text{weakly in } H^{4,0}(\Omega)$$

as $\delta \downarrow 0$. Since $\Delta^2\widetilde{w}_\varepsilon + \frac{1}{\varepsilon}p(\widetilde{w}_\varepsilon) = \widetilde{u}_\varepsilon$ we obtain $\widetilde{w}_\varepsilon = w_\varepsilon(\widetilde{u}_\varepsilon)$.

Furthermore,

$$\mathcal{J}_{\varepsilon,\delta}(u_{\varepsilon,\delta}, w_{\varepsilon,\delta}) \geq \|w_{\varepsilon,\delta} - w_0\|_0^2 + \|u_{\varepsilon,\delta}\|_0^2 .$$

Consequently, by the weak convergence of the sequence $\{(u_{\varepsilon,\delta}, w_{\varepsilon,\delta})\}$ as $\delta \downarrow 0$, we get

$$\liminf_{\delta\downarrow 0} \mathcal{J}_{\varepsilon,\delta}(u_{\varepsilon,\delta}, w_{\varepsilon,\delta}) \geq \|\widetilde{w}_\varepsilon - w_0\|_0^2 + \|\widetilde{u}_\varepsilon\|_0^2 \equiv$$

$$\equiv \|w_\varepsilon(\widetilde{u}_\varepsilon) - w_0\|_0^2 + \|\widetilde{u}_\varepsilon\|_0^2 .$$

On the other hand, by (1.25),

$$\limsup_{\delta\downarrow 0} \mathcal{J}_{\varepsilon,\delta}(u_{\varepsilon,\delta}, w_{\varepsilon,\delta}) \leq \|w_\varepsilon(u_\varepsilon) - w_0\|_0^2 + \|u_\varepsilon\|_0^2 .$$

This proves that $j_{\varepsilon\delta} \to j_\varepsilon$ as $\delta \downarrow 0$. Moreover, $\widetilde{u}_\varepsilon$ solves problem (1.17)–(1.18). Thus, we have established the following fact.

Theorem 1.3. *There exist subsequences, without loss of generality denoted by* $\{w_{\varepsilon,\delta}\}$, $\{u_{\varepsilon,\delta}\}$, *such that*

$$u_{\varepsilon,\delta} \to \widetilde{u}_\varepsilon \quad \text{weakly in } L^2(\Omega),$$

$$w_{\varepsilon,\delta} \to \widetilde{w}_\varepsilon \quad \text{weakly in } H_0^2(\Omega),$$

$$j_{\varepsilon\delta} \to j_\varepsilon,$$

as $\delta \downarrow 0$, *where* $\widetilde{u}_\varepsilon$ *is a solution to problem (1.17), (1.18).*

2 Optimal shape of obstacles

2.1 Cost functionals with norms

Let $\Omega \subset \mathbb{R}^2$ be a bounded domain with smooth boundary Γ. For $\phi \in \Psi$, we introduce the following convex set,

$$K_\phi = \{\varphi \in H_0^2(\Omega) | \ \varphi \geq \phi \quad \text{in } \Omega\}$$

and consider the variational inequality

$$w \equiv w(\phi) \in K_\phi : \quad \int_\Omega \Delta w \Delta(\varphi - w) dx \geq \int_\Omega f(\varphi - w) dx \quad \forall\, \varphi \in K_\phi . \tag{2.1}$$

Here $f \in L^2(\Omega)$ is a given function. We assume that $\Psi \subset H^2(\Omega)$ is a closed and convex subset, and $\phi < 0$ on Γ for every $\phi \in \Psi$.

For variational inequality (2.1), consider the following optimization problem. Find an element $\psi \in \Psi$ such that

$$\mathcal{J}(\psi) \leq \mathcal{J}(\phi) \quad \forall\, \phi \in \Psi, \tag{2.2}$$

where

$$\mathcal{J}(\phi) = \|w(\phi) - w_0\|_0^2 + \|\phi\|_2^2 ,$$

and $w_0 \in L^2(\Omega)$ is given.

We begin by examining solvability of problem (2.2). It can be verified that for any $\phi \in \Psi$ there exists a unique $w(\phi) \in K_\phi$ which solves variational inequality (2.1).

Let $\{\phi_m\}_{m=1}^\infty$ be a minimizing sequence for problem (2.2). There exists an element $\psi \in \Psi$ such that for a subsequence, without loss of generality denoted by $\{\phi_m\}$,

$$\begin{aligned} \phi_m &\to \psi \quad \text{weakly in } H^2(\Omega), \\ \phi_m &\to \psi \quad \text{uniformly in } \bar{\Omega} . \end{aligned}$$

Obviously, $\psi \in \Psi$ since the set $\Psi \subset H^2(\Omega)$ is weakly closed. The uniform convergence of ϕ_m on $\bar{\Omega}$ follows from the Sobolev imbedding theorem since $H^2(\Omega) \subset C(\bar{\Omega})$ with compact imbedding.

We now show that for any element $\bar{\varphi} \in K_\psi$ there exists a sequence $\{\varphi_m\}_{m=1}^\infty$, $\varphi_m \in K_{\phi_m}$, $\varphi_m \to \bar{\varphi}$ strongly in $H_0^2(\Omega)$ as $m \to \infty$. Such a sequence can be constructed in the following way.

By the uniform convergence of ϕ_m to ψ, we have $|\phi_m(x) - \psi(x)| < \frac{1}{m}$ for all $x \in \Omega$. Putting $\widetilde{\varphi}_m = \bar{\varphi} + \frac{1}{m}$, we get $\widetilde{\varphi}_m(x) \geq \phi_m(x)$ since $\bar{\varphi}(x) \geq \psi(x)$.

On the other hand, there exists an open neighbourhood $\mathcal{O}$ of Γ such that $\psi(x) \leq -\delta < 0$ for all $x \in \Omega_0 \equiv \mathcal{O} \cap \bar{\Omega}$ and some $\delta > 0$. Therefore, $\phi_m(x) < -\frac{\delta}{2}$ for all $x \in \Omega_0$ and all m large enough.

Moreover, there exists a sequence $\{\chi_m\}_{m=1}^\infty$ with the following properties: $\text{spt}\chi_m \subset \Omega_0$, $\chi_m(x) = \frac{1}{m}$ and $\nabla \chi_m(x) = 0$ on the boundary Γ, $|\partial^\alpha \chi_m / \partial x^\alpha| \leq \frac{c}{m}$ in Ω_0 for all $\alpha = (\alpha_1, \alpha_2)$, where α_1 and α_2 are integers, $|\alpha| = \alpha_1 + \alpha_2 \leq 2$ and for all m, with the constant c independent of m. Setting $\varphi_m = \widetilde{\varphi}_m - \chi_m$, we obtain the required sequence, ie.,

$$\varphi_m \to \bar{\varphi} \quad \text{strongly in } H_0^2(\Omega),$$

$$\varphi_m(x) \geq \phi_m(x) \quad \text{for all } x \in \Omega .$$

Consider the variational inequality

$$\begin{aligned} w(\phi_m) \in K_{\phi_m} : \int_\Omega \Delta w(\phi_m) \Delta(\varphi - w(\phi_m)) dx \\ \geq \int_\Omega f(\varphi - w(\phi_m)) dx \quad \forall\, \varphi \in K_{\phi_m} \end{aligned} \tag{2.3}$$

for $\varphi = \varphi_m$. The sequence $\{\varphi_m\}$ is bounded in $H_0^2(\Omega)$, and therefore

$$\|\Delta w(\phi_m)\|_0 \leq c$$

with a constant c independent of m.

The sequence $\{w(\phi_m)\}$ is bounded in $H_0^2(\Omega)$, hence there exists an element $\widetilde{w}$ and a subsequence, without loss of generality denoted by $\{w(\phi_m)\}$, such that

$$w(\phi_m) \to \widetilde{w} \quad \text{weakly in } H_0^2(\Omega) .$$

Given $\bar{\varphi} \in K_\psi$, let $\varphi_m \in K_{\phi_m}$ be a sequence such that $\varphi_m \to \bar{\varphi}$ strongly in $H_0^2(\Omega)$ as $m \to \infty$. Inequality (2.3) can be rewritten as

$$\int_\Omega \Delta w(\phi_m) \Delta \varphi_m dx \geq \|\Delta w(\phi_m)\|_0^2 + \int_\Omega f(\varphi_m - w(\phi_m))dx \ .$$

In view of the strong convergence of the sequence φ_m and the lower semicontinuity of the L^2 norm, we can pass to the lower limit in this inequality. This leads to the following inequality

$$\widetilde{w} \in K_\psi : \int_\Omega \Delta \widetilde{w} \Delta(\bar{\varphi} - \widetilde{w})dx \geq \int_\Omega f(\bar{\varphi} - \widetilde{w})dx \quad \forall\, \bar{\varphi} \in K_\psi \ ,$$

and hence $\widetilde{w} = w(\psi)$. Since

$$j = \liminf_{m \to \infty} \mathcal{J}(\phi_m) \geq \mathcal{J}(\psi) \geq j \ ,$$

ψ solves (2.1)–(2.2).

We now consider the following family of optimization problems which depend on $\varepsilon > 0$,

$$\inf_{\phi \in \Psi} \mathcal{J}_\varepsilon(\phi) \tag{2.4}$$

with $\mathcal{J}_\varepsilon(\phi) = \|w_\varepsilon(\phi) - w_0\|_0^2 + \|\phi\|_2^2$, where

$$w_\varepsilon(\phi) \in H_0^2(\Omega) \ : \quad \Delta^2 w_\varepsilon(\phi) + \frac{1}{\varepsilon}\mathcal{P}_\phi(w_\varepsilon(\phi)) = f \ . \tag{2.5}$$

Here $\mathcal{P}_\phi(w_\varepsilon(\phi))$ is a penalty term corresponding to the inequality constraints $w_\varepsilon(\phi) \geq \phi$,

$$\mathcal{P}_\phi(v)(x) = \begin{cases} v(x) - \phi(x), & \text{if } v(x) - \phi(x) < 0 \\ 0, & \text{if } v(x) - \phi(x) \geq 0 \ . \end{cases}$$

Before proving an existence result for problem (2.4)–(2.5), let us note that for any $\varepsilon > 0$ and $\phi \in \Psi$, there exists a solution to equation (2.5). Moreover, $w_\varepsilon(\phi) \in H^{4,0}(\Omega) = H^4(\Omega) \cap H_0^2(\Omega)$.

Let ϕ_m be a minimizing sequence for problem (2.4), (2.5). The sequence ϕ_m is bounded, therefore there exists an element ϕ_ε such that $\phi_m \to \phi_\varepsilon$ weakly in $H^2(\Omega)$ and uniformly in $\bar{\Omega}$ as $m \to \infty$. For a given element $\bar{\varphi} \in K_{\phi_\varepsilon}$ a sequence $\bar{\varphi}_m \in K_{\phi_m}$ can be constructed such that $\bar{\varphi}_m \to \bar{\varphi}$ strongly in $H_0^2(\Omega)$ as $m \to \infty$. Multiplying

$$\Delta^2 w_\varepsilon(\phi_m) + \frac{1}{\varepsilon}\mathcal{P}_{\phi_m}(w_\varepsilon(\phi_m)) = f \tag{2.6}$$

by $w_\varepsilon(\phi_m) - \bar{\varphi}_m$, and taking into account that

$$\int_\Omega \mathcal{P}_{\phi_m}(w_\varepsilon(\phi_m))(w_\varepsilon(\phi_m) - \bar{\varphi}_m)dx \geq 0$$

we obtain the estimate

$$\|\Delta w_\varepsilon(\phi_m)\|_0 \le c$$

which holds uniformly with respect to m. Hence, there exist an element $\widetilde{w}_\varepsilon \in H_0^2(\Omega)$ and a subsequence, without loss of generality denoted by $\{w_\varepsilon(\phi_m)\}_{m=1}^{\infty}$, such that $w_\varepsilon(\phi_m) \to \widetilde{w}_\varepsilon$ weakly in $H_0^2(\Omega)$ and strongly in $L^2(\Omega)$ as $m \to \infty$.

Passing now to the limit in (2.6), as $m \to \infty$, we get $\widetilde{w}_\varepsilon = w_\varepsilon(\phi_\varepsilon)$. Since

$$j_\varepsilon = \inf_{\phi \in \Psi} \mathcal{J}_\varepsilon(\phi),$$

and the sequences ϕ_m, $w_\varepsilon(\phi_m)$ converge weakly as $m \to \infty$, it follows that

$$j_\varepsilon = \liminf_{m \to \infty} \mathcal{J}_\varepsilon(\phi_m) \ge \mathcal{J}_\varepsilon(\phi_\varepsilon) \ge j_\varepsilon$$

which means that ϕ_ε is a solution to problem (2.4)–(2.5).

The following theorem establishes a relationship between solutions of problem (2.1)–(2.2) and those of problem (2.4)–(2.5).

Theorem 2.1. *There exist subsequences, without loss of generality denoted by* $\{\phi_\varepsilon\}$ *and* $\{w_\varepsilon(\phi_\varepsilon)\}$, *such that*

$$\begin{aligned}
&\phi_\varepsilon \to \psi \quad \text{weakly in } H^2(\Omega),\\
&w_\varepsilon(\phi_\varepsilon) \to w(\psi) \quad \text{weakly in } H_0^2(\Omega),\\
&j_\varepsilon \to j,
\end{aligned}$$

as $\varepsilon \downarrow 0$, *where* ψ *is a solution to problem (2.1)–(2.2).*

Proof. For any $\phi \in \Psi$ we have

$$\mathcal{J}_\varepsilon(\phi_\varepsilon) \le \mathcal{J}_\varepsilon(\phi)\,. \tag{2.7}$$

Multiplying equation (2.5) by $w_\varepsilon(\phi) - \bar{\varphi}$, $\bar{\varphi} \in K_\phi$, in a standard way we get the estimate

$$\|\Delta w_\varepsilon(\phi)\|_0 \le c$$

which holds uniformly with respect to $\varepsilon > 0$. Hence, $\mathcal{J}_\varepsilon(\phi)$ is bounded from above for all $\varepsilon > 0$, and, in view of (2.7), the sequence $\{\phi_\varepsilon\}$ is bounded uniformly with respect to $\varepsilon > 0$,

$$\|\phi_\varepsilon\|_2 \le c\,.$$

By this, there exists $\widetilde{\phi} \in H^2(\Omega)$ such that for a subsequence, without loss of generality denoted by $\{\phi_\varepsilon\}$, $\phi_\varepsilon \to \widetilde{\phi}$ weakly in $H^2(\Omega)$.

Take $\bar{\varphi} \in K_{\widetilde{\phi}}$. As before, a family $\{\bar{\varphi}_\varepsilon\} \in K_{\phi_\varepsilon}$ exists such that $\bar{\varphi}_\varepsilon \to \bar{\varphi}$ strongly in $H_0^2(\Omega)$ as $\varepsilon \downarrow 0$. Since the non-negative penalty terms can be neglected, the multiplication of

$$\Delta^2 w_\varepsilon(\phi_\varepsilon) + \frac{1}{\varepsilon}\mathcal{P}_{\phi_\varepsilon}(w_\varepsilon(\phi_\varepsilon)) = f$$

by $w_\varepsilon(\phi_\varepsilon) - \bar{\varphi}_\varepsilon$ gives

$$\int_\Omega \Delta w_\varepsilon(\phi_\varepsilon)\Delta(\bar{\varphi}_\varepsilon - w_\varepsilon(\phi_\varepsilon))dx \geq \int_\Omega f(\bar{\varphi}_\varepsilon - w_\varepsilon(\phi_\varepsilon))dx \ . \tag{2.8}$$

From (2.8) we obtain the estimate

$$\|w_\varepsilon(\phi_\varepsilon)\|_2 \leq c$$

which is uniform with respect to $\varepsilon > 0$. Hence, there exists an element $\widetilde{w} \in H_0^2(\Omega)$ such that for a subsequence, without loss of generality denoted by $\{w_\varepsilon(\phi_\varepsilon)\}$, $w_\varepsilon(\phi_\varepsilon) \to \widetilde{w}$ weakly in $H_0^2(\Omega)$.

Passing to the upper limit in (2.8), we get the variational inequality

$$\widetilde{w} \in K_{\widetilde{\phi}} : \quad \int_\Omega \Delta\widetilde{w}\Delta(\bar{\varphi} - \widetilde{w})dx \geq \int_\Omega f(\bar{\varphi} - \widetilde{w})dx \quad \forall\, \bar{\varphi} \in K_{\widetilde{\phi}} \ .$$

Thus, $\widetilde{w} = w(\widetilde{\phi})$. By the weak convergence of the sequences $\{\phi_\varepsilon\}$ and $\{w_\varepsilon(\phi_\varepsilon)\}$,

$$\liminf_{\varepsilon\downarrow 0} j_\varepsilon = \liminf_{\varepsilon\downarrow 0} \mathcal{J}_\varepsilon(\phi_\varepsilon) \geq \mathcal{J}(\widetilde{\phi}) \ . \tag{2.9}$$

On the other hand, for $\psi \in \Psi$ which solves problem (2.1)–(2.2), we have $\mathcal{J}_\varepsilon(\psi) \to \mathcal{J}(\psi)$ as $\varepsilon \downarrow 0$, and therefore,

$$j_\varepsilon \leq \mathcal{J}_\varepsilon(\psi) \to \mathcal{J}(\psi) \ .$$

Hence

$$\limsup_{\varepsilon\downarrow 0} j_\varepsilon \leq \mathcal{J}(\psi) \ . \tag{2.10}$$

From inequalities (2.9), (2.10) it follows that $\widetilde{\phi}$ solves problem (2.1)–(2.2) and $j_\varepsilon \to j$. This completes the proof of Theorem 2.1. □

We now concentrate on certain approximation of problem (2.1)–(2.2) with a parameter $\delta > 0$.

Denote $\mathcal{U} = \Psi \times H^{4,0}(\Omega)$. Let $\mathcal{J}_{\varepsilon,\delta}(\cdot,\cdot)$ be a class of functionals defined on $H^2(\Omega) \times H^{4,0}(\Omega)$ of the form

$$\mathcal{J}_{\varepsilon,\delta}(\phi,\varphi) = \|\varphi - w_0\|_0^2 + \|\phi\|_2^2 + \frac{1}{\delta}\|\Delta^2\varphi + \frac{1}{\varepsilon}\mathcal{P}_\phi(\varphi) - f\|_0^2 \ .$$

Consider the following family of approximate optimization problems

$$\inf_{(\phi,\varphi)\in\mathcal{U}} \mathcal{J}_{\varepsilon,\delta}(\phi,\varphi) \ . \tag{2.11}$$

We first examine the solvability of (2.11). To this end, we show that the functional $\mathcal{J}_{\varepsilon,\delta}(\cdot,\cdot)$ is coercive and weakly lower semicontinuous.

Since

$$\|\mathcal{P}_\phi(\varphi)\|_0 \le \|\varphi\|_0 + \|\phi\|_0 \quad \forall\, (\phi,\varphi) \in L^2(\Omega)\times L^2(\Omega),$$
$$\|\Delta^2\varphi\|_0 \ge c\|\varphi\|_4 \quad \forall\, \varphi \in H^{4,0}(\Omega)$$

we have

$$\mathcal{J}_{\varepsilon,\delta}(\phi,\varphi) \to \infty \quad \text{for } \|\phi\|_2 + \|\varphi\|_4 \to \infty\,.$$

We now show the property that for any sequence $\{(\phi_k,\varphi_k)\}_{k=1}^\infty$, $(\phi_k,\varphi_k) \to (\phi,\varphi)$ weakly in $H^2(\Omega)\times H^{4,0}(\Omega)$, there exists a subsequence, without loss of generality denoted by $\{(\phi_k,\varphi_k)\}$, such that

$$\liminf_{k\to\infty} \mathcal{J}_{\varepsilon,\delta}(\phi_k,\varphi_k) \ge \mathcal{J}_{\varepsilon,\delta}(\phi,\varphi)\,.$$

Indeed, since ε,δ are fixed, and for a subsequence of the sequence $\{(\phi_k,\varphi_k)\}$, without loss of generality denoted by $\{(\phi_k,\varphi_k)\}$, $(\phi_k,\varphi_k) \to (\phi,\varphi)$ strongly in $L^2(\Omega)\times L^2(\Omega)$, we have

$$\mathcal{P}_{\phi_k}(\varphi_k) \to \mathcal{P}_\phi(\varphi) \quad \text{strongly in } L^2(\Omega),$$

and our property immediately follows from the lower semicontinuity of the norms in the spaces $L^2(\Omega)$ and $H^2(\Omega)$.

Let $(\phi_{\varepsilon,\delta}\,, w_{\varepsilon,\delta})$ solve problem (2.11), and let ε be fixed. Since $\frac{1}{\delta}\|\Delta^2 w_\varepsilon + \frac{1}{\varepsilon}\mathcal{P}_{\phi_\varepsilon}(w_\varepsilon) - f\|_0^2 = 0$, we obtain

$$j_{\varepsilon,\delta} = \mathcal{J}_{\varepsilon,\delta}(\phi_{\varepsilon,\delta}\,, w_{\varepsilon,\delta}) \le \mathcal{J}_{\varepsilon,\delta}(\phi_\varepsilon, w_\varepsilon) = \mathcal{J}_\varepsilon(\phi_\varepsilon) = j_\varepsilon\,. \tag{2.12}$$

From (2.12) it follows that

$$\|\phi_{\varepsilon,\delta}\|_2 \le c_1\,, \quad \|w_{\varepsilon,\delta}\|_0 \le c_2\,,$$
$$\|\Delta^2 w_{\varepsilon,\delta} + \frac{1}{\varepsilon}\mathcal{P}_{\phi_{\varepsilon,\delta}}(w_{\varepsilon,\delta}) - f\|_0 \le c_3\delta^{\frac{1}{2}}\,.$$

Hence, there exist subsequences, without loss of generality denoted by $\{\phi_{\varepsilon,\delta}\}$ and $\{w_{\varepsilon,\delta}\}$, such that

$$\phi_{\varepsilon,\delta} \to \widetilde{\phi}_\varepsilon \quad \text{weakly in } H^2(\Omega),$$
$$w_{\varepsilon,\delta} \to \widetilde{w}_\varepsilon \quad \text{weakly in } H^{4,0}(\Omega),$$
$$(\phi_{\varepsilon,\delta}, w_{\varepsilon,\delta}) \to (\widetilde{\phi}_\varepsilon, \widetilde{w}_\varepsilon) \quad \text{strongly in } L^2(\Omega)\times L^2(\Omega)$$

as $\delta \to 0$. Since

$$\Delta^2\widetilde{w}_\varepsilon + \frac{1}{\varepsilon}\mathcal{P}_{\widetilde{\phi}_\varepsilon}(\widetilde{w}_\varepsilon) = f,$$

we get $\widetilde{w}_\varepsilon = w_\varepsilon(\widetilde{\phi}_\varepsilon)$. By (2.12),

$$\limsup_{\delta\downarrow 0} j_{\varepsilon,\delta} \le j_\varepsilon\,. \tag{2.13}$$

On the other hand, by the weak convergence of the sequences $\{\phi_{\varepsilon,\delta}\}$, $\{w_{\varepsilon,\delta}\}$ (or their subsequences) we have

$$\liminf_{\delta\downarrow 0} \mathcal{J}_{\varepsilon,\delta}(\phi_{\varepsilon,\delta}\, , w_{\varepsilon,\delta}) \geq \liminf_{\delta\downarrow 0}(\|w_{\varepsilon,\delta} - w_0\|_0^2 + \\ + \|\phi_{\varepsilon,\delta}\|_2^2) \geq \|\widetilde{w}_\varepsilon - w_0\|_0^2 + \|\widetilde{\phi}_\varepsilon\|_2^2 \geq j_\varepsilon$$

and consequently, in view of (2.13), $\widetilde{\phi}_\varepsilon$ solves problem (2.4)–(2.5) and

$$\lim_{\delta\downarrow 0} j_{\varepsilon,\delta} = j_\varepsilon \ .$$

Therefore, the following result holds true.

Theorem 2.2. *There exist subsequences, without loss of generality denoted by $\{\phi_{\varepsilon,\delta}\}$, $\{w_{\varepsilon,\delta}\}$ such that*

$$\phi_{\varepsilon,\delta} \to \phi_\varepsilon \quad \text{weakly in } H^2(\Omega),$$
$$w_{\varepsilon,\delta} \to w_\varepsilon(\phi_\varepsilon) \quad \text{weakly in } H^{4,0}(\Omega),$$
$$j_{\varepsilon,\delta} \to j_\varepsilon,$$

as $\delta \to 0$, ϕ_ε is a solution to problem (2.4)–(2.5).

2.2 Cost functionals with measures

Consider the following optimal control problem: find $\psi \in \Psi$ such that

$$\mathcal{J}(\psi) \leq \mathcal{J}(\phi) \quad \forall\, \phi \in \Psi, \tag{2.14}$$

where

$$\mathcal{J}(\phi) = \mu_\phi(\Omega) + \|\phi\|_2^2\ ,$$
$$\mu_\phi \equiv \Delta^2 w(\phi) - f\ .$$

The set Ψ has the properties listed in Section 2.1.

We first study the existence of a solution to problem (2.14), (2.1).

Let $\{\phi_k\}$ be a minimizing sequence. The sequence is bounded, and hence, without loss of generality, we assume that $\phi_k \to \phi$ weakly in $H^2(\Omega)$ and uniformly in $\bar{\Omega}$ as $k \to \infty$. By the same arguments as in Section 2.1, the sequence $\{w(\phi_k)\}$ of solutions to variational inequality (2.1) is bounded in $H_0^2(\Omega)$ which allow us to conclude that $w(\phi_k) \to \widetilde{w}$ weakly in $H_0^2(\Omega)$ as $k \to \infty$. Moreover. $\widetilde{w} = w(\phi)$, ie. $\widetilde{w}$ is a solution to variational inequality (2.1).

We now show the weak-(∗) convergence of the sequence of measures $\mu_k \equiv \mu_{\phi_k}$. To this end, it is enough to prove that for every compact $B \subset \Omega$ the values $\mu_k(B)$ are bounded uniformly with respect to $k = 1, 2, \ldots$

For any $\xi \in C_0^\infty(\Omega)$, $\xi \equiv 1$ on B, $\xi \geq 0$, we have

$$\mu_k(B) \leq \int \xi d\mu_k = \int_\Omega (\Delta\xi \Delta w(\phi_k) - \xi f)dx \leq c,$$

where c is independent of $k = 1, 2, \dots$ Hence,

$$\mu_k \equiv \Delta^2 w(\phi_k) - f \to \mu_\phi \equiv \Delta^2 w(\phi) - f \quad \text{weakly-}(*) \ .$$

The measure μ_ϕ is obtained as the limit, since the sequence $\{w(\phi_k)\}$ is bounded in $H_0^2(\Omega)$. Thus

$$\Delta^2 w(\phi_k) - f \to \Delta^2 w(\phi) - f \quad \text{weakly in } H^{-2}(\Omega) \ .$$

By the weak-$(*)$ convergence of the sequence $\{\mu_k\}$,

$$\liminf_{k\to\infty} \mu_k(\Omega) \geq \mu_\phi(\Omega) \ .$$

Hence

$$j = \liminf_{k\to\infty} \mathcal{J}(\phi_k) \geq \mathcal{J}(\phi) \geq j \ .$$

Therefore, ϕ is a solution to problem (2.14), (2.1).

Consider a family of control problems depending on the parameter $\varepsilon > 0$.

Find $\psi_\varepsilon \in \Psi$ such that

$$\begin{aligned} &\mathcal{J}_\varepsilon(\psi_\varepsilon) \leq \mathcal{J}_\varepsilon(\phi) \quad \forall\, \phi \in \Psi, \\ &\mathcal{J}_\varepsilon(\phi) = \mu_{\phi,\varepsilon}(\Omega) + \|\phi\|_2^2 \ . \end{aligned} \tag{2.15}$$

Here

$$\mu_{\phi,\varepsilon}(\Omega) = \int_\Omega (\Delta^2 w_\varepsilon(\phi) - f)dx,$$

and $w_\varepsilon(\phi)$ is a solution of the equation

$$w_\varepsilon(\phi) \in H_0^2(\Omega) \ : \quad \Delta^2 w_\varepsilon(\phi) + \frac{1}{\varepsilon}\mathcal{P}_\phi(w_\varepsilon(\phi)) = f. \tag{2.16}$$

It should be noted that for any $\varepsilon > 0$, $w_\varepsilon(\phi) \in H^{4,0}(\Omega)$, and hence $\mu_{\phi,\varepsilon}(\Omega)$ is well defined. As in Section 2.1, we can prove the solvability of problem (2.15)–(2.16). A solution to this problem can be regarded as an approximation of a solution to problem (2.14), (2.1).

We introduce the notation

$$\begin{aligned} j &= \inf_{\phi\in\Psi} \mathcal{J}(\phi), \\ j_\varepsilon &= \inf_{\phi\in\Psi} \mathcal{J}_\varepsilon(\phi) \ . \end{aligned}$$

The following result holds true.

Theorem 2.3. *There exist subsequences, without loss of generality denoted by* $\{\psi_\varepsilon\}$, $\{w_\varepsilon(\psi_\varepsilon)\}$ *such that*

$$\psi_\varepsilon \to \psi \quad \textit{weakly in } H^2(\Omega),$$
$$w_\varepsilon(\psi_\varepsilon) \to w(\psi) \quad \textit{weakly in } H_0^2(\Omega),$$
$$j_\varepsilon \to j$$

as $\varepsilon \downarrow 0$. *Here* ψ *is a solution to problem (2.14), (2.1).*

Proof. We begin by showing that for any $\phi \in \Psi$

$$\mathcal{J}_\varepsilon(\phi) \to \mathcal{J}(\phi) \ ,$$

as $\varepsilon \downarrow 0$.

Multiplying the nonlinear equation

$$\Delta^2 w_\varepsilon(\phi) + \frac{1}{\varepsilon}\mathcal{P}_\phi(w_\varepsilon(\phi)) = f$$

by $w_\varepsilon(\phi) - \varphi$, $\varphi \in K_\phi$ we obtain that $w_\varepsilon(\phi)$ is bounded in the norm of $H_0^2(\Omega)$ uniformly with respect to $\varepsilon > 0$.

Passing to subsequences, if necessary, we obtain that $w_\varepsilon(\phi) \to \widetilde{w}$ weakly in $H_0^2(\Omega)$ and uniformly in $\bar{\Omega}$. Moreover, $\widetilde{w}$ is a solution to the variational inequality

$$w \in K_\phi \ : \quad \int_\Omega \Delta w \Delta(\varphi - w)dx \geq \int_\Omega f(\varphi - w)dx \quad \forall\, \varphi \in K_\phi,$$

and hence $\widetilde{w} = w(\phi)$.

Let

$$\mu_{\phi,\varepsilon} \equiv \Delta^2 w_\varepsilon(\phi) - f \to \mu_\phi \equiv \Delta^2 w(\phi) - f \quad \text{weakly-}(*) \ .$$

By the Sobolev imbedding theorem, $w(\phi)$ is a continuous function in $\bar{\Omega}$ and vanishes on the boundary Γ. Besides, $\mathcal{P}_\phi(w_\varepsilon(\phi))(x) = 0$ for $x \in \Omega$ such that $w_\varepsilon(\phi)(x) > \phi(x)$. Since $\phi(x) < 0$ in an open neighbourhood of Γ and the sequence $\{w_\varepsilon(\phi)\}$ converges uniformly to $w(\phi)$, there exists a compact $B \subset \Omega$,

$$\mu_{\phi,\varepsilon}(\Omega \setminus B) = 0 \ . \tag{2.17}$$

We can assume that $\mu_\phi(\partial B) = 0$, and hence $\mu_{\phi,\varepsilon}(B) \to \mu_\phi(B)$ as $\varepsilon \downarrow 0$ (for details see Chapter 1, Section 1.10). Taking into account (2.17),

$$\mu_{\phi,\varepsilon}(\Omega) = \mu_{\phi,\varepsilon}(B) \to \mu_\phi(B) = \mu_\phi(\Omega) \ .$$

Therefore,

$$\mathcal{J}_\varepsilon(\phi) \equiv \mu_{\phi,\varepsilon}(\Omega) + \|\phi\|_2^2 \to \mathcal{J}(\phi) \equiv \mu_\phi(\Omega) + \|\phi\|_2^2 \ .$$

Let ψ_ε solve problem (2.15)–(2.16). Then

$$\mu_{\psi_\varepsilon,\varepsilon}(\Omega) + \|\psi_\varepsilon\|_2^2 \leq \mu_{\phi,\varepsilon}(\Omega) + \|\phi\|_2^2$$

for any $\phi \in \Psi$. From what has already been proved it follows that the right-hand side of the last inequality is bounded uniformly with respect to ε. Hence, the sequence $\{\psi_\varepsilon\}$ is bounded in $H^2(\Omega)$ for $\varepsilon > 0$ and there exists an element $\widetilde{\phi}$, $\psi_\varepsilon \to \widetilde{\phi}$ weakly in $H^2(\Omega)$ and uniformly in $\bar{\Omega}$ as $\varepsilon \downarrow 0$.

It follows from the equation

$$\Delta^2 w_\varepsilon(\psi_\varepsilon) + \mathcal{P}_{\psi_\varepsilon}(w_\varepsilon(\psi_\varepsilon)) = f, \tag{2.18}$$

that the sequence $\{w_\varepsilon(\psi_\varepsilon)\}$ is bounded in $H_0^2(\Omega)$. In fact, for a given sequence $\{\bar{\varphi}_\varepsilon\}$, $\bar{\varphi}_\varepsilon \in K_{\psi_\varepsilon}$, $\bar{\varphi}_\varepsilon \to \bar{\varphi} \in K_{\widetilde{\phi}}$ strongly in $H_0^2(\Omega)$, equation (2.18) multiplied by $w_\varepsilon(\psi_\varepsilon) - \bar{\varphi}_\varepsilon$ yields the required estimate.

Therefore, we can assume that

$$\begin{aligned} &w_\varepsilon(\psi_\varepsilon) \to \widetilde{w} \quad \text{weakly in } H_0^2(\Omega), \\ &\mu_{\psi_\varepsilon,\varepsilon} \equiv \Delta^2 w_\varepsilon(\psi_\varepsilon) - f \to \mu_{\widetilde{\phi}} \equiv \Delta^2 \widetilde{w} - f \quad \text{weakly-}(*), \end{aligned}$$

for some $\widetilde{w} \in H_0^2(\Omega)$.

From equation (2.18) it follows that $\widetilde{w}$, $\widetilde{\phi}$ satisfy the variational inequality

$$\int_\Omega \Delta\widetilde{w}\Delta(\varphi - \widetilde{w})dx \geq \int_\Omega f(\varphi - \widetilde{w})dx \quad \forall\, \varphi \in K_{\widetilde{\phi}}\,,$$

and hence $\widetilde{w} = w(\widetilde{\phi})$, and the measure $\mu_{\widetilde{\phi}}$ is well defined.

The weak-$(*)$ convergence of the sequence $\{\mu_{\psi_\varepsilon,\varepsilon}\}$ implies, in particular, that

$$\liminf_{\varepsilon\downarrow 0} \mu_{\psi_\varepsilon,\varepsilon}(\Omega) \geq \mu_{\widetilde{\phi}}(\Omega)\ .$$

This entails that

$$\liminf_{\varepsilon\downarrow 0} j_\varepsilon \geq \mu_{\widetilde{\phi}}(\Omega) + \|\widetilde{\phi}\|_2^2\ . \tag{2.19}$$

By taking ψ which solves problem (2.14), (2.1), we get

$$j_\varepsilon \leq \mathcal{J}_\varepsilon(\psi) \to \mathcal{J}(\psi) = j,$$

and hence

$$\limsup_{\varepsilon\downarrow 0} j_\varepsilon \leq \mathcal{J}(\psi)\ . \tag{2.20}$$

From (2.19)–(2.20) it follows that $\widetilde{\phi}$ is a solution to problem (2.14), (2.1) and $j_\varepsilon \to j$ as $\varepsilon \downarrow 0$. This completes the proof. □

2.3 Finite set of pointwise restrictions

Let us consider a particular case where the convex set $\mathcal{K}$ is defined by a finite number of poitwise constraints

$$\mathcal{K} = \{\varphi \in H_0^2(\Omega) | \varphi(x_i) \geq C_i \ , \quad i = 1, 2, \ldots, \ell\} \ .$$

Constants C_i are given and points x_i are fixed. A construction composed of a finite set of rigid rods located at points x_i is an example of an obstacle of such a form. The state of static equilibrium of the plate with the obstacle is given as a solution to the following variational inequality

$$w_C \in \mathcal{K} \ : \quad \int_\Omega \Delta w_C \Delta(\varphi - w_C) dx \geq \int_\Omega f(\varphi - w_C) dx \quad \forall\, \varphi \in \mathcal{K} \ . \tag{2.21}$$

The existence and uniqueness of solutions to (2.21) follow by standard arguments.

The variational inequality is equivalent to the minimization of the corresponding quadratic functional over the set $\mathcal{K}$. In the same way as before we introduce a class of optimization problems

$$\inf_{C^1 \leq C \leq C^2} \|w_C - w_0\|_0^2 \ .$$

Here C^i, $i = 1, 2$, are given vectors, w_0 is a given function in $L^2(\Omega)$. There exists a solution to the above minimization problem.

3 Other cost functionals

In this section we consider some optimization problems for plates with cost functionals other than those studied in Sections 1, 2. In particular, our analysis includes the case of a cost functional with an integral on the boundary Γ which allows us to take into account the traces on the boundary of the displacement of the plate or of its normal derivative.

Let $\Omega \subset \mathbb{R}^2$ be a bounded domain with smooth boundary Γ, $w \equiv w_u$ be a solution to the variational inequality

$$w \in \mathcal{K} \ : \quad \int_\Omega \Delta w \Delta(\varphi - w) dx \geq \int_\Omega u(\varphi - w) dx \quad \forall\, \varphi \in \mathcal{K} \ . \tag{3.1}$$

Here $u \in \mathcal{U}$ is a given element, $\mathcal{U} \subset L^2(\Omega)$ is a closed and convex set,

$$\mathcal{K} = \left\{\varphi \in H^{2,0}(\Omega) \ \middle| \ \varphi(x) \geq \psi(x), \quad x \in \Omega\right\} \ ,$$

where ψ is a given function such that $\mathcal{K}$ is nonempty, $H^{2,0}(\Omega) = H^2(\Omega) \cap H_0^1(\Omega)$.

Let $w_* \in L^2(\Gamma)$ be a given function. We consider the cost functional

$$\mathcal{J}(u) = \int_\Gamma \left(\frac{\partial w_u}{\partial \nu} - w_*\right)^2 d\Gamma + \int_\Omega u^2 dx \ .$$

The first problem we examine in this section is to find an element $u \in \mathcal{U}$ such that

$$\mathcal{J}(u) \leq \mathcal{J}(v) \quad \forall\, v \in \mathcal{U} \ . \tag{3.2}$$

There exists a solution to problem (3.2), and a family of optimization problems approximating (3.2) can be constructed in the same way as in the preceding sections. The existence of a solution to (3.2) can be proved in much the same way as in Section 2. However, we should point out here a slight modification.

For a minimizing sequence $u^n \in \mathcal{U}$, $n = 1, 2, \ldots$, the uniform estimate

$$\|w^n\|_2 \leq c$$

holds for $w^n \equiv w_{u^n}$. The functions u^n and w^n satisfy the variational inequality

$$\begin{gathered} w^n \in \mathcal{K} \ : \\ \int_\Omega \Delta w^n (\Delta \bar{w} - \Delta w^n) dx \geq \int_\Omega u^n (\bar{w} - w^n) dx \quad \forall\, \bar{w} \in \mathcal{K} \ . \end{gathered} \tag{3.3}$$

The sequence w^n is bounded in $H^{2,0}(\Omega)$. Hence, by letting $n \to \infty$

$$w_\nu^n \to w_\nu \quad \text{strongly in} \quad L^2(\Gamma) \ . \tag{3.4}$$

We can assume that

$$\begin{gathered} w^n \to w \quad \text{weakly in} \quad H^{2,0}(\Omega) \quad \text{and strongly in} \quad L^2(\Omega) \ , \\ u^n \to u \quad \text{weakly in} \quad L^2(\Omega) \ , \quad u \in \mathcal{U} \end{gathered} \tag{3.5}$$

as $n \to \infty$. The weak convergence of u^n results from the boundedness of u^n in $L^2(\Omega)$. Without loss of generality we can use the same notations for subsequences. Passing to the limit in (3.3) as $n \to \infty$, we get $w = w(u)$. By (3.4), u solves problem (3.2).

Let us consider the optimal control problem governed by the equation with a penalty operator. We seek an element $u \in \mathcal{U}$ such that

$$\mathcal{J}_\varepsilon(u) \leq \mathcal{J}_\varepsilon(\bar{u}) \quad \forall\, \bar{u} \in \mathcal{U} \ , \tag{3.6}$$

where

$$\mathcal{J}_\varepsilon(u) = \|w_\nu - w_*\|^2_{L^2(\Gamma)} + \|u\|^2 \ ,$$

and the function w solves the equation

$$\Delta^2 w + \frac{1}{\varepsilon}\, p(w) = u \ , \tag{3.7}$$

$$w = \Delta w = 0 \quad \text{on} \quad \Gamma \ . \tag{3.8}$$

The function p is defined, as earlier, by the formula $p(w) = -(w - \psi)^-$. Two questions arise. The first question is the following. What can be said about the solvability of problem (3.6) for fixed $\varepsilon > 0$? The second question concerns relationships between solutions of (3.6) and those of (3.2).

We now discuss the second question. Let u^ε be a solution to problem (3.6), (3.7), (3.8) corresponding to ε and let w^ε be defined by (3.7), (3.8) for $u = u^\varepsilon$. For any element $v \in \mathcal{U}$,

$$\mathcal{J}_\varepsilon(u^\varepsilon) \leq \mathcal{J}_\varepsilon(v) \ . \tag{3.9}$$

We now demonstrate the uniform boundedness with respect to ε of the right-hand side of (3.9). This will yield the boundedness of $u^\varepsilon \in L^2(\Omega)$.

To this end, consider the equation

$$\Delta^2 w^\varepsilon(v) + \frac{1}{\varepsilon}\, p\big(w^\varepsilon(v)\big) = v \ , \tag{3.10}$$

$$w^\varepsilon(v) = \Delta w^\varepsilon(v) = 0 \quad \text{on} \quad \Gamma \ . \tag{3.11}$$

Multiplying (3.10) by $w^\varepsilon(v) - \bar{w}$, $\bar{w} \in \mathcal{K}$, and using the monotonicity of the operator p, we obtain

$$\int_\Omega \Delta w^\varepsilon(v)\big(\Delta w^\varepsilon(v) - \Delta\bar{w}\big)dx \leq \int_\Omega v\big(w^\varepsilon(v) - \bar{w}\big)dx \ ,$$

and hence

$$\|\Delta w^\varepsilon(v)\|^2 \leq c \ , \tag{3.12}$$

with the constant c independent of ε. Inequality (3.12) implies the boundedness of $w^\varepsilon(v)$ in $H^{2,0}(\Omega)$, and consequently, $w^\varepsilon_\nu(v)$ are bounded in $L^2(\Gamma)$. As mentioned above, this implies the boundedness of u^ε in $L^2(\Omega)$. Without loss of generality, we can assume that

$$u^\varepsilon \to \tilde{u} \quad \text{weakly in} \quad L^2(\Omega) \tag{3.13}$$

as $\varepsilon \to 0$. It should be noted here that the multiplication of (3.10) by $w^\varepsilon(v) - \bar{w}$ and the integration over Ω actually reduces to the substitution of $w^\varepsilon(v) - \bar{w}$ as test functions into the corresponding relation defining solutions of problem (3.10), (3.11).

Multiplying the equation

$$\Delta^2 w^\varepsilon + \frac{1}{\varepsilon}\, p(w^\varepsilon) = u^\varepsilon \ , \tag{3.14}$$

$$w^\varepsilon = \Delta w^\varepsilon = 0 \quad \text{on} \quad \Gamma \tag{3.15}$$

by $w^\varepsilon - \bar{w}$, $\bar{w} \in \mathcal{K}$, and integrating over Ω, one easily gets the estimate

$$\|\Delta w^\varepsilon\|^2 \leq c$$

which holds uniformly with respect to ε. This implies the boundedness of w^ε in the space $H^{2,0}(\Omega)$. Choosing a subsequence, if necessary, the following convergence holds

$$\begin{aligned} w^\varepsilon \to \widetilde{w} \quad &\text{weakly in} \quad H^{2,0}(\Omega), \quad \text{strongly in} \quad L^2(\Omega) \ , \\ w^\varepsilon_\nu \to \widetilde{w}_\nu \quad &\text{strongly in} \quad L^2(\Gamma) \end{aligned} \tag{3.16}$$

as $\varepsilon \to 0$. The multiplication of (3.14) by $\bar{w} - w^\varepsilon$, $\bar{w} \in \mathcal{K}$, and integration over Ω imply

$$\int_\Omega \Delta w^\varepsilon (\Delta \bar{w} - \Delta w^\varepsilon) dx \geq \int_\Omega u^\varepsilon (\bar{w} - w^\varepsilon) dx \ .$$

Taking into account (3.13), (3.16), we can pass to the limit here that readily gives

$$\widetilde{w} \in \mathcal{K} \ :$$

$$\int_\Omega \Delta \widetilde{w} (\Delta \bar{w} - \Delta \widetilde{w}) dx \geq \int_\Omega \widetilde{u} (\bar{w} - \widetilde{w}) dx \quad \forall \, \bar{w} \in \mathcal{K} \ ,$$

and $\widetilde{w} = w(\widetilde{u})$. Consequently, the weak lower semicontinuity of $\| \cdot \|$ and $\| \cdot \|_{L^2(\Gamma)}$ results in

$$\underline{\lim}\ \mathcal{J}_\varepsilon(u^\varepsilon) \geq \mathcal{J}(\widetilde{u}) \ . \tag{3.17}$$

On the other hand, for a fixed $v \in \mathcal{U}$,

$$w_\nu^\varepsilon(v) \to w_\nu(v) \quad \text{strongly in} \quad L^2(\Gamma)$$

as $\varepsilon \to 0$. Hence,

$$\mathcal{J}_\varepsilon(v) \to \mathcal{J}(v) \ , \quad \varepsilon \to 0 \ .$$

Therefore, by taking u which solves problem (3.2), (3.1), the inequality

$$\mathcal{J}_\varepsilon(u^\varepsilon) \leq \mathcal{J}_\varepsilon(u)$$

implies

$$\overline{\lim}\ \mathcal{J}_\varepsilon(u^\varepsilon) \leq \mathcal{J}(u) \ . \tag{3.18}$$

Comparing (3.17) with (3.18), we see that $\widetilde{u}$ solves problem (3.2), (3.1) and $\mathcal{J}_\varepsilon(u^\varepsilon) \to \mathcal{J}(\widetilde{u})$. In this way we establish the following theorem.

Theorem 3.1. *From the sequence $u^\varepsilon, w^\varepsilon, \mathcal{J}_\varepsilon(u^\varepsilon)$ one can choose a subsequence (without loss of generality we use the same notation for subsequences) such that*

$$u^\varepsilon \to u \quad \textit{weakly in} \quad L^2(\Omega) \ ,$$

$$w^\varepsilon \to w \quad \textit{weakly in} \quad H^{2,0}(\Omega) \ ,$$

$$\mathcal{J}_\varepsilon(u^\varepsilon) \to \mathcal{J}(u) \ ,$$

as $\varepsilon \to 0$, where u solves problem (3.2), (3.1) and w corresponds to u.

Finally, we shall consider one more optimal control problem. Denote

$$\mathcal{K}_\varphi = \{ w \in H^{2,0}(\Omega) \mid w(x) \geq \varphi(x) \ , \quad x \in \Omega \} \ ,$$

where $\varphi \in \Phi$, $\Phi \subset H^2(\Omega)$ is a closed convex set whose every element φ satisfies the inequality $\varphi < 0$ on Γ.

Introduce the variational inequality

$$w(\varphi) \in \mathcal{K}_\varphi \ : \quad \int_\Omega \Delta w(\varphi)\big(\Delta \bar{w} - \Delta w(\varphi)\big) dx \geq \\ \geq \int_\Omega f\big(\bar{w} - w(\varphi)\big) dx \quad \forall\, \bar{w} \in \mathcal{K}_\varphi \ . \tag{3.19}$$

The function $f \in L^2(\Omega)$ is given. For the cost functional

$$\mathcal{J}(\varphi) = \|w_\nu(\varphi) - w_*\|^2_{L^2(\Gamma)} + \|\varphi\|_2^2$$

we seek a solution to the optimal control problem

$$\inf_{\varphi \in \Phi} \mathcal{J}(\varphi) \ . \tag{3.20}$$

The properties of solutions to (3.20), (3.19) will be discussed somewhat later.

Now we shall consider another problem. Let the cost functional have the form

$$\mathcal{J}_\varepsilon(\varphi) = \|w^\varepsilon_\nu(\varphi) - w_*\|^2_{L^2(\Gamma)} + \|\varphi\|_2^2 \ ,$$

where $w^\varepsilon(\varphi)$ solves the equation

$$\Delta^2 w^\varepsilon(\varphi) + \frac{1}{\varepsilon}\, p_\varphi\big(w^\varepsilon(\varphi)\big) = f \ , \tag{3.21}$$

$$w^\varepsilon(\varphi) = \Delta w^\varepsilon(\varphi) = 0 \quad \text{on} \quad \Gamma \tag{3.22}$$

with a positive parameter $\varepsilon > 0$, $p_\varphi(w) = -(w - \varphi)^-$.

The problem is to minimize $\mathcal{J}_\varepsilon(\varphi)$ over Φ, ie.,

$$\inf_{\varphi \in \Phi} \mathcal{J}_\varepsilon(\varphi) \ . \tag{3.23}$$

The theorem below characterizes the properties of these optimal control problems.

Theorem 3.2. *The following statements are valid:*

1. There exists a solution to problem (3.20), (3.19).

2. There exists a solution to problem (3.23), (3.21), (3.22) for any fixed $\varepsilon > 0$.

3. Let φ^ε *be a solution to problem (3.23), (3.21), (3.22) for given* $\varepsilon > 0$ *and* w^ε *defined by (3.21), (3.22) for* $\varphi = \varphi^\varepsilon$. *Then, one can choose a subsequence (without loss of generality we can use the same notation for subsequences)* φ^ε, w^ε, $\mathcal{J}_\varepsilon(\varphi^\varepsilon)$ *such that as* $\varepsilon \to 0$

$$\varphi^\varepsilon \to \varphi \quad \textit{weakly in} \quad H^2(\Omega) \ ,$$

$$w^\varepsilon \to w \quad \textit{weakly in} \quad H^{2,0}(\Omega) \ ,$$

$$\mathcal{J}_\varepsilon(\varphi^\varepsilon) \to \mathcal{J}(\varphi) \ .$$

Here φ *is a solution to problem (3.20), (3.19) and* w *corresponds to* φ.

4 Plastic hinge on the boundary

We shall consider an optimal control problem with boundary conditions corresponding to the plastic hinge. A shape of a punch will be a governed function. Our objective is to prove a solvability theorem and to state some results on the convergence of solutions when parameters of the problem tend to zero or to infinity.

4.1 Cost functionals with displacements

Let $\Omega \subset \mathbb{R}^2$ be a bounded domain with the smooth boundary Γ; $H^{2,0}(\Omega) = H^2(\Omega) \cap H_0^1(\Omega)$. Consider the functional defined on $H^{2,0}(\Omega)$

$$\Pi(w) = \frac{1}{2} B(w,w) + q \int_\Gamma |w_\nu| d\Gamma \ , \quad q = \text{const} > 0 \ .$$

As in the preceeding section, w_ν denotes the derivative along the outer normal ν to the boundary Γ,

$$B(w,w) = \int_\Omega \{w_{x_1x_1}^2 + w_{x_2x_2}^2 + 2\sigma w_{x_1x_1} w_{x_2x_2}$$
$$+ 2(1-\sigma) w_{x_1x_2}^2\} dx \ , \quad \sigma = \text{const} \ , \quad 0 < \sigma < \frac{1}{2} \ .$$

Let us denote by Ψ a nonempty closed and convex set in $H^2(\Omega)$ whose every element ψ satisfies the inequality $\psi < 0$ on Γ. For any $\psi \in \Psi$, we introduce a convex and closed set in $H^{2,0}(\Omega)$,

$$\mathcal{K}_\psi = \{w \in H^{2,0}(\Omega) | w(x) \geq \psi(x) \ , \quad x \in \Omega\} \ .$$

For a fixed $\psi \in \Psi$, a unique solution to the problem

$$\begin{aligned} w(\psi) \in \mathcal{K}_\psi \ : \quad & B(w(\psi), \bar{w} - w(\psi)) \\ & + q \int_\Gamma |\bar{w}_\nu| d\Gamma - q \int_\Gamma |w_\nu(\psi)| d\Gamma \geq 0 \quad \forall\, \bar{w} \in \mathcal{K}_\psi \end{aligned} \tag{4.1}$$

exists. This inequality is equivalent to the problem of minimizing Π over the set $\mathcal{K}_\psi$.

Let $w_* \in L^2(\Omega)$ be a fixed element and

$$\mathcal{J}_q(\psi) = \|w(\psi) - w_*\|_0^2 + \|\psi\|_2^2$$

be the cost functional. The optimal control problem consists in finding an element $\psi \in \Psi$ such that

$$\mathcal{J}_q(\psi) \leq \mathcal{J}_q(\bar{\psi}) \quad \forall\, \bar{\psi} \in \Psi \ . \tag{4.2}$$

For a fixed element $\psi \in \Psi$, inequality (4.1) describes the equilibrium state of a plate having a contact with a punch. The minimization of the functional $\mathcal{J}_q$ expresses the fact that we are looking for the best punch shape. The function $w(\psi)$ represents the displacement of the plate. The first boundary condition for $w(\psi)$ depends on the choice of the space $H^{2,0}(\Omega)$. The second one will be discussed below. It corresponds to the plastic hinge located on the boundary. The corresponding bilinear form $B(w,u)$ was defined in Section 6.2, Chapter 2.

We begin by proving the following solvability result.

Theorem 4.1. *There exists a solution to problem (4.2).*

Proof. Let $\psi^n \in \Psi$ be a minimizing sequence. This sequence is clearly bounded in $H^2(\Omega)$. Choosing a subsequence, if necessary, one can assume that

$$\psi^n \to \psi \quad \text{weakly in} \quad H^2(\Omega) \ ,$$
$$\psi^n \to \psi \quad \text{uniformly in} \quad \bar{\Omega}$$

as $n \to \infty$. The properties of the set Ψ obviously ensure that $\psi \in \Psi$. It was proved in Section 2 that for every $\bar{w} \in \mathcal{K}_\psi$ there exists a sequence $\bar{w}^n \in \mathcal{K}_{\psi^n}$ strongly converging to $\bar{w}$.

Let $\bar{w}^n \in \mathcal{K}_{\psi^n}$ converge to $\bar{w} \in \mathcal{K}_\psi$. By the inequality

$$B\big(w(\psi^n), \bar{w}^n - w(\psi^n)\big) + q\int_\Gamma \big(|\bar{w}_\nu^n| - |w_\nu(\psi^n)|\big)d\Gamma \geq 0 \tag{4.3}$$

we have

$$B\big(w(\psi^n), w(\psi^n)\big) \leq c \ .$$

Taking into account the boundary condition for $w(\psi^n)$, it follows that

$$\|w(\psi^n)\|_2 \leq c \ .$$

Hence, one can assume that there exists a subsequence such that

$$w(\psi^n) \to \widetilde{w} \quad \text{weakly in} \quad H^{2,0}(\Omega) \ ,$$
$$w_\nu(\psi^n) \to \widetilde{w}_\nu \quad \text{strongly in} \quad L^1(\Gamma)$$

as $n \to \infty$. Then, inequality (4.3) takes the form

$$\begin{aligned} &B\big(w(\psi^n), \bar{w}^n\big) + q\int_\Gamma |\bar{w}_\nu^n| d\Gamma - q\int_\Gamma |w_\nu(\psi^n)| d\Gamma \\ &\geq B\big(w(\psi^n), w(\psi^n)\big) \ . \end{aligned} \tag{4.4}$$

By the weak lower semicontinuity of $B(\cdot,\cdot)$,

$$\widetilde{w} \in \mathcal{K}_\psi \ : \quad B(\widetilde{w}, \bar{w} - \widetilde{w}) + q\int_\Gamma |\bar{w}_\nu| d\Gamma - q\int_\Gamma |\widetilde{w}_\nu| d\Gamma \geq 0 \quad \forall\, \bar{w} \in \mathcal{K}_\psi \ ,$$

and consequently $\widetilde{w} = w(\psi)$. Moreover,

$$j^q \equiv \inf_{\bar{\psi} \in \Psi} \mathcal{J}_q(\bar{\psi}) = \lim_{n\to\infty} \mathcal{J}_q(\psi^n) \geq \mathcal{J}_q(\psi) \geq j^q \ .$$

Therefore, ψ solves problem (4.2). The theorem is proved. □

Let us discuss the boundary conditions imposed on the solution. For the sake of simplicity, assume that the domain Ω is simply connected. Consider the curve obtained by moving the vector $-\delta\nu$ along Γ, $\delta = \text{const} > 0$. Points lying on this curve correspond to the ends of the vector $-\delta\nu$. Accordingly, the curves $\Gamma_{2\delta}$, $\Gamma_{\lambda\delta}$ are obtained by moving the vectors $-2\delta\nu$, $-\lambda\delta\nu$ along Γ, $\lambda = \text{const} > 0$. Assuming that δ is small enough, there is a one-to-one mapping between Γ and $\Gamma_{2\delta}$, and moreover, the inequalities $w(\psi) > \psi$, $\psi < 0$ are satisfied in the domain Ω_2. The domain Ω_2 is located between Γ and $\Gamma_{2\delta}$. Let $\varphi \in C_0^\infty(\Omega_2)$. Take $\bar{w} = w(\psi) + \lambda\varphi \in \mathcal{K}_\psi$, where λ is small enough, and substitute it into inequality (4.1). We have $\lambda B\big(w(\psi), \varphi\big) \geq 0$. Consequently,

$$\Delta^2 w(\psi) = 0$$

holds in Ω_2 in the sense of distributions.

Let Ω_1 be the domain bounded by the curve Γ_δ,

$$Q_\psi = \{\bar{w} \in \mathcal{K}_\psi | \bar{w} = w(\psi) \quad \text{on} \quad \Omega_1\} .$$

Here $w(\psi)$ solves variational inequality (4.1). Substituting $\bar{w} \in Q_\psi$ into (4.1) the inequality

$$\begin{aligned} \int_{\Omega\setminus\bar{\Omega}_1} &\{w_{x_1x_1}(\psi)\big(\bar{w}_{x_1x_1} - w_{x_1x_1}(\psi)\big) + w_{x_2x_2}(\psi)\big(\bar{w}_{x_2x_2} - w_{x_2x_2}(\psi)\big) \\ &+\sigma w_{x_1x_1}(\psi)\big(\bar{w}_{x_2x_2} - w_{x_2x_2}(\psi)\big) + \sigma w_{x_2x_2}(\psi)\big(\bar{w}_{x_1x_1} - w_{x_1x_1}(\psi)\big) \\ &\quad + 2(1-\sigma)w_{x_1x_2}(\psi)\big(\bar{w}_{x_1x_2} - w_{x_1x_2}(\psi)\big)\}dx \\ &\quad +q\int_\Gamma |\bar{w}_\nu| d\Gamma - q\int_\Gamma |w_\nu(\psi)| d\Gamma \geq 0 \quad \forall\, \bar{w} \in Q_\psi \end{aligned} \tag{4.5}$$

results. The first term in the left-hand side of this relation is equal to $B\big(w(\psi), \bar{w} - w(\psi)\big)$. Note that the integration is taken over $\Omega\setminus\bar{\Omega}_1$ since $\bar{w} = w(\psi)$ on Ω_1. By Green's formula applied to the domain $\Omega\setminus\bar{\Omega}_1$ (see Section 6, Chapter 2) and by the equalities $\bar{w} = w(\psi)$, $\bar{w}_\nu = w_\nu(\psi)$ on Γ_δ, $w(\psi) = 0$ on Γ, from (4.5) we have

$$\begin{aligned} &\langle\, M\big(w(\psi)\big), \bar{w}_\nu - w_\nu(\psi)\,\rangle_{1/2} \\ &+ q\int_\Gamma |\bar{w}_\nu| d\Gamma - q\int_\Gamma |w_\nu(\psi)| d\Gamma \geq 0 \quad \forall\, \bar{w} \in Q_\psi \ . \end{aligned} \tag{4.6}$$

Here $M\big(w(\psi)\big) = \sigma\Delta w(\psi) + (1-\sigma)\dfrac{\partial^2 w(\psi)}{\partial\nu^2}$, $M\big(w(\psi)\big) \in H^{-1/2}(\Gamma)$, the brackets $\langle\, \cdot, \cdot\,\rangle_{1/2}$ denote the duality pairing between $H^{-1/2}(\Gamma)$ and $H^{1/2}(\Gamma)$. Recall that $\Delta^2 w(\psi) = 0$ holds in $\Omega\setminus\bar{\Omega}_1$.

We now show that for any $\varphi \in H^{1/2}(\Gamma)$ there exists $\bar{w} \in Q_\psi$ such that $\bar{w}_\nu = \varphi$ on Γ. For the pair $(0, \varphi) \in H^{3/2}(\Gamma) \times H^{1/2}(\Gamma)$, one can easily find $\widetilde{w} \in H^2(\Omega)$ such that

$$\widetilde{w}\big|_\Gamma = 0 \ , \quad \widetilde{w}_\nu\big|_\Gamma = \varphi \ .$$

There exists a function χ with the properties $\chi = 1$ on Γ, $|\chi| \leq 1$, $\chi \equiv 0$ in the domain bounded by $\Gamma_{\lambda\delta}$. The parameter λ is chosen in such a way that $\chi\widetilde{w} > \psi$ in the domain located between Γ and $\Gamma_{\lambda\delta}$.

Furthermore, there exists a smooth function ξ equal to one in $\bar{\Omega}_1$ and to zero outside $\bar{\Omega}_1 \cup \widetilde{\Omega}_1$, where $\widetilde{\Omega}_1$ is the outer small neighbourhood of the curve Γ_δ, $0 \leq \xi \leq 1$. Then $\xi w(\psi) \in Q_\psi$, $\xi w(\psi) + \chi\widetilde{w} \in Q_\psi$, and

$$\frac{\partial}{\partial\nu}\left(\xi w(\psi) + \chi\widetilde{w}\right) = \varphi \quad \text{on} \quad \Gamma$$

which entails that inequality (4.6) can be rewritten as

$$\langle\, M\big(w(\psi)\big), \varphi - w_\nu(\psi)\,\rangle_{1/2}$$
$$+ q\int_\Gamma |\varphi| d\Gamma - q\int_\Gamma |w_\nu(\psi)| d\Gamma \geq 0 \quad \forall\, \varphi \in H^{1/2}(\Gamma) \ .$$

Substituting here $\varphi = 0$, $\varphi = 2w_\nu(\psi)$, and next replacing φ by $\pm\lambda\varphi$, $\lambda > 0$, we get

$$\big|\langle\, M\big(w(\psi)\big), \varphi\,\rangle_{1/2}\big| \leq q\int_\Gamma |\varphi| d\Gamma \quad \forall\, \varphi \in H^{1/2}(\Gamma) \ ,$$

$$\langle\, M\big(w(\psi)\big), w_\nu(\psi)\,\rangle_{1/2} + q\int_\Gamma \big|w_\nu(\psi)\big| d\Gamma = 0 \ .$$

Provided solutions are sufficiently smooth we can put the above relations in the following form

$$\big|M\big(w(\psi)\big)\big| \leq q \ , \quad M\big(w(\psi)\big) w_\nu(\psi) + q|w_\nu(\psi)| = 0 \ .$$

These relations may also be written as

$$\big|M\big(w(\psi)\big)\big| < q \ \Rightarrow\ w_\nu(\psi) = 0 \ ,$$
$$M\big(w(\psi)\big) = q \ \Rightarrow\ w_\nu(\psi) \leq 0 \ ,$$
$$M\big(w(\psi)\big) = -q \ \Rightarrow\ w_\nu(\psi) \geq 0 \ .$$

The above boundary condition describes a plastic hinge and depends on the parameter q. The passage to the limit as $q \to 0$, and/or $q \to \infty$ is interesting for applications. Pertaining results are Theorem 4.2 and Theorem 4.3.

In our considerations the equilibrium state of the plate is described by the variational inequality. Convergence of solutions in the case where there is no control was studied in (Naumann, 1975) for the equilibrium state described by equations.

Let us consider the following optimal control problem. Let $w(\psi)$ solve the variational inequality

$$w(\psi) \in \mathcal{K}_\psi \ : \ B(w(\psi), \bar{w} - w(\psi)) \geq 0 \quad \forall\, \bar{w} \in \mathcal{K}_\psi \ , \tag{4.7}$$

and

$$\mathcal{J}_0(\psi) = \|w(\psi) - w_*\|_0^2 + \|\psi\|_2^2.$$

Find an element $\psi \in \Psi$ such that

$$\mathcal{J}_0(\psi) \leq \mathcal{J}_0(\bar{\psi}) \quad \forall\, \bar{\psi} \in \Psi \ . \tag{4.8}$$

Put $j = \inf\limits_{\psi \in \Psi} \mathcal{J}_0(\psi)$. A solution to (4.2) will be denoted by ψ^q to indicate the dependence on q. Accordingly, $w(\psi^q)$ solves (4.1).

Theorem 4.2. *From the sequence ψ^q one can choose a subsequence (without loss of generality we can use the same notation as for the sequence) such that*

$$\psi^q \to \psi \quad \textit{weakly in} \quad H^2(\Omega) \ ,$$

$$w(\psi^q) \to w(\psi) \quad \textit{weakly in} \quad H^{2,0}(\Omega) \ ,$$

$$j^q \to j^0$$

as $q \to 0$. The function $w(\psi)$ satisfies the boundary condition

$$M(w(\psi)) = 0 \quad \textit{on} \quad \Gamma \ .$$

Here, ψ solves (4.8) and $w(\psi)$ is defined by variational inequality (4.7).

Proof. Let $\psi \in \Psi$ be a fixed element. From (4.1) it follows that the estimate

$$\|w(\psi)\|_2 \leq c$$

holds uniformly with respect to $q \leq q_0$. By the inequality

$$\mathcal{J}_q(\psi^q) \leq \mathcal{J}_q(\psi) \ ,$$

one gets

$$\|\psi^q\|_2 \leq c \ .$$

Choosing a subsequence, if necessary, we have

$$\psi^q \to \widetilde{\psi} \quad \text{weakly in} \quad H^2(\Omega)$$

as $q \to 0$. For any element $\bar{w} \in \mathcal{K}_{\widetilde{\psi}}$, there exists a sequence $\bar{w}^q \in \mathcal{K}_{\psi^q}$ possessing the property that

$$\bar{w}^q \to \bar{w} \quad \text{strongly in} \quad H^{2,0}(\Omega)$$

as $q \to 0$. Substituting for every q the functions $\bar{w}^q$ into inequality (4.1) as test elements we get

$$\|w(\psi^q)\|_2 \leq c \ .$$

Passing to subsequences, if necessary, we have

$$w(\psi^q) \to \widetilde{w} \quad \text{weakly in} \quad H^{2,0}(\Omega)$$

as $q \to 0$. Since $\|w_\nu(\psi^q)\|_{L^1(\Gamma)}$, $\|\bar{w}_\nu^q\|_{L^1(\Gamma)}$ are uniformly bounded with respect to $q \le q_0$ we can pass to the limit in the inequality

$$B\big(w(\psi^q),\ \bar{w}^q - w(\psi^q)\big) + q\int_\Gamma |\bar{w}_\nu^q| d\Gamma - q\int_\Gamma |w_\nu(\psi^q)| d\Gamma \ge 0$$

as $q \to 0$ which gives

$$\widetilde{w} \in \mathcal{K}_{\widetilde{\psi}} \ :\ B(\widetilde{w}, \bar{w} - \widetilde{w}) \ge 0 \quad \forall\, \bar{w} \in \mathcal{K}_{\widetilde{\psi}} \ . \tag{4.9}$$

Therefore,

$$\underline{\lim}\ j^q = \underline{\lim}\ \mathcal{J}_q(\psi^q) \ge \mathcal{J}_0(\widetilde{\psi}) \ . \tag{4.10}$$

On the other hand, for a fixed $\psi \in \Psi$, there exists a subsequence $\mathcal{J}_q(\psi)$ such that

$$\mathcal{J}_q(\psi) \to \mathcal{J}_0(\psi) \ , \quad q \to 0 \ . \tag{4.11}$$

Indeed, if $w^q(\psi)$ is a solution of variational inequality (4.1), then

$$\|w^q(\psi)\|_2 \le c$$

holds uniformly with respect to $q \le q_0$. Assuming that

$$w^q(\psi) \to w \quad \text{strongly in} \quad L^2(\Omega), \quad \text{weakly in} \quad H^{2,0}(\Omega) \ ,$$

$$w_\nu^q(\psi) \to w_\nu \quad \text{strongly in} \quad L^1(\Gamma)$$

we can pass to the limit in the inequality

$$B(w^q(\psi), \bar{w} - w^q(\psi)) + q\int_\Gamma \big(|\bar{w}_\nu| - |w_\nu^q(\psi)|\big) d\Gamma \ge 0$$

as $q \to 0$. This leads to the inequality

$$w \in \mathcal{K}_\psi \ :\ B(w, \bar{w} - w) \ge 0 \quad \forall\, \bar{w} \in \mathcal{K}_\psi$$

which implies that $w = w(\psi)$. This proves (4.11).

Now, let ψ be a solution to problem (4.8). Then

$$j^q \le \mathcal{J}_q(\psi) \to \mathcal{J}_0(\psi) \ .$$

Consequently,

$$\overline{\lim}\ j^q \le \mathcal{J}_0(\psi) \ . \tag{4.12}$$

By inequality (4.9), we have $\widetilde{w} = w(\widetilde{\psi})$. Then, by (4.10), (4.12), $\widetilde{w}$ is a solution to problem (4.8) and $j^q \to j^0$.

Let us discuss the boundary conditions imposed on Γ. From (4.9) it follows that

$$\langle\, M(\widetilde{w}), \varphi\, \rangle_{1/2} = 0 \quad \forall\, \varphi \in H^{1/2}(\Gamma) \ ,$$

and consequently

$$M(\widetilde{w}) = 0 \quad \text{on} \quad \Gamma \ .$$

This proves the theorem. □

We now investigate convergence of solutions as $q \to \infty$. To this end we introduce one more minimization problem with the cost functional

$$\mathcal{J}_\infty(\psi) = \|w(\psi) - w_*\|_0^2 + \|\psi\|_2^2 \ ,$$

where the function $w(\psi)$ solves the variational inequality

$$\begin{aligned} & w(\psi) \in \mathcal{K}_\psi \cap H_0^2(\Omega) \ : \\ & B(w(\psi), \bar{w} - w(\psi)) \geq 0 \quad \forall\, \bar{w} \in \mathcal{K}_\psi \cap H_0^2(\Omega) \ . \end{aligned} \tag{4.13}$$

We want to find an element $\psi \in \Psi$ such that

$$\mathcal{J}_\infty(\psi) \leq \mathcal{J}_\infty(\bar{\psi}) \quad \forall\, \bar{\psi} \in \Psi \ . \tag{4.14}$$

This problem is obviously solvable. As in the previous considerations, we denote by ψ^q a solution to problem (4.2). Accordingly, $w(\psi^q)$ is defined by (4.1). Put

$$j^\infty = \inf_{\psi \in \Psi} \mathcal{J}_\infty(\psi) \ .$$

Theorem 4.3. *From the sequence ψ^q one can choose a subsequence such that*

$$\psi^q \to \psi \quad \textit{weakly in} \quad H^2(\Omega) \ ,$$

$$w(\psi^q) \to w(\psi) \quad \textit{weakly in} \quad H^{2,0}(\Omega) \ ,$$

$$j^q \to j^\infty \ ,$$

as $q \to \infty$ where ψ solves problem (4.14) and $w(\psi)$ is defined by variational inequality (4.13). The function $w(\psi)$ satisfies the boundary condition

$$w_\nu(\psi) = 0 \quad \textit{on} \quad \Gamma \ .$$

Proof. Fix $\psi \in \Psi$. We rewrite variational inequality (4.1) as

$$\begin{aligned} & B\big(w^q(\psi), w^q(\psi)\big) + q \int_\Gamma |w_\nu^q(\psi)| d\Gamma \\ & \quad \leq B\big(\bar{w}, w^q(\psi)\big) + q \int_\Gamma |\bar{w}_\nu| d\Gamma \ . \end{aligned} \tag{4.15}$$

By taking $\bar{w} \in \mathcal{K}_\psi \cap H_0^2(\Omega)$, it follows that $\|w^q(\psi)\|_2 \leq c$ holds uniformly with respect to q. This entails the boundedness of $\mathcal{J}_q(\psi)$ with respect to q, and hence by the inequality $\mathcal{J}_q(\psi^q) \leq \mathcal{J}_q(\psi)$, one gets

$$\|\psi^q\|_2 \leq c \ .$$

Without loss of generality, we can assume that

$$\psi^q \to \widetilde{\psi} \quad \text{weakly in} \quad H^2(\Omega)$$

as $q \to \infty$. As in the proof of Theorem 4.1, for any element $\bar{w} \in \mathcal{K}_{\widetilde{\psi}} \cap H_0^2(\Omega)$, there exists a sequence $\bar{w}^q \in \mathcal{K}_{\psi^q} \cap H_0^2(\Omega)$ such that

$$\bar{w}^q \to \bar{w} \quad \text{strongly in} \quad H_0^2(\Omega) \ .$$

From the inequality

$$B\big(w(\psi^q), \bar{w}^q - w(\psi^q)\big) - q \int_\Gamma \big|w_\nu(\psi^q)\big| d\Gamma \geq 0 \tag{4.16}$$

we get the estimate

$$\|w(\psi^q)\|_2 \leq c$$

which holds uniformly with respect to q. As above, we can now assume that

$$w(\psi^q) \to \widetilde{w} \quad \text{weakly in} \quad H^{2,0}(\Omega)$$

as $q \to \infty$. With (4.16) we conclude that

$$\|w_\nu(\psi^q)\|_{L^1(\Gamma)} \to 0$$

as $q \to \infty$. Assuming additionally that $w_\nu(\psi^q) \to \widetilde{w}_\nu$ strongly in $L^1(\Gamma)$, we get $\widetilde{w} \in H_0^2(\Omega)$.

By inequality (4.16),

$$w(\psi^q) \in \mathcal{K}_{\psi^q} \ : \ B\big(w(\psi^q), \bar{w}^q - w(\psi^q)\big) \geq 0 \ .$$

Passing to the upper limit we get

$$\widetilde{w} \in \mathcal{K}_{\widetilde{\psi}} \cap H_0^2(\Omega) \ :$$
$$B(\widetilde{w}, \bar{w} - \widetilde{w}) \geq 0 \quad \forall\, \bar{w} \in \mathcal{K}_{\widetilde{\psi}} \cap H_0^2(\Omega) \ .$$

This, in particular, entails that $\widetilde{w} = w(\widetilde{\psi})$, and hence

$$\underline{\lim}\, j^q = \underline{\lim}\, \mathcal{J}_q(\psi^q) \geq \mathcal{J}_\infty(\widetilde{\psi}) \ . \tag{4.17}$$

We now show that, for an arbitrary $\psi \in \Psi$, passing to a subsequence in $\mathcal{J}_q(\psi)$, if necessary, we have

$$\mathcal{J}_q(\psi) \to \mathcal{J}_\infty(\psi) \ , \quad q \to \infty \ .$$

In fact, replacing in inequality (4.1) $\bar{w}$ with an element from $\mathcal{K}_\psi \cap H_0^2(\Omega)$ for some solution $w^q(\psi)$ we get the estimates

$$\|w^q(\psi)\|_2 \leq c \ , \quad q\|w_\nu^q(\psi)\|_{L^1(\Gamma)} \leq c$$

which hold uniformly with respect to q.

We can assume that

$$w^q(\psi) \to w \quad \text{weakly in} \quad H^{2,0}(\Omega) \ , \quad \text{strongly in} \quad L^2(\Omega) \ ,$$
$$w_\nu^q(\psi) \to w_\nu \quad \text{strongly in} \quad L^1(\Gamma)$$

as $q \to \infty$. Then $w \in \mathcal{K}_\psi \cap H_0^2(\Omega)$. Letting $q \to \infty$ we can pass to the limit in the inequality

$$B\big(w^q(\psi), \bar{w} - w^q(\psi)\big) + q \int_\Gamma |\bar{w}_\nu| d\Gamma \geq 0$$

with $\bar{w} \in \mathcal{K}_\psi \cap H_0^2(\Omega)$ substituted as a test function. As a result the relation

$$w \in \mathcal{K}_\psi \cap H_0^2(\Omega) \ :$$
$$B(w, \bar{w} - w) \geq 0 \quad \forall\, \bar{w} \in \mathcal{K}_\psi \cap H_0^2(\Omega)$$

follows. Thus, $\mathcal{J}_q(\psi) \to \mathcal{J}_\infty(\psi)$. Consequently, by taking ψ which solves problem (4.14),

$$j^q \leq \mathcal{J}_q(\psi) \to \mathcal{J}_\infty(\psi) \ .$$

It follows that

$$\overline{\lim}\ j^q \leq \mathcal{J}_\infty(\psi) \ . \tag{4.18}$$

Hence, by (4.17), (4.18), we conclude that $\widetilde{\psi}$ solves problem (4.14) and $j^q \to j^\infty$. This proves the theorem. □

4.2 Cost functionals with measures

In this section we examine other cost functionals. Observe firstly that for a solution $w(\psi)$ of (4.1) and for any element $\varphi \in C_0^\infty(\Omega)$, $\varphi \geq 0$, we have $w(\psi) + \varphi \in \mathcal{K}_\psi$, and for $\bar{w} = w(\psi) + \varphi$,

$$B\big(w(\psi), \varphi\big) \geq 0 \ .$$

This means that the distribution $\mu_\psi \equiv \Delta^2 w(\psi)$ is positive and can be regarded as a nonnegative measure defined on the domain Ω. This measure characterizes the work of interaction forces.

Let

$$\mathcal{J}_q(\psi) = \mu_\psi(\Omega) + \|\psi\|_2^2 \ .$$

The optimal control problem we shall analyse consists in finding an element $\psi \in \Psi$ such that

$$\mathcal{J}_q(\psi) \leq \mathcal{J}_q(\bar{\psi}) \quad \forall\, \bar{\psi} \in \Psi \ . \tag{4.19}$$

We begin by showing that problem (4.19) has a solution. To this end we consider a minimizing sequence $\psi^n \in \Psi$. By its boundedness, we can assume that

$$\psi^n \to \psi \quad \text{weakly in} \quad H^2(\Omega), \ \ \psi \in \Psi \ ,$$
$$\psi^n \to \psi \quad \text{uniformly in} \quad \bar{\Omega}$$

as $n \to \infty$. As in the proof of Theorem 4.1, this convergence entails the boundedness of the solutions $w(\psi^n)$ of (4.1) in the space $H^{2,0}(\Omega)$. Choosing, if necessary, a subsequence, we assume that

$$w(\psi^n) \to \widetilde{w} \quad \text{weakly in} \quad H^{2,0}(\Omega) \ ,$$
$$w_\nu(\psi^n) \to \widetilde{w}_\nu \quad \text{strongly in} \quad L^1(\Gamma)$$

as $n \to \infty$. As before, we can show that $\widetilde{w} = w(\psi)$ which implies that $\widetilde{w}$ solves the problem

$$\widetilde{w} \in \mathcal{K}_\psi \ :$$

$$B(\widetilde{w}, \bar{w} - \widetilde{w}) + q\int_\Gamma |\bar{w}_\nu| d\Gamma - q\int_\Gamma |\widetilde{w}| d\Gamma \geq 0 \quad \forall\, \bar{w} \in \mathcal{K}_\psi \ .$$

We now prove that the sequence of measures $\mu_{\psi^n} \equiv \Delta^2 w(\psi^n)$ converges weakly-(*). In fact, the functions $\Delta^2 w(\psi^n)$ are bounded in $H^{-2}(\Omega) \equiv (H_0^2(\Omega))^*$.

Therefore,

$$\langle\, \Delta^2 w(\psi^n), \varphi \,\rangle \to \langle\, \Delta^2 \widetilde{w}, \varphi \,\rangle$$

as $n \to \infty$ for any fixed $\varphi \in H_0^2(\Omega)$. The brackets $\langle\, \cdot, \varphi \,\rangle$ denote the value of the distribution at the point φ. Thus, the measures μ_{ψ^n} weakly-(*) converge to μ_ψ. Consequently,

$$\underline{\lim}\ \mu_{\psi^n}(\Omega) \geq \mu_\psi(\Omega)$$

and

$$j^q \equiv \inf_{\psi \in \Psi} \mathcal{J}_q(\psi) = \underline{\lim}_{n \to \infty}\ \mathcal{J}_q(\psi^n) \geq \mathcal{J}_q(\psi) \geq j^q \ .$$

This means that ψ solves problem (4.19).

We now consider another optimal control problem. Define the cost functional

$$\mathcal{J}_0(\psi) = \mu_\psi(\Omega) + \|\psi\|_2^2 \ .$$

In this case a function $w(\psi)$ can be found which satisfies the variational inequality

$$w(\psi) \in \mathcal{K}_\psi \ : \ B\big(w(\psi), \bar{w} - w(\psi)\big) \geq 0 \quad \forall\, \bar{w} \in \mathcal{K}_\psi \ . \tag{4.20}$$

and defines the measure $\mu_\psi \equiv \Delta^2 w(\psi)$.

The optimal control problem is formulated as follows. Find an element $\psi \in \Psi$ such that

$$\mathcal{J}_0(\psi) \leq \mathcal{J}_0(\bar{\psi}) \quad \forall\, \bar{\psi} \in \Psi \ . \tag{4.21}$$

Put $j^0 = \inf_{\psi \in \Psi} \mathcal{J}_0(\psi)$. Denote by ψ^q a solution to (4.19). Let $w(\psi^q)$ be defined by variational inequality (4.1) for $\psi = \psi^q$.

Theorem 4.4. *From the sequence ψ^q one can choose a subsequence such that as $q \to 0$*

$$\psi^q \to \psi \quad \textit{weakly in} \quad H^2(\Omega) \ ,$$

$$w(\psi^q) \to w(\psi) \quad \textit{weakly in} \quad H^{2,0}(\Omega) \ ,$$

$$j^q \to j^0 \ ,$$

and, moreover, the function $w(\psi)$ satisfies the boundary condition

$$M\big(w(\psi)\big) = 0 \quad \textit{on} \quad \Gamma \ .$$

Here ψ solves problem (4.21), $w(\psi)$ corresponds to ψ and is defined by (4.20); $j^q = \inf_{\psi \in \Psi} \mathcal{J}_q(\psi)$.

Proof. We show that for any fixed $\psi \in \Psi$ a subsequence of $\mathcal{J}_q(\psi)$ exists such that

$$\mathcal{J}_q(\psi) \to \mathcal{J}_0(\psi) \ , \quad q \to 0 \ .$$

For a given ψ, the corresponding solution to inequality (4.1) will be denoted by $w^q(\psi)$. Since the estimate

$$\|w^q(\psi)\|_2 \le c$$

holds uniformly with respect to $q \le q_0$ we can assume that as $q \to 0$

$$w^q(\psi) \to \widetilde{w} \quad \text{weakly in} \quad H^{2,0}(\Omega) \ ,$$

$$w^q_\nu(\psi) \to \widetilde{w}_\nu \quad \text{strongly in} \quad L^1(\Gamma) \ .$$

Now we can pass to the limit in the inequality

$$B\big(w^q(\psi), \bar{w} - w^q(\psi)\big) + q\int_\Gamma |\bar{w}_\nu| d\Gamma - q\int_\Gamma |w^q_\nu(\psi)| d\Gamma \ge 0 \quad \forall\, \bar{w} \in \mathcal{K}_\psi$$

as $q \to 0$. As a result, we get

$$\widetilde{w} \in \mathcal{K}_\psi \ : \ B(\widetilde{w}, \bar{w} - \widetilde{w}) \ge 0 \quad \forall\, \bar{w} \in \mathcal{K}_\psi \ .$$

Thus, $\widetilde{w} = w(\psi)$, and $\mu^q_\psi \equiv \Delta^2 w^q(\psi) \to \mu_\psi \equiv \Delta^2 w(\psi)$ weakly-(*).

Moreover,

$$\mu^q_\psi(\Omega) \to \mu_\psi(\Omega)$$

as $q \to 0$.

Let us verify this. The function $\widetilde{w}$ is continuous on $\bar{\Omega}$ and equal to zero on Γ. Moreover, $\psi < 0$ in some neighbourhood of Γ. We can assume that $w^q(\psi) \to \widetilde{w}$ uniformly in $\bar{\Omega}$. Thus, there exists a compact $B \subset \Omega$ such that $\mu^q_\psi(\Omega \backslash B) = 0$ for $q > 0$ small enough. Without loss of generality, we can assume that the measure $\mu_\psi(\partial B)$ of the boundary is equal to zero. This implies that $\mu^q_\psi(B) \to \mu_\psi(B)$ (see Chapter 1, Section 1.10) and hence

$$\mu^q_\psi(\Omega) = \mu^q_\psi(B) \to \mu_\psi(B) = \mu_\psi(\Omega) \ .$$

So, by letting $q \to 0$ we conclude that

$$\mathcal{J}_q(\psi) \equiv \mu^q_\psi(\Omega) + \|\psi\|_2^2 \to \mathcal{J}_0(\psi) \equiv \mu_\psi(\Omega) + \|\psi\|_2^2 \ .$$

For every fixed $\psi \in \Psi$ the inequality

$$\mathcal{J}_q(\psi^q) \le \mathcal{J}_q(\psi)$$

holds and its right-hand side is bounded uniformly with respect to q. Hence

$$\|\psi^q\|_2 \le c \ .$$

Let

$$\psi^q \to \widetilde{\psi} \quad \text{weakly in} \quad H^2(\Omega) \quad \text{and uniformly in} \quad \bar{\Omega} \ .$$

We substitute into the variational inequality

$$\begin{aligned} & B(w(\psi^q), \bar{w}^q - w(\psi^q)) \\ & + q \int_\Gamma |\bar{w}^q_\nu| d\Gamma - q \int_\Gamma |w_\nu(\psi^q)| d\Gamma \geq 0 \quad \forall\, \bar{w}^q \in \mathcal{K}_{\psi^q} \end{aligned} \tag{4.22}$$

a sequence $\bar{w}^q \in \mathcal{K}_{\psi^q}$ strongly converging in $H^{2,0}(\Omega)$ to $\bar{w} \in \mathcal{K}_{\widetilde{\psi}}$. This gives us the estimation

$$\|w(\psi^q)\|_2 \leq c \tag{4.23}$$

which holds uniformly with respect to q. Passing now to a subsequence, if necessary, we have

$$\begin{aligned} & w(\psi^q) \to \widetilde{w} \quad \text{weakly in} \quad H^{2,0}(\Omega) \ , \\ & \mu_{\psi^q} \equiv \Delta^2 w(\psi^q) \to \mu_{\widetilde{\psi}} \equiv \Delta^2 \widetilde{w} \quad \text{weakly-}(*) \end{aligned} \tag{4.24}$$

as $q \to 0$.

Take a sequence $\bar{w}^q \in \mathcal{K}_{\psi^q}$ such that $\bar{w}^q \to \bar{w}$ strongly in $H^{2,0}(\Omega)$. Passing to the limit in (4.22) as $q \to 0$ we see that the functions $\widetilde{w}, \widetilde{\psi}$ satisfy the inequality

$$\widetilde{w} \in \mathcal{K}_{\widetilde{\psi}} \ : \ B(\widetilde{w}, \bar{w} - \widetilde{w}) \geq 0 \quad \forall\, \bar{w} \in \mathcal{K}_{\widetilde{\psi}} \ .$$

Thus, $\widetilde{w} = w(\widetilde{\psi})$. This justifies the definition of the measure in (4.24).

According to (4.24), we have

$$\underline{\lim}\, \mu_{\psi^q}(\Omega) \geq \mu_{\widetilde{\psi}}(\Omega)$$

and hence

$$\underline{\lim}\, j^q \geq \mu_{\widetilde{\psi}}(\Omega) + \|\widetilde{\psi}\|_2^2 \ . \tag{4.25}$$

Let ψ be a solution to (4.21). By the above arguments,

$$j^q \leq \mathcal{J}_q(\psi) \to \mathcal{J}_0(\psi) = j^0 \ ,$$

that is

$$\overline{\lim}\, j^q \leq \mathcal{J}_0(\psi) \ . \tag{4.26}$$

Using (4.25), (4.26) we see that $\widetilde{\psi}$ solves problem (4.21) and $j^q \to j^0$. Moreover, the boundary condition $M\big(w(\psi)\big) = 0$ is satisfied which completes the proof. □

We conclude this section with a result for the case where $q \to \infty$. Consider the variational inequality

$$\begin{gathered} w(\psi) \in \mathcal{K}_\psi \cap H_0^2(\Omega) \ : \\ B\big(w(\psi), \bar{w} - w(\psi)\big) \geq 0 \quad \forall\, \bar{w} \in \mathcal{K}_\psi \cap H_0^2(\Omega) \end{gathered} \tag{4.27}$$

and define the cost functional

$$\mathcal{J}_\infty(\psi) = \mu_\psi(\Omega) + \|\psi\|_2^2 \ , \quad \mu_\psi \equiv \Delta^2 w(\psi) \ .$$

Find an element $\psi \in \Psi$ such that

$$\mathcal{J}_\infty(\psi) \leq \mathcal{J}_\infty(\bar{\psi}) \quad \forall\, \bar{\psi} \in \Psi \ . \tag{4.28}$$

As before, we use the notation $j^\infty = \inf\limits_{\psi \in \Psi} \mathcal{J}_\infty(\psi)$. Let ψ^q be a solution to problem (4.19) and $w(\psi^q)$ be determined by (4.1).

Theorem 4.5. *From the sequence ψ^q one can choose a subsequence such that*

$$\psi^q \to \psi \quad \textit{weakly in} \quad H^2(\Omega) \ ,$$

$$w(\psi^q) \to w(\psi) \quad \textit{weakly in} \quad H^{2,0}(\Omega) \ ,$$

$$j^q \to j^\infty$$

as $q \to \infty$. Here, ψ is a solution to problem (4.28) and we can find $w(\psi)$ satisfying (4.27), such that

$$w_\nu(\psi) = 0 \quad on \quad \Gamma \ .$$

The idea of proof is the same as that of the previous theorem and therefore we omit it here.

5 Optimal control problem for a beam

We consider an optimal choice of the punch shape in contact problems of a beam and a rigid punch. The beam points have both the vertical and horizontal displacements w, u. As we know the nonpenetration condition in this case is more complicated than that considered in the previous section.

Let us consider a beam having a unit length; the axis x is directed along the middle line of the beam, the axis z has the orthogonal direction. The equation $z = \varphi(x)$ describes the punch shape. We write the nonpenetration condition (compare Section 3, Chapter 2) as follows

$$w(x) - u(x)\varphi_x(x) \geq \varphi(x) \ . \tag{5.1}$$

The boundary conditions are of the form

$$u = w = w_x = 0 \quad \text{when} \quad x = 0, 1 \; .$$

The energy functional for the beam has the form

$$\Pi(u, w) = \frac{1}{2}\int_0^1 (u_x^2 + w_{xx}^2)dx - \int_0^1 (f_1 u + f_2 w)dx \; . \tag{5.2}$$

Here f_1, f_2 are given exterior forces. Let $H(0,1) = H_0^1(0,1) \times H_0^2(0,1)$.

We consider the equilibrium state of the beam in contact with the punch. Denoting

$$\mathcal{K}_\varphi = \{(u, w) \in H(0,1) \big| w - u\varphi_x \geq \varphi\}$$

we can formulate the problem as the following variational inequality

$$\begin{aligned}(u, w) \in \mathcal{K}_\varphi \; : \; & \int_0^1 \{w_{xx}(\bar{w}_{xx} - w_{xx}) + u_x(\bar{u}_x - u_x)\}dx \\ & \geq \int_0^1 \{f_1(\bar{u} - u) + f_2(\bar{w} - w)\}dx \quad \forall\, (\bar{u}, \bar{w}) \in \mathcal{K}_\varphi \; .\end{aligned} \tag{5.3}$$

We use the cost functional of the form

$$\mathcal{J}(\varphi) = \|w - w_0\|_0^2 + \|u - u_0\|_0^2 + \|\varphi\|_2^2 \; , \tag{5.4}$$

where $\|\cdot\|_s$ is the norm in $H^s(0,1)$, the functions w, u are defined by (5.3) for a given φ. The functions w_0, u_0 are fixed.

Assume that $\Phi \subset H^2(0,1)$ is a convex and closed set and $\varphi(0) < -\delta$, $\varphi(1) < -\delta$, $\delta = \text{const} > 0$, for every $\varphi \in \Phi$.

The optimal control problem consists in the minimization of the functional $\mathcal{J}(\varphi)$ over Φ,

$$\inf_{\varphi \in \Phi} \mathcal{J}(\varphi) \; . \tag{5.5}$$

The main result of this section is formulated as follows.

Theorem 5.1. *There exists a solution to problem (5.5), (5.3).*

Proof. Let φ^n be a minimizing sequence. This sequence is bounded in the space $H^2(0,1)$, and, therefore, without loss of generality, we assume that

$$\varphi^n \to \varphi \quad \text{weakly in} \quad H^2(0,1) \; , \quad \text{strongly in} \quad C^1[0,1] \tag{5.6}$$

as $n \to \infty$. For every φ^n, there exists a solution to the variational inequality

$$\begin{aligned}(u^n, w^n) \in \mathcal{K}_{\varphi^n} \; : \; & \int_0^1 \{w_{xx}^n(\bar{w}_{xx}^n - w_{xx}^n) + u_x^n(\bar{u}_x^n - u_x^n)\}dx \\ & \geq \int_0^1 \{f_1(\bar{u}^n - u^n) + f_2(\bar{w}^n - w^n)\}dx \quad \forall\, (\bar{u}^n, \bar{w}^n) \in \mathcal{K}_{\varphi^n} .\end{aligned} \tag{5.7}$$

This inequality enables us to derive a priori estimates for (u^n, w^n).

We begin by proving the following preliminary statement. For any $(\bar{u}, \bar{w}) \in \mathcal{K}_\varphi$ there exists a sequence $(\bar{u}^n, \bar{w}^n) \in \mathcal{K}_{\varphi^n}$ with the property

$$(\bar{u}^n, \bar{w}^n) \to (\bar{u}, \bar{w}) \quad \text{strongly in} \quad H(0,1) \ .$$

Here φ is defined by (5.6).

To see this, we recall that the sequence $(\bar{u}^n, \bar{w}^n)$ satisfies the inequality

$$\bar{w}^n - \bar{u}^n \varphi_x^n \geq \varphi^n \ , \tag{5.8}$$

φ takes the negative values at $x = 0, 1$, and φ^n converges to φ strongly in $C^1[0,1]$. There exists $\varepsilon > 0$ such that the inequalities

$$\varphi^n < -\frac{2\delta}{3} \ , \quad |\bar{u}\varphi_x^n| < \frac{\delta}{4} \ , \quad |\bar{w}| < \frac{\delta}{4} \tag{5.9}$$

hold in the intervals $(0, \varepsilon)$, $(1 - \varepsilon, 1)$.

Take a nonnegative function $\psi \in H_0^2(0,1)$ such that

$$\psi \equiv 1 \quad \text{on} \quad (\varepsilon, 1 - \varepsilon) \ . \tag{5.10}$$

Since

$$|\varphi^n(x) - \varphi(x)| < \frac{1}{2n} \ , \quad |\varphi_x^n(x) - \varphi_x(x)| \ |\bar{u}(x)| < \frac{1}{2n} \tag{5.11}$$

the sequence

$$(\bar{u}^n, \bar{w}^n) = (\bar{u}, \bar{w} + \frac{1}{n}\psi) \ ,$$

satisfies (5.8). In other words, we have $(\bar{u}^n, \bar{w}^n) \in \mathcal{K}_{\varphi^n}$. In fact, the functions $\bar{u}, \bar{w}$ satisfy the inequality $\bar{w} - \bar{u}\varphi_x \geq \varphi$. Hence, in view of (5.9), (5.10), (5.11), we get

$$\bar{w} + \frac{1}{n}\psi - \bar{u}\varphi_x^n \geq \varphi^n \ ,$$

which entails (5.8). In this way we get the required sequence $(\bar{u}^n, \bar{w}^n)$.

Substituting into (5.7) the elements $(\bar{u}^n, \bar{w}^n)$ as test functions, we obtain the inequality

$$\|w_{xx}^n\|_0^2 + \|u_x^n\|_0^2 \leq c$$

which holds uniformly with respect to n. This implies that w^n and u^n are bounded in $H_0^2(0,1)$ and $H_0^1(0,1)$, respectively. Without loss of generality, it can be assumed that

$$(u^n, w^n) \to (u, w) \quad \text{weakly in} \quad H(0,1)$$

as $n \to \infty$. We put inequality (5.7) in the form

$$\begin{aligned} \int_0^1 (w_{xx}^n \bar{w}_{xx}^n + u_x^n \bar{u}_x^n) dx &\geq \|w_{xx}^n\|_0^2 + \|u_x^n\|_0^2 \\ &+ \int_0^1 f_1(\bar{u}^n - u^n) dx + \int_0^1 f_2(\bar{w}^n - w^n) dx \ . \end{aligned} \tag{5.12}$$

As above, the elements of the strongly converging sequence $(\bar{u}^n, \bar{w}^n)$ can be substituted into (5.12). Passing to the lower limit in both sides of (5.12), we get

$$(u,w) \in \mathcal{K}_\varphi \ : \ \int_0^1 \{w_{xx}(\bar{w}_{xx} - w_{xx}) + u_x(\bar{u}_x - u_x)\}dx \geq \int_0^1 f_1(\bar{u} - u)dx + \int_0^1 f_2(\bar{w} - w)dx \ .$$

Obviously, this inequality holds for every $(\bar{u}, \bar{w}) \in \mathcal{K}_\varphi$. Thus, $u = u(\varphi)$, $w = w(\varphi)$. With this we have

$$\inf_{\bar{\varphi} \in \Phi} \mathcal{J}(\bar{\varphi}) = \underline{\lim} \ \mathcal{J}(\varphi^n) \geq \mathcal{J}(\varphi) \ .$$

Hence, the constructed element $\varphi \in \Phi$ solves optimal control problem (5.5), (5.3). This completes the proof. □

6 Optimal control problem for a fourth-order variational inequality

In this section we study an optimal control problem governed by the variational inequality for the operator

$$w \to \varepsilon \Delta^2 w - \varphi(\|\nabla w\|^2)\Delta w \ , \quad \varepsilon > 0 \ ,$$

with a nonnegative continuous function φ. Our goal is to prove the convergence of solutions of optimal control problems when ε tends to zero. The set of admissible displacements forms a control region. In applications, we get the convergence from a plate to a membrane.

6.1 Fourth-order operator

We formulate the problem as follows. Let $\Omega \subset \mathbb{R}^2$ be a bounded domain with smooth boundary Γ, and $\Psi \subset H^2(\Omega)$ be a closed, convex set, $H^{2,0}(\Omega) \equiv H^2(\Omega) \cap H_0^1(\Omega)$. For every $\psi \in \Psi$ we define the set

$$\mathcal{K}_\psi^2 = \{w \in H^{2,0}(\Omega) \mid w(x) \geq \psi(x) \ , \quad x \in \Omega\} \ .$$

Each element $\psi \in \Psi$ satisfies the inequality $\psi < 0$ on Γ. For a fixed $\psi \in \Psi$ and $\varepsilon > 0$, we can solve the variational inequality

$$w \in \mathcal{K}_\psi^2 \ : \ \varepsilon \int_\Omega \Delta w(\Delta \bar{w} - \Delta w)dx + \varphi(\|\nabla w\|^2) \int_\Omega \nabla w(\nabla \bar{w} - \nabla w)dx \geq \int_\Omega f(\bar{w} - w)dx \quad \forall \, \bar{w} \in \mathcal{K}_\psi^2 \ . \tag{6.1}$$

Some additional conditions providing the existence and uniqueness of solution to (6.1) will be formulated below.

Let us consider the cost functional

$$\mathcal{J}_\varepsilon(\psi) = \|w(\psi) - w_0\|^2 + \|\psi\|_2^2 \ ,$$

where $w(\psi) \equiv w$ is a solution to (6.1) and $w_0 \in L^2(\Omega)$ is a given element. The dependence of $w(\psi)$ on ε is not indicated for the sake of simplicity.

For any fixed $\varepsilon > 0$, we formulate the following optimal control problem. Find $\psi \in \Psi$ such that

$$\mathcal{J}_\varepsilon(\psi) \leq \mathcal{J}_\varepsilon(\bar{\psi}) \quad \forall\, \bar{\psi} \in \Psi \ . \tag{6.2}$$

The state of the system is defined by variational inequality (6.1), the cost functional depends on solutions of (6.1) and ψ.

We organize this section as follows. At the beginning some complementary statements concerning the operator

$$w \to -\varphi(\|\nabla w\|^2)\Delta w$$

will be proved. Next we establish a solvability result for problem (6.2). In Section 6.2 we study problem (6.2) with $\varepsilon = 0$. The main result of the present section concerns the convergence of solutions to (6.2) to the solution of the problem corresponding to $\varepsilon = 0$ as $\varepsilon \to 0$. This proof is given in Section 6.3.

Properties of the above operator depend on ε. In Lemma 6.1 we consider a more general operator

$$B(w) = (-1)^n \varphi(\|\nabla^n w\|^2)\Delta^n w \ ,$$

where $\nabla^n = \Delta^{\frac{n}{2}}$ for n even, $\nabla^n = \nabla\big(\Delta^{\frac{n-1}{2}}\big)$ for n odd.

Let $H^{-n}(\Omega)$ be the dual space to $H_0^n(\Omega)$. Obviously, the operator B acts from $H_0^n(\Omega)$ into $H^{-n}(\Omega)$.

Lemma 6.1. *The operator B from $H_0^n(\Omega)$ into $H^{-n}(\Omega)$ is monotone if and only if the function $\sqrt{s}\,\varphi(s)$ is nondecreasing.*

Proof. We start with the proof of the sufficiency part. Let $w_1, w_2 \in H_0^n(\Omega)$. By the Cauchy inequality

$$\begin{aligned}
L(w_1, w_2) &\equiv \big(B(w_1) - B(w_2)\big)(w_1 - w_2) \\
&\geq \varphi\big(\|\nabla^n w_1\|^2\big)\|\nabla^n w_1\|^2 + \varphi\big(\|\nabla^n w_2\|^2\big)\|\nabla^n w_2\|^2 \\
&\quad - \|\nabla^n w_1\|\,\|\nabla^n w_2\|\big(\varphi(\|\nabla^n w_1\|^2) + \varphi(\|\nabla^n w_2\|^2)\big) \\
&= \big(\sqrt{s_1} - \sqrt{s_2}\big)\big(\sqrt{s_1}\,\varphi(s_1) - \sqrt{s_2}\,\varphi(s_2)\big) \geq 0,
\end{aligned}$$

where $s_i = \|\nabla^n w_i\|^2$, $i = 1, 2$, which is the required property.

We prove the necessity by contradiction. Assuming that the condition is not satisfied we can take s_1, s_2, δ such that

$$\sqrt{s_1}\,\varphi(s_1) - \sqrt{s_2}\,\varphi(s_2) = \delta\ , \quad s_1 > s_2\ , \quad \delta < 0$$

and choose functions w_1, w_2 such that $\|\nabla^n w_1\|^2 = s_1$, $w_2 = \sqrt{\frac{s_2}{s_1}}\, w_1$. By this

$$L(w_1, w_2) = \left(\sqrt{s_1} - \sqrt{s_2}\right)\left(\sqrt{s_1}\,\varphi(s_1) - \sqrt{s_2}\,\varphi(s_2)\right) < 0\ .$$

which contradicts the monotonicity of B. Lemma 6.1 is proved. □

It is worth observing that $B(w)$ is the gradient of the functional

$$w \to \frac{1}{2}\int_0^{\|\nabla^n w\|^2} \varphi(s)ds$$

at w. Provided that the function $\sqrt{s}\,\varphi(s)$ is nondecreasing, the operator B is monotone, and hence the functional is convex.

Consider now a minimization problem which is equivalent to variational inequality (6.1). On the space $H^{2,0}(\Omega)$ we define the functional

$$\Pi_\varepsilon(w) = \frac{1}{2}\,\varepsilon\|\Delta w\|^2 + \frac{1}{2}\int_0^{\|\nabla w\|^2} \varphi(s)ds - \int_\Omega fwdx\ .$$

The problem is

$$\inf_{w\in\mathcal{K}^2_\psi} \Pi_\varepsilon(w)\ . \tag{6.3}$$

Lemma 6.2. *Let $\sqrt{s}\,\varphi(s)$ be a nondecreasing function. Then problem (6.3) has a unique solution.*

Proof. Let A_ε be an operator acting from $H^{2,0}(\Omega)$ into its dual space,

$$A_\varepsilon(w)(\bar{w}) = \int_\Omega \left(\varepsilon\Delta w - \varphi(\|\nabla w\|^2)w\right)\Delta\bar{w}dx - \int_\Omega f\bar{w}dx\ .$$

It is easy to verify that $A_\varepsilon(w)$ is the gradient of Π_ε at the point w. Thus, the problem of minimization of Π_ε over the set $\mathcal{K}^2_\psi$ is equivalent to the variational inequality

$$w \in \mathcal{K}^2_\psi\ :\ A_\varepsilon(w)(\bar{w} - w) \geq 0\quad \forall\, \bar{w} \in \mathcal{K}^2_\psi\ .$$

This inequality coincides with (6.1).

We may note here that the operator A_ε is semicontinuous. More precisely, for any fixed u, v, w the function $A_\varepsilon(u + sv)(w)$ of the variable s is continuous. This fact, though not used in the proof of Lemma 6.2 will be used later on. It is also clear, that the functional Π_ε is coercive, ie.,

$$\Pi_\varepsilon(w) \to \infty\ , \quad \|w\|_2 \to \infty\ .$$

Moreover, this functional is weakly lower semicontinuous, and the set $\mathcal{K}^2_\psi$ is convex and closed in $H^{2,0}(\Omega)$. Hence, a solution to problem (6.3) exists.

We now demonstrate the uniqueness of a solution. Assuming that there are two different solutions w_1 and w_2 we substitute into (6.1) the function w_1 as $\bar{w}$ and next w_2 as $\bar{w}$. Summing up the resulting inequalities, we get

$$\varepsilon\|\Delta w_1 - \Delta w_2\|^2 + \int_\Omega (\varphi(\|\nabla w_1\|^2)\nabla w_1 - \varphi(\|\nabla w_2\|^2)\nabla w_2)(\nabla w_1 - \nabla w_2)dx \le 0 \ .$$

The second term is nonnegative by Lemma 6.1, which completes the proof of the lemma. □

We can now formulate an existence theorem for optimal control problem (6.2).

Theorem 6.1. *Let the function $\sqrt{s}\,\varphi(s)$ be nondecreasing. Then a solution to (6.2) exists.*

Proof. A minimizing sequence ψ_n is obviously bounded in $H^2(\Omega)$. Choosing a subsequence, if necessary, we have

$$\psi_n \to \psi \quad \text{weakly in} \quad H^2(\Omega) \quad \text{and uniformly in} \quad \bar{\Omega} \tag{6.4}$$

as $n \to \infty$. Clearly, $\psi \in \Psi$. Furthemore, for every $\psi_n \in \Psi$ we can form the set $\mathcal{K}^2_{\psi_n}$ and find a solution to the variational inequality

$$w_n \in \mathcal{K}^2_{\psi_n} \quad : \quad A_\varepsilon(w_n)(\bar{w}_n - w_n) \ge 0 \quad \forall\, \bar{w}_n \in \mathcal{K}^2_{\psi_n} \ . \tag{6.5}$$

Let $\psi \in \Psi$ be given by (6.4). We shall use the result proved in Section 2 that for any $\bar{w} \in \mathcal{K}^2_\psi$ there exists a sequence $\bar{w}_n \in \mathcal{K}^2_{\psi_n}$ strongly converging to $\bar{w}$.

Choose an element $\bar{w} \in \mathcal{K}^2_\psi$ and construct a sequence $\bar{w}_n \in \mathcal{K}^2_{\psi_n}$ such that

$$\bar{w}_n \to \bar{w} \quad \text{strongly in} \quad H^{2,0}(\Omega)$$

as $n \to \infty$. By inequality (6.5),

$$\Pi_\varepsilon(w_n) \le \Pi_\varepsilon(\bar{w}_n) \quad \forall\, \bar{w}_n \in \mathcal{K}^2_{\psi_n} \ .$$

Substituting $\bar{w}_n$, we get the estimate

$$\|w_n\|_2 \le c$$

which holds uniformly with respect to n (recall that ε is fixed). By this estimate, we can assume that

$$w_n \to w \quad \text{weakly in} \quad H^{2,0}(\Omega) \tag{6.6}$$

as $n \to \infty$. In view of the monotonicity and semicontinuity of the operator A_ε acting from the space $H^{2,0}(\Omega)$ into its dual space, inequality (6.5) can be rewritten as

$$w_n \in \mathcal{K}^2_{\psi_n} \quad : \quad A_\varepsilon(\bar{w}_n)(\bar{w}_n - w_n) \geq 0 \quad \forall\, \bar{w}_n \in \mathcal{K}^2_{\psi_n} \ .$$

This, in turn, can be rewritten as

$$\varepsilon \int_\Omega \Delta\bar{w}_n(\Delta\bar{w}_n - \Delta w_n)dx + \varphi(\|\nabla\bar{w}_n\|^2) \int_\Omega \nabla\bar{w}_n(\nabla\bar{w}_n - \nabla w_n)dx$$
$$\geq \int_\Omega f(\bar{w}_n - w_n)dx \quad \forall\, \bar{w}_n \in \mathcal{K}^2_{\psi_n} \ .$$

By (6.6), we can pass to the limit as $n \to \infty$ provided $\bar{w}_n$ is the strongly converging sequence constructed above. This gives

$$w \in \mathcal{K}^2_\psi \quad : \quad \varepsilon \int_\Omega \Delta\bar{w}(\Delta\bar{w} - \Delta w)dx$$
$$+ \varphi(\|\nabla\bar{w}\|^2) \int_\Omega \nabla\bar{w}(\nabla\bar{w} - \nabla w)dx \geq \int_\Omega f(\bar{w} - w)dx \quad \forall\, \bar{w} \in \mathcal{K}^2_\psi.$$

As before, this inequality can be written in form (6.1). Hence, for ψ defined by (6.4) and w defined by (6.6) we have $w = w(\psi)$. Therefore,

$$d \equiv \inf_{\bar{\psi} \in \Psi} \mathcal{J}_\varepsilon(\bar{\psi}) = \varliminf_{n \to \infty} \mathcal{J}_\varepsilon(\psi_n) \geq \mathcal{J}_\varepsilon(\psi) \geq d \ .$$

This proves that the function ψ solves problem (6.2). □

6.2 Second-order operator

In the space $H^1_0(\Omega)$ we consider the convex and closed set

$$\mathcal{K}^1_\psi = \{w \in H^1_0(\Omega) \mid w(x) \geq \psi(x) \quad \text{a.e. in} \quad \Omega\} \ .$$

On the space $H^1_0(\Omega)$ we define the functional

$$\Pi_0(w) = \frac{1}{2}\int_0^{\|\nabla w\|^2} \varphi(s)ds - \int_\Omega fwdx \ ,$$

having the derivative $A_0(w)$ at w. The derivative is equal to $-\varphi(\|\nabla w\|^2)\Delta w - f$, and its value at $\bar{w}$ can be found by the formula

$$A_0(w)(\bar{w}) = \varphi(\|\nabla w\|^2) \int_\Omega \nabla w \nabla\bar{w} dx - \int_\Omega f\bar{w}dx \ .$$

For any fixed $\psi \in \Psi$, we consider the problem

$$w \in \mathcal{K}^1_\psi \quad : \quad \varphi(\|\nabla w\|^2) \int_\Omega \nabla w(\nabla\bar{w} - \nabla w)dx$$
$$\geq \int_\Omega f(\bar{w} - w)dx \quad \forall\, \bar{w} \in \mathcal{K}^1_\psi \ . \tag{6.7}$$

As above, let $\sqrt{s}\,\varphi(s)$ be a nondecreasing function. According to Lemma 6.1 the functional Π_0 is convex. Hence, the minimization problem

$$\inf_{w\in\mathcal{K}_\psi^1} \Pi_0(w)$$

is equivalent to variational inequality (6.7). The sufficient conditions for solvability of (6.7) are contained in the following result.

Lemma 6.3. *Let $\sqrt{s}\,\varphi(s)$ be a strictly increasing function and $\sqrt{s}\,\varphi(s) \to \infty$ as $s \to \infty$. Then, a solution of variational inequality (6.7) exists and is unique.*

Proof. We first show the coercivity of the functional Π_0. In view of the formula

$$\Pi_0(w) - \Pi_0(0) = \int_0^1 A_0(sw)(w)ds \ ,$$

we have

$$\Pi_0(w) = \int_0^{1/2} \big(A_0(sw) - A_0(0)\big)(w)ds$$

$$+ \int_{1/2}^1 A_0(sw)(w)ds + \frac{1}{2}\,A_0(0)(w) \ .$$

By Lemma 6.1, the first term of this relation is nonnegative, the second one is equal to $\frac{1}{2}\,A_0(s_0w)(w)$, $s_0 \in \left[\frac{1}{2}, 1\right]$. Thus,

$$\Pi_0(w) \geq \frac{1}{2}\,A_0(s_0w)(w) - c\|\nabla w\|$$

$$\geq \frac{1}{2}\,\|\nabla w\|\big(\varphi(\|s_0\nabla w\|^2)\|s_0\nabla w\| - 2c\big) \to \infty \ , \quad \|\nabla w\| \to \infty \ .$$

We note also that the functional Π_0 is weakly lower semicontinuous on $H_0^1(\Omega)$. Consequently, a solution to our minimization problem exists. Hence, a solution to (6.7) exists.

Let us prove the uniqueness of a solution. Assuming the existence of two solutions w_1, w_2, we substitute w_1 into (6.7) as $\bar{w}$, and next w_2 as $\bar{w}$. First, suppose that $s_1 \neq s_2$, where $s_i = \|\nabla w_i\|^2$, $i = 1, 2$. Summing up the inequalities, we find that

$$\varphi(\|\nabla w_1\|^2)\int_\Omega \nabla w_1(\nabla w_1 - \nabla w_2)dx$$

$$+\, \varphi(\|\nabla w_2\|^2)\int_\Omega \nabla w_2(\nabla w_2 - \nabla w_1)dx \leq 0 \ .$$

By the Cauchy inequality, the left-hand side of the above inequality can be bounded by the quantity

$$\varphi(s_1)s_1 + \varphi(s_2)s_2 - \sqrt{s_1 s_2}\,\big(\varphi(s_1) + \varphi(s_2)\big)$$
$$= \big(\sqrt{s_1} - \sqrt{s_2}\big)\big(\sqrt{s_1}\,\varphi(s_1) - \sqrt{s_2}\,\varphi(s_2)\big) > 0$$

which leads to a contradiction. If $s_1 = s_2$, then

$$\int_\Omega \nabla w_1 \nabla w_2 dx < \|\nabla w_1\| \, \|\nabla w_2\| \ ,$$

and again we get a contradiction. This proves the lemma. □

For a given $\psi \in \Psi$, let $w = w(\psi)$ be a solution to variational inequality (6.7). We formulate an optimal control problem governed by (6.7) with the cost functional

$$\mathcal{J}_0(\psi) = \|w(\psi) - w_0\|^2 + \|\psi\|_2^2 \ .$$

The problem consists in finding an element $\psi \in \Psi$ such that

$$\mathcal{J}_0(\psi) \leq \mathcal{J}_0(\bar{\psi}) \quad \forall \, \bar{\psi} \in \Psi \ . \tag{6.8}$$

The main purpose of this section is to prove solvability of problem (6.8). We need two auxiliary lemmas.

Let us denote by $H_0^{1,p}(\Omega)$ the space of functions, having the first derivatives integrable with the power $p > 2$ and vanishing on the boundary Γ. Note that these functions are continuous in $\bar{\Omega}$.

Let $w \in \mathcal{K}_\psi^1$. There exists a sequence w_n such that $w_n \in \mathcal{K}_\psi^1 \cap H_0^{1,p}(\Omega)$ and

$$w_n \to w \quad \text{strongly in} \quad H_0^1(\Omega)$$

as $n \to \infty$. To see this we choose a sequence of smooth functions $\widetilde{w}_n$ with compact supports converging strongly in $H_0^1(\Omega)$ to w, and define $w_n = \max\,\{\widetilde{w}_n, \psi\}$.

Lemma 6.4. *For every fixed $w \in \mathcal{K}_\psi^1$ there exists a sequence $w_n \in \mathcal{K}_\psi^2$ such that for $n \to \infty$*

$$w_n \to w \quad \textit{strongly in} \quad H_0^1(\Omega) \ . \tag{6.9}$$

Proof. We begin by proving that a sequence $w_n \in \mathcal{K}_\psi^2$ satisfying (6.9) can be chosen for $w \in \mathcal{K}_\psi^1 \cap H_0^{1,p}(\Omega)$. To this end we consider a domain Ω_1, $\bar{\Omega} \subset \Omega_1$. We extend the function w by assigning it the value zero in an outer neighbourhood of Ω. The function ψ is also extended in this neighbourhood. Since $w \in H_0^{1,p}(\Omega)$ we can exploit the continuity of w in $\bar{\Omega}$. The notation $(w)_\varepsilon$ will be used to denote the mollifiers of w calculated for a smooth kernel. For every n we take ε_n such that

$$(\bar{w})_{\varepsilon_n}(x) + \frac{1}{n} \geq (\bar{\psi})_{\varepsilon_n}(x) + \frac{1}{n} \geq \psi(x) \tag{6.10}$$

in Ω, where $\bar{w}, \bar{\psi}$ are the above extensions. Let us consider in $\Omega_1 \backslash \Omega$ the sequence $v_n = (\bar{w})_{\varepsilon_n} + \dfrac{1}{n}$. It is clear that

$$\|v_n\|_{1,\Omega_1 \backslash \Omega} \to 0 \ , \quad n \to \infty \ . \tag{6.11}$$

To avoid misunderstandings we indicate the domain here. In general, the functions $(\bar{w})_{\varepsilon_n}$ are defined only for $x \in \Omega_1$ such that dist $(x, \partial\Omega_1) \leq \varepsilon_n$. Without loss of generality, we can assume that for sufficiently large n the functions $(\bar{w})_{\varepsilon_n}$ are equal to zero when dist $(x, \partial\Omega_1) \geq \varepsilon_n$. Let us choose a domain Ω_2, $\bar{\Omega}_2 \subset \Omega$, and extend the functions v_n in Ω assuming that v_n vanish in Ω_2 (see Mikhailov, 1976). For these extensions $\bar{v}_n$ the estimate

$$\|\bar{v}_n\|_{1,\Omega_1\setminus\Omega_2} \leq c\,\|v_n\|_{1,\Omega_1\setminus\Omega} \tag{6.12}$$

holds. In particular, it can be assumed that $\psi < \delta < 0$ and $w > \dfrac{\delta}{2}$ in some neighbourhood Ω_0 of the boundary Γ, $\Omega_2 = \Omega\setminus\bar{\Omega}_0$. From (6.12) it follows that

$$\|\bar{v}_n\|_{1,\Omega\setminus\Omega_2} \leq c\,\|v_n\|_{1,\Omega_1\setminus\Omega} \ .$$

By (6.11), the right-hand side of this inequality converges to zero as $n \to \infty$. On the other hand, by the continuity of $\bar{w}$,

$$\|(\bar{w})_{\varepsilon_n}\|_{C(\overline{\Omega_1\setminus\Omega})} \to 0\ , \quad n \to \infty \tag{6.13}$$

in the topology of $C(\overline{\Omega_1\setminus\Omega})$. We can assume that the extensions $\bar{v}_n$ satisfy the inequality (see Mikhailov, 1976)

$$\|\bar{v}_n\|_{C(\overline{\Omega\setminus\Omega_2})} \leq c\|v_n\|_{C(\overline{\Omega_1\setminus\Omega})}$$

uniformly with respect to n. By (6.13), the right-hand side converges to zero as $n \to \infty$. Thus, the extensions possess the following properties

$$\begin{gathered} \bar{v}_n(x) = (\bar{w})_{\varepsilon_n}(x) + \frac{1}{n}\ , \quad x \in \Gamma\ , \\ \|\bar{v}_n\|_{1,\Omega\setminus\Omega_2} \to 0\ , \quad \|\bar{v}_n\|_{C(\overline{\Omega\setminus\Omega_2})} \to 0\ . \end{gathered} \tag{6.14}$$

From (6.10) it follows that

$$(\bar{w})_{\varepsilon_n}(x) + \frac{1}{n} \geq \psi(x)\ , \quad x \in \Omega\ .$$

The right-hand side of this inequality is bounded from above in Ω_0 by a negative constant δ, the left-hand side converges to w uniformly with respect to x in the same neighbourhood. We recall that $w > \dfrac{\delta}{2}$ in Ω_0. Hence, taking into account (6.14) for $\bar{v}_n$, one can assume that

$$(\bar{w})_{\varepsilon_n}(x) + \frac{1}{n} - \bar{v}_n(x) \geq \psi(x)\ , \quad x \in \Omega\ .$$

In Ω we define

$$w_n = (\bar{w})_{\varepsilon_n} + \frac{1}{n} - \bar{v}_n\ . \tag{6.15}$$

Sequence (6.15) satisfies all the required conditions. In fact, $w_n \in \mathcal{K}_\psi^2$ and $w_n \to w$ strongly in $H_0^1(\Omega)$.

To complete the proof we have to get rid of the assumption $w \in H_0^{1,p}(\Omega)$. Let $w \in \mathcal{K}_\psi^1$. By the remark we made before stating Lemma 6.4, it is possible to construct a sequence $\widetilde{w}_n \in \mathcal{K}_\psi^1 \cap H_0^{1,p}(\Omega)$ strongly converging to w in $H_0^1(\Omega)$. On the other hand, the statement proved above allows us to choose a sequence from $\mathcal{K}_\psi^2$ strongly converging to $\widetilde{w}_n$ in $H_0^1(\Omega)$. This completes the proof. □

We now show that every element $w \in \mathcal{K}_\psi^1$ can be approximated by a sequence from $\mathcal{K}_{\psi_n}^1$ when ψ_n converges to ψ in a certain sense.

Lemma 6.5. *Let $\psi_n \in \Psi$ and $\psi_n \to \psi$ weakly in $H^2(\Omega)$, strongly in $H^1(\Omega)$. Then for any fixed element $w \in \mathcal{K}_\psi^1$ there exists a sequence $w_n \in \mathcal{K}_{\psi_n}^1$ such that as $n \to \infty$*

$$w_n \to w \quad \textit{strongly in} \quad H_0^1(\Omega) \ .$$

To prove this it is enough to observe that the required sequence can be given by the formula $w_n = \max\{w, \psi_n\}$.

Let us now return to optimal control problem (6.8) and prove its solvability. Lemma 6.5 will be used directly. Lemma 6.4 will be used in Section 6.3 to justify a passage to the limit as $\varepsilon \to 0$.

Theorem 6.2. *Let $\sqrt{s}\,\varphi(s)$ be a strictly increasing function such that $\sqrt{s}\,\varphi(s) \to \infty$ as $s \to \infty$. Then a solution to problem (6.8) exists.*

Proof. By the boundedness of a minimizing sequence ψ_n in the space $H^2(\Omega)$, we can assume that

$$\psi_n \to \psi \quad \text{weakly in} \quad H^2(\Omega) \ , \quad \text{strongly in} \quad H^1(\Omega) \tag{6.16}$$

as $n \to \infty$. For every ψ_n we can define the corresponding set $\mathcal{K}_{\psi_n}^1$ and solve the variational inequality

$$w_n \in \mathcal{K}_{\psi_n}^1 \quad : \quad A_0(w_n)(\bar{w}_n - w_n) \geq 0 \quad \forall\, \bar{w}_n \in \mathcal{K}_{\psi_n}^1 \ . \tag{6.17}$$

Let $\psi \in \Psi$ satisfy (6.16). We choose an arbitrary element $\bar{w} \in \mathcal{K}_\psi^1$ and construct, by Lemma 6.5, a sequence $\bar{w}_n \in \mathcal{K}_{\psi_n}^1$ such that

$$\bar{w}_n \to \bar{w} \quad \text{strongly in} \quad H_0^1(\Omega) \tag{6.18}$$

as $n \to \infty$. Inequality (6.17) is equivalent to

$$\Pi_0(w_n) \leq \Pi_0(\bar{w}_n) \quad \forall\, \bar{w}_n \in \mathcal{K}_{\psi_n}^1 \ .$$

Substituting here the elements of the sequence constructed above as $\bar{w}_n$ we see that the right-hand side is bounded uniformly with respect to n and, therefore, by the coercivity of Π_0,

$$\|\nabla w_n\| \le c \ .$$

Choosing a subsequence, if necessary, we have

$$w_n \to w \quad \text{weakly in} \quad H_0^1(\Omega) \tag{6.19}$$

as $n \to \infty$. Rewriting (6.17) as

$$\varphi(\|\nabla \bar{w}_n\|^2) \int_\Omega \nabla \bar{w}_n (\nabla \bar{w}_n - \nabla w_n) dx$$
$$\ge \int_\Omega f(\bar{w}_n - w_n) dx \quad \forall\, \bar{w}_n \in \mathcal{K}^1_{\psi_n} \ .$$

we see that, by (6.18), (6.19), we can pass to the limit here. Thus, the inequality

$$w \in \mathcal{K}^1_\psi \quad : \quad A_0(\bar{w})(\bar{w} - w) \ge 0 \quad \forall\, \bar{w} \in \mathcal{K}^1_\psi$$

holds which is equivalent to (6.7), and hence $w = w(\psi)$. Finally,

$$d \equiv \inf_{\bar{\psi} \in \Psi} \mathcal{J}_0(\bar{\psi}) = \underline{\lim}\ \mathcal{J}_0(\psi_n) \ge \mathcal{J}_0(\psi) \ge d \ .$$

This proves that ψ solves problem (6.8). □

6.3 The passage to the limit

The aim of this section is to prove the main result of Section 6 concerning the convergence of solutions of (6.2) to a solution of (6.8) as $\varepsilon \to 0$.

Let ψ_ε be a solution to problem (6.2) and $w_\varepsilon = w(\psi_\varepsilon)$ be defined by (6.1). Assume that $j_\varepsilon = \mathcal{J}_\varepsilon(\psi_\varepsilon)$, and $j_0 = \mathcal{J}_0(\psi)$, where ψ solves optimal control problem (6.8). With this we can formulate the following result.

Theorem 6.3. *Let $\sqrt{s}\,\varphi(s)$ be a strictly increasing function such that $\sqrt{s}\,\varphi(s) \to \infty$ as $s \to \infty$. Then, from the sequence $\psi_\varepsilon, w_\varepsilon$ one can choose a subsequence (with the same notations) such that as $\varepsilon \to 0$*

$$\psi_\varepsilon \to \psi \quad \textit{weakly in} \quad H^2(\Omega) \ ,$$

$$w_\varepsilon \to w \quad \textit{weakly in} \quad H_0^1(\Omega) \ ,$$

$$j_\varepsilon \to j_0 \ ,$$

where ψ is a solution to problem (6.8), and $w = w(\psi)$ is defined by variational inequality (6.7).

Proof. Let $\psi \in \Psi$. Obviously,

$$\mathcal{J}_\varepsilon(\psi_\varepsilon) \leq \mathcal{J}_\varepsilon(\psi) \tag{6.20}$$

for any ε. The variational inequality

$$w_\varepsilon(\psi) \in \mathcal{K}^2_\psi \quad : \quad A_\varepsilon\big(w_\varepsilon(\psi)\big)\big(\bar{w} - w_\varepsilon(\psi)\big) \geq 0 \quad \forall\, \bar{w} \in \mathcal{K}^2_\psi$$

yields an a priori estimate for $w_\varepsilon(\psi)$. In fact, this inequality can be rewritten as

$$\Pi_\varepsilon\big(w_\varepsilon(\psi)\big) \leq \Pi_\varepsilon(\bar{w}) \quad \forall\, \bar{w} \in \mathcal{K}^2_\psi \ .$$

Here the right-hand side is bounded uniformly with respect to $\varepsilon \leq \varepsilon_0$ for a fixed $\bar{w} \in \mathcal{K}^2_\psi$. Therefore,

$$\frac{1}{2}\,\varepsilon\|\Delta w_\varepsilon(\psi)\|^2 + \frac{1}{2}\int_0^{\|\nabla w_\varepsilon(\psi)\|^2} \varphi(s)ds - \int_\Omega f w_\varepsilon(\psi)dx \leq c \ .$$

The arguments used in Lemma 6.3 to prove the coercivity of Π_0 allow us to derive the estimate

$$\|\nabla w_\varepsilon(\psi)\| \leq c$$

with the constant c independent of ε. This means that $\mathcal{J}_\varepsilon(\psi)$ is bounded uniformly with respect to ε.

From (6.20) it follows that

$$\|\psi_\varepsilon\|_2 \leq c \ .$$

We choose a subsequence of ψ_ε, still denoted by ψ_ε, such that

$$\psi_\varepsilon \to \widetilde{\psi} \quad \text{weakly in} \quad H^2(\Omega) \quad \text{and uniformly in} \quad \bar{\Omega}$$

as $\varepsilon \to 0$. According to what we have observed in the proof of Theorem 6.1, for any fixed element $\bar{w} \in \mathcal{K}^2_{\widetilde{\psi}}$, we can choose a sequence $\bar{w}_\varepsilon \in \mathcal{K}^2_{\psi_\varepsilon}$ strongly converging to $\bar{w}$ in $H^{2,0}(\Omega)$. Substituting into the variational inequality

$$\begin{gathered} w_\varepsilon(\psi_\varepsilon) \in \mathcal{K}^2_{\psi_\varepsilon} \quad : \\ A_\varepsilon\big(w_\varepsilon(\psi_\varepsilon)\big)\big(\bar{w}_\varepsilon - w_\varepsilon(\psi_\varepsilon)\big) \geq 0 \quad \forall\, \bar{w}_\varepsilon \in \mathcal{K}^2_{\psi_\varepsilon} \end{gathered} \tag{6.21}$$

the elements $\bar{w}_\varepsilon$ as test functions we get

$$\Pi_\varepsilon\big(w_\varepsilon(\psi_\varepsilon)\big) \leq \Pi_\varepsilon(\bar{w}_\varepsilon)$$

which holds uniformly with respect to ε. This gives the estimation

$$\varepsilon\|\Delta w_\varepsilon(\psi_\varepsilon)\|^2 + \|\nabla w_\varepsilon(\psi_\varepsilon)\| \leq c$$

which holds uniformly with respect to $\varepsilon \leq \varepsilon_0$. Choosing a subsequence, if necessary,

$$\begin{aligned} w_\varepsilon(\psi_\varepsilon) &\to \widetilde{w} \quad \text{weakly in} \quad H^1_0(\Omega) \ , \\ \varepsilon w_\varepsilon(\psi_\varepsilon) &\to 0 \quad \text{weakly in} \quad H^{2,0}(\Omega) \end{aligned}$$

as $\varepsilon \to 0$. Inequality (6.21) can be rewritten as

$$A_\varepsilon(\bar{w}_\varepsilon)\big(\bar{w}_\varepsilon - w_\varepsilon(\psi_\varepsilon)\big) \geq 0 \quad \forall\, \bar{w}_\varepsilon \in \mathcal{K}^2_{\psi_\varepsilon} \ .$$

Let us substitute here the above elements $\bar{w}_\varepsilon$ strongly converging to $\bar{w} \in \mathcal{K}^2_{\widetilde{\psi}}$. After passing to the limit the inequality takes the form

$$\begin{aligned} \widetilde{w} \in \mathcal{K}^1_{\widetilde{\psi}} \ : \ \varphi(\|\nabla \bar{w}\|^2) \int_\Omega \nabla \bar{w}(\nabla \bar{w} - \nabla \widetilde{w}) dx \\ \geq \int_\Omega f(\bar{w} - \widetilde{w}) dx \quad \forall\, \bar{w} \in \mathcal{K}^2_{\widetilde{\psi}} \ . \end{aligned} \tag{6.22}$$

By Lemma 6.4, for a fixed element $w \in \mathcal{K}^1_{\widetilde{\psi}}$, a sequence $w_n \in \mathcal{K}^2_{\widetilde{\psi}}$ can be constructed such that

$$w_n \to w \quad \text{strongly in} \quad H^1_0(\Omega) \ .$$

This enables us to substitute into (6.22) the elements $\bar{w}$ from $\mathcal{K}^1_{\widetilde{\psi}}$. Hence, the functions $\widetilde{w}, \widetilde{\psi}$ satisfy the inequality

$$\begin{aligned} \widetilde{w} \in \mathcal{K}^1_{\widetilde{\psi}} \ : \ \varphi(\|\nabla \bar{w}\|^2) \int_\Omega \nabla \bar{w}(\nabla \bar{w} - \nabla \widetilde{w}) dx \\ \geq \int_\Omega f(\bar{w} - \widetilde{w}) dx \quad \forall\, \bar{w} \in \mathcal{K}^1_{\widetilde{\psi}} \ . \end{aligned}$$

This means that $\widetilde{w} = w(\widetilde{\psi})$, and therefore, it is easy to evaluate $\mathcal{J}_0(\widetilde{\psi})$. Indeed, in view of the weak semicontinuity of the norm in $L^2(\Omega)$,

$$\underline{\lim}\ j_\varepsilon = \underline{\lim}\ \mathcal{J}_\varepsilon(\psi_\varepsilon) \geq \mathcal{J}_0(\widetilde{\psi}) \ . \tag{6.23}$$

We now show that for a fixed $\psi \in \Psi$ we can choose a subsequence of $\mathcal{J}_\varepsilon(\psi)$, still denoted by $\mathcal{J}_\varepsilon(\psi)$, such that

$$\mathcal{J}_\varepsilon(\psi) \to \mathcal{J}_0(\psi)$$

as $\varepsilon \to 0$. Indeed, by the inequality

$$\begin{aligned} w_\varepsilon(\psi) \in \mathcal{K}^2_\psi \ : \\ A_\varepsilon\big(w_\varepsilon(\psi)\big)\big(\bar{w} - w_\varepsilon(\psi)\big) \geq 0 \quad \forall\, \bar{w} \in \mathcal{K}^2_\psi, \end{aligned} \tag{6.24}$$

we easily obtain the estimate

$$\varepsilon\|\Delta w_\varepsilon(\psi)\|^2 + \|\nabla w_\varepsilon(\psi)\| \leq c$$

which holds uniformly with respect to ε. This implies that passing to a subsequence of the sequence $w_\varepsilon(\psi)$, if necessary, we get

$$\begin{aligned} w_\varepsilon(\psi) \to w \quad \text{weakly in} \quad H^1_0(\Omega) \ , \ \text{strongly in} \quad L^2(\Omega) \ , \\ \varepsilon w_\varepsilon(\psi) \to 0 \quad \text{weakly in} \quad H^{2,0}(\Omega) \ , \end{aligned} \tag{6.25}$$

as $\varepsilon \to 0$, and, moreover, $w \in \mathcal{K}_\psi^1$. Using the above arguments and putting inequality (6.24) in the equivalent form, we can pass to the limit as $\varepsilon \to 0$. The result can be written as

$$w \in \mathcal{K}_\psi^1 \ : \ \varphi(\|\nabla \bar{w}\|^2) \int_\Omega \nabla \bar{w}(\nabla \bar{w} - \nabla w)dx$$
$$\geq \int_\Omega f(\bar{w} - w)dx \quad \forall\, \bar{w} \in \mathcal{K}_\psi^2 \ .$$

As previously, this inequality holds for every $\bar{w} \in \mathcal{K}_\psi^1$. Hence $w = w(\psi)$. Thus, in view of the strong convergence of $w_\varepsilon(\psi)$ in $L^2(\Omega)$,

$$\mathcal{J}_\varepsilon(\psi) \equiv \|w_\varepsilon(\psi) - w_0\|^2 + \|\psi\|_2^2 \to \mathcal{J}_0(\psi) \ ,$$

which establishes the required convergence.

Let ψ be a solution to optimal control problem (6.8). Then

$$j_\varepsilon = \mathcal{J}_\varepsilon(\psi_\varepsilon) \leq \mathcal{J}_\varepsilon(\psi) \to \mathcal{J}_0(\psi) \ , \quad \varepsilon \to 0 \ .$$

Consequently,

$$\overline{\lim}\ j_\varepsilon \leq \mathcal{J}_0(\psi) \ .$$

Combining this with (6.23) we conclude that $\widetilde{\psi}$ solves (6.8) and $j_\varepsilon \to j_0$. This proves our theorem. □

Observe that the assumptions imposed on the function $\sqrt{s}\,\varphi(s)$ can be weakened. Actually it is enough to assume that $\sqrt{s}\,\varphi(s)$ is strictly increasing and

$$\frac{1}{\sqrt{\tau}} \int_0^\tau \varphi(s)ds \to \infty \ , \quad \tau \to \infty \ .$$

To see this we should return to the part of the proof of Lemma 6.3 where the coercivity of Π_0 is shown. The proof can be modified by noting first that

$$\Pi_0(w) \geq \frac{1}{2} \int_0^{\|\nabla w\|^2} \varphi(s)ds - c\|\nabla w\|$$
$$= \frac{1}{2}\sqrt{\tau}\left(\frac{1}{\sqrt{\tau}} \int_0^\tau \varphi(s)ds - 2c\right) \to \infty \ , \quad \tau \equiv \|\nabla w\|^2 \to \infty \ .$$

Since $\sqrt{s}\,\varphi(s)$ is strictly increasing a solution to (6.7) is unique. In the case where $f \equiv 0$, we can weaken the hypotheses imposed on φ even further. In fact, assume that

a) $\sqrt{s}\,\varphi(s)$ is nondecreasing,

b) $\varphi(s) > 0$, $s > 0$.

From b) it follows that the variational inequality

$$w \in \mathcal{K}^1_\psi \ : \ \varphi(\|\nabla w\|^2) \int_\Omega \nabla w(\nabla \bar{w} - \nabla w)dx \geq 0 \quad \forall\, \bar{w} \in \mathcal{K}^1_\psi$$

has a unique solution. This fact can be verified directly both in the case where $0 \in \mathcal{K}^1_\psi$, and in the case where $0 \notin \mathcal{K}^1_\psi$. Moreover, the above inequality coincides with the following one

$$w \in \mathcal{K}^1_\psi \ : \ \int_\Omega \nabla w(\nabla \bar{w} - \nabla w)dx \geq 0 \quad \forall\, \bar{w} \in \mathcal{K}^1_\psi \ . \tag{6.26}$$

When proving Theorem 6.2, we have used the coercivity of $\Pi_0(w)$ which enables us to estimate solutions $w_n \equiv w(\psi_n)$ uniformly with respect to n. In the present case the corresponding estimation follows directly from the inequalities

$$\varphi(\|\nabla w_n\|^2) \int_\Omega \nabla w_n(\nabla \bar{w}_n - \nabla w_n)dx \geq 0 \quad \forall\, \bar{w}_n \in \mathcal{K}^1_{\psi_n} \ .$$

Thus, we conclude that for $f \equiv 0$ and a fixed function φ satisfying a) - b), from the sequence of solutions of (6.2) we can choose a subsequence which converges to a solution of (6.8). Passing to the limit we get variational inequality (6.26) which does not depend on φ.

This independence on φ which occurs after passing to the limit is an important property.

As indicated at the beginning of the section, a passage to the limit as $\varepsilon \to 0$ corresponds to the convergence of a plate to a membrane when the thickness of the plate tends to zero. We have to note that the model considered in this section corresponds to a plate subjected to stretching forces in the plane x. In particular, for $\varphi(s) = s$ we get Berger's operator

$$w \to \varepsilon \Delta^2 w - \|\nabla w\|^2 \Delta w$$

(see Grigolyuk, Kulikov, 1981). The case where $\varphi(s) = 1$ corresponds to the linear model of the plate.

7 The case of two punches

We discuss the optimal control problem governed by a variational inequality for the operator

$$w \to \varepsilon^3 \Delta^2 w - \Delta w \ , \quad \varepsilon > 0 \ .$$

For a given $\varepsilon > 0$, the problem corresponds to a plate, and for $\varepsilon = 0$ it corresponds to a membrane. The function w satisfies a bilateral restriction. The set of admissible displacements depends on ε.

7.1 Optimal control for a plate

Let the domain Ω and its boundary satisfy the same conditions as in the preceeding section. Moreover, Ψ_1, Ψ_2 are convex and closed sets in $H^2(\Omega)$ such that

$$\psi_1 \le -\delta_0 \ , \quad \psi_2 \ge \delta \quad \text{on} \quad \Gamma \ ; \quad \psi_1(x) \le \psi_2(x) - \delta \ , \quad x \in \Omega$$

for any $\psi_i \in \Psi_i$, $\delta_0 = \text{const} > 0$, $\delta = \text{const} > 0$. Assume that $\delta \ge \delta_0$. In the space $H^{2,0}(\Omega) \equiv H^2(\Omega) \cap H_0^1(\Omega)$ define the convex and closed set

$$\mathcal{K}^2_{\psi,\varepsilon} = \left\{ w \in H^{2,0}(\Omega) \ \middle| \ \psi_1(x) \le w(x) \le \psi_2(x) - \varepsilon \ , \quad x \in \Omega \right\} ,$$

where $\varepsilon < \delta$. The pair (ψ_1, ψ_2) is denoted by ψ.

Consider the variational inequality

$$\begin{aligned} w \in \mathcal{K}^2_{\psi,\varepsilon} \ : \ \varepsilon^3 \int_\Omega \Delta w(\Delta \bar{w} - \Delta w) dx \\ + \int_\Omega \nabla w(\nabla \bar{w} - \nabla w) dx \ge 0 \quad \forall \, \bar{w} \in \mathcal{K}^2_{\psi,\varepsilon} \ . \end{aligned} \tag{7.1}$$

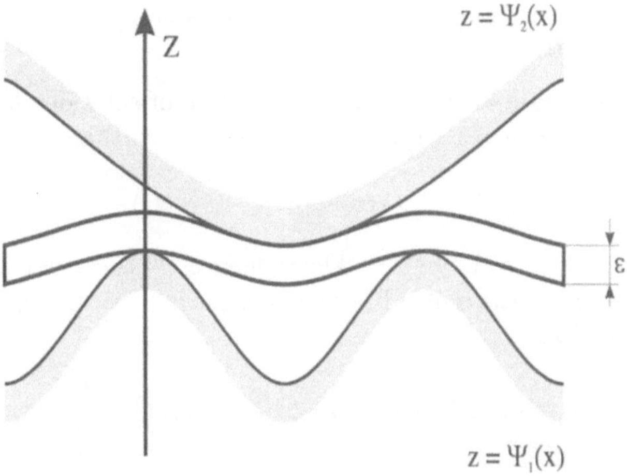

Fig. 3. Optimal control for a plate.

For a fixed $\varepsilon > 0$, inequality (7.1) has a unique solution $w = w(\psi)$. This inequality describes the equilibrium state of the plate between two punches. The punch shapes are described by the equations $z = \psi_i(x)$, $i = 1, 2$, w is the plate displacement (see Fig. 3). Let $\Psi = \Psi_1 \times \Psi_2$ and w_0 be a given function. The optimal control problem which will be analysed in this section is a minimization problem over the set Ψ.

Define the cost functional

$$\mathcal{J}_\varepsilon(\psi) = \|w(\psi) - w_0\|^2 + \|\psi\|_2^2 \ .$$

We want to find an element $\psi \in \Psi$ such that

$$\mathcal{J}_\varepsilon(\psi) \le \mathcal{J}_\varepsilon(\bar{\psi}) \quad \forall\, \bar{\psi} \in \Psi \ . \tag{7.2}$$

In this problem ε is assumed to be fixed.

We begin by proving the solvability of the optimal control problem under consideration.

Theorem 7.1. *There exists a solution to problem (7.2).*

Proof. Let $\psi^n \in \Psi$ be a minimizing sequence. This sequence is bounded in the space $H^2(\Omega) \times H^2(\Omega)$, and hence passing to a subsequence, if necessary, we have

$$\psi^n \to \psi \quad \text{weakly in} \quad H^2(\Omega) \times H^2(\Omega) \quad \text{and uniformly in} \quad \bar{\Omega} \tag{7.3}$$

as $n \to \infty$. For every $\psi^n \in \Psi$ there exists a solution to the variational inequality

$$\begin{aligned} w^n \in \mathcal{K}^2_{\psi^n,\varepsilon} \ : \ \varepsilon^3 \int_\Omega \Delta w^n (\Delta \bar{w}^n - \Delta w^n) dx \\ + \int_\Omega \nabla w^n (\nabla \bar{w}^n - \nabla w^n) dx \ge 0 \quad \forall\, \bar{w}^n \in \mathcal{K}^2_{\psi^n,\varepsilon} \ . \end{aligned} \tag{7.4}$$

We now prove that from (7.4) an a priori estimate of solutions w^n follows if for every $\bar{w} \in \mathcal{K}^2_{\psi,\varepsilon}$ there exists a sequence $\bar{w}^n \in \mathcal{K}^2_{\psi^n,\varepsilon}$ such that

$$\bar{w}^n \to \bar{w} \quad \text{strongly in} \quad H^{2,0}(\Omega)$$

as $n \to \infty$, and ψ is defined by (7.3). The sequence $\bar{w}^n$ is constructed below.

Note first that, by the imbedding theorems, $\bar{w}$ is a continuous function on $\bar{\Omega}$. Consider the closed set

$$M_1 = \left\{ x \in \Omega \ \middle| \ \psi_1(x) \le \bar{w}(x) \le \psi_1(x) + \frac{\delta_0}{8} \right\} \ ,$$

and denote by M_1^ν its ν-neighbourhood. Analogously, let

$$M_2 = \left\{ x \in \Omega \ \middle| \ \psi_2(x) - \varepsilon - \frac{\delta_0}{8} \le \bar{w}(x) \le \psi_2(x) - \varepsilon \right\} \ ,$$

and M_2^ν be its ν-neighbourhood.

We now construct functions $\xi_1, \xi_2 \in H^2(\Omega)$ such that

$$\xi_1(x) = \begin{cases} 1 \ , & x \in M_1 \\ 0 \ , & x \notin M_1^\nu \ , \end{cases} \qquad \xi_2(x) = \begin{cases} -1 \ , & x \in M_2 \\ 0 \ , & x \notin M_2^\nu \ , \end{cases}$$

$$0 \le \xi_1(x) \le 1 \ , \quad -1 \le \xi_2(x) \le 0 \ .$$

We choose ν small enough to ensure that

$$\psi_1(x) \leq \bar{w}(x) \leq \psi_1(x) + \frac{\delta_0}{4} \ ,$$

$$\psi_2(x) - \varepsilon - \frac{\delta_0}{4} \leq \bar{w}(x) \leq \psi_2(x) - \varepsilon$$

on the sets M_1^ν, M_2^ν. By the uniform convergence of the sequence ψ^n,

$$\left|\psi_i^n(x) - \psi_i(x)\right| < \frac{1}{n} \ , \quad x \in \Omega \ , \quad i = 1, 2 \ .$$

Hence, the sequence

$$\bar{w}^n = \bar{w} + \frac{\delta_0}{4n}\,\xi_1 + \frac{\delta_0}{4n}\,\xi_2$$

has the required properties. Precisely, for n sufficiently large

$$\psi_1^n(x) \leq \bar{w}^n(x) \leq \psi_2^n(x) - \varepsilon \ , \quad x \in \Omega \ ,$$

and $\bar{w}^n$ strongly converge in $H^{2,0}(\Omega)$ to the function $\bar{w}$. The sequence $\bar{w}_n$ is appropriate for (7.4).

Consider variational inequality (7.4). Let $\psi = (\psi_1, \psi_2)$ be the limit defined by (7.3). Basing on what we have said we can choose an arbitrary function $\bar{w} \in \mathcal{K}^2_{\psi,\varepsilon}$ and construct a sequence $\bar{w}^n \in \mathcal{K}^2_{\psi^n,\varepsilon}$ strongly converging to $\bar{w}$. Substituting $\bar{w}^n$ into (7.4) the inequality

$$\varepsilon^3 \|\Delta w^n\|^2 + \|\nabla w^n\|^2 \leq c \tag{7.5}$$

results with the constant independent of n. It follows from (7.5) that w^n are bounded in $H^{2,0}(\Omega)$, and therefore, one can assume that

$$w^n \to w \quad \text{weakly in} \quad H^{2,0}(\Omega)$$

as $n \to \infty$. Supposing that the elements of the constructed strongly converging sequence are chosen as $\bar{w}^n$ in variational inequality (7.4) it follows that

$$\int_\Omega (\varepsilon^3 \Delta w^n \Delta \bar{w}^n + \nabla w^n \nabla \bar{w}^n) dx \geq \varepsilon^3 \|\Delta w^n\|^2 + \|\nabla w^n\|^2 \ ,$$

which is the required a priori estimate for solutions w^n. By the weak lower semicontinuity of the norm in the space $L^2(\Omega)$ we can pass to the limit which gives

$$w \in \mathcal{K}^2_{\psi,\varepsilon} \ : \ \varepsilon^3 \int_\Omega \Delta w(\Delta \bar{w} - \Delta w) dx$$

$$+ \int_\Omega \nabla w(\nabla \bar{w} - \nabla w) dx \geq 0 \quad \forall\, \bar{w} \in \mathcal{K}^2_{\psi,\varepsilon} \ .$$

This implies that $w = w(\psi)$. Consequently,

$$j_\varepsilon \equiv \inf_{\bar{\psi} \in \Psi} \mathcal{J}_\varepsilon(\bar{\psi}) = \varliminf_{n \to \infty} \ \mathcal{J}_\varepsilon(\psi^n) \geq \mathcal{J}_\varepsilon(\psi) \geq j_\varepsilon \ .$$

Hence, the function ψ is a solution to problem (7.2). This completes the proof.

□

7.2 Optimal control for a membrane

For any $\psi \in \Psi$ we introduce

$$\mathcal{K}^1_\psi = \{w \in H^1_0(\Omega) \mid \psi_1(x) \leq w(x) \leq \psi_2(x) \quad \text{a. e. in} \quad \Omega\} .$$

The variational inequality

$$w \in \mathcal{K}^1_\psi \ : \ \int_\Omega \nabla w(\nabla \bar{w} - \nabla w) dx \geq 0 \quad \forall\, \bar{w} \in \mathcal{K}^1_\psi \tag{7.6}$$

is uniquely solvable. Let

$$\mathcal{J}_0(\psi) = \|w(\psi) - w_0\|^2 + \|\psi\|_2^2 ,$$

where $w = w(\psi)$ solves (7.6).

In this section we consider the optimal control problem of minimizing the functional $\mathcal{J}_0$ over Ψ, ie. the problem is to find $\psi \in \Psi$ such that

$$\mathcal{J}_0(\psi) \leq \mathcal{J}_0(\bar{\psi}) \quad \forall\, \bar{\psi} \in \Psi . \tag{7.7}$$

Theorem 7.2. *There exists a solution to problem (7.7).*

Proof. Let $\psi^n \in \Psi$ be a minimizing sequence. This sequence is bounded in the space $H^2(\Omega) \times H^2(\Omega)$, and hence passing to a subsequence, if necessary, we have

$$\psi_i^n \to \psi_i \text{ weakly in } H^2(\Omega) \text{ strongly in } H^1(\Omega) \text{ and uniformly in } \bar{\Omega} \ , \quad i = 1, 2$$

as $n \to \infty$. We denote the pair (ψ_1, ψ_2) by ψ and prove that ψ solves the problem under consideration.

We claim that for any $\bar{w} \in \mathcal{K}^1_\psi$ there exists a sequence $\bar{w}^n \in \mathcal{K}^1_{\psi^n}$ such that

$$\bar{w}^n \to \bar{w} \quad \text{strongly in } H^1_0(\Omega) .$$

Indeed, one can construct a sequence of smooth functions $\hat{w}^n$ with compact supports strongly converging to $\bar{w}$. Let $\widetilde{w}^n = \max \{\psi_1^n, \hat{w}^n\}$. Then, $\widetilde{w}^n \to \max \{\psi_1, \bar{w}\}$, $\widetilde{w}^n \geq \psi_1^n$. Put $\bar{w}^n = \min \{\psi_2^n, \widetilde{w}^n\}$. We have $\bar{w}^n \to \min \{\psi_2, \bar{w}\}$, and $\psi_1^n \leq \bar{w}^n \leq \psi_2^n$. Thus, the required sequence exists.

Consider the variational inequality

$$w^n \in \mathcal{K}^1_{\psi^n} \ : \ \int_\Omega \nabla w^n(\nabla \bar{w}^n - \nabla w^n) dx \geq 0 \quad \forall\, \bar{w}^n \in \mathcal{K}^1_{\psi^n} . \tag{7.8}$$

By repeating the above reasoning, for any element $\bar{w} \in \mathcal{K}^1_\psi$, one can construct a sequence $\bar{w}^n \in \mathcal{K}^1_{\psi^n}$ strongly converging to $\bar{w}$. Substituting this sequence into (7.8) we get the estimate

$$\|\nabla w^n\|^2 \leq c$$

which holds uniformly which respect to n. Without loss of generality, we assume that

$$w^n \to w \quad \text{weakly in} \quad H_0^1(\Omega)$$

as $n \to \infty$.

After passing to the limit in (7.8), the relation

$$w \in \mathcal{K}_\psi^1 \ : \ \int_\Omega \nabla w(\nabla \bar{w} - \nabla w)dx \geq 0 \quad \forall\, \bar{w} \in \mathcal{K}_\psi^1$$

results, and so $w = w(\psi)$.

Now it is easy to see that the function $\psi \in \Psi$ solves problem (7.7). Indeed,

$$j_0 \equiv \inf_{\bar{\psi} \in \Psi} \mathcal{J}_0(\bar{\psi}) = \underline{\lim}\ \mathcal{J}_0(\psi^n) \geq \mathcal{J}_0(\psi) \geq j_0 \ .$$

This completes the proof. □

7.3 The passage to the limit

In this section we examine the behaviour of solutions to problem (7.2) as $\varepsilon \to 0$.

Let ψ^ε be a solution to (7.2), and $w^\varepsilon = w(\psi^\varepsilon)$ be the corresponding solution of variational inequality (7.1),

$$j_\varepsilon = \mathcal{J}_\varepsilon(\psi^\varepsilon) \ , \quad j_0 = \inf_{\psi \in \Psi} \mathcal{J}_0(\psi) \ .$$

Theorem 7.3. *From the sequence of solutions $\{\psi^\varepsilon, w^\varepsilon\}$ we can choose a subsequence (the notation of the subsequence is the same) such that as $\varepsilon \to 0$*

$$\psi^\varepsilon \to \psi \quad \textit{weakly in} \quad H^2(\Omega) \times H^2(\Omega) \ ,$$

$$w^\varepsilon \to w \quad \textit{weakly in} \quad H_0^1(\Omega) \ ,$$

$$j_\varepsilon \to j_0 \ .$$

Here ψ is a solution to problem (7.7), and $w = w(\psi)$ is defined by variational inequality (7.6).

Proof. Fix $\psi \in \Psi$. We have

$$\mathcal{J}_\varepsilon(\psi^\varepsilon) \leq \mathcal{J}_\varepsilon(\psi) \ . \tag{7.9}$$

We shall prove that the right-hand side of this inequality is bounded in ε, $\varepsilon \leq \varepsilon_0$. To this end it is enough to prove that $w^\varepsilon(\psi)$ are bounded in the space $L^2(\Omega)$. We now demonstrate the boundedness of $w^\varepsilon(\psi)$. Consider the variational inequality

$$\begin{aligned} w^\varepsilon(\psi) \in \mathcal{K}_{\psi,\varepsilon}^2 \ : \ \varepsilon^3 \int_\Omega \Delta w^\varepsilon(\psi)\big(\Delta \bar{w}^\varepsilon - \Delta w^\varepsilon(\psi)\big)dx + \\ + \int_\Omega \nabla w^\varepsilon(\psi)\big(\nabla \bar{w}^\varepsilon - \nabla w^\varepsilon(\psi)\big)dx \geq 0 \quad \forall\, \bar{w}^\varepsilon \in \mathcal{K}_{\psi,\varepsilon}^2 \ . \end{aligned} \tag{7.10}$$

According to the definition of $\mathcal{K}^2_{\psi,\varepsilon}$ trial functions satisfy the inequalities

$$\psi_1(x) \le \bar{w}^\varepsilon(x) \le \psi_2(x) - \varepsilon \ , \quad x \in \Omega \ .$$

Construct a function $\bar{w}$ belonging to $\mathcal{K}^2_{\psi,\varepsilon}$ for every $\varepsilon \le \varepsilon_0$. It is enough to take $\bar{w}$ satisfying $\psi_1 \le \bar{w} \le \psi_2 - \varepsilon$ and belonging to the space $H^{2,0}(\Omega)$. This holds if ε_0 is sufficiently small and $\varepsilon_0 < \delta$. Then, for every $\varepsilon \le \varepsilon_0$, $\bar{w} \in \mathcal{K}^2_{\psi,\varepsilon}$.

Substituting this function into inequality (7.10), we get the estimate

$$\varepsilon^3 \|\Delta w^\varepsilon(\psi)\|^2 + \|\nabla w^\varepsilon(\psi)\|^2 \le c \ ,$$

which holds uniformly with respect to $\varepsilon \le \varepsilon_0$. This entails that the right-hand side of (7.9) is bounded for $\varepsilon \le \varepsilon_0$. Therefore,

$$\|\psi^\varepsilon\|_2 \le c$$

with the constant c independent of $\varepsilon \le \varepsilon_0$. Passing to subsequences, if necessary, we have

$$\psi^\varepsilon \to \widetilde{\psi} \quad \text{weakly in} \quad H^2(\Omega) \times H^2(\Omega) \quad \text{and uniformly in} \quad \bar{\Omega}$$

as $\varepsilon \to 0$.

We now prove that from the variational inequality

$$\begin{aligned} w^\varepsilon \in \mathcal{K}^2_{\psi^\varepsilon,\varepsilon} \ : \ & \varepsilon^3 \int_\Omega \Delta w^\varepsilon (\Delta \bar{w}^\varepsilon - \Delta w^\varepsilon) dx \\ + & \int_\Omega \nabla w^\varepsilon (\nabla \bar{w}^\varepsilon - \nabla w^\varepsilon(\psi)) dx \ge 0 \quad \forall\, \bar{w}^\varepsilon \in \mathcal{K}^2_{\psi^\varepsilon,\varepsilon} \end{aligned} \tag{7.11}$$

it follows the estimate,

$$\varepsilon^3 \|\Delta w^\varepsilon\|^2 + \|\nabla w^\varepsilon\|^2 \le c \tag{7.12}$$

which holds uniformly with respect to $\varepsilon \le \varepsilon_0$. Take any fixed element $\bar{w} \in H^{2,0}(\Omega)$ satisfying the inequalities $\widetilde{\psi}_1(x) \le \bar{w}(x) \le \widetilde{\psi}_2(x)$ and construct a sequence $\bar{w}^\varepsilon \in H^{2,0}(\Omega)$ such that

$$\psi_1^\varepsilon(x) \le \bar{w}^\varepsilon(x) \le \psi_2^\varepsilon(x) - \varepsilon \ , \quad \bar{w}^\varepsilon \to \bar{w} \quad \text{strongly in} \ H^{2,0}(\Omega) \ .$$

As in the proof of Theorem 7.1 we assume that

$$\left|\psi_i^\varepsilon(x) - \widetilde{\psi}_i(x)\right| < \varepsilon \ , \quad x \in \Omega \ , \quad i = 1,2 \ .$$

Introduce the closed sets

$$M_1 = \left\{ x \in \Omega \mid \widetilde{\psi}_1(x) \le \bar{w}(x) \le \widetilde{\psi}_1(x) + \frac{\delta_0}{8} \right\} \ ,$$

$$M_2 = \left\{ x \in \Omega \mid \widetilde{\psi}_2(x) - \frac{\delta_0}{8} \le \bar{w}(x) \le \widetilde{\psi}_1(x) \right\} \ .$$

Let M_1^ν, M_2^ν be their ν-neighbourhoods such that

$$\widetilde{\psi}_1 \le \bar{w} \le \widetilde{\psi}_1 + \frac{\delta_0}{4} \ , \quad \widetilde{\psi}_2 - \frac{\delta_0}{4} \le \bar{w} \le \widetilde{\psi}_2 \ ,$$

respectively. There exist two functions $\xi_1, \xi_2 \in H^2(\Omega)$ satisfying the conditions

$$\xi_1(x) = \begin{cases} 1 \ , & x \in M_1 \\ 0 \ , & x \notin M_1^\nu \ , \end{cases} \qquad \xi_2(x) = \begin{cases} -1 \ , & x \in M_2 \\ 0 \ , & x \notin M_2^\nu \ , \end{cases}$$

$$0 \le \xi_1 \le 1 \ , \quad -1 \le \xi_2 \le 0 \ .$$

For small ε the function $\bar{w}^\varepsilon = \bar{w} + \dfrac{\delta_0\sqrt{\varepsilon}}{4}\,\xi_1 + \dfrac{\delta_0\sqrt{\varepsilon}}{4}\,\xi_2$ satisfies the inequalities

$$\psi_1^\varepsilon(x) \le \bar{w}^\varepsilon(x) \le \psi_2^\varepsilon(x) - \varepsilon \ , \quad x \in \Omega \ ,$$

and strongly converges to $\bar{w}$ in the space $H^{2,0}(\Omega)$. Substituting the functions $\bar{w}^\varepsilon$ into variational inequality (7.11) we get the required estimate (7.12).

We assume that for $\varepsilon \to 0$

$$w^\varepsilon \to w \quad \text{weakly in} \quad H_0^1(\Omega) \ ,$$

$$\varepsilon^3 w^\varepsilon \to 0 \quad \text{weakly in} \quad H^{2,0}(\Omega) \ .$$

We can now pass to the limit in (7.11) as $\varepsilon \to 0$. To this end we rewrite (7.11) as

$$\int_\Omega \left(\varepsilon^3 \Delta w^\varepsilon \Delta \bar{w}^\varepsilon + \nabla w^\varepsilon \nabla \bar{w}^\varepsilon\right) dx \ge \varepsilon^3 \|\Delta w^\varepsilon\|^2 + \|\nabla w^\varepsilon\|^2 \ .$$

Substituting the elements of the above sequence $\bar{w}^\varepsilon$ we pass to the lower limit and get the variational inequality

$$w \in \mathcal{K}^1_{\widetilde{\psi}} \ : \ \int_\Omega \nabla w (\nabla \bar{w} - \nabla w) dx \ge 0 \quad \forall\, \bar{w} \in \mathcal{K}^1_{\widetilde{\psi}} \cap H^2(\Omega) \ . \tag{7.13}$$

In the sequel we shall prove (see Lemma 7.1) that functions from $\mathcal{K}^1_{\widetilde{\psi}}$ can be approximated by functions from $\mathcal{K}^1_{\widetilde{\psi}} \cap H^2(\Omega)$. This implies that inequality (7.13) holds for every $\bar{w} \in \mathcal{K}^1_{\widetilde{\psi}}$, and consequently $w = w(\widetilde{\psi})$. By the weak lower semicontinuity of the norm $\|\cdot\|$, one easily obtains

$$\underline{\lim}\ j_\varepsilon = \underline{\lim}\ \mathcal{J}_\varepsilon(\psi^\varepsilon) \ge \mathcal{J}_0(\widetilde{\psi}) \ . \tag{7.14}$$

We shall need the following statement. For a fixed $\psi \in \Psi$, from the sequence $\mathcal{J}_\varepsilon(\psi)$, a subsequence can be chosen such that

$$\mathcal{J}_\varepsilon(\psi) \to \mathcal{J}_0(\psi) \ , \quad \varepsilon \to 0 \ .$$

To verify this, let us note that from the variational inequality

$$\begin{aligned} w^\varepsilon(\psi) \in \mathcal{K}^2_{\psi,\varepsilon} \ : \ \varepsilon^3 \int_\Omega \Delta w^\varepsilon(\psi)\big(\Delta \bar{w}^\varepsilon - \Delta w^\varepsilon(\psi)\big)dx \\ + \int_\Omega \nabla w^\varepsilon(\psi)\big(\nabla \bar{w}^\varepsilon - \nabla w^\varepsilon(\psi)\big)dx \geq 0 \quad \forall\, \bar{w}^\varepsilon \in \mathcal{K}^2_{\psi,\varepsilon} \end{aligned} \tag{7.15}$$

we get the estimate

$$\varepsilon^3 \|\Delta w^\varepsilon(\psi)\|^2 + \|\nabla w^\varepsilon(\psi)\|^2 \leq c$$

which holds uniformly with respect to $\varepsilon \leq \varepsilon_0$. Thus, passing to a subsequence, if necessary, we have

$$w^\varepsilon(\psi) \to w \quad \text{weakly in} \quad H^1_0(\Omega) \ , \quad \text{strongly in} \quad L^2(\Omega) \ ,$$

$$\varepsilon^3 w^\varepsilon(\psi) \to 0 \quad \text{weakly in} \quad H^{2,0}(\Omega) \ ,$$

as $\varepsilon \to 0$, and $w \in \mathcal{K}^1_\psi$.

As before, we can now pass to the limit as $\varepsilon \to 0$ and we get

$$w \in \mathcal{K}^1_\psi \ : \ \int_\Omega \nabla w(\nabla \bar{w} - \nabla w)dx \geq 0 \quad \forall\, \bar{w} \in \mathcal{K}^1_\psi \cap H^2(\Omega) \ .$$

According to Lemma 7.1, this inequality holds for every element $\bar{w} \in \mathcal{K}^1_\psi$, and therefore, $w = w(\psi)$. In view of the strong convergence of $w^\varepsilon(\psi)$ to w in $L^2(\Omega)$,

$$\mathcal{J}_\varepsilon(\psi) \equiv \|w^\varepsilon(\psi) - w_0\|^2 + \|\psi\|_2^2 \to \mathcal{J}_0(\psi) \ .$$

Taking ψ which solves problem (7.7),

$$j_\varepsilon = \mathcal{J}_\varepsilon(\psi^\varepsilon) \leq \mathcal{J}_\varepsilon(\psi) \ ,$$

and by the previous arguments

$$\overline{\lim}\ j_\varepsilon \leq \mathcal{J}_0(\psi) \ .$$

Taking into account (7.14) we conclude that $\widetilde{\psi}$ solves (7.7) and $j_\varepsilon \to j_0$. This proves the theorem. □

We now establish the result used to prove Theorem 7.3.

Lemma 7.1. *For an arbitrary function* $\bar{w} \in \mathcal{K}^1_\psi$ *there exists a sequence* $\bar{w}^n \in \mathcal{K}^1_\psi \cap H^2(\Omega)$ *such that as* $n \to \infty$

$$\bar{w}^n \to \bar{w} \quad \textit{strongly in} \quad H^1_0(\Omega) \ . \tag{7.16}$$

Proof. Let $H^{1,p}_0(\Omega)$ denote the space of functions having the first generalized derivatives, integrable with the power $p > 2$ and equal to zero on Γ. Elements of this space are continuous on $\bar{\Omega}$.

Note first that there exists a sequence $\bar{w}^n \in \mathcal{K}^1_\psi \cap H^{1,p}_0(\Omega)$ satisfying condition (7.16). In fact, if $\hat{w}^n \in C^\infty_0(\Omega)$ and $\hat{w}^n \to \bar{w}$ strongly in $H^1_0(\Omega)$, the sequence $\bar{w}^n = \min\left\{\psi_2, \max\left\{\psi_1, \hat{w}^n\right\}\right\}$ converges to the function $\bar{w}$ in $H^1_0(\Omega)$ and has the required properties.

Thus, it is enough to prove that for every function $\bar{w} \in \mathcal{K}^1_\psi \cap H^{1,p}_0(\Omega)$ there exists a sequence $\bar{w}^n \in \mathcal{K}^1_\psi \cap H^2(\Omega)$ satisfying condition (7.16).

As in the proof of Theorem 7.1, we construct functions $\xi_1, \xi_2 \in H^2(\Omega)$ with the only difference that now $\varepsilon = 0$. The sequence $\hat{w}^n = \bar{w} + \dfrac{\delta_0}{4n}\,\xi_1 + \dfrac{\delta_0}{4n}\,\xi_2$ converges to $\bar{w}$ in $H^1_0(\Omega)$. It is essential that there is a nonzero "gap" between $\hat{w}^n$ and the functions ψ_1, ψ_2, or in other words,

$$\psi_1(x) < \hat{w}^n(x) < \psi_2(x) \ , \quad x \in \Omega \ .$$

We now extend the functions $\hat{w}^n$ beyond Ω by assigning them the value zero and introduce the mollifiers $(\hat{w}^n)_{\varepsilon_n}$ with a smooth kernel. By choosing ε_n sufficiently small

$$\|(\hat{w}^n)_{\varepsilon_n} - \hat{w}^n\|_1 < \frac{1}{n} \ ,$$

and

$$\psi_1(x) < (\hat{w}^n)_{\varepsilon_n}(x) < \psi_2(x) \ , \quad x \in \Omega \ . \tag{7.17}$$

Consider now the sequence $(\hat{w}^n)_{\varepsilon_n}$ in the outer neighbourhood Ω_1 of the boundary Γ. These functions can be extended to Ω in such a way that the extensions denoted by v_n vanish in $\Omega \backslash \Omega_0$. The domain Ω_0 is chosen as an inner neighbourhood of the boundary Γ, where

$$|\bar{w}| < \frac{\nu}{2} \ , \quad \psi_1 < -\nu \ , \quad \psi_2 > \nu \tag{7.18}$$

with some constant $\nu > 0$. Since the functions ξ_1, ξ_2 have compact supports, one can assume that $\hat{w}^n = \bar{w}$ in Ω_0. In particular, this entails that for n large enough the functions $(\hat{w}^n)_{\varepsilon_n}$ coincide with $(\bar{w})_{\varepsilon_n}$ on Ω_1. For the extensions v_n,

$$\|v_n\|_{1,\Omega} \le c\,\|(\bar{w})_{\varepsilon_n}\|_{1,\Omega_1} \ , \tag{7.19}$$

$$\max_{\Omega} |v_n| \le c \max_{\Omega_1} |(\bar{w})_{\varepsilon_n}| \tag{7.20}$$

with the constants c independent of n (see Mikhailov, 1976). From the construction it follows that the right-hand sides of inequalities (7.19), (7.20) tend to zero as $n \to \infty$. Therefore, $\bar{w}^n = (\hat{w}^n)_{\varepsilon_n} - v_n$ is the required sequence. Indeed, bearing in mind (7.18)–(7.20), and by (7.17), we have

$$\psi_1(x) \le \bar{w}^n(x) \le \psi_2(x) \ .$$

Moreover, the sequence $\bar{w}^n$ strongly converges to $\bar{w}$. This proves the lemma. □

8 Optimal control of stretching forces

We consider again the variational inequality examined in Section 6. We shall investigate an optimal control problem of stretching forces. Our main interest now is to prove existence theorems for the problems relating to a plate and a membrane and to analyse relationships between the two problems.

8.1 Optimal control for a plate

Let a domain $\Omega \subset \mathbb{R}^2$ satisfy the conditions of Section 6, and $\psi \in H^2(\Omega)$ be a given function, $\psi|_\Gamma < 0$. The set of admissible displacements of the plate is

$$\mathcal{K}_2 = \{w \in H^{2,0}(\Omega) \mid w(x) \ge \psi(x) \ , \quad x \in \Omega\} \ .$$

Consider the variational inequality

$$\begin{aligned} \varepsilon \int_\Omega \Delta w(\Delta\bar{w} - \Delta w)dx + \varphi(\|\nabla w\|^2) \int_\Omega \nabla w(\nabla\bar{w} - \nabla w)dx \\ \ge \int_\Omega f(\bar{w} - w)dx \quad \forall\, \bar{w} \in \mathcal{K}_2 \ . \end{aligned} \tag{8.1}$$

As before, we assume that $f \in L^2(\Omega)$. Additional conditions imposed on φ will be formulated below.

Let Φ be a convex and closed set in $H^1(0,\infty)$ and $\varphi(x) \ge 0$ for every $\varphi \in \Phi$. Consider the cost functional

$$E_\varepsilon(\varphi) = \|w(\varphi) - w_0\| + \|\varphi\|_1 \ , \quad \varphi \in \Phi \ .$$

Here $w(\varphi)$ is a solution to variational inequality (8.1) corresponding to φ, $w_0 \in L^2(\Omega)$ is a given element, $\|\cdot\|_s$ is the norm in $H^s(\Omega)$ or in $H^s(0,\infty)$, $\|\cdot\|_0 \equiv \|\cdot\|$. The optimal control problem is formulated as follows. Find an element $\varphi \in \Phi$ such that

$$E_\varepsilon(\varphi) \le E_\varepsilon(\bar{\varphi}) \quad \forall\, \bar{\varphi} \in \Phi \ . \tag{8.2}$$

We fix $\varepsilon > 0$. The dependence of the solutions on the parameter ε will be discussed later on.

Let us assume that $\sqrt{s}\ \varphi(s)$ is a nondecreasing function for each element $\varphi \in \Phi$. As shown in Section 6, in this case $w \to -\varphi(\|\nabla w\|^2)\Delta w$ is a monotone operator from $H_0^1(\Omega)$ into $H^{-1}(\Omega)$. The energy functional in this section is denoted by Π_ε^φ and

$$\Pi_\varepsilon^\varphi(w) = \frac{\varepsilon}{2}\ \|\Delta w\|^2 + \frac{1}{2}\int_0^{\|\nabla w\|^2} \varphi(s)ds - \langle\ f, w\ \rangle\ .$$

Inequality (8.1) can be written as

$$w \in \mathcal{K}_2\ :\ \partial\Pi_\varepsilon^\varphi(w)(\bar{w} - w) \geq 0 \quad \forall\, \bar{w} \in \mathcal{K}_2\ .$$

Here $\partial\Pi_\varepsilon^\varphi(w)$ is the derivative of the functional Π_ε^φ at the point w.

Theorem 8.1. *Let the elements of the set Φ satisfy the conditions formulated above. Then optimal control problem (8.2) has a solution.*

Proof. Let $\varphi_n \in \Phi$ be a minimizing sequence. Since φ_n are bounded in $H^1(0,\infty)$, passing to a subsequence, if necessary, we can write that $\varphi_n \to \varphi$ weakly in $H^1(0,\infty)$. A solution of the problem

$$w_n \in \mathcal{K}_2\ :\ \partial\Pi_\varepsilon^{\varphi_n}(w_n)(\bar{w} - w_n) \geq 0 \quad \forall\, \bar{w} \in \mathcal{K}_2 \tag{8.3}$$

exists for every n. For a fixed $\bar{w} \in \mathcal{K}_2$, from (8.3) it follows that

$$\Pi_\varepsilon^{\varphi_n}(w_n) \leq \Pi_\varepsilon^{\varphi_n}(\bar{w}) \leq c$$

with the constant c independent of n. Taking into account that φ_n are nonnegative, from the last inequality we easily get the following estimate

$$\|\Delta w_n\|^2 \leq c\ .$$

Recall that now ε is fixed. By this estimate, w_n are bounded in $H^{2,0}(\Omega)$. Passing to a subsequence, if necessary, we have $w_n \to w$ weakly in $H^{2,0}(\Omega)$ and strongly in $H_0^1(\Omega)$.

Let $\|\nabla w_n\|^2 \leq \alpha$. Then $\varphi_n \to \varphi$ uniformly in $[0,\alpha]$ which is a consequence of the compact imbedding of $H^1(0,\alpha)$ into $C[0,\alpha]$. We can now pass to the limit in (8.3). Actually, $\varphi_n(\|\nabla w_n\|^2) \to \varphi(\|\nabla w\|^2)$, $\underline{\lim}\ \|\Delta w_n\|^2 \geq \|\Delta w\|^2$. Therefore, the function w satisfies the inequality

$$w \in \mathcal{K}_2\ :\ \partial\Pi_\varepsilon^\varphi(w)(\bar{w} - w) \geq 0 \quad \forall\, \bar{w} \in \mathcal{K}_2\ , \tag{8.4}$$

and hence $w = w(\varphi)$. By the lower semicontinuity of the norm,

$$\inf_{\bar{\varphi}\in\Phi} E_\varepsilon(\bar{\varphi}) = \underline{\lim}_{n\to\infty}\ E_\varepsilon(\varphi_n) \geq E_\varepsilon(\varphi) \geq \inf_{\bar{\varphi}\in\Phi} E_\varepsilon(\bar{\varphi})\ .$$

This means that the function φ minimizes the functional E_ε over Φ. The theorem is proved. □

8.2 Optimal control for a membrane

In the space $H_0^1(\Omega)$ consider the closed and convex set

$$\mathcal{K}_1 = \{w \in H_0^1(\Omega) \mid w(x) \geq \psi(x) \quad \text{a.e. in} \quad \Omega\}$$

and the variational inequality

$$\begin{gathered} w \in \mathcal{K}_1 \quad : \\ \varphi(\|\nabla w\|^2) \langle \nabla w, \nabla \bar{w} - \nabla w \rangle \geq \langle f, \bar{w} - w \rangle \quad \forall \, \bar{w} \in \mathcal{K}_1 \ . \end{gathered} \tag{8.5}$$

Assume that $\sqrt{s}\ \varphi(s)$ is a strongly increasing function for each element $\varphi \in \Phi$. Moreover, the convergence $\sqrt{s}\ \varphi(s) \to \infty$, $s \to \infty$, is uniform with respect to $\varphi \in \Phi$.

Under these assumptions, for a fixed $\varphi \in \Phi$, there exists a unique solution to inequality (8.5) (see Section 6). The minimization problem of Π_0^φ over $\mathcal{K}_1$ is equivalent to variational inequality (8.5), as in the case of variational inequality (8.1).

Consider now an optimal control problem. The cost functional is of the form

$$E_0(\varphi) = \|w(\varphi) - w_0\| + \|\varphi\|_1 \ ,$$

where $w(\varphi)$ solves variational inequality (8.5). We seek an element $\varphi \in \Phi$ such that

$$E_0(\varphi) \leq E_0(\bar{\varphi}) \quad \forall \, \bar{\varphi} \in \Phi \ . \tag{8.6}$$

Theorem 8.2. *Assume that Φ is a convex and closed set of nonnegative functions in $H^1(0,\infty)$ such that $s \to \sqrt{s}\varphi(s)$ is a strictly increasing function for any $\varphi \in \Phi$ and $\sqrt{s}\varphi(s) \to \infty$, $s \to \infty$, uniformly with respect to $\varphi \in \Phi$. Then there exists a solution to (8.6).*

Proof. Let $\varphi_n \in \Phi$ be a minimizing sequence. Without loss of generality, we assume that $\varphi_n \to \varphi$ weakly in $H^1(0,\infty)$. By this, the variational inequality

$$\begin{gathered} w_n \in \mathcal{K}_1 \quad : \\ \varphi_n(\|\nabla w_n\|^2) \langle \nabla w_n, \nabla \bar{w} - \nabla w_n \rangle \geq \langle f, \bar{w} - w_n \rangle \quad \forall \, \bar{w} \in \mathcal{K}_1 \end{gathered} \tag{8.7}$$

is solvable for every n. Equivalently,

$$w_n \in \mathcal{K}_1 \ : \ \Pi_0^{\varphi_n}(w_n) \leq \Pi_0^{\varphi_n}(\bar{w}) \quad \forall \, \bar{w} \in \mathcal{K}_1 \ . \tag{8.8}$$

We show that $\Pi_0^\varphi(w)$ is coercive and this property holds uniformly with respect to $\varphi \in \Phi$. Actually, we see that

$$\Pi_0^\varphi(w) - \Pi_0^\varphi(0) = \int_0^1 \partial \Pi_0^\varphi(sw)(w) ds \ .$$

Therefore,

$$\Pi_0^\varphi(w) = \int_0^{1/2} \left(\partial\Pi_0^\varphi(sw) - \partial\Pi_0^\varphi(0)\right)(w)ds$$
$$+ \frac{1}{2}\,\partial\Pi_0^\varphi(0)(w) + \int_{1/2}^1 \partial\Pi_0^\varphi(sw)(w)ds\ .$$

According to Lemma 6.1 the first term of the right-hand side is nonnegative, the second one is equal to $-\frac{1}{2}\langle f, w\rangle$, and the third one is $\frac{1}{2}\partial\Pi_0^\varphi(\bar{s}w)(w)$, $\bar{s} \in \left[\frac{1}{2}, 1\right]$.

Consequently,

$$\Pi_0^\varphi(w) \geq \frac{1}{2}\,\partial\Pi_0^\varphi(\bar{s}w)(w) - \frac{1}{2}\,\langle f, w\rangle$$
$$\geq \frac{1}{2}\,\|\nabla w\|\left(\varphi(\bar{s}^2\|\nabla w\|^2)\|\bar{s}\nabla w\| - c\right) \to +\infty$$

as $\|\nabla w\| \to \infty$, uniformly with respect to $\varphi \in \Phi$. For a fixed $\bar{w}$ in (8.8), we obtain that $\Pi_0^{\varphi_n}(w_n) \leq c$ with the constant c independent of n. By the coercivity of the functional Π_0^φ, we conclude that there exists a constant α independent of n, such that

$$\|\nabla w_n\|^2 \leq \alpha\ .$$

As before, we assume that $\varphi_n \to \varphi$ strongly in $C[0,\alpha]$ and $w_n \to w$ weakly in $H_0^1(\Omega)$. Note that $w \in \mathcal{K}_1$.

After passing to the limit in (8.7) inequality (8.8) which is equivalent to (8.7) takes the form

$$\frac{1}{2}\int_0^{\|\nabla w_n\|^2} \varphi_n(s)ds - \langle f, w_n\rangle \leq \frac{1}{2}\int_0^{\|\nabla \bar{w}\|^2} \varphi_n(s)ds - \langle f, \bar{w}\rangle\ .$$

On the other hand, by the above arguments

$$\underline{\lim} \int_0^{\|\nabla w_n\|^2} \varphi_n(s)ds \geq \int_0^{\|\nabla w\|^2} \varphi(s)ds\ .$$

Thus, after passing to the lower limit in both sides of (8.8), we obtain

$$\Pi_0^\varphi(w) \leq \Pi_0^\varphi(\bar{w}) \quad \forall\, \bar{w} \in \mathcal{K}_1\ ,$$

which is equivalent to the inequality

$$w \in \mathcal{K}_1\ :\quad \varphi(\|\nabla w\|^2)\,\langle \nabla w, \nabla\bar{w} - \nabla w\rangle \geq \langle f, \bar{w} - w\rangle \quad \forall\, \bar{w} \in \mathcal{K}_1\ .$$

This gives that $w = w(\varphi)$ which completes the proof. □

8.3 Transition from a plate to a membrane

Here we assume that the elements $\varphi \in \Phi$ satisfy the same conditions as in Section 8.2. By Lemma 6.4, for any $\bar{w} \in \mathcal{K}_1$ there exists a sequence $\bar{w}^n \in \mathcal{K}_2$ strongly converging in $H_0^1(\Omega)$ to $\bar{w}$.

Let φ_ε be a solution to problem (8.2) and the corresponding $w(\varphi_\varepsilon)$ be determined by variational inequality (8.1). Relationships between solutions of optimal control problems (8.2) and (8.6) can be characterized by the following result.

Theorem 8.3. *There exists a subsequence φ_ε (without loss of generality we can use the same notation for subsequences as for sequences) such that*

$$\varphi_\varepsilon \to \varphi \quad \text{weakly in} \quad H^1(0,\infty) ,$$

$$w(\varphi_\varepsilon) \to w \quad \text{weakly in} \quad H_0^1(\Omega) ,$$

$$E_\varepsilon(\varphi_\varepsilon) \to E_0(\varphi)$$

as $\varepsilon \to 0$. Here, φ solves problem (8.6), w corresponds to φ and is defined by variational inequality (8.5).

Proof. Fix $\varphi \in \Phi$. For every ε

$$E_\varepsilon(\varphi_\varepsilon) \le E_\varepsilon(\varphi) . \tag{8.9}$$

We show that solutions $w_\varepsilon(\varphi)$ to variational inequality (8.1) corresponding to φ are bounded uniformly with respect to ε. This implies, in particular, that the right-hand side of (8.9) is bounded.

The variational inequality

$$w_\varepsilon(\varphi) \in \mathcal{K}_2 \ :$$

$$\partial \Pi_\varepsilon^\varphi(w_\varepsilon(\varphi))(\bar{w} - w_\varepsilon(\varphi)) \ge 0 \quad \forall\, \bar{w} \in \mathcal{K}_2$$

is equivalent to

$$\Pi_\varepsilon^\varphi(w_\varepsilon(\varphi)) \le \Pi_\varepsilon^\varphi(\bar{w}) \quad \forall\, \bar{w} \in \mathcal{K}_2 .$$

Hence, the relation

$$\frac{\varepsilon}{2}\, \|\Delta w_\varepsilon(\varphi)\|^2 + \frac{1}{2} \int_0^{\|\nabla w_\varepsilon(\varphi)\|^2} \varphi(s)ds - \langle\, f, w_\varepsilon(\varphi)\,\rangle \le c$$

holds uniformly with respect to ε. Consequently,

$$\|\nabla w_\varepsilon(\varphi)\|^2 \le \alpha .$$

Therefore, $E_\varepsilon(\varphi)$ are bounded uniformly with respect to ε, and from (8.9), it follows that

$$\|\varphi_\varepsilon\|_1 \le c .$$

Passing to a subsequence, if necessary, $\varphi_\varepsilon \to \varphi$ weakly in $H^1(0,\infty)$.

Let $\bar{w} \in \mathcal{K}_2$. From the inequality

$$w_\varepsilon(\varphi_\varepsilon) \in \mathcal{K}_2 \ : \quad \partial \Pi_\varepsilon^{\varphi_\varepsilon}(w_\varepsilon(\varphi_\varepsilon))(\bar{w} - w_\varepsilon(\varphi_\varepsilon)) \geq 0 \quad \forall\, \bar{w} \in \mathcal{K}_2 \tag{8.10}$$

an estimate for $w_\varepsilon(\varphi_\varepsilon)$ follows. Indeed, this inequality is equivalent to

$$\Pi_\varepsilon^{\varphi_\varepsilon}(w_\varepsilon(\varphi_\varepsilon)) \leq \Pi_\varepsilon^{\varphi_\varepsilon}(\bar{w}) \ . \tag{8.11}$$

So

$$\varepsilon \|\Delta w_\varepsilon(\varphi_\varepsilon)\|^2 + \|\nabla w_\varepsilon(\varphi_\varepsilon)\| \leq \sqrt{\alpha}$$

with some constant α independent of ε. Passing to a subsequence, if necessary,

$$w_\varepsilon(\varphi_\varepsilon) \to w \quad \text{weakly in} \quad H_0^1(\Omega) \ , \quad w \in \mathcal{K}_1 \ ,$$

$$\varepsilon w_\varepsilon(\varphi_\varepsilon) \to 0 \quad \text{weakly in} \quad H^{2,0}(\Omega)$$

as $\varepsilon \to 0$. Assuming additionally that $\varphi_\varepsilon \to \varphi$ uniformly on $[0, \alpha]$, we can pass in (8.10) to the limit as $\varepsilon \to 0$. In fact, from (8.11) it follows that

$$\frac{1}{2} \int_0^{\|\nabla w_\varepsilon(\varphi_\varepsilon)\|^2} \varphi_\varepsilon(s) ds - \langle\, f, w_\varepsilon(\varphi_\varepsilon)\,\rangle \leq \Pi_\varepsilon^{\varphi_\varepsilon}(\bar{w}) \ .$$

After passing to the lower limit in both sides with a fixed $\bar{w} \in \mathcal{K}_2$ we have

$$\frac{1}{2} \int_0^{\|\nabla w\|^2} \varphi(s) ds - \langle\, f, w\,\rangle \leq \Pi_0^{\varphi}(\bar{w}) \ . \tag{8.12}$$

By Lemma 6.4, inequality (8.12) holds for every $\bar{w} \in \mathcal{K}_1$. Therefore,

$$\varphi(\|\nabla w\|^2)\, \langle\, \nabla w, \nabla \bar{w} - \nabla w\,\rangle \geq \langle\, f, \bar{w} - w\,\rangle \quad \forall\, \bar{w} \in \mathcal{K}_1 \ .$$

This inequality entails that $w = w(\varphi)$ and, consequently,

$$\underline{\lim}\ E_\varepsilon(\varphi_\varepsilon) \geq E_0(\varphi) \ . \tag{8.13}$$

On the other hand, for any fixed element $\varphi \in \Phi$, we have $E_\varepsilon(\varphi) \to E_0(\varphi)$. In fact, from the variational inequality

$$w_\varepsilon(\varphi) \in \mathcal{K}_2 \ : \quad \partial \Pi_\varepsilon^{\varphi}(w_\varepsilon(\varphi))(\bar{w} - w_\varepsilon(\varphi)) \geq 0 \quad \forall\, \bar{w} \in \mathcal{K}_2 \tag{8.14}$$

we derive the estimate

$$\varepsilon \|\Delta w_\varepsilon(\varphi)\|^2 + \|\nabla w_\varepsilon(\varphi)\| \leq c \ ,$$

which holds uniformly with respect to ε. Passing to a subsequence, if necessary,

$$w_\varepsilon(\varphi) \to \widetilde{w} \quad \text{weakly in} \quad H_0^1(\Omega) \ , \quad \text{strongly in} \quad L^2(\Omega) \ ,$$

$$\varepsilon w_\varepsilon(\varphi) \to 0 \quad \text{weakly in} \quad H^{2,0}(\Omega)$$

as $\varepsilon \to 0$. Obviously, $\widetilde{w} \in \mathcal{K}_1$. Now, as in (8.10), we can pass to the limit in (8.14) by letting $\varepsilon \to 0$, and we obtain

$$\widetilde{w} \in \mathcal{K}_1 \;:\; \varphi(\|\nabla\widetilde{w}\|^2)\,\langle\, \nabla\widetilde{w}, \nabla\bar{w} - \nabla\widetilde{w}\,\rangle \geq \langle\, f, \bar{w} - \widetilde{w}\,\rangle \quad \forall\, \bar{w} \in \mathcal{K}_1 \;.$$

It follows that $\widetilde{w} = w(\varphi)$, and

$$E_\varepsilon(\varphi) \equiv \|w_\varepsilon(\varphi) - w_0\| + \|\varphi\|_1 \to E_0(\varphi) \;.$$

Taking $\widetilde{\varphi}$ which solves optimal control problem (8.6), we have

$$E_\varepsilon(\varphi_\varepsilon) \leq E_\varepsilon(\widetilde{\varphi}) \;.$$

The right-hand side of this inequality, by the previous considerations, converges to $E_0(\widetilde{\varphi})$, and therefore,

$$\overline{\lim}\; E_\varepsilon(\varphi_\varepsilon) \leq E_0(\widetilde{\varphi}) \;.$$

Bearing in mind (8.13), we complete the proof of the theorem. □

9 Extreme shapes of cracks in a plate

We consider the contact problem between a plate and a rigid punch under the assumption that a plate has a vertical crack. The shape of the crack may change. Our problem is to find among all admissible shapes of the crack an extreme one. This can be done by considering a functional defined on the set of all admissible shapes of the crack and reducing the problem to the minimization of this functional. For any shape of the crack the displacements of the plate can be defined in a unique way. Our main goal in this section is to prove the existence of extreme shapes of cracks.

Let $\Omega \subset \mathbb{R}^2$ be a bounded domain with smooth boundary Γ, $(x, y) \in \Omega$. The shapes of cracks are described by the equations $y = \delta\phi(x)$, $x \in [a, b]$, where δ is a parameter. We assume that the axis z is orthogonal to the plane x, y and $z = \psi(x, y)$ describes the punch shape.

Let Γ_δ be the graph of the function $y = \delta\phi(x)$, $x \in [a, b]$, $\Omega_\delta = \Omega \setminus \Gamma_\delta$. For every δ the middle surface of the plate occupies the domain Ω_δ. We assume that $\Gamma_\delta \subset \Omega$ for $\delta \in [0, 1]$. In particular, $\Omega_0 = \Omega \setminus \Gamma_0$. The nonpenetration condition is expressed by the inequality $w_\delta(x, y) \geq \psi(x, y)$, $(x, y) \in \Omega_\delta$, where $w_\delta(x, y)$ is the plate displacement for a given δ.

We shall start with analyzing the behaviour of solutions for $\delta \downarrow 0$. The minimization problem of finding an extreme shape of the crack will be formulated below. Let $H_0^{2,0}(\Omega_\delta)$ be the Sobolev space of functions having the derivatives up to the second order square integrable in Ω_δ and vanishing on the outer boundary Γ, together with the first order derivatives

$$H_0^{2,0}(\Omega_\delta) = \{\varphi \in L^2(\Omega_\delta) | D^\alpha\varphi \in L^2(\Omega_\delta),\ |\alpha| \leq 2, \\ \varphi = 0 \text{ and } \frac{\partial\varphi}{\partial n} = 0 \text{ on } \Gamma\},$$

where $\alpha = (\alpha_1\ , \alpha_2)$, $\alpha_i = 0, 1, 2$.

We introduce a bilinear form $B_\delta : H_0^{2,0}(\Omega_\delta) \times H_0^{2,0}(\Omega_\delta) \to \mathbb{R}$.

$$B_\delta(v,\varphi) = \int_{\Omega_\delta} (v_{xx}\varphi_{xx} + v_{yy}\varphi_{yy} + \sigma(v_{xx}\varphi_{yy} + v_{yy}\varphi_{xx}) + 2(1-\sigma)v_{xy}\varphi_{xy})d\Omega_\delta,$$

and the set $\mathcal{K}_\psi(\Omega_\delta)$ of admissible displacements of the plate,

$$\mathcal{K}_\psi(\Omega_\delta) = \{\varphi \in H_0^{2,0}(\Omega_\delta) | \varphi(x,y) \geq \psi(x,y), \quad (x,y) \in \Omega_\delta\},$$

where $0 < \sigma < \frac{1}{2}$ is a constant, $\psi \in H^2(\Omega)$, $\psi_{|\Gamma} < 0$.

The contact problem between a plate and a punch can be formulated in the variational form

$$\min_{\varphi \in \mathcal{K}_\psi(\Omega_\delta)} \left\{ \frac{1}{2} B_\delta(\varphi,\varphi) - \langle f,\varphi\rangle_\delta \right\}, \tag{9.1}$$

where $\langle f,\varphi\rangle_\delta = \int_{\Omega_\delta} f\varphi d\Omega_\delta$, $f \in L^2(\Omega)$.

For every fixed δ this problem has a unique solution w_δ and is equivalent to the variational inequality

$$w_\delta \in \mathcal{K}_\psi(\Omega_\delta) \; : \quad B_\delta(w_\delta, \varphi - w_\delta) \geq \langle f, \varphi - w_\delta\rangle_\delta \quad \forall\, \varphi \in \mathcal{K}_\psi(\Omega_\delta). \tag{9.2}$$

First of all we shall demonstrate the boundedness of solutions to (9.2) in the space $H_0^{2,0}(\Omega_\delta)$ with the norm $\|\cdot\|_{2,\Omega_\delta}$.

We take any fixed element $\varphi \in H_0^2(\Omega)$, such that $\varphi \geq \psi$ in Ω. By taking the restriction $\varphi = \varphi_{|\Omega_\delta}$ to Ω_δ, we get $\varphi \geq \psi$ in Ω_δ, or equivalently, $\varphi = \varphi_{|\Omega_\delta} \in \mathcal{K}_\psi(\Omega_\delta)$.

Substituting the functions $\varphi_{|\Omega_\delta}$ into (9.2) as test functions we get

$$B_\delta(w_\delta, w_\delta) \leq c,$$

with a constant c independent of δ. Hence, we get an estimate for the solutions w_δ,

$$\|w_\delta\|_{2,\Omega_\delta} \leq c\,. \tag{9.3}$$

In order to study the behavior of solutions for $\delta \downarrow 0$, we establish a one-to-one mapping between Ω_δ and Ω_0 and we find the limit of the solutions in the fixed domain Ω_0.

In addition to the above assumptions, it will be assumed that $\phi \in H_0^3(a,b)$. Choose the domains Ω_1, Ω_2 such that $\bar{\Omega}_1 \subset \Omega_2$, $\bar{\Omega}_2 \subset \Omega$ and $\Gamma_\delta \subset \Omega_1$ for all sufficiently small δ. Let h be a function such that $h \equiv 1$ on Ω_1, $h \in C_0^\infty(\Omega_2)$.

We apply a transformation of independent variables

$$\tilde{x} = x, \quad \tilde{y} = y - \delta\phi h, \tag{9.4}$$

where the function ϕ is extended beyond (a,b) by putting its values equal to zero. For δ small enough the Jacobian of this transformation is strictly positive.

Let $u_\delta(\widetilde{x},\widetilde{y}) \equiv w_\delta(x,y)$, $f_\delta(\widetilde{x},\widetilde{y}) \equiv f(x,y)$. In the variables $\widetilde{x},\widetilde{y}$, inequality (9.2) can be written as

$$B^\delta(u_\delta, \varphi - u_\delta) \geq \int_{\Omega_0} f^\delta(\varphi - u_\delta) d\Omega_0 \quad \forall\, \varphi \in \mathcal{K}_{\psi_\delta}(\Omega_0)\,, \tag{9.5}$$

where

$$\begin{aligned}
&B^\delta(u_\delta,\xi_\delta) = B_\delta(w_\delta,\xi),\\
&\mathcal{K}_{\psi_\delta}(\Omega_0) = \{\varphi \in H_0^{2,0}(\Omega_0) | \varphi(\widetilde{x},\widetilde{y}) \geq \psi_\delta(\widetilde{x},\widetilde{y}), \quad (\widetilde{x},\widetilde{y}) \in \Omega_0\},\\
&\xi_\delta(\widetilde{x},\widetilde{y}) \equiv \xi(x,y),\ \psi_\delta(\widetilde{x},\widetilde{y}) \equiv \psi(x,y),\\
&f^\delta = f_\delta \lambda_\delta^2,\ \lambda_\delta = (1-\delta\phi h_y)^{-\frac{1}{2}}\ .
\end{aligned}$$

Estimate (9.3) in variables $\widetilde{x},\widetilde{y}$ takes the form

$$\|u_\delta\|_{2,\Omega_0} \leq c \tag{9.6}$$

with the constant c independent of δ, $\delta \leq \delta_0$. Observe that $\psi_\delta \to \psi$ uniformly on Ω. By this, for each element $\bar{\varphi} \in \mathcal{K}_\psi(\Omega_0)$ there exists a sequence $\bar{\varphi}_\delta \in \mathcal{K}_{\psi_\delta}(\Omega_0)$ such that

$$\bar{\varphi}_\delta \to \bar{\varphi} \quad \text{strongly in } H_0^{2,0}(\Omega_0)\ .$$

This result is analogous to the corresponding result of Section 2 of Chapter 4. By (9.6), from the sequence $\{u_\delta\}_{\delta>0}$ one can choose a subsequence (without loss of generality we can use the same notation for subsequences as for sequences) such that for $\delta \downarrow 0$

$$u_\delta \to u \quad \text{weakly in } H_0^{2,0}(\Omega_0)\ . \tag{9.7}$$

For any element $\bar{\varphi} \in \mathcal{K}_\psi(\Omega_0)$, by the above considerations, we can choose a subsequence $\bar{\varphi}_\delta \in \mathcal{K}_{\psi_\delta}(\Omega_0)$ strongly converging to $\bar{\varphi}$ in $H_0^{2,0}(\Omega_0)$. By the strong convergence $\lambda_\delta \to 1$ in $L^2(\Omega_0)$, we can pass to the upper limit in (9.5). Therefore,

$$u \in \mathcal{K}_\psi(\Omega_0)\ :\quad B(u, \bar{\varphi} - u) \geq \langle f, \bar{\varphi} - u\rangle_0 \quad \forall\, \bar{\varphi} \in \mathcal{K}_\psi(\Omega_0)\,, \tag{9.8}$$

which means that the limit u satisfies the corresponding variational inequality. Hence, the following statement is true.

Theorem 9.1. *There exists a subsequence of u_δ, with $u_\delta(\widetilde{x},\widetilde{y}) = w_\delta(x,y)$, still denoted by u_δ, satisfying (9.7) and such that the limit of this subsequence satisfies (9.8).*

On the basis of this statement we can formulate the problem of finding extreme shapes of cracks. For this purpose let the graph of a function $y = \phi(x)$ describe the shape of the crack, where $\phi \in \Phi$ and $\Phi \subset H_0^3(a,b)$ is a bounded, closed and convex set. For a given $\phi \in \Phi$, a solution to problem (9.2) is denoted by w_ϕ. The function $w_\phi \in \mathcal{K}_\psi(\Omega_\phi)$ satisfies the inequality

$$B_\phi(w_\phi, \varphi - w_\phi) \geq \langle f, \varphi - w_\phi\rangle_\phi \quad \forall\, \varphi \in \mathcal{K}_\psi(\Omega_\phi)\,, \tag{9.9}$$

where

$$\mathcal{K}_\psi(\Omega_\phi) = \{\varphi \in H_0^{2,0}(\Omega_\phi) | \varphi(x,y) \geq \psi(x,y), \quad (x,y) \in \Omega_\phi = \Omega \setminus \Gamma_\phi\}$$

and Γ_ϕ is the graph of the function $\phi(x)$.

Let $w_0 \in L^2(\Omega)$ be a given function. We formulate an optimization problem of finding the shape $y = \phi(x)$ of a crack, $\phi \in \Phi$, as the maximization of the difference between w_ϕ and w_0. The following cost functional is used

$$\mathcal{J}(\phi) = \|w_\phi - w_0\|_{0,\Omega_\phi} ,$$

where $\|\cdot\|_{0,\Omega_\phi}$ is the norm in $L^2(\Omega_\phi)$, and the maximization problem takes the form

$$\sup_{\phi \in \Phi} \mathcal{J}(\phi). \tag{9.10}$$

Let us prove the existence of a solution to (9.10). Choose a maximizing sequence $\phi_m \in \Phi$ with the property $\mathcal{J}(\phi_m) \to \sup_{\phi\in\Phi} \mathcal{J}(\phi)$. Due to the requirements imposed on Φ the sequence ϕ_m is bounded in $H_0^3(a,b)$. Without loss of generality we can write

$$\phi_m \to \phi \quad \text{weakly in } H_0^3(a,b) .$$

For every ϕ_m the displacement $w_m = w_{\phi_m}$ of the plate can be found as a solution to the variational inequality

$$w_m \in \mathcal{K}_\psi(\Omega_{\phi_m}) : \; B_m(w_m, \varphi - w_m) \geq \langle f, \varphi - w_m \rangle_m \quad \forall \varphi \in \mathcal{K}_\psi(\Omega_{\phi_m}) \tag{9.11}$$

Let the domains Ω_1, Ω_2 and the function h be as above. Moreover, let $\Gamma_{\phi_m} \subset \Omega_1$ for all m. Assuming that ϕ_1 , ϕ_2 , ..., are extended by assigning them the value zero outside the interval (a,b), the following transformation of independent variables

$$\widetilde{x} = x, \quad \widetilde{y} = y + h(\phi - \phi_m) \tag{9.12}$$

can be applied. In view of the imbedding theorem, the inequality $|\phi_{xx} - \phi_{mxx}| < \frac{1}{m}$ holds on (a,b) which allows us to rewrite transformation (9.12) in the form $\widetilde{x} = x, \;\; \widetilde{y} = y - \frac{1}{m}\zeta_m h$, where $\zeta_m = m(\phi_m - \phi)$ are bounded functions in $C^2([a,b])$. The Jacobian $1 - \frac{1}{m}\zeta_m h_y$ of this transformation is positive for m sufficiently large.

Let $u_m(\widetilde{x},\widetilde{y}) \equiv w_m(x,y)$. Using the variables $\widetilde{x}$, $\widetilde{y}$, we can rewrite inequality (9.11) in the form

$$B^m(\Omega_\phi; u_m, \varphi - u_m) \geq \langle f^m, \varphi - u_m \rangle_\phi \quad \forall \varphi \in \mathcal{K}_{\psi_m}(\Omega_\phi) , \tag{9.13}$$

where

$$B^m(\Omega_\phi; u_m, \varphi) \equiv B_m(w_m, \xi), \quad \varphi(\widetilde{x},\widetilde{y}) = \xi(x,y),$$
$$f^m(\widetilde{x},\widetilde{y}) = \lambda_m^2 f(x,y),$$
$$\psi_m(\widetilde{x},\widetilde{y}) = \psi(x,y),$$
$$\lambda_m = (1 - \frac{1}{m}\zeta_m h_y)^{-\frac{1}{2}}.$$

In the space $H_0^{2,0}(\Omega_\phi)$

$$\|u_m\|_{2,\Omega_\phi} \leq c$$

with the constant c independent of m. Choosing a subsequence, if necessary, we have

$$u_m \to u \quad \text{weakly in } H_0^{2,0}(\Omega_\phi) \text{ and strongly in } L^2(\Omega_\phi) \ .$$

Moreover, for any element $\varphi \in \mathcal{K}_\psi(\Omega_\phi)$ there exists a subsequence $\varphi_m \in \mathcal{K}_{\psi_m}(\Omega_\phi)$, $\varphi_m \to \varphi$ strongly in $H_0^{2,0}(\Omega_\phi)$. Now we can pass to the limit in (9.13) as $m \to \infty$ and get

$$u \in \mathcal{K}_\psi(\Omega_\phi) \ : \quad B_\phi(u, \varphi - u) \geq \langle f, \varphi - u \rangle_\phi \quad \forall \, \varphi \in \mathcal{K}_\psi(\Omega_\phi) \ .$$

This means that the constructed function u corresponds to the crack defined by ϕ and to the punch shape $z = \psi$, i.e. $u = u_\phi$.

Let $w_{0m}(\widetilde{x}, \widetilde{y}) \equiv w_0(x, y)$. Then

$$\begin{aligned} &\sup_{\eta \in \Phi} \mathcal{J}(\eta) = \overline{\lim}\, \mathcal{J}(\phi_m) = \\ &= \overline{\lim}\, \|w_m - w_0\|_{0,\Omega_{\phi_m}} = \overline{\lim}\, \|\lambda_m(u_m - w_{0m})\|_{0,\Omega_\phi} = \\ &= \|u_\phi - w_0\|_{0,\Omega_\phi} = \mathcal{J}(\phi) \leq \sup_{\eta \in \Phi} \mathcal{J}(\eta) \ . \end{aligned}$$

Consequently, the function ϕ is a solution to problem (9.10). □

To conclude this section we observe that by using a similar method it is possible to prove the existence of a shape of the crack which minimizes the functional $\mathcal{J}$ on Φ. Moreover, stresses in the plate can also be minimized. This leads to the following problem.

Let $w_1 \in H^1(\Omega)$ be a given function. Consider the cost functional

$$\mathcal{J}_1(\phi) = \|w_\phi - w_1\|_{1,\phi} \ ,$$

where $\|\cdot\|_{1,\phi}$ is the norm in $H^1(\Omega_\phi)$ and w_ϕ is a solution to (9.9) for a given ϕ. We seek a solution to the problem

$$\inf_{\phi \in \Phi} \mathcal{J}_1(\phi) \ .$$

Taking into account our considerations it can be easily seen that the above problem has a solution.

10 Extreme shapes of unilateral cracks

We consider the equilibrium state of a plate with both the vertical and horizontal displacements of the plate. As in the previous section, the plate is assumed to have a vertical crack with a variable shape. The nonpenetration condition imposed on crack faces is formulated in the form of an inequality. An extreme shape of the crack is sought among all admissible cracks with fixed ends. It means that the displacements of the plate should have the maximal deflection from a given function.

In other words, we define a cost functional on a set of admissible crack shapes and we aim at maximizing this functional. Cracks with interior and boundary ends are considered. For both cases we prove existence theorems. Crack lengths are assumed to be limited.

10.1 Interior cracks

Let the crack ends be situated at the points $x = a$, $x = b$ of the axis x. The crack shape is described by the equation $y = \delta\psi(x)$ with a fixed function $\psi(x)$ and a parameter δ. We assume that the middle surface of the plate occupies a domain $\Omega_\delta = \Omega\backslash\Gamma_\delta$, where Γ_δ is the graph of the function $y = \delta\psi(x)$, and $\Omega \subset \mathbb{R}^2$ is a bounded domain with smooth boundary Γ.

We first study the dependence of solutions on the parameter δ. The problem of extreme crack shape will be formulated later on.

Let w^δ and $W^\delta = (w_1^\delta, w_2^\delta)$ be the vertical and the horizontal displacements of a plate, respectively. We shall make use of bilinear forms B_δ and b_δ. The first one was defined in the previous section and the second one is defined by the formula

$$b_\delta(W,U) = \int_{\Omega_\delta} \{\varepsilon_{11}(W)\varepsilon_{11}(U) + \varepsilon_{22}(W)\varepsilon_{22}(U)$$

$$+ \sigma\varepsilon_{11}(W)\varepsilon_{22}(U) + \sigma\varepsilon_{22}(W)\varepsilon_{11}(U) + 2(1-\sigma)\varepsilon_{12}(W)\varepsilon_{12}(U)\}d\Omega_\delta \ .$$

Here $\varepsilon(W)$ is a strain tensor, $\varepsilon_{ij}(W) = \frac{1}{2}\Big(\frac{\partial w_i}{\partial x_j} + \frac{\partial w_j}{\partial x_i}\Big)$, $x_1 \equiv x$, $x_2 \equiv y$, $0 < \sigma < \frac{1}{2}$. The exterior forces are denoted by f and $F = (f_1, f_2)$, respectively. Let

$$\langle\, f, w \,\rangle_\delta = \int_{\Omega_\delta} fw d\Omega_\delta \ , \quad \langle\, F, W \,\rangle_\delta = \int_{\Omega_\delta} FW d\Omega_\delta \ ,$$

$$f, F \in L^2(\Omega) \ .$$

Assume that crack faces may be pulled apart, ie. crack is open, in this case the contact set is unknown. Let $\nu^\delta = (\nu_1^\delta, \nu_2^\delta)$ be the unit normal vector to the curve $y = \delta\psi(x)$ (sec Fig. 4).

The displacement W^δ evaluated on the crack face corresponding to the positive direction of ν^δ is denoted by W_+^δ. By W_-^δ we denote W^δ evaluated on the opposite crack face, $[W^\delta] = W_+^\delta - W_-^\delta$. The nonpenetration condition has the form

$$[W^\delta]\nu^\delta \geq 0 \quad \text{on} \quad \Gamma_\delta \ . \tag{10.1}$$

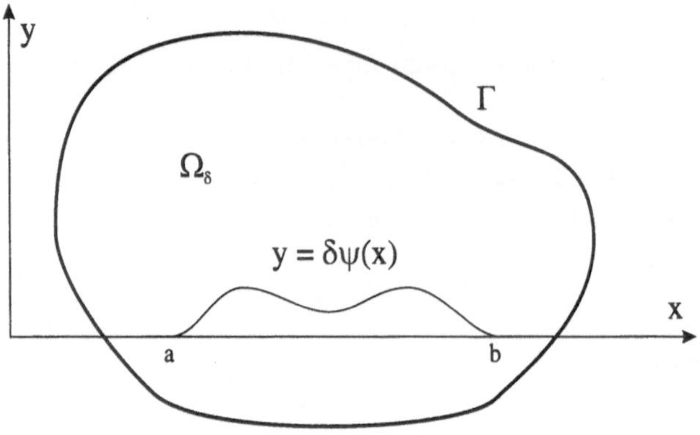

Fig. 4. Cracks in a plate.

We should note that in general the displacements w^δ evaluated on the opposite crack faces are different. For the sake of simplicity the following boundary conditions are considered on the external boundary,

$$w^\delta = \frac{\partial w^\delta}{\partial n} = W^\delta = 0 \quad \text{on} \quad \Gamma \ . \tag{10.2}$$

Let $H_0^{2,0}(\Omega_\delta)$ be the completion in $H^2(\Omega_\delta)$ of smooth functions vanishing near Γ and $\mathcal{K}_\delta^1(\Omega_\delta)$ be the set of functions W^δ from $H^1(\Omega_\delta)$ satisfying (10.1), (10.2). The problem of finding the equilibrium state of the plate having a given crack shape can be formulated as a variational problem. Namely, we seek $w^\delta \in H_0^{2,0}(\Omega_\delta)$ and $W^\delta \in \mathcal{K}_\delta^1(\Omega_\delta)$ that minimize the energy functionals

$$\frac{1}{2}\, B_\delta(w,w) - \langle\, f, w \,\rangle_\delta \ ,$$

$$\frac{1}{2}\, b_\delta(W,w) - \langle\, F, W \,\rangle_\delta$$

over the sets $H_0^{2,0}(\Omega_\delta)$ and $\mathcal{K}_\delta^1(\Omega_\delta)$, respectively. These variational problems are, respectively, equivalent to

$$B_\delta(w^\delta, \bar{w}) = \langle\, f, \bar{w} \,\rangle_\delta \quad \forall\, \bar{w} \in H_0^{2,0}(\Omega_\delta) \ , \tag{10.3}$$

$$b_\delta(W^\delta, \bar{W} - W^\delta) \geq \langle\, F, \bar{W} - W^\delta \,\rangle_\delta \quad \forall\, \bar{W} \in \mathcal{K}_\delta^1(\Omega_\delta) \ . \tag{10.4}$$

It is worth observing that the functions w^δ and W^δ can be found from (10.3), (10.4) independently.

To study the behaviour of solutions as $\delta \to 0$, we establish a one-to-one mapping of Ω_δ on the fixed domain Ω_0 using a transformation of independent variables.

Assume that $\psi \in H_0^3(a,b)$. Extend the function ψ beyond the segment (a,b) by assigning it the value zero.

Let Ω_1, Ω_2 be domains such that $\bar{\Omega}_1 \subset \Omega_2$, $\bar{\Omega}_2 \subset \Omega$, $\Gamma_\delta \subset \Omega_1$ for δ small enough. We can choose a function $h \in C_0^\infty(\Omega_2)$ such that $h \equiv 1$ on Ω_1. Consider the transformation of the independent variables

$$\widetilde{x} = x \ , \ \widetilde{y} = y - \delta\psi h \tag{10.5}$$

with the positive Jacobian $q_\delta = 1 - \delta\psi h_y$ for all δ, $|\delta| \leq \delta_0$. This transformation maps Ω_δ onto Ω_0. Denote $u^\delta(\widetilde{x},\widetilde{y}) \equiv w^\delta(x,y)$, $U^\delta(\widetilde{x},\widetilde{y}) \equiv W^\delta(x,y)$. Equation (10.3) and inequality (10.4) can be written in the new variables. The structure of relations obtained in this way depends on transformation (10.5). For example, for the second derivatives of w^δ we have

$$w^\delta_{xx} = u^\delta_{\underset{xx}{\sim}} - 2\delta u^\delta_{\underset{xy}{\sim}}(h\psi)_x + \delta^2 u^\delta_{\underset{yy}{\sim}}(h\psi)^2_x - \delta u^\delta_{\underset{y}{\sim}}(h\psi)_{xx} \ . \tag{10.6}$$

The right-hand side is represented as the sum of $u^\delta_{\underset{xx}{\sim}}$ and the terms proportional to δ^i, $i \geq 1$. In the same way we can transform (10.3), (10.4).

We now derive an a priori estimate for solutions w^δ. We substitute w^δ into (10.3) as a test function $\bar{w}$ which gives

$$B_\delta(w^\delta, w^\delta) \leq \|f\|_{0,\Omega_\delta} \|w^\delta\|_{0,\Omega_\delta} \ , \tag{10.7}$$

where $\|\cdot\|_{s,\Omega_\delta}$ is the norm in the space $H^s(\Omega_\delta)$. It follows that $\|w^\delta\|_{2,\Omega_\delta} \leq c$ uniformly with respect to δ. Hence,

$$\|u^\delta\|_{2,\Omega_0} \leq c \tag{10.8}$$

with the constant c independent of δ. Equation (10.3) can be rewritten as

$$B_0^\delta(u^\delta, \bar{u}) = \langle\, f^\delta, \bar{u}\,\rangle_0 \quad \forall\, \bar{u} \in H_0^{2,0}(\Omega_0) \ . \tag{10.9}$$

where $B_0^\delta(u^\delta,\bar{u}) = B_\delta(w^\delta,\bar{w})$, $f^\delta = q_\delta^{-1} f_\delta$, $f_\delta(\widetilde{x},\widetilde{y}) = f(x,y)$. As in the case of variational inequality (10.4) it is easy to derive the following relation

$$\begin{gathered} U^\delta \in \mathcal{K}_\delta^1(\Omega_0) \ : \\ b_0^\delta(U^\delta, \bar{U} - U^\delta) \geq \langle\, F^\delta, \bar{U} - U^\delta\,\rangle_0 \quad \forall\, \bar{U} \in \mathcal{K}_\delta^1(\Omega_0) \ . \end{gathered} \tag{10.10}$$

Here $b_0^\delta(U^\delta,\bar{U}) = b_\delta(W^\delta,\bar{W})$, $F^\delta = q_\delta^{-1} F_\delta$, $F_\delta(\widetilde{x},\widetilde{y}) = F(x,y)$. To obtain the estimate of U^δ we note that the first Korn inequality

$$\sum_{i,j=1}^{2} \|\varepsilon_{ij}(U^\delta)\|^2_{0,\Omega_0} \geq c\, \|U^\delta\|^2_{1,\Omega_0}$$

holds in Ω_0, and the inequality $\dfrac{3}{4} < q_\delta < \dfrac{5}{4}$ holds for δ small enough. Now, it is easily seen that zero belongs to $\mathcal{K}_\delta^1(\Omega_0)$. Hence, by taking $\bar{U} = 0$ in (10.10) we get the estimation

$$\|U^\delta\|_{1,\Omega_0} \leq c \tag{10.12}$$

which holds uniformly with respect to δ, $|\delta| \leq \delta_0$.

In order to pass to the limit in (10.10) as $\delta \to 0$, we need the following auxiliary statement. Namely, for any element $\bar{U} \in \mathcal{K}_0^1(\Omega_0)$ there exists a sequence $\bar{U}^\delta \in \mathcal{K}_\delta^1(\Omega_0)$ such that $\bar{U}^\delta \to \bar{U}$ strongly in $H^1(\Omega_0)$.

To see this, take $\bar{U} = (\bar{u}_1, \bar{u}_2)$ and put $\bar{U}^\delta = \bar{U} + (0, \delta\psi_x h\bar{u}_1)$, where the function h is defined as in (10.5). The function $\bar{U}^\delta$ possesses all the required properties. In fact, since $\bar{U} \in \mathcal{K}_0^1(\Omega_0)$ we have $[\bar{u}_2] \geq 0$ on Γ_0. Moreover, the inclusion $\bar{U}^\delta \in \mathcal{K}_\delta^1(\Omega_0)$ yields $[\bar{u}_1^\delta](-\delta\psi_x) + [\bar{u}_2^\delta] \geq 0$ on Γ. Clearly, the function $\bar{U}^\delta$ constructed above satisfies the latter inequality. Obviously, $\bar{U}^\delta$ converges to $\bar{U}$. This proves the statement.

By (10.8), (10.12), one can choose a subsequence u^δ, U^δ (still denoted as the sequence) such that as $\delta \to 0$

$$u^\delta \to u \quad \text{weakly in} \quad H_0^{2,0}(\Omega_0) \ ,$$

$$U^\delta \to U \quad \text{weakly in} \quad H^1(\Omega_0) \ .$$

Observe that $q_\delta \to 1$ strongly in $L^2(\Omega)$. Hence, in the limit equation (10.9) takes the form

$$B_0(u, \bar{u}) = \langle\, f, \bar{u}\,\rangle_0 \quad \forall\, \bar{u} \in H_0^{2,0}(\Omega_0) \ . \tag{10.13}$$

Now, take a fixed element $\bar{U} \in \mathcal{K}_0^1(\Omega_0)$ and build a sequence $\bar{U}^\delta \in \mathcal{K}_\delta^1(\Omega_0)$ strongly converging to $\bar{U}$. Passing to the upper limit in both sides of (10.10) the inequality

$$b_0(U, \bar{U} - U) \geq \langle\, F, \bar{U} - U\,\rangle_0 \quad \forall\, \bar{U} \in \mathcal{K}_0^1(\Omega_0) \tag{10.14}$$

follows, and $U \in \mathcal{K}_0^1(\Omega_0)$. In this way we have proved the following statement.

Theorem 10.1. *There exists a subsequence of solutions $w^\delta = u^\delta$, $W^\delta = U^\delta$ to (10.3), (10.4), still denoted by $w^\delta = u^\delta$, $W^\delta = U^\delta$, weakly convergent to a solution u, U of (10.13), (10.14), respectively.*

This result can be used to find extreme shapes of cracks. The problem is formulated as follows. Let Ψ be a convex, bounded and closed set in the space $H_0^3(a,b)$. Each element $\psi \in \Psi$ defines the crack shape $y = \psi(x)$, $x \in [a,b]$, and generates the solution w_ψ, W_ψ of problem (10.3), (10.4). Denote by Γ_ψ the graph of the function $y = \psi(x)$, $\Omega_\psi = \Omega\backslash\Gamma_\psi$. The functions w_0, W_0 belong to the space $L^2(\Omega)$. The problem consists in finding an element $\psi \in \Psi$ maximizing the functional

$$\mathcal{J}(\psi) = \|w_\psi - w_0\|_{0,\Omega_\psi} + \|W_\psi - W_0\|_{0,\Omega_\psi} \ .$$

Thus, we have to find a crack shape which maximizes the deviation of (w_ψ, W_ψ) from (w_0, W_0). The space of functions having properties analogous to $H_0^{2,0}(\Omega_\delta)$ is now denoted by $H_0^{2,0}(\Omega_\psi)$, and the set of functions W satisfying the inequality $[W]\nu^\psi \geq 0$ on Γ_ψ and vanishing on Γ is denoted by $\mathcal{K}_\psi^1(\Omega_\psi)$. Here $\nu^\psi = \dfrac{(-\psi_x, 1)}{\sqrt{1+\psi_x^2}}$

is the unit normal vector to Γ_ψ. For any $\psi \in \Psi$ the functions w_ψ and W_ψ are uniquely determined by the relations

$$B_\psi(w_\psi, \bar{w}) = \langle f, \bar{w} \rangle_\psi \quad \forall \bar{w} \in H_0^{2,0}(\Omega_\psi) \ , \tag{10.15}$$

$$b_\psi(W_\psi, \bar{W} - W_\psi) \geq \langle F, \bar{W} - W_\psi \rangle_\psi \quad \forall \bar{W} \in \mathcal{K}_\psi^1(\Omega_\psi) \ . \tag{10.16}$$

In (10.15), (10.16) we add ψ in the notation to indicate that the integration is taken over the domain Ω_ψ. Determining the extreme crack shape is equivalent to solving the problem

$$\sup_{\psi \in \Psi} \mathcal{J}(\psi) \ . \tag{10.17}$$

Theorem 10.2. *There exists a solution $\psi \in \Psi$ to problem (10.17).*

Proof. Let ψ^n be a maximizing sequence. By the properties of Ψ, the sequence is bounded in $H_0^3(a,b)$. Without loss of generality we can assume that $\psi^n \to \psi$ weakly in $H_0^3(a,b)$. Due to imbedding theorems, the inequalities $|\psi_{xx}^n - \psi_{xx}| < \frac{1}{n}$ hold in (a,b).

Moreover, for every ψ^n the solution $w^n \equiv w_{\psi^n}$ and $W^n \equiv W_{\psi^n}$ can be found from the relations

$$B_{\psi^n}(w^n, \bar{w}) = \langle f, \bar{w} \rangle_{\psi^n} \quad \forall \bar{w} \in H_0^{2,0}(\Omega_{\psi^n}) \ , \tag{10.18}$$

$$b_{\psi^n}(W^n, \bar{W} - W^n) \geq \langle F, \bar{W} - W^n \rangle_{\psi^n} \quad \forall \bar{W} \in \mathcal{K}_{\psi^n}^1(\Omega_{\psi^n}) \ . \tag{10.19}$$

Let us choose domains Ω_1, Ω_2 and a function h as before, assuming that $\Gamma_{\psi^n} \subset \Omega_1$ for all n. Consider the transformation

$$\widetilde{x} = x \ , \quad \widetilde{y} = y + (\psi - \psi^n)h \ .$$

The functions ψ and ψ^n are extended by assigning them the value zero beyond the interval (a,b). Our transformation can be written in the form $\widetilde{x} = x$, $\widetilde{y} = y - \frac{1}{n}\,\xi_n h$, where the functions $\xi_n = n(\psi^n - \psi)$ are bounded in $C^2[a,b]$. In this way we get a one-to-one mapping between Ω_{ψ^n} and Ω_ψ with the positive Jacobian $q_n = 1 - \frac{1}{n}\,\xi_n h_y$.

The rest of the proof is similar to that of Theorem 10.1. We rewrite (10.18), (10.19) using the variables $\widetilde{x}, \widetilde{y}$ and pass to the limit as $n \to \infty$ using corresponding a priori estimates. Relations (10.18), (10.19) can be written in the form

$$B_n^\psi(u^n, u) = \langle f^n, \bar{u} \rangle_\psi \quad \forall \bar{u} \in H_0^{2,0}(\Omega_\psi) \ , \tag{10.20}$$

$$b_n^\psi(U^n, \bar{U} - U^n) \geq \langle F^n, \bar{U} - U^n \rangle_\psi \quad \forall \bar{U} \in \mathcal{K}_{\psi^n}^1(\Omega_\psi) \ . \tag{10.21}$$

A priori estimates take the form

$$\|u^n\|_{2,\Omega_\psi} \leq c \ , \quad \|U^n\|_{1,\Omega_\psi} \leq c$$

with the constants independent of n. Let us choose a subsequence u^n, U^n (still denoted as the sequence) weakly converging to $u \in H_0^{2,0}(\Omega_\psi)$ and $U \in \mathcal{K}_\psi^1(\Omega_\psi)$, respectively. It can be assumed that u^n, $U^n \to u$, U strongly in $L^2(\Omega_\psi)$.

Notice that for any fixed $\bar{U} \in \mathcal{K}_\psi^1(\Omega_\psi)$ there exists a sequence $\bar{U}^n \in \mathcal{K}_{\psi^n}^1(\Omega_\psi)$ such that $\bar{U}^n \to \bar{U}$ strongly in $H^1(\Omega_\psi)$. This fact can be proved in the same way as the above statement concerning the strong convergence of $\bar{U}^\delta$. Consequently, we can pass to the limit in (10.20), (10.21) as $n \to \infty$

$$B_\psi(u, \bar{u}) = \langle f, \bar{u} \rangle_\psi \quad \forall\, \bar{u} \in H_0^{2,0}(\Omega_\psi) \ ,$$

$$b_\psi(U, \bar{U} - U) \geq \langle F, \bar{U} - U \rangle_\psi \quad \forall\, \bar{U} \in \mathcal{K}_\psi^1(\Omega_\psi) \ .$$

This means that the functions u, U correspond to the crack shape $y = \psi(x)$, ie., $u = u_\psi$, $U = U_\psi$. Let $w_{0n}(\widetilde{x}, \widetilde{y}) \equiv w_0(x, y)$, $W_{0n}(\widetilde{x}, \widetilde{y}) \equiv W_0(x, y)$. By the above arguments

$$\begin{aligned}
\sup_{\bar{\psi} \in \Psi} \mathcal{J}(\bar{\psi}) &= \overline{\lim}\ \mathcal{J}(\psi^n) = \overline{\lim}\ \{\|w^n - w_0\|_{0,\Omega_{\psi^n}} + \|W^n - W_0\|_{0,\Omega_{\psi^n}}\} \\
&= \overline{\lim}\ \{\|q_n^{-1/2}(u^n - w_{0n})\|_{0,\Omega_\psi} + \|q_n^{-1/2}(U^n - W_{0n}\|_{0,\Omega_\psi}\} \\
&= \|u_\psi - w_0\|_{0,\Omega_\psi} + \|U_\psi - W_0\|_{0,\Omega_\psi} = \mathcal{J}(\psi) \ .
\end{aligned}$$

Consequently, the constructed function ψ maximizes the functional $\mathcal{J}$ over Ψ and, thus, solves problem (10.17). The theorem is proved. □

An interesting observation is that there exists a crack shape such that the deviation of (w_ψ, W_ψ) from (w_0, W_0) is minimal. This problem can be formulated as

$$\inf_{\psi \in \Psi} \mathcal{J}(\psi) \ . \tag{10.22}$$

10.2 Boundary cracks

Let again the function $y = \delta\psi(x)$ describe a crack shape. The points $(0,0)$, $(b,0)$ are the crack ends. We assume that $(0,0) \in \Gamma$ and $\psi \in H_0^4(0,b)$. As before, $\Omega_\delta = \Omega \backslash \Gamma_\delta$, Γ_δ is the graph of the function $y = \delta\psi$. The vertical and horizontal displacements can be derived from the relations

$$\begin{aligned}
&w^\delta \in H_0^{2,0}(\Omega_\delta) \ , \quad W^\delta \in \mathcal{K}_\delta^1(\Omega_\delta) \ : \\
&B_\delta(w^\delta, \bar{w}) = \langle f, \bar{w} \rangle_\delta \quad \forall\, \bar{w} \in H_0^{2,0}(\Omega_\delta) \ ,
\end{aligned} \tag{10.23}$$

$$b_\delta(W^\delta, \bar{W} - W^\delta) \geq \langle F, \bar{W} - W^\delta \rangle_\delta \quad \forall\, \bar{W} \in \mathcal{K}_\delta^1(\Omega_\delta) \ . \tag{10.24}$$

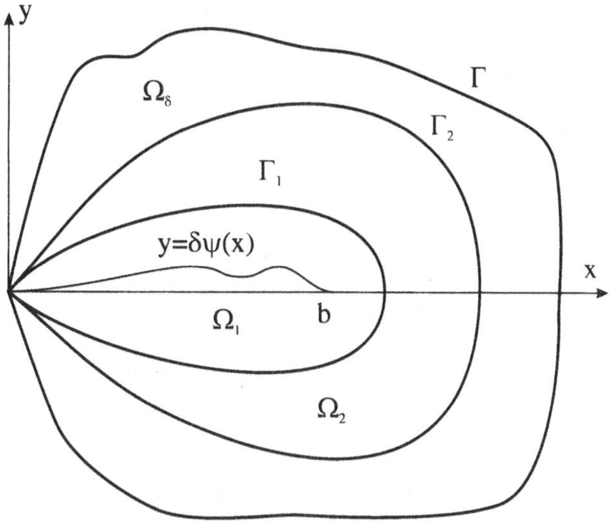

Fig. 5. Boundary cracks.

In what follows we shall assume that the tangent to the boundary Γ at the point $(0, 0)$ does not coincide with the tangent to the curve $y = \delta\psi$ at the same point.

Properties of solutions will be first analysed as $\delta \to 0$. Choose domains $\Omega_1 \subset \Omega_2$, $\Omega_2 \subset \Omega$ with the boundaries Γ_1, Γ_2 as shown in Fig. 5. Moreover, h is a function such that $h \equiv 1$ in Ω_1, $h \equiv 0$ beyond Ω_2, $h \in C^\infty(\Omega)$.

Such a choice of the domains is possible if the boundaries Γ_2, Γ_1 near $x = 0$ are described by the equations $y = \alpha_2 x$, $y = \alpha_1 x$, $y = \beta_2 x$, $y = \beta_1 x$, $\beta_2 > 0$, $\beta_1 < 0$, α_i, β_i – const, $\alpha_2 > \beta_2 > \beta_1 > \alpha_1$, respectively.

To demonstrate this, we introduce the domains I - V near $x = 0$ as indicated in Fig. 6. In this case $h \equiv 0$ in the domains I, V; $h \equiv 1$ in the domain III;

$$h(x, y) = \exp\left\{1 - \frac{(\alpha_2 - \beta_2)^2 x^2}{(\alpha_2 - \beta_2)^2 x^2 - (y - \beta_2 x)^2}\right\} \quad \text{in the domain} \quad II \ ;$$

$$h(x, y) = \exp\left\{1 - \frac{(\alpha_1 - \beta_1)^2 x^2}{(\alpha_1 - \beta_1)^2 x^2 - (y - \beta_1 x)^2}\right\} \quad \text{in the domain} \quad IV \ .$$

Since $\psi \in H_0^4(0, b)$ we have $\psi(x) = o(x^2)$, $\psi_x(x) = o(x)$, $\psi_{xx}(x) = o(1)$. Thus, the function ψh and its derivatives up to the second order are bounded as $x \to 0$, $(x, y) \in \Omega$. We consider the transformation of independent variables which maps Ω_δ onto Ω_0

$$\tilde{x} = x \ , \quad \tilde{y} = y - \delta\psi h \ . \tag{10.25}$$

The Jacobian of this transformation is positive for small δ, $|\delta| \le \delta_0$. The remaining part of the proof is analogous to the corresponding part of the proof of

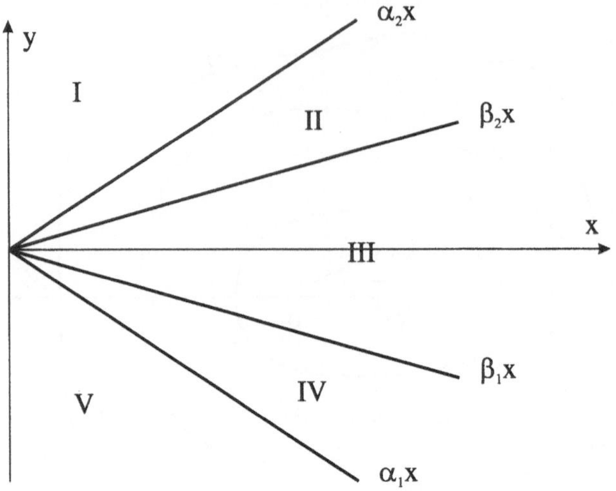

Fig. 6. Choice of auxiliary domains.

Theorem 10.1. In consequence, relations (10.23), (10.24) can be written using the variables $\widetilde{x}, \widetilde{y}$, and a priori estimates for $u^\delta = w^\delta$, $U^\delta = W^\delta$ follow

$$\|u^\delta\|_{2,\Omega_0} + \|U^\delta\|_{1,\Omega_0} \le c \ .$$

Having these estimates we can pass to the limit as $\delta \to 0$. This proves that the following statement holds true.

Theorem 10.3. *From the sequence $\{u^\delta, U^\delta\}$ one can choose a subsequence such that as $\delta \to 0$*

$$u^\delta \to u \quad \textit{weakly in} \quad H_0^{2,0}(\Omega_0) \ ,$$
$$U^\delta \to U \quad \textit{weakly in} \quad H^1(\Omega_0) \ ,$$

moreover, the functions u, U satisfy the relations

$$B_0(u, \bar{u}) = \langle\, f, \bar{u} \,\rangle_0 \quad \forall\, \bar{u} \in H_0^{2,0}(\Omega_0) \ ,$$

$$U \in \mathcal{K}_0^1(\Omega_0) \ : \ b_0(U, \bar{U} - U) \ge \langle\, F, \bar{U} - U \,\rangle_0 \quad \forall\, \bar{U} \in \mathcal{K}_0^1(\Omega_0) \ .$$

Let us turn to the extreme crack shapes problem. As before, we assume that $y = \psi(x)$ is the equation of the crack, the points $(0,0)$, $(b,0)$ are the crack ends, $(0,0) \in \Gamma$, $(b,0) \in \Omega$. For any $\psi \in \Psi$ the solution to the problem

$$w_\psi \in H_0^{2,0}(\Omega_\psi) \ , \quad W_\psi \in \mathcal{K}_\psi^1(\Omega_\psi) \ ,$$

$$B_\psi(w_\psi, \bar{w}) = \langle\, f, \bar{w} \,\rangle_\psi \quad \forall\, \bar{w} \in H_0^{2,0}(\Omega_\psi) \ ,$$
$$b_\psi(W_\psi, \bar{W} - W_\psi) \ge \langle\, F, \bar{W} - W_\psi \,\rangle_\psi \quad \forall\, \bar{W} \in \mathcal{K}_\psi^1(\Omega_\psi)$$

can be found. It is assumed that Ψ is a nonempty, closed, bounded and convex set in $H_0^4(0,b)$. Define the cost functional

$$\mathcal{J}(\psi) = \|w_\psi - w_0\|_{0,\Omega_\psi} + \|W_\psi - W_0\|_{0,\Omega_\psi} .$$

We want to find an element $\psi \in \Psi$ such that

$$\mathcal{J}(\psi) \geq \mathcal{J}(\bar{\psi}) \quad \forall \bar{\psi} \in \Psi .$$

It is also assumed that for any $\psi \in \Psi$ the graph $y = \psi(x)$ have no common points with the boundary Γ when $x \neq 0$.

It can be shown that there exists a function $\psi \in \Psi$ maximizing $\mathcal{J}$ over Ψ. To do this we can choose a maximizing sequence ψ^n. Without loss of generality we can assume that $\psi^n \to \psi$ weakly in $H_0^4(0,b)$, $\left|\psi^n_{xx} - \psi_{xx}\right| < \frac{1}{n}$ on $(0,b)$.

A function h can be constructed as that in the proof of Theorem 10.3 provided that appropriate domains Ω_1, Ω_2 are chosen. We also apply the same transformation of independent variables. The Jacobian of this transformation in positive for large n. The remaining part of the proof coincides with the corresponding part of the proof of Theorem 10.2. □

Observe that the minimization problem

$$\inf_{\psi \in \Psi} \mathcal{J}(\psi)$$

has a solution.

10.3 A more precise nonpenetration condition

In this section we formulate the nonpenetration condition taking into account the dependence of horizontal displacements on the distance from the middle surface. As before, a bounded domain $\Omega \subset \mathbb{R}^2$ is assumed to have smooth boundary Γ. A trace of the crack shape on the plane x, y is described by the function $y = \delta\psi$, where $x \in [0,1]$, δ is a positive parameter, $(x,y) \in \Omega$. The middle surface of the plate occupies the domain $\Omega_\delta = \Omega \backslash \Gamma_\delta$; Γ_δ is the graph of the function $y = \delta\psi(x)$. The horizontal and vertical displacements of the middle surface points are denoted by $W^\delta = (w_1^\delta, w_2^\delta)$ and w^δ, respectively, $\chi^\delta = (W^\delta, w^\delta)$.

Consider the energy functional

$$\Pi_\delta(\chi) = \frac{1}{2}\, B_\delta(w,w) + \frac{1}{2}\, b_\delta(W,W) - \langle\, F, \chi\, \rangle_\delta .$$

The bilinear forms b_δ, B_δ have been already used in the previous sections. Here $F = (f_1, f_2, f_3)$ is the vector of external forces belonging to $L^2(\Omega)$. Recall that the plate has a nonzero thickness. In particular, the displacements of the plate points depend on the distance from the middle surface. Actually the model of the plate considered throughout this book is characterized by the following dependence of the horizontal displacements on z, $w_i^z = w_i - z w_{x_i}$, $i = 1, 2$. It is assumed that $z = 0$ corresponds to the middle surface.

In consequence, we should be more precise in formulating the nonpenetration condition. To do this, we formulate the following condition

$$\big[W^\delta - z\nabla w^\delta\big]\nu^\delta \geq 0 \quad \text{on} \quad \Gamma_\delta \ , \quad |z| \leq \Lambda \ . \tag{10.26}$$

Here $2\Lambda = \text{const}$ is the thickness of the plate. As before, the plate is assumed to be clamped on its external boundary

$$W^\delta = w^\delta = \frac{\partial w^\delta}{\partial n} = 0 \quad \text{on} \quad \Gamma \ .$$

Let $H(\Omega_\delta) = H_0^{1,0}(\Omega_\delta) \times H_0^{1,0}(\Omega_\delta) \times H_0^{2,0}(\Omega_\delta)$, where the spaces $H_0^{s,0}(\Omega_\delta)$ are defined in the previous sections.

With

$$\mathcal{K}_\delta(\Omega_\delta) = \big\{(W, w) \in H(\Omega_\delta) \ \big|\ (W, w) \quad \text{satisfy (10.26)}\big\}$$

being a closed and convex set the equilibrium state of the plate is described by the following variational problem

$$\inf_{\chi \in \mathcal{K}_\delta(\Omega_\delta)} \Pi_\delta(\chi) \ .$$

This problem is equivalent to the variational inequality

$$\chi^\delta \in \mathcal{K}_\delta(\Omega_\delta) \ : \ \langle\, \Pi'_\delta(\chi^\delta), \bar\chi - \chi^\delta \,\rangle \geq 0 \quad \forall\, \bar\chi \in \mathcal{K}_\delta(\Omega_\delta) \ , \tag{10.27}$$

where $\Pi'_\delta(\chi^\delta)$ is the derivative of Π_δ at the point χ^δ. It is easily seen that a solution $\chi^\delta = (W^\delta, w^\delta)$ to problem (10.27) exists for every fixed δ, and inequality (10.27) can be written as

$$B_\delta(w^\delta, \bar w - w^\delta) + b_\delta(W^\delta, \bar W - W^\delta) \geq \langle\, F, \bar\chi - \chi^\delta \,\rangle_\delta \quad \forall\, \bar\chi \in \mathcal{K}_\delta(\Omega_\delta) \ . \tag{10.28}$$

Let $\psi \in H_0^3(0,1)$. In the same way as in Section 10.1 we choose domains Ω_1, Ω_2 and a nonnegative function h with the required properties. The transformation of independent variables

$$\widetilde{x} = x \ , \quad \widetilde{y} = y - \delta\psi h$$

maps Ω_δ onto Ω_0 with the positive Jacobian $q_\delta = 1 - \delta\psi h_y$. Denote $u^\delta(\widetilde{x}, \widetilde{y}) = w^\delta(x, y)$, $U^\delta(\widetilde{x}, \widetilde{y}) = W^\delta(x, y)$, $\omega^\delta = (U^\delta, u^\delta)$.

It is seen that restriction (10.26) can be rewritten as

$$\big[U^\delta - z(u^\delta_{\widetilde{x}} - \delta\psi_x u^\delta_{\widetilde{y}}, u^\delta_{\widetilde{y}})\big](-\delta\psi_x, 1) \geq 0 \quad \text{on} \quad \Gamma_0 \ , \quad |z| \leq \Lambda \ . \tag{10.29}$$

The set of functions from $H(\Omega_0)$ satisfying (10.29) is denoted by $\mathcal{K}_\delta(\Omega_0)$. Inequality (10.28) is transformed to the form

$$\begin{aligned} \omega^\delta \in \mathcal{K}_\delta(\Omega_0) \ : \ & B_0^\delta(u^\delta, \bar{u} - u^\delta) + b_0^\delta(U^\delta, \bar{U} - U^\delta) \\ & \geq \langle F^\delta, \bar{\omega} - \omega^\delta \rangle_0 \quad \forall \, \bar{\omega} \in \mathcal{K}_\delta(\Omega_0) \ . \end{aligned} \tag{10.30}$$

It is clear that $0 \in \mathcal{K}_\delta(\Omega_0)$ and $\frac{3}{4} < q_\delta < \frac{5}{4}$ for all δ small enough. We substitute $\bar{\omega} = 0$ into (10.30) as a test function. Taking into account the estimate

$$B_0(u, u) \geq c\|u\|_{2,\Omega_0}^2$$

and the first Korn inequality, both valid in Ω_0, we get the estimate

$$\|U^\delta\|_{1,\Omega_0} + \|u^\delta\|_{2,\Omega_0} \leq c \tag{10.31}$$

which holds uniformly with respect to $\delta \leq \delta_0$. It turns out that every element from $\mathcal{K}_0(\Omega_0)$ can be approximated by elements from $\mathcal{K}_\delta(\Omega_0)$.

Lemma 10.1. *For every element* $(u_1, u_2, u) \in \mathcal{K}_0(\Omega_0)$ *there exists a sequence* $(u_1^\delta, u_2^\delta, u^\delta) \in \mathcal{K}_\delta(\Omega_0)$ *such that as* $\delta \to 0$

$$(u_1^\delta, u_2^\delta, u^\delta) \to (u_1, u_2, u) \quad \textit{strongly in} \quad H(\Omega_0) \ .$$

Proof. Since $(u_1, u_2, u) \in \mathcal{K}_0(\Omega_0)$ we have

$$[u_2] - z[u_{\tilde{y}}] \geq 0 \quad \text{on} \quad \Gamma_0, \quad |z| \leq \Lambda \ .$$

The function $u_{\tilde{x}} - \delta\psi_x u_{\tilde{y}}$ belongs to the space $H_0^{1,0}(\Omega_0)$. Hence, its traces on the lines $\tilde{y} = 0^+$, $\tilde{y} = 0^-$ are elements of $H^{\frac{1}{2}}(\tilde{y} = 0^\pm)$. The difference of these traces belongs to $H^{\frac{1}{2}}(\tilde{y} = 0)$ and coincides with $[u_{\tilde{x}} - \delta\psi_x u_{\tilde{y}}]$ on Γ_0.

Choose an extension of this difference from the space $H^1(\mathbb{R}^2)$ and denote it by Q. Consequently, the restriction of the function $h\Lambda|\delta\psi_x Q|$ to Ω is an element of $H_0^1(\Omega)$. In Ω_0 we define

$$(u_1^\delta, u_2^\delta, u^\delta) = (u_1, u_2, u) + (0, \delta\psi_x u_1 + h\Lambda\, |\delta\psi_x Q| \, , \, 0) \ .$$

We first show that $(u_1^\delta, u_2^\delta, u^\delta) \in \mathcal{K}_\delta(\Omega_0)$. To this end we notice that the boundary conditions on Γ are fulfilled. Hence, it suffices to prove (10.29). From the above considerations it follows that

$$\Lambda\, |\delta\psi_x Q| \geq -\delta z\psi_x[u_{\tilde{x}} - \delta\psi_x u_{\tilde{y}}] \quad \forall \, z, \ |z| \leq \Lambda$$

holds in Γ_0. Hence, on Γ_0 we have

$$\begin{aligned} & [u_1](-\delta\psi_x) + [u_2] + \delta\psi_x[u_1] + \Lambda\, |\delta\psi_x Q| \\ & + \delta z\psi_x[u_{\tilde{x}} - \delta\psi_x u_{\tilde{y}}] - z[u_{\tilde{y}}] \geq [u_2] - z[u_{\tilde{y}}] \geq 0 \ , \quad |z| \leq \Lambda \ , \end{aligned}$$

which means that inequality (10.29) holds. The strong convergence of the sequence $(u_1^\delta, u_2^\delta, u^\delta)$ to (u_1, u_2, u) in $H(\Omega_0)$ is obvious. This proves the lemma. □

By (10.31), one can choose a subsequence from the sequence ω^δ (still denoted by ω^δ) such that as $\delta \to 0$

$$\omega^\delta \to \omega \quad \text{weakly in} \quad H(\Omega_0) \ .$$

Let us take any fixed element $\bar{\omega} \in \mathcal{K}_0(\Omega_0)$ and construct a sequence $\bar{\omega}^\delta \in \mathcal{K}_\delta(\Omega_0)$ strongly converging in $H(\Omega_0)$ to $\bar{\omega}$. By the strong convergence $F_\delta \to F$, $q_\delta \to 1$ in $L^2(\Omega)$ we can pass to the limit (10.30) as $\delta \to 0$ in (10.30) and get

$$B_0(u, \bar{u} - u) + b_0(U, \bar{U} - U) \geq \langle\, F, \bar{\omega} - \omega \,\rangle_0 \quad \forall\, \bar{\omega} \in \mathcal{K}_0(\Omega_0) \ . \tag{10.32}$$

This inequality means that the function $\omega = (U, u)$ corresponds to the crack shape $\widetilde{y} = 0$.

We have proved the following result.

Theorem 10.4. *From the sequence $\chi^\delta = \omega^\delta$ of solutions to problem (10.27) one can choose a subsequence weakly converging in $H(\Omega_0)$ to a solution ω of problem (10.32).*

Now we are in a position to prove the existence of an extreme crack shape. The problem is formulated as follows. Let Ψ be a closed convex and bounded set in $H_0^3(0,1)$. Each element $\psi \in \Psi$ describes a crack shape. The space of functions analogous to $H(\Omega_\delta)$ is denoted by $H(\Omega_\psi)$. The nonpenetration condition in this case has the form

$$[W_\psi - z \nabla w_\psi] \nu_\psi \geq 0 \quad \text{on} \quad \Gamma_\psi,\ |z| \leq \Lambda \ . \tag{10.33}$$

Here (W_ψ, w_ψ) is the displacement, Γ_ψ is the graph of the function $y = \psi(x)$, $\nu_\psi = (-\psi_x, 1)(1 + \psi_x^2)^{-\frac{1}{2}}$ is the normal to Γ_ψ. We denote by $\mathcal{K}_\psi(\Omega_\psi)$ the set of functions from $H(\Omega_\psi)$ satisfying (10.33). The solution $\chi_\psi = (W_\psi, w_\psi)$ can be determined by the variational inequality

$$\chi_\psi \in \mathcal{K}_\psi(\Omega_\psi) \ : \ \langle\, \Pi'_\psi(\chi_\psi), \bar{\chi} - \chi_\psi \,\rangle \geq 0 \quad \forall\, \bar{\chi} \in \mathcal{K}_\psi(\Omega_\psi) \ .$$

Let the functions W_0, w_0 belong to the space $L^2(\Omega)$. We introduce the cost functional

$$\mathcal{J}(\psi) = \|W_\psi - W_0\|_{0,\Omega_\psi} + \|w_\psi - w_0\|_{0,\Omega_\psi}$$

and consider the optimal control problem

$$\sup_{\psi \in \Psi} \mathcal{J}(\psi) \ . \tag{10.34}$$

Theorem 10.5. *There exists a solution to problem (10.34).*

Proof. Let ψ^n be a maximizing sequence. It is bounded in the space $H_0^3(0,1)$. Choosing a subsequence, if necessary, we can assume that $\psi^n \to \psi$ weakly in

$H_0^3(0,1)$ as $n \to \infty$. In view of the imbedding theorems, $|\psi_{xx}^n - \psi_{xx}| < \dfrac{1}{n}$ on $(0,1)$. For each n a unique solution $\chi^n = (W^n, w^n)$ can be found such that

$$\chi^n \in \mathcal{K}_{\psi^n}(\Omega_{\psi^n}) \ : \ \langle \Pi'_{\psi^n}(\chi^n), \bar{\chi} - \chi^n \rangle \geq 0 \quad \forall \bar{\chi} \in \mathcal{K}_{\psi^n}(\Omega_{\psi^n}) \ . \tag{10.35}$$

We choose domains Ω_1, Ω_2 and a nonnegative function h as above assuming $\Gamma_{\psi^n} \subset \Omega_1$ for all n and consider the transformation of independent variables

$$\widetilde{x} = x \ , \quad \widetilde{y} = y + (\psi - \psi^n)h \ .$$

As before, the functions ψ, ψ^n are extended by assigning them the value zero outside the interval $(0,1)$. In this way we obtain a one-to-one mapping between Ω_{ψ^n} and Ω_ψ with the positive Jacobian $q_n = 1 + (\psi - \psi^n)h_y$ for all sufficiently large n. The further reasonings are similar to those used to prove the convergence of W^δ, w^δ.

Let $\chi^n(x,y) = \omega^n(\widetilde{x}, \widetilde{y})$. Inequality (10.35) can be rewritten in the variables $\widetilde{x}, \widetilde{y}$

$$\begin{aligned} \omega^n \in \mathcal{K}_{\psi^n}(\Omega_\psi) \ : \ & B_n^\psi(u^n, \bar{u} - u^n) + b_n^\psi(U^n, \bar{U} - U^n) \\ & \geq \langle F^n, \bar{\omega} - \omega^n \rangle_\psi \quad \forall \bar{\omega} \in \mathcal{K}_{\psi^n}(\Omega_\psi) \ , \end{aligned} \tag{10.36}$$

where

$$F^n = q_n^{-1} F_n, \quad F_n(\widetilde{x}, \widetilde{y}) = F(x,y) \ ,$$

$$\mathcal{K}_{\psi^n}(\Omega_\psi) = \{(U,u) \in H(\Omega_\psi) \ |$$

$$\left[U - z\left(u_{\widetilde{x}} + (\psi_n - \psi_x^n)u_{\widetilde{y}}, u_{\widetilde{y}}\right)\right](-\psi_x^n, 1) \geq 0 \quad \text{on} \quad \Gamma_\psi, \ |z| \leq \Lambda\} \ .$$

By the above convergence of ψ^n, all the derivatives of $n(\psi - \psi^n)h$ up to the second order appearing in (10.36) are bounded. A priori estimates of the solutions are as follows

$$\|U^n\|_{1,\Omega_\psi} + \|u^n\|_{2,\Omega_\psi} \leq c$$

with a constant c independent of n. Choosing a subsequence, if necessary,

$$\omega^n \to \omega \quad \text{weakly in} \quad H(\Omega_\psi) \ , \quad \text{strongly in} \quad L^2(\Omega_\psi)$$

as $n \to \infty$. In order to pass to the limit in (10.36) we need the fact that for every fixed $\bar{\omega} \in \mathcal{K}_\psi(\Omega_\psi)$ there exists a sequence $\bar{\omega}^n \in \mathcal{K}_{\psi^n}(\Omega_\psi)$ such that

$$\bar{\omega}^n \to \bar{\omega} \quad \text{strongly in} \quad H(\Omega_\psi) \ .$$

This fact can be proved as in Lemma 10.1. Hence, from (10.36) it follows that

$$\omega \in \mathcal{K}_\psi(\Omega_\psi) \ : \ \langle \Pi'_\psi(\omega), \bar{\omega} - \omega \rangle \geq 0 \quad \forall \bar{\omega} \in \mathcal{K}_\psi(\Omega_\psi) \ ,$$

ie. $\omega = \omega(\psi)$. Now, by the same arguments as those used to complete the proof of Theorem 10.2 we obtain the assertion of Theorem 10.5. □

The same reasoning allows us to prove the existence of a solution to the problem

$$\inf_{\psi \in \Psi} \mathcal{J}(\psi) .$$

Let us consider the case where the crack ends belong to the external boundary Γ. We assume that the points $(0,0) \in \Gamma$ and $(0,1) \in \Omega$ correspond to the crack ends. As before, $\Omega_\delta = \Omega \backslash \Gamma_\delta$, Γ_δ is the graph of the function $y = \delta\psi(x)$. The function ψ belongs to the space $H_0^4(0,1)$. Moreover, the tangent to the boundary Γ at the point $(0,0)$ and the tangent to the graph $y = \delta\psi(x)$ at the same point are assumed to be different. The displacements of the middle surface of the plate can be determined by the inequality

$$\chi^\delta \in \mathcal{K}_\delta(\Omega_\delta) \ : \ \langle \Pi'_\delta(\chi^\delta), \bar{\chi} - \chi^\delta \rangle \geq 0 \quad \forall \, \bar{\chi} \in \mathcal{K}_\delta(\Omega_\delta) . \tag{10.37}$$

We first investigate the behaviour of the solution χ^δ as $\delta \to 0$. We choose domains Ω_1, Ω_2 with boundaries and a smooth function $h \geq 0$ such that $h = 1$ in Ω_1, $h \in C^\infty(\Omega)$, $\Omega_1 \subset \Omega_2$, $\Omega_2 \subset \Omega$, $h = 0$ beyond Ω_2, $\{\Gamma_\delta \backslash (0,0)\} \subset \Omega_1$. Such a choice is possible if the boundaries near the point $(0,0)$ are of the form $y = \alpha_2 x$, $y = \alpha_1 x$, $y = \beta_2 x$, $y = \beta_1 x$, $\beta_2 > 0$, $\beta_1 < 0$, α_i, β_i – const, $\alpha_2 > \beta_2 > \beta_1 > \alpha_1$ (see Section 10.2).

As before, the transformation of the variables x, y has the form $\widetilde{x} = x$, $\widetilde{y} = y - \delta\psi h$. Since $\psi \in H_0^4(0,1)$ we have $\psi(x) = o(x^2)$, $\psi_x(x) = o(x)$, $\psi_{xx}(x) = o(1)$ near $x = 0$, so that the function ψh and its derivatives up to the second order are bounded when $x \to 0$, $(x,y) \in \Omega$. The argumentation is analogous to that of Section 10.1. In this way we obtain the following result.

Theorem 10.6. *From the sequence ω^δ one can choose a subsequence (still denoted as the original sequence) such that as $\delta \to 0$*

$$\omega^\delta \to \omega \quad \textit{weakly in} \quad H(\Omega_0) ,$$

where ω is a solution to the problem

$$\omega \in \mathcal{K}_0(\Omega_0) \ : \ \langle \Pi'_0(\omega), \bar{\omega} - \omega \rangle \geq 0 \quad \forall \, \bar{\omega} \in \mathcal{K}_0(\Omega_0) .$$

In the same way we can study the existence of extreme crack shapes. Let $y = \psi(x)$ be the equation of the crack shape, the points $(0,0)$ and $(0,1)$ are the crack ends, $(0,0) \in \Gamma$, $(0,1) \in \Omega$. The convex closed and bounded set $\Psi \subset H_0^4(0,1)$ is chosen in such a way that the assumption concerning the discrepancy of the tangents at the point $(0,0)$ is fulfilled.

The formulation of the problem is as follows. Maximize the cost functional

$$\sup_{\psi \in \Psi} \mathcal{J}(\psi) , \tag{10.38}$$

where, as before,

$$\mathcal{J}(\psi) = \|W_\psi - W_0\|_{0,\Omega_\psi} + \|w_\psi - w_0\|_{0,\Omega_\psi} \ .$$

A solution to problem (10.38) exists. We omit the proof here.

To conclude let us observe that by using analogous arguments one can prove the existence of extreme crack shapes in the case where both crack ends belong to the external boundary: $(0,0) \in \Gamma$, $(0,1) \in \Gamma$.

11 Optimal control in elastoplastic problems

In Section 8 of the previous chapter we have considered an elastoplastic contact problem for the Timoshenko model of a beam. In particular, an existence result was proved for fixed punch shapes. Here we shall study an optimal control problem.

We recall basic assumptions used to prove the existence theorem. We assume that there exists a solution to the equation

$$-m^0_{xx} = f \ ,$$

on the interval $(0,1)$ such that $(1+\kappa)m^0 \in \mathcal{K}$, $\kappa > 0$, where

$$\mathcal{K} = \{m \in H^1_0(0,1) \mid |m(x)| \le c_0 \quad \text{on} \quad (0,1)\}$$

and f belongs to $L^2(0,1)$. It is also assumed that

$$\mathcal{B}_\psi = \{w \in H^1_0(0,1) \mid \psi_1(x) \le w(x) \le \psi_2(x) \quad \text{on} \quad (0.1)\}$$

is nonempty.

Punch shapes are described by the functions ψ_1, ψ_2 which belong to $H^1(0,1)$, $\psi = (\psi_1, \psi_2)$. These assumptions ensure the existence of a solution to the problem

$$w \in \mathcal{B}_\psi \ , \quad m \in \mathcal{K} \ , \quad \xi \in H^{-1}(0,1) \ , \tag{11.1}$$

$$\langle\, m_{xx} + f, \bar{w} - w \,\rangle \le 0 \quad \forall\, \bar{w} \in \mathcal{B}_\psi \ , \tag{11.2}$$

$$m - m_{xx} + w_{xx} + \xi = 0 \ , \tag{11.3}$$

$$\langle\, \xi, \bar{m} - m \,\rangle \le 0 \quad \forall\, \bar{m} \in \mathcal{K} \ . \tag{11.4}$$

In this section we adopt an additional assumption that two functions $\psi^0_1, \psi^0_2 \in H^1(0,1)$ are given such that $\psi^0_1 \le \psi^0_2 - c_1$ on $(0,1)$, $c_1 = \text{const} > 0$; $\psi^0_1 \le -c_1$, $\psi^0_2 \ge c_1$ when $x = 0, 1$, and $\Psi_1, \Psi_2 \subset H^1(0,1)$ are two closed, convex, bounded sets such that

$$\psi_1(x) \le \psi^0_1(x) \quad \forall\, \psi_1 \in \Psi_1 \ ; \quad \psi_2(x) \ge \psi^0_2(x) \quad \forall\, \psi_2 \in \Psi_2 \ .$$

It is easily seen that this assumption ensures the existence of a function w^0 which belongs to $\mathcal{B}_\psi$ for all $\psi \in \Psi_1 \times \Psi_2$.

Consider the cost functional

$$\mathcal{J}(\psi, w) = \|w - w_0\| \ ,$$

where $\psi \in \Psi_1 \times \Psi_2$, $w_0 \in L^2(0,1)$ is a fixed element, the function w corresponds to the function ψ and is determined by (11.1)–(11.4).

The optimal control problem consists in minimizing the functional $\mathcal{J}$ over the set $\Psi_1 \times \Psi_2 \times \mathcal{B}_\psi$

$$\mathcal{J}(\psi, w) \to \min \ . \tag{11.5}$$

Let us observe that the functional is defined for a pair "control-state". This is caused by the nonuniqueness of the function w. Our aim is to prove an existence theorem for problem (11.5).

Theorem 11.1. *A solution to optimal control problem (11.5) exists.*

Proof. Let (ψ^n, w^n) be a minimizing sequence. Due to the boundedness of Ψ_1, Ψ_2 we can assume that

$$\psi_i^n \to \psi_i \quad \text{weakly in} \quad H^1(0,1) \quad \text{and uniformly on} \quad [0,1]$$

as $n \to \infty$. For every n, one can find a solution to the problem

$$w^n \in \mathcal{B}_{\psi^n} \ , \quad m^n \in \mathcal{K} \ , \quad \xi^n \in H^{-1}(0,1) \ , \tag{11.6}$$

$$\langle\, m_{xx}^n + f \ , \ \bar{w} - w^n \,\rangle \le 0 \quad \forall\, \bar{w} \in \mathcal{B}_{\psi^n} \ , \tag{11.7}$$

$$m^n - m_{xx}^n + w_{xx}^n + \xi^n = 0 \ , \tag{11.8}$$

$$\langle\, \xi^n, \bar{m} - m^n \,\rangle \le 0 \quad \forall\, \bar{m} \in \mathcal{K} \ . \tag{11.9}$$

Substitute $\bar{w} = w^0$ into (11.7) as a test function, and multiply (11.8) by $m^n - m_0$. Bearing in mind (11.9),

$$\|m_x^n\|^2 + \|m^n\|^2 + \langle\, \xi^n, m^n - m^0 \,\rangle \le c$$

with the constant c independent of n. Summing up (11.9) with the inequality $\langle\, \xi^n, m^n - m^0 \,\rangle \le c$, the relation $\langle\, \xi^n, \bar{m} - m^0 \,\rangle \le 0$ follows.

Assuming that $\bar{m} = m^0 + m^1$, $m^1 \in H_0^1(0,1)$, $|m^1| \le \dfrac{c_0 \kappa}{1+\kappa}$, the inequality

$$\langle\, \xi^n, m^1 \,\rangle \le c \quad \forall\, m^1 \in H_0^1(0,1) \ , \quad |m^1| \le \frac{c_0 \kappa}{1+\kappa}$$

results. Consequently, the following estimate, uniform with respect to n,

$$\|\xi^n\|_{H^{-1}(0,1)} \le c \tag{11.10}$$

holds. From (11.8) and (11.10) it is seen that

$$\|w^n\|_1 \le c \ .$$

By the obtained estimates, we have

$$w^n, m^n \to w, m \quad \text{weakly in} \quad H_0^1(0,1) \ ,$$

$$\xi^n \to \xi \quad \text{weakly in} \quad H^{-1}(0,1)$$

as $n \to \infty$. Let us now prove the following fact. For every function $\bar{w} \in \mathcal{B}_\psi$ there exists a sequence $\bar{w}^n \in \mathcal{B}_{\psi^n}$ such that $\bar{w}^n \to \bar{w}$ strongly in $H_0^1(0,1)$. Consider the set

$$Q_1 = \left\{x \in (0,1) \mid \psi_1(x) \le \bar{w}(x) \le \psi_1(x) + \frac{c_1}{8}\right\}$$

and its ν-neighbourhood Q_1^ν. Let

$$Q_2 = \left\{x \in (0,1) \mid \psi_2(x) - \frac{c_1}{8} \le \bar{w}(x) \le \psi_2(x)\right\} \ ,$$

and Q_2^ν be its ν-neighbourhood. We construct two smooth functions

$$\mu_1(x) = \begin{cases} 1 \ , & x \in Q_1 \\ 0 \ , & x \notin Q_1^\nu \ , \end{cases} \qquad \mu_2(x) = \begin{cases} -1 \ , & x \in Q_2 \\ 0 \ , & x \notin Q_2^\nu \ , \end{cases}$$

$$0 \le \mu_1(x) \le 1 \ , \quad -1 \le \mu_2(x) \le 0 \ ,$$

assuming that ν is sufficiently small. Then, on the sets Q_1^ν, Q_2^ν the inequalities

$$\psi_1 \le \bar{w} \le \psi_1 + \frac{c_1}{4} \ , \quad \psi_2 - \frac{c_1}{4} \le \bar{w} \le \psi_2$$

hold. In view of the uniform convergence of ψ_i^n,

$$\left|\psi_i^n(x) - \psi_i(x)\right| < \frac{1}{n^2} \ , \quad x \in [0,1] \ .$$

The sequence $\bar{w}^n = \bar{w} + \frac{c_1}{4n}\mu_1 + \frac{c_1}{4n}\mu_2$ possesses the required properties. Namely, $\bar{w}^n \to \bar{w}$ strongly in $H_0^1(0,1)$ and $\psi_1^n(x) \le \bar{w}^n(x) \le \psi_2^n(x)$. Substituting this sequence into (11.9) as a test function and taking into account the convergence of w^n, m^n, ξ^n, ψ^n, we can pass to the limit in (11.6)–(11.9) as in Section 8, Chapter 3. In this way we get the variational inequality

$$w \in \mathcal{B}_\psi \ , \quad m \in \mathcal{K} \ , \quad \xi \in H^{-1}(0,1) \ ,$$

$$\langle\, m_{xx} + f, \bar{w} - w \,\rangle \le 0 \quad \forall\, \bar{w} \in \mathcal{B}_\psi \ ,$$

$$m - m_{xx} + w_{xx} + \xi = 0 \ ,$$

$$\langle\, \xi, \bar{m} - m \,\rangle \le 0 \quad \forall\, \bar{m} \in \mathcal{K} \ .$$

These relations show that $w = w(\psi)$, $m = m(\psi)$, $\xi = \xi(\psi)$. In view of the weak lower semicontinuity of the functional $\mathcal{J}$

$$j = \varliminf_{n\to\infty} \mathcal{J}(\psi^n, w^n) \ge \mathcal{J}(\psi, w) \ge j \ .$$

Here $j = \inf\limits_{(\bar{\psi},\bar{w})} \mathcal{J}(\bar{\psi}, \bar{w})$. This implies that the pair (ψ, w) is a solution to optimal control problem (11.5). This completes the proof. □

Remark. 11.1. The established property relating to the existence of the strongly converging sequence $\bar{w}^n \in \mathcal{B}_{\psi^n}$ is similar to that proved in Section 7 in the two-dimensional case.

12 The case of vertical and horizontal displacements

In this section we shall study an optimal control problem for the Timoshenko model of an elastoplastic beam in the case of both vertical and horizontal displacements of beam points.

In Chapter 3, Section 9, the corresponding existence theorem has been proved. Denote by Φ a convex, closed and bounded set in $H^3(I)$ such that $\varphi(0) < 0$, $\varphi(1) < 0$ for every element $\varphi \in \Phi$. For any $\varphi \in \Phi$ a solution to problem (9.7)–(9.10) of Chapter 3 exists, and the measure ν_φ can be constructed.

Introduce the notations

$$\mathcal{B}_\varphi = \left\{(v, w) \mid v, w \in H_0^1(I),\ \ w - v\varphi_x \geq \varphi \quad \text{on} \quad I\right\} ,$$

$$\mathcal{K} = \left\{m \in H_0^1(I) \mid |m| \leq c^* \quad \text{on} \quad I\right\} .$$

Define the cost functional

$$\mathcal{J}(\varphi) = \nu_\varphi(I) .$$

Our optimal control problem is formulated as follows. Find an element $\varphi \in \Phi$ such that

$$\mathcal{J}(\varphi) = \inf_{\bar{\varphi} \in \Phi} \mathcal{J}(\bar{\varphi}) . \tag{12.1}$$

Theorem 12.1. *A solution to optimal control problem (12.1) exists.*

Proof. Let φ^i be a minimizing sequence. This sequence is bounded in the space $H^3(I)$. For every i, a solution v^i, w^i, m^i, ξ^i exists such that

$$(v^i, w^i) \in \mathcal{B}_{\varphi^i} , \quad m^i \in \mathcal{K} , \quad \xi^i \in H^{-1}(I) , \tag{12.2}$$

$$\langle\, m^i_{xx} + f, \bar{w}^i - w^i \,\rangle + \langle\, v^i_{xx} + g, \bar{v}^i - v^i \,\rangle \leq 0 \quad \forall\, (\bar{v}^i, \bar{w}^i) \in \mathcal{B}_{\varphi^i} , \tag{12.3}$$

$$-w^i_{xx} = m^i - m^i_{xx} + \xi^i , \tag{12.4}$$

$$\langle\, \xi^i, \bar{m} - m^i \,\rangle \leq 0 \quad \forall\, \bar{m} \in \mathcal{K} . \tag{12.5}$$

The measure ν_{φ^i} is defined by the formula $\nu_{\varphi^i} = -\sqrt{1 + \varphi_x^{i2}}\, (m^i_{xx} + f)$. Without loss of generality, we assume that as $i \to \infty$

$$\varphi^i \to \varphi \quad \text{weakly in} \quad H^3(I) \ , \ \text{strongly in} \quad C^2(\bar{I}) \ .$$

We shall prove that for every fixed element $(\bar{v}, \bar{w}) \in \mathcal{B}_\varphi$ there exists a sequence $(\bar{v}^i, \bar{w}^i) \in \mathcal{B}_{\varphi^i}$ such that as $i \to \infty$

$$\bar{v}^i, \bar{w}^i \to \bar{v}, \bar{w} \quad \text{strongly in} \quad H_0^1(I) \ .$$

We choose $\beta > 0$ possessing the properties $\varphi(0) < -\beta$, $\varphi(1) < -\beta$. It can be assumed that the inequalities

$$\varphi^i < -\frac{2}{3}\beta\ , \quad |\bar{v}\varphi^i_x| < \frac{\beta}{4}\ , \quad |\bar{w}^i| < \frac{\beta}{4}$$

are satisfied on the intervals $(0, \mu)$, $(1-\mu, 1)$. Here $\mu > 0$ is a sufficiently small number. A nonnegative function $\psi \in H^1_0(I)$ can be chosen such that $\psi \equiv 1$ on $(\mu, 1-\mu)$. Taking into account the convergence of φ^i,

$$|\varphi^i - \varphi| < \frac{1}{2i}\ , \quad |\varphi^i_x - \varphi_x| \cdot |\bar{v}| < \frac{1}{2i}$$

on I. The sequence

$$(\bar{v}^i, \bar{w}^i) \equiv (\bar{v}, \bar{w} + \frac{1}{i}\,\psi)$$

satisfies the inequality

$$\bar{w}^i - \bar{v}^i\varphi^i_x \geq \varphi^i\ .$$

Thus, the constructed sequence possesses the required properties.

Let us substitute the elements $(\bar{v}^i, \bar{w}^i)$ of the constructed sequence into (12.3). We multiply (12.4) by $m - m^0$. By this, from (12.3), (12.4) we derive the a priori estimate

$$\|v^i_x\| + \|w^i_x\| + \|m^i_x\| + \|\xi^i\|_{H^{-1}(I)} \leq c$$

which is uniform with respect to i. The estimate of ξ^i is obtained in the same way as that used to evaluate $\lambda^{-1}p(m)$ in (9.17), Chapter 3. We assume that as $i \to \infty$

$$v^i, w^i, m^i \to v, w, m \quad \text{weakly in} \quad H^1_0(I)\ ,$$

$$v^i \to v \quad \text{strongly in} \quad L^2(I)\ ,$$

$$\xi^i \to \xi \quad \text{weakly in} \quad H^{-1}(I)\ .$$

By this, after passing to the limit in the inequality $w^i - v^i\varphi^i_x \geq \varphi^i$ we get $w - v\varphi_x \geq \varphi$.

We can pass to the limit in (12.2)–(12.5) as $i \to \infty$ in much the same way as in Theorem 9.1, Chapter 3 when $\delta \to 0$. Now, we choose in (12.3) a sequence $(\bar{v}^i, \bar{w}^i) \in \mathcal{B}_{\varphi^i}$ strongly converging to a fixed function $(\bar{v}, \bar{w}) \in \mathcal{B}_\varphi$. From (12.2)–(12.5) we obtain the required relations

$$(v, w) \in \mathcal{B}_\varphi\ , \quad m \in \mathcal{K}\ , \quad \xi \in H^{-1}(I)\ ,$$

$$\langle\, m_{xx} + f, \bar{w} - w \,\rangle + \langle\, v_{xx} + g, \bar{v} - v \,\rangle \leq 0 \quad \forall\, (\bar{v}, \bar{w}) \in \mathcal{B}_\varphi\ ,$$

$$-w_{xx} = m - m_{xx} + \xi\ ,$$

$$\langle\, \xi, \bar{m} - m \,\rangle \leq 0 \quad \forall\, \bar{m} \in \mathcal{K}\ .$$

This implies that $v = v(\varphi)$, $w = w(\varphi)$, $m = m(\varphi)$, $\xi = \xi(\varphi)$.

To complete the proof one has to show that the constructed element φ solves optimal control problem (12.1). Indeed, by the convergence properties established above

$$m^i_{xx} + f \to m_{xx} + f \quad \text{weakly in} \quad H^{-1}(I)$$

as $i \to \infty$. Consequently,

$$\sqrt{1+\varphi_x^{i2}}\ (m^i_{xx} + f) \to \sqrt{1+\varphi_x^2}\ (m_{xx} + f) \quad \text{weakly in} \quad H^{-1}(I) \ .$$

It follows that the measures converge, ie.,

$$\nu_{\varphi^i} \to \nu_\varphi \quad \text{weakly-}(*) \ .$$

The last property implies that for every fixed $\psi \in C_0(I)$,

$$\int_0^1 \psi d\nu_{\varphi^i} \to \int_0^1 \psi d\nu_\varphi$$

as $i \to \infty$. The convergence of measures entails

$$\underline{\lim}\ \nu_{\varphi^i}(I) \geq \nu_\varphi(I) \ .$$

Finally, by the weak lower semicontinuity, we complete the proof. □

Chapter 5
Sensitivity Analysis

In this chapter we study different aspects of sensitivity analysis of nonlinear problems, as eg., Gateaux differentiability of solutions to variational inequalities with respect to perturbations of the right-hand side and coefficients of the corresponding partial differential operator, differentiability of solutions to nonlinear problems with respect to deformations of the geometrical domain of integration, and so on.

For a class of shape optimization problems we investigate local stability. We make use of the displacement derivative (defined in Section 2.4) when examining the sensitivity of systems defined on moving surfaces and we apply the results to the shell equations. We also discuss the problem of an optimal choice of interfaces for a parabolic equation.

We start with a recollection of results on the directional differentiability of the metric projection onto closed convex subsets of a Hilbert space. The material and displacement derivatives, specific tools for shape sensitivity analysis, are presented in Section 2.

Section 3 is entirely devoted to sensitivity analysis of fourth-order elliptic variational inequalities. These inequalities are used to model eg. elastic plates with obstacles. In the Sobolev space $H_0^2(\Omega)$ an explicit formula for the directional derivative of the metric projection onto the convex set

$$K = \{f \in H_0^2(\Omega) \mid f \geq \psi\}$$

is derived. With the help of this formula we establish differentiability of solutions to obstacle problems for the Kirchhoff plate and investigate shape design sensitivity of obstacle problems.

In Section 4 we examine obstacle problems for the simply supported Kirchhoff plate and give an explicit formula for the tangent cone at any point of convex set defined by unilateral conditions. In Section 5 sensitivity results are given for variational inequalities involving vector functions.

In Section 6 we propose an approach to sensitivity analysis of shape optimization problems. This approach is based on the fact that local solutions to regularized problems can be represented as fixed points of metric projections onto the sets of admissible graphs. For a class of shape optimization problems involving second order elliptic equations we prove local stability of optimal graphs.

In Section 7 we formulate a domain optimization problem for the heat equation and propose a method for sensitivity analysis of solutions with respect to perturbations of interfaces. Finally, in Section 8 the displacement derivative is used to derive formulae for the first and second order shape derivatives for a class of shape functionals for optimization problems of thin shells.

1 Properties of metric projections in Hilbert spaces

We start with a recollection of results concerning directional differentiability of metric projections.

Let H be a separable Hilbert space. Assume we are given a bilinear form

$$a(\cdot,\cdot) : H \times H \mapsto \mathbb{R}$$

which is coercive and continuous, ie.

$$\begin{aligned} &a(v,v) \geq \alpha \|v\|_H^2 \quad \forall v \in H, \\ &|a(v,z)| \leq M\|v\|_H\|z\|_H \quad \forall v,z \in H, \end{aligned} \tag{1.1}$$

where $\alpha > 0$ and M are constants.
For the sake of simplicity, we assume that the bilinear form is symmetric

$$a(v,z) = a(z,v) \quad \forall v,z \in H \ . \tag{1.2}$$

Let $K \subset H$ be a convex and closed subset of H. By $P_K(f)$ we denote the a-projection in H of an element $f \in H$ onto K. The element $y = P_K(f)$ minimizes the quadratic functional

$$I(v) = \frac{1}{2}a(v-f, v-f) \tag{1.3}$$

over the set K. Therefore, $y \in K$ solves the variational inequality

$$a(y-f, v-f) \geq 0 \quad \forall v \in K \ . \tag{1.4}$$

The mapping $P_K : H \mapsto K \subset H$ is Lipschitz continuous, ie.,

$$\|P_K(f_1) - P_K(f_2)\|_H \leq \frac{M}{\alpha}\|f_1 - f_2\|_H \quad \forall f_1, f_2 \in H \ . \tag{1.5}$$

By a generalization of Rademacher's theorem (Mignot 1976), there exists a dense subset $\Upsilon \subset H$ such that P_K is Gateaux differentiable on Υ, ie. for any $f \in \Upsilon$ we can find a linear continuous mapping $P_K'(\cdot) = P_K'(f,\cdot) : H \mapsto H$ such that

$$\forall h \in H : P_K(f + \varepsilon h) = P_K(f) + \varepsilon P_K'(h) + o(\varepsilon) \ , \tag{1.6}$$

where $\|o(\varepsilon)\|_H/\varepsilon \to 0$ as $\varepsilon \downarrow 0$.

Definition 1.1. The mapping P_K is *conically differentiable* at $f \in H$ if there exists a continuous mapping

$$\begin{aligned} &\mathcal{Q} : H \mapsto H, \\ &\mathcal{Q}(\alpha h) = \alpha \mathcal{Q}(h) \quad \text{for all } \alpha > 0 \text{ and } h \in H \end{aligned} \tag{1.7}$$

such that for $\varepsilon > 0$, ε small enough,

$$\forall h \in H : P_K(f + \varepsilon h) = P_K(f) + \varepsilon \mathcal{Q}(h) + o(\varepsilon), \tag{1.8}$$

where $\|o(\varepsilon)\|_H/\varepsilon \to 0$ as $\varepsilon \to 0$, uniformly with respect to h on compact subsets of H.

Remark. 1.1. In view of (1.8) conical differentiability implies the differentiability in the sense of Hadamard.

Now we recall some classical concepts helpful in investigating the mapping $\mathcal{Q}$ defined by (1.8).

For a given $y \in K$, the set

$$C_K(y) = \{\phi \in H | \exists \varepsilon > 0 \text{ such that } y + \varepsilon\phi \in K\} \tag{1.9}$$

is the radial cone. In general, the cone $C_K(y)$ is convex but not closed. Its closure in H, called the tangent cone, is denoted by $T_K(y)$, $T_K(y) = \mathrm{cl}(C_K(y))$.

The normal cone to K at $y \in K$, denoted by $N_K(y)$, is given as

$$N_K(y) = \{\phi \in H | a(\phi, z - y) \leq 0 \quad \forall z \in K\} \ . \tag{1.10}$$

The normal cone is convex and closed. For a given $f \in H$, and $y = P_K(f)$,

$$S_K(f) = \{v \in T_K(y) | a(f - y, v) = 0\} \ , \tag{1.11}$$

is a convex and closed cone.

Consider a continuous mapping

$$f : [0, \delta) \mapsto H$$

which is right-differentiable at 0, ie. there exists an element $f'(0) \in H$ such that

$$\lim_{s \downarrow 0} \|(f(s) - f(0))/s - f'(0)\|_H = 0 \ .$$

Put

$$\begin{aligned} &y(s) = P_K(f(s)) \quad s \in [0, \delta), \\ &\varrho(s) = (y(s) - y(0))/s \ . \end{aligned}$$

From (1.5) it clearly follows that

$$\|\varrho(s))\|_H \leq M/\alpha \quad \forall s \in (0, \delta) \ .$$

It can be shown that

$$\varrho \in S_K(f(0))$$

for every weak limit ϱ of the function $s \mapsto \varrho(s)$ at $s = 0^+$. For the proof we refer the reader to (Mignot, 1976, Theorem 1.1).

The orthogonal subspace to the element $f - y \in H$, $[f-y]^\perp$, is given as

$$[f-y]^\perp = \{\phi \in H | a(y-f,\phi) = 0\} \ .$$

Denote $(\varphi, v) = a(\varphi, v)$ for all $\varphi, v \in H$, $\|\varphi\|_H = a(\varphi,\varphi)^{\frac{1}{2}}$.

Proposition 1.1. *Let $K \subset H$ be a closed and convex subset of a Hilbert space H. Then for any $f \in H$, and $w \in C_K(u) \cap [f-u]^\perp$, where $u = P_K(f)$, we have*

$$P_K(f+tw) = P_K(f) + tw \ . \tag{1.12}$$

Therefore,

$$P_K(f+tw) = P_K(f) + tw + o(t) \tag{1.13}$$

for all $w \in \mathrm{cl}(C_K(u) \cap [f-y]^\perp)$.

Theorem 1.1. *Let $f \in H$ and $u = P_K(f)$. If for any $w \in S_K(u)$,*

$$P_K(f+tw) = P_K(f) + tw + o(t) \ ,$$

then for $t > 0$, t small enough,

$$\forall h \in H : P_K(f+th) = P_K(f) + tP_S(h) + o(t), \tag{1.14}$$

where P_S is the metric projection in H onto $S_K(u)$.

Theorem 1.1 remains valid for a non-symmetric, coercive bilinear form $a(\cdot,\cdot)$ provided H is a so-called Dirichlet space (Mignot, 1976).

From Theorem 1.1 and Proposition 1.1 it follows that the metric projection onto a convex and closed set $K \subset H$ is conically differentiable at $f \in H$, if

$$S_K(f) = \mathrm{cl}(\{v \in C_K(y) | a(f-y,v) = 0\}), \tag{1.15}$$

where $y = P_K(f)$, and the cone $S_K(f)$ is defined by (1.11). In such a case we have (1.8) with $Q = P_S$, ie., for $t > 0$, t small enough

$$P_K(g+th) = P_K(g) + tP_S(h) + o(t) \ .$$

Observe that for any element $f \in H$,

$$\mathrm{cl}(\{v \in C_K(y) | a(f-y,v) = 0\}) \subset S_K(f)$$

and

$$\mathrm{cl}(\{v \in C_K(y) | a(f-y,v) = 0\}) \neq S_K(f) \ .$$

Corollary 1.1. *Let $f : [0, \delta) \mapsto H$ be right-differentiable in the norm of H at $s = 0$ and let $K \subset H$ be a convex and closed set. Assume that*

$$T_K(f) \cap [f - g]^{\perp} = \mathrm{cl}(C_K(f) \cap [f - g]^{\perp}),$$

where $f = f(0)$, $g = P_K(f(0))$. Then for $s > 0$, s small enough,

$$P_K(f(s)) = P_K(f(0)) + sP_{S_K(f(0))}(f'(0)) + o(s), \tag{1.16}$$

where $\|o(s)\|_H / s \to 0$ as $s \to 0$. □

From (1.14) it follows that the projection P_K is conically differentiable at $f = f(0) \in H$. Moreover, we have

$$Q(h) = P_{S_K(f)}(h) \quad \forall h \in H \ . \tag{1.17}$$

Note that in general

$$Q(h) \neq -Q(-h) \ .$$

Now we give a result on the directional differentiability of the metric projection onto the unit ball in a Hilbert space H. By $\|f\|$ we denote the norm of $f \in H$.

Proposition 1.2. *In a Hilbert space H the metric projection P_K onto the convex set*

$$K = \{v \in H \mid \|v\| \leq 1\} \tag{1.18}$$

has the form

$$P_K(f) = \begin{cases} f/\|f\|, & \text{if } \|f\| > 1; \\ f, & \text{otherwise.} \end{cases} \tag{1.19}$$

If $\|f\| = 1$, then for $\varepsilon > 0$, ε small enough

$$\forall v \in H \ : \ P_K(f + \varepsilon v) = P_K(f) + \varepsilon P_S(v) + o(\varepsilon). \tag{1.20}$$

where

$$S = T_K(f) = \{v \in H \mid (f, v) \leq 0\} \ . \tag{1.21}$$

For $f \notin K$ the metric projection is differentiable ie., for ε small enough

$$\forall v \in H \ : \ P_K(f + \varepsilon v) = P_K(f) + \varepsilon Rv + o(\varepsilon), \tag{1.22}$$

where $R : H \mapsto H$ is a linear mapping of the form

$$\forall v \in H \ : \ Rv = \|f\|^{-1} P_S v \tag{1.23}$$

and S is a linear subspace,

$$S = T_K(P_K(f)) \cap [f - P_K(f)]^{\perp} = \{v \in H \mid (f, v) = 0\} \ . \tag{1.24}$$

Definition 1.2. The set K is *polyhedric* at $f \in K$, if for any $g \in H$ such that $f = P_K(g)$ we have

$$T_K(f) \cap [f-g]^\perp = \text{cl}(C_K(f) \cap [f-g]^\perp) . \tag{1.25}$$

Let $K \subset H$ be a convex and closed subset in a Hilbert space H, and let $\langle \cdot, \cdot \rangle$ denote the duality pairing between H' and H, where H' denotes the dual of H.

Consider the following family of variational inequalities depending on a parameter $t \in [0, \delta)$, $\delta > 0$,

$$y_t \in K : \qquad a_t(y_t, \varphi - y_t) \geq \langle f_t, \varphi - y_t \rangle \qquad \forall \varphi \in K . \tag{1.26}$$

Let $y_t = \mathcal{P}_t(f_t)$ be a solution to (1.26).

Theorem 1.2. *Assume that*

(i) the bilinear form $a_t : H \times H \mapsto \mathbb{R}$ *is coercive and uniformly continuous with respect to* $t \in [0, \delta)$, $\mathcal{A}_t \in \mathcal{L}(H; H')$ *is a linear operator defined by* $a_t(\phi, \varphi) = \langle \mathcal{A}_t \phi, \varphi \rangle$ $\forall \phi, \varphi \in H$, *and there exists* $\mathcal{A}' \in \mathcal{L}(H; H')$ *such that*

$$\mathcal{A}_t = \mathcal{A}_0 + t\mathcal{A}' + o(t) \quad \text{in } \mathcal{L}(H; H') , \tag{1.27}$$

(ii) for $t > 0$, t *small enough,*

$$f_t = f_0 + tf' + o(t) \quad \text{in } H', \tag{1.28}$$

where $f_t, f_0, f' \in H'$,

(iii) $K \subset H$ *is convex and closed, and solutions to the variational inequality*

$$\Pi f = \mathcal{P}_0(f) \in K : \quad a_0(\Pi f, \varphi - \Pi f) \geq \langle f, \varphi - \Pi f \rangle \quad \forall \varphi \in K \tag{1.29}$$

satisfy the following differential stability relation

$$\forall h \in H : \qquad \Pi(f_0 + \varepsilon h) = \Pi f_0 + \varepsilon \Pi' h + o(\varepsilon) \quad \text{in } H \tag{1.30}$$

for $\varepsilon > 0$, ε *small enough, where the mapping* $\Pi' : H' \mapsto H$ *is continuous and positively homogeneous.*

Then solutions to variational inequality (1.26) are right-differentiable with respect to t *at* $t = 0$, *ie. for* $t > 0$, t *small enough,*

$$y_t = y_0 + ty' + o(t) \quad \text{in } H,$$

where

$$y' = \Pi'(f' - \mathcal{A}' y_0) .$$

2 Shape sensitivity analysis

The shape sensitivity analysis for classes of shape functionals defined on geometrical domains and surfaces is presented in this section. The material and shape derivatives for classes of functions are introduced. The displacement derivates for functions defined on moving surfaces are defined. The first and second order differentiable shape functionals are considered and results on the structure of first and second derivatives of shape functionals are established.

2.1 Material derivatives

Let $\Omega \subset \mathbb{R}^N$ be a domain with smooth boundary $\Gamma = \partial\Omega$.

Following (Sokołowski, Zolesio, 1992), we assume that $\Omega \subset D$ is bounded, where D is a fixed bounded domain, ∂D is piecewise smooth, and $\partial\Omega = \Gamma$ is C^k (ie. Ω is a domain of class C^k). Let V be a vector field, it is assumed that

$$V(.,.) \in C(0, \delta; C^k(\overline{D}; \mathbb{R}^N)) \tag{2.1}$$

where $\delta > 0$ is a given constant, $k \geq 1$.

Assume that V is an admissible vector field provided that $\langle V, n\rangle_{\mathbb{R}^N} = 0$ on ∂D, except for the singular points of ∂D where $V = 0$.

Consider a family of domains $\{\Omega_t\} \subset \mathbb{R}^N, t \in [0, \delta)$,

$$\begin{aligned} \Omega_t &= T_t(V)(\Omega) \\ &= \{x \in \mathbb{R}^N \mid \exists X \in \Omega \text{ such that } x(0) = X, x = x(t)\} \end{aligned} \tag{2.2}$$

where $x(t) \in \mathbb{R}^N$, $t \in [0, \delta)$, is a unique solution to the system

$$\begin{aligned} &\frac{dx}{dt} = V(t, x(t)), \quad t \in (0, \delta), \\ &x(0) = X\ . \end{aligned} \tag{2.3}$$

Let $y(\Omega) \in W^{s,p}(\Omega)$, $s \in [0, k]$, $1 \leq p < +\infty$.

Remark. 2.1. By a result in (Sokołowski, Zolesio, 1992, Theorem 2.3), there exists $\varepsilon > 0$ such that

$$y(\Omega_t) \circ T_t(V) \in W^{s,p}(\Omega) \quad \text{for} \quad 0 \leq t < \varepsilon\ . \tag{2.4}$$

Definition 2.1. An element $\dot{y}(\Omega; V) \in W^{s,p}(\Omega)$ is the material derivative of $y(\Omega) \in W^{s,p}(\Omega)$ in the direction of a vector field $V \in C(0, \varepsilon; C^k(D; \mathbb{R}^N))$ if

$$\dot{y}(\Omega; V) = \lim_{t \to 0} \frac{1}{t}(y(\Omega_t) \circ T_t(V) - y(\Omega))$$

provided the limit exists. This limit can be taken in the sense of the strong or weak convergence in $W^{s,p}(\Omega)$. Respectively, $\dot{y}(\Omega;V)$ is called the strong (s,p) derivative, or the weak (s,p) derivative.

Remark. 2.2. We consider problems where $y(\Omega) \in \mathcal{W}(\Omega)$ and $\mathcal{W}(\Omega)$ is a Banach space. We say that $\dot{y}(\Omega;V)$ is the weak (strong) material derivative of y in $\mathcal{W}(\Omega)$ if $\frac{1}{t}(y(\Omega_t)\circ T_t - y(\Omega))$ weakly (strongly) converges to $\dot{y}(\Omega;V)$ in $\mathcal{W}(\Omega)$ as $t \downarrow 0$.

We have the following formula for the material derivative

$$\dot{y}(\Omega;V) = y'(\Omega;V) + \nabla y(\Omega)\cdot V(0) ,$$

where $y'(\Omega;V)$ denotes the shape derivative.

Example. Let $Y \in W^{m,p}(D)$, for some $m \in \mathbb{N}$, $p \geq 1$, and $y(\Omega) = Y|_\Omega$ be the restriction of Y to Ω. Hence, $y(\Omega) \in W^{m,p}(\Omega)$ and $y(\Omega_t)\circ T_t = (Y\circ T_t)|_\Omega \in W^{m,p}(\Omega)$. The following proposition describes regularity of the mapping $t \mapsto (y\circ T_t)|_\Omega$.

Proposition 2.1. *Let $Y \in W^{m,p}(D)$, $m \in \mathbb{N}$, $m \leq k$, where the integer $k \geq 1$ determines the regularity of a vector field V. Then*
(i) for $m \geq 1$ the mapping $\Omega \mapsto Y|_\Omega$ has the strong (m,p) material derivative,

$$\dot{y}(\Omega;V) = \nabla Y|_\Omega \cdot V(0) ,$$

(ii) for $m \leq 0$ the mapping $\Omega \mapsto Y|_\Omega$ has the weak (m,p) material derivative,

$$\dot{y}(\Omega;V) = \nabla Y|_\Omega \cdot V(0) .$$

2.2 Material derivatives on the boundary Γ

Now we define the material derivative for $y(\Gamma) \in W^{r,p}(\Gamma)$ under the assumption that $y(\Gamma)$ is well-defined for all the boundaries Γ of domains Ω of class C^k. This means that for a given domain Ω with boundary Γ and for

$$\Gamma_t = T_t(V)(\Gamma) \quad \text{with } V \in C(0,\varepsilon;C^k(\mathbb{R}^N;\mathbb{R}^N))$$

$y(\Gamma_t)$ is a well-defined element of the space $W^{r,p}(\Gamma_t)$.

Definition 2.2. We say that

$$\dot{y} = \dot{y}(\Gamma;V) \in W^{r,p}(\Gamma)$$

is the *weak (strong)* (r,p) *material derivative* of an element $y(\Gamma)$ at Γ, in the direction of a vector field V, if

$$\frac{1}{t}(y(\Gamma_t)\circ T_t(V) - y(\Gamma)) \to \dot{y}(\Gamma;V),$$

as $t \to 0$ provided the limit of the left-hand side exists in the weak (strong) convergence of the space $W^{r,p}(\Gamma)$.

Proposition 2.2. *Assume that $y(\Omega)$ posseses the weak (strong) material derivative $\dot{y}(\Omega;V)$ at Ω in the direction of a field V. Then for $s > \frac{1}{p}$ there exists the weak (strong) $(s-\frac{1}{p},p)$ material derivative $\dot{y}(\Gamma;V)$ of $y(\Gamma)$, $y(\Gamma) = y(\Omega)|_\Gamma$, and*

$$\dot{y}(\Gamma;V) = \dot{y}(\Omega;V)|_\Gamma \in W^{s-\frac{1}{p},p}(\Gamma)\ .$$

For a vector field V tangent to Γ an explicit formula for the material derivative can be given.

Proposition 2.3. *Let $\mathcal{V} \in C^k(D;\mathbb{R}^N)$ be a vector field such that $\langle \mathcal{V}, n\rangle_{\mathbb{R}^N} = 0$ on Γ, and the weak (s,p) material derivative $\dot{y}(\Omega;\mathcal{V})$ of $y(\Omega)$ exist. If the derivative $\dot{y}(\Omega;V)$ exists in all admissible directions V, then for the vector field $\mathcal{V}$ we have*

$$\dot{y}(\Omega;\mathcal{V}) = \nabla y(\Omega)\cdot \mathcal{V} \in W^{s,p}(\Omega)\ .$$

The same formula holds for the material derivatives on boundaries which are called boundary material derivatives.

Proposition 2.4. *Let $\dot{y}(\Gamma;V)$ be the weak (r,p) material derivative of $y(\Gamma)$. If the material derivative $\dot{y}(\Gamma;V)$ exists for any admissible direction*

$$V \in C(0,\varepsilon;C^k(D;\mathbb{R}^N))$$

such that $\langle V, n\rangle_{\mathbb{R}^N} = 0$ on $\Gamma = \partial\Omega$, then

$$\dot{y}(\Gamma;V) = \nabla_\Gamma y(\Gamma)\cdot V \in W^{r,p}(\Omega)\ .$$

Here ∇_Γ denotes the tangential gradient on Γ; in the particular case of a function y defined in an open neighbourhood of Γ,

$$\nabla_\Gamma y = \nabla y - \frac{\partial y}{\partial n}n\ .$$

2.3 Shape derivatives on the boundary Γ

Let D be a fixed bounded domain in $\mathbb{R}^N$. For any domain Ω in D with boundary Γ of class C^k, $k \geq 2$, consider an element $z(\Gamma)$ of a Sobolev space $W(\Gamma)$. Throughout this section we assume that

(i) the weak material derivative $\dot{z}(\Gamma; V)$ exists in $W(\Gamma)$,
(ii) $\nabla_\Gamma z(\Gamma) \cdot V(0)$ belongs to the space $W(\Gamma)$ for all vector fields $V \in C(0, \varepsilon; C^k(D; \mathbb{R}^N))$, where $k \geq 3$ is a fixed integer.

Definition 2.3. The *shape derivative* of $z(\Gamma)$ in the direction of V is an element of $W(\Gamma)$ defined by the formula

$$z'(\Gamma; V) = \dot{z}(\Gamma; V) - \nabla_\Gamma z(\Gamma) \cdot V(0) . \tag{2.5}$$

Proposition 2.5. *Assume that $V \mapsto \dot{z}(\Omega; V)$ is a continuous mapping from $C(0, \varepsilon; C^k(D; \mathbb{R}^N))$ into $W(\Gamma)$. Then*

$$z'(\Gamma; V) = z'(\Gamma; V(0)) . \tag{2.6}$$

Proposition 2.6. *Assume that $V \to \dot{z}(\Gamma; V)$ is a linear and continuous mapping from $C(0, \varepsilon; C^k(D; \mathbb{R}^N))$ into $W(\Gamma)$. If V_1 and V_2 are two vector fields in $C(0, \varepsilon; C^k(D; \mathbb{R}^N))$, and*

$$\langle V_1(0), n \rangle_{\mathbb{R}^N} = \langle V_2(0), n \rangle_{\mathbb{R}^N} \quad \text{on } \Gamma,$$

then

$$z'(\Gamma; V_1) = z'(\Gamma; V_2) .$$

In particular, if $z(\Gamma) = y(\Omega)|_\Gamma$, then

$$z'(\Gamma; V) = y'(\Omega; V)|_\Gamma + \frac{\partial y}{\partial n}(\Omega) \langle V(0), n \rangle_{\mathbb{R}^N} .$$

Remark. 2.3. In the next section we introduce the displacement derivative $\frac{\delta}{dt}$ and in Section 8 we apply this concept to shape sensitivity analysis for elliptic partial differential equations defined on moving surfaces S_t, $t \in [0, \delta]$, transported to a given reference domain by means of regular parametrizations. Instead of the boundary shape derivatives on moving surfaces, we consider an equivalent displacement derivative defined on reference domains.

2.4 Displacement derivatives on S

In this section and in Section 8, $\mathbf{a}, \mathbf{u}, \mathbf{v}, \mathbf{b}, \mathbf{l}$ denote vectors in $\mathbb{R}^3$ or vector functions with values in $\mathbb{R}^3$.

Let $\mathcal{O}$ be a bounded domain of $\mathbb{R}^2$ with boundary $\partial\mathcal{O}$. By $S \subset \mathbb{R}^3$ we denote the image of $\mathcal{O} \subset \mathbb{R}^2$ under a regular mapping (parametrization) $\varphi : \mathbb{R}^2 \mapsto \mathbb{R}^3$,

$$\begin{cases} \varphi : \ (\xi^1, \xi^2) \in \overline{\mathcal{O}} \subset \mathbb{R}^2 \mapsto \varphi(\xi^1, \xi^2) \in \overline{S} \subset \mathbb{R}^3, \\ \overline{S} = \varphi(\overline{\mathcal{O}}) \ . \end{cases} \tag{2.7}$$

We define two local bases in $\overline{S}$:

i) *the covariant basis* $(\mathbf{a}_i, \ i = 1, 2, 3)$

$$\mathbf{a}_\alpha = \varphi_{,\alpha} = \frac{\partial \varphi}{\partial \xi^\alpha}, \qquad \mathbf{a}_3 = \frac{\mathbf{a}_1 \times \mathbf{a}_2}{\|\mathbf{a}_1 \times \mathbf{a}_2\|_{\mathbb{R}^3}} \ ; \tag{2.8}$$

ii) *the contravariant basis* $(\mathbf{a}^i, \ i = 1, 2, 3)$

$$\mathbf{a}^\beta \cdot \mathbf{a}_\alpha = \begin{cases} 1, & \text{if } \alpha = \beta \\ 0, & \text{if } \alpha \neq \beta \end{cases} \quad ; \qquad \mathbf{a}^3 = \mathbf{a}_3 \ . \tag{2.9}$$

From now on small greek letter indices take values 1 and 2 while small latin indices take values 1, 2 and 3. We use Einstein's summation convention for repeated indices at higher and lower positions.

To the covariant and contravariant bases we assign the first $(a_{\alpha\beta}, a^{\alpha\beta})$ and the second $(b_{\alpha\beta}, b_\alpha^\beta, b^{\alpha\beta})$ fundamental forms of the middle surface, respectively,

$$\begin{cases} a_{\alpha\beta} = \mathbf{a}_\alpha \cdot \mathbf{a}_\beta \ ; & a^{\alpha\beta} = \mathbf{a}^\alpha \cdot \mathbf{a}^\beta \ ; \quad a = \det(a_{\alpha\beta}) \\ b_{\alpha\beta} = \mathbf{a}_3 \cdot \mathbf{a}_{\alpha,\beta} \ ; & b_\alpha^\beta = a^{\beta\lambda} b_{\lambda\alpha} \ ; \qquad b^{\alpha\beta} = a^{\alpha\lambda} b_\lambda^\beta. \end{cases} \tag{2.10}$$

It is also convenient to introduce the covariant derivatives

$$u_{\alpha|\beta} = u_{\alpha,\beta} - \Gamma_{\alpha\beta}^\lambda u_\lambda \ ; \qquad u_{3|\lambda} = u_{3,\alpha} \ ; \qquad u_{3|\alpha\beta} = u_{3,\alpha\beta} - \Gamma_{\alpha\beta}^\lambda u_{3,\lambda} \ , \tag{2.11}$$

where the Christoffel symbols are given as

$$\Gamma_{\alpha\beta}^\lambda = \mathbf{a}^\lambda \cdot \mathbf{a}_{\alpha,\beta} = \Gamma_{\beta\alpha}^\lambda \ . \tag{2.12}$$

The same notation will be used for the surface S_t.

The family of surfaces S_t. By t we denote a real parameter which belongs to $[0, \delta]$, $\delta > 0$. Let S_t be a family of surfaces defined as the images of the reference plane domain $\mathcal{O}$ under regular mappings φ_t :

$$\begin{cases} \varphi_t : \ (\xi^1, \xi^2) \in \overline{\mathcal{O}} \subset \mathbb{R}^2 \mapsto \varphi_t(\xi^1, \xi^2) \in \overline{S_t} \subset \mathbb{R}^3 \\ \overline{S_t} = \varphi_t(\overline{\mathcal{O}}) \ . \end{cases} \tag{2.13}$$

For $t = 0$ we recover the original surface S, ie.,

$$\varphi_0 = \varphi, \qquad \text{and} \qquad S_0 = S.$$

It can be shown that

$$\frac{\partial \varphi_t}{\partial t} = c_t^\alpha \mathbf{a}_{t\alpha} + w_t \mathbf{a}_{t3}.$$

For the proof of this fact we refer the reader to (Kosinski, 1986), (formulae (1.13),(1.16) and (1.17)). For convected parametrizations, ie. when $c_t^\alpha \equiv 0$ we have

$$\frac{\partial \varphi_t}{\partial t} = w_t \mathbf{a}_{t3} \ , \tag{2.14}$$

where w_t is the normal speed of the surface S_t, and $\mathbf{a}_{t3}$ is the unit normal vector to S_t. In particular, $\mathbf{a}_{03} = \mathbf{a}_3$.

For any parametrization of the surface S_t we have

$$\frac{\partial \varphi_t}{\partial t} = c_t^\alpha \mathbf{a}_{t\alpha} + \frac{\delta \varphi_t}{\delta t},$$

where $\frac{\delta}{\delta t}$ denotes the displacement derivative defined below (Definition 2.4).

A mapping $T_t = T_t(V) : \mathbb{R}^3 \mapsto \mathbb{R}^3$ assigns the surface S_t to the reference surface S, ie.,

$$S_t = T_t(S)$$

where V is a given vector field.

Definition 2.4. Let $z_t \ : \mathcal{O} \mapsto \mathbb{R}$, $t \in [0, \delta)$ be a family of functions and let φ_t, $t \in [0,\delta)$, be a family of regular mappings, $\overline{S}_t = \varphi_t(\overline{\mathcal{O}})$. The *displacement derivative* $\frac{\delta z_t}{\delta t}$ for this family is

$$\frac{\delta z_t}{\delta t} = \frac{\partial z_t}{\partial t} - z_{t,\alpha} a^{\alpha\beta} \varphi_{t,\beta} \cdot \frac{\partial \varphi_t}{\partial t} \ .$$

Remark. 2.4. For

$$\varphi_t \equiv T_t \circ \varphi = T_t(V) \circ \varphi, \quad t \in [0,\delta)$$

we have

$$\frac{\partial \varphi_t}{\partial t} = \left[\frac{\partial T_t}{\partial t} \circ T_t^{-1} \right] \circ \varphi = V(t) \circ \varphi_t = V(t, \varphi_t) \ ,$$

therefore

$$\frac{\delta z_t}{\delta t} = \frac{\partial z_t}{\partial t} - z_{t,\alpha} a^{\alpha\beta} \varphi_{t,\beta} \cdot V(t,\varphi_t) \ .$$

It is easily seen that the displacement derivative of φ_t takes the form

$$\frac{\delta \varphi_t}{\delta t} = \langle V(t,\varphi_t), \mathbf{a}_{3t} \rangle_{\mathbb{R}^3} \mathbf{a}_{3t} = w_t \mathbf{a}_{3t} \ .$$

Now we determine the displacement derivative of the restriction of a function to S. Let $\psi : [0,\delta) \times D \mapsto \mathbb{R}$. Denoting $u_t \equiv \psi(t,\varphi_t)$, we get

$$\frac{\partial u_t}{\partial t}(\xi) = \frac{\partial \psi}{\partial t}(t, \varphi_t(\xi)) + \langle \nabla_x \psi(t, \varphi_t(\xi)), V(t, \varphi_t(\xi)) \rangle_{\mathbb{R}^3} \ ,$$
$$\frac{\delta u_t}{\delta t}(\xi) = \frac{\partial \psi}{\partial t}(t, \varphi_t(\xi)) + \langle \nabla_x \psi(t, \varphi_t(\xi)), \mathbf{a}_{t3} \rangle_{\mathbb{R}^3} w_t(\xi) \ .$$

If $z'(S;V)$ is the boundary shape derivative of $z(S_t) = \psi_{|S_t}$ in the direction of the vector field $V(t) = \frac{\partial T_t}{\partial t} \circ T_t^{-1}$, the following relation is obtained

$$\frac{\delta u_t}{\delta t}_{|t=0}(\xi) = z'(S;V)(\varphi(\xi)) \ .$$

For $t > 0$ we use the following notation,

$$\frac{\delta u_t}{\delta t}(\xi) = z'(S_t;V(t)) \circ \varphi_t \ .$$

It is clear that if $\frac{\partial \psi}{\partial t} = \frac{\partial \psi}{\partial n} \equiv 0$ on S and $u_t \equiv \psi(t, \varphi_t)$, then $\frac{\delta u_t}{\delta t} = 0$.

Remark. 2.5. Given mappings φ_t , φ, we can define the mapping

$$T_t : S \mapsto S_t$$

of the form $T_t = \varphi_t \circ \varphi^{-1}$, and the vector field

$$V(t,x) = \left[\frac{\partial T_t}{\partial t} \circ T_t^{-1}\right](x) = \left[\frac{\partial \varphi_t}{\partial t} \circ \varphi_t^{-1}\right](x) \quad x \in S_t \ .$$

The normal and tangent components of the field V are given by

$$\begin{aligned} &w(t,x) = \langle V(t,x), n(t,x)\rangle_{\mathbb{R}^3} \quad x \in S_t \ , \\ &V_\tau(t,x) = V(t,x) - w(t,x)n(t,x), \\ &w_t(\xi) = w(t, \varphi_t(\xi)), \\ &n(t,x) = \mathbf{a}_{t3}(\varphi_t^{-1}(x)), \quad x \in S_t \ . \end{aligned}$$

Therefore, without any loss of generality we assume that there is given a mapping φ and an admissible vector field V such that $\varphi_t \equiv T_t(V) \circ \varphi$.

A tangent vector on $\partial S_t = \{x \in \mathbb{R}^3 | x = \varphi(\xi), \quad \xi \in \partial \mathcal{O}\}$ is denoted by $\mathbf{b}_t(\xi)$; $\mathbf{l}_t(\xi)$ is the unit vector normal to ∂S_t such that $\langle \mathbf{l}_t(\xi), \mathbf{b}_t(\xi)\rangle_{\mathbb{R}^3} = 0$. For $t = 0$ we denote $\mathbf{b}(\xi) = \mathbf{b}_0(\xi), \mathbf{l}(\xi) = \mathbf{l}_0(\xi)$, respectively; $\mathbf{l}(t,x) \equiv \mathbf{l}_t(\varphi_t^{-1}(\xi))$ for $x = \varphi_t(\xi) \in \partial S_t$, $\mathbf{b}(t,x)$ is defined in the same way for $x \in \partial S_t$.

We refer to (Kosinski, 1986) for properties of second order displacement derivatives. Such derivatives are used, eg. in Section 8.5.

2.5 Derivatives of shape functionals

Suppose we are given an open set D in $\mathbb{R}^N$, a measurable subset Ω of D, an admissible vector field $V \in C(0,\varepsilon; C^k(\overline{D};\mathbb{R}^N))$, $k \geq 1$, and an associated transformation $T_t(V)$ from $\overline{D}$ onto $\overline{D}$.

Let $J(\Omega)$ be a well defined functional for any measurable subset Ω of D. Assume that $\Omega_t = T_t(V)(\Omega)$, $t \in [0,\delta)$, is a family of deformations of Ω. The set Ω_t is a measurable subset of D for any $t \in [0,\delta)$.

Definition 2.5. For an admissible vector field $V \in C(0,\varepsilon; C^k(\overline{D};\mathbb{R}^N))$, the Eulerian derivative of the domain functional $J(\Omega)$ at Ω in the direction of V is the limit

$$dJ(\Omega;V) = \lim_{t\downarrow 0}(J(\Omega_t) - J(\Omega))/t, \tag{2.15}$$

where

$$\Omega_t = T_t(V)(\Omega) \ .$$

Definition 2.6. A functional $J(\Omega)$ is *shape differentiable* (or simply differentiable) at Ω if

(i) there exists the Eulerian derivative $dJ(\Omega;V)$ for all directions V,
(ii) the mapping $V \mapsto dJ(\Omega;V)$ is a linear and continuous mapping from $C(0,\varepsilon;C^k(D;\mathbb{R}^N))$ into $\mathbb{R}$.

Gradients of shape differentiable functionals can be characterized as follows.

Theorem 2.1. *Let $J(\Omega)$ be a shape differentiable functional at every domain Ω of class C^k, $\Omega \subset D$. Assume that $\Omega \subset D$ is a domain with boundary of class C^{k-1}. There exists a scalar distribution*

$$g(\Gamma) \in \mathcal{D}^{-k}(\Gamma)$$

such that the gradient of the functional J at Ω, $G(\Omega) \in \mathcal{D}^{-k}(\Omega;\mathbb{R}^N)$, with $\mathrm{spt}G(\Omega) \in \Gamma$, *is given by*

$$G(\Omega) = {}^*\gamma_\Gamma(g\cdot n), \tag{2.16}$$

where $\gamma_\Gamma \in \mathcal{L}(\mathcal{D}(\overline{D};\mathbb{R}^N),\mathcal{D}(\Gamma;\mathbb{R}^N))$ is the trace operator and ${}^*\gamma_\Gamma$ is the transpose of γ_Γ, n is a unit normal vector on Γ directed into the exterior of Ω.

From (2.16) it follows that

$$dJ(\Omega;V) = \langle g, V\cdot n\rangle_{\mathcal{D}^{-k}(\Gamma)\times\mathcal{D}^k(\Gamma)},$$

where $V\cdot n = \langle V(0,x), n(x)\rangle_{\mathbb{R}^N}$, $x \in \Gamma$.

In general, $g = g(\Omega) \in \mathcal{D}^{-k}(\Gamma)$. However, for some classes of shape functionals it can be assumed that $g(\Omega)$ is an integrable function on Γ and then

$$dJ(\Omega;V) = \int_\Gamma g(x)\langle V(0,x), n(x)\rangle_{\mathbb{R}^N} d\Gamma \tag{2.17}$$

and we shall denote

$$\mathfrak{D}J(\Omega;V(0)) = \int_\Gamma g(x)\langle V(0,x), n(x)\rangle_{\mathbb{R}^N} d\Gamma \ .$$

Example. For the shape functional

$$J(\Omega) = \frac{1}{2}\int_\Omega (y(\Omega) - z_d)^2 dx, \quad z_d \in H_0^1(\Omega),$$

where $y(\Omega) \in H_0^1(\Omega)$, $-\Delta y(\Omega) = f(\Omega)$ in Ω, $f(\Omega) \in L^2(\Omega)$, it follows that

$$dJ(\Omega; V) = \int_\Omega f'(\Omega; V)p(\Omega)dx + \int_{\partial\Omega} \frac{\partial y}{\partial n}(\Omega)\frac{\partial p}{\partial n}(\Omega)\langle V(0), n\rangle d\Gamma$$

with $p(\Omega) \in H_0^1(\Omega)$, $\Delta p(\Omega) = y(\Omega) - z_d$ in Ω. For $f(\Omega) = f_{|\Omega}$ we have $f'(\Omega; V) = 0$ and

$$\mathfrak{D}J(\Omega; V(0)) = \int_{\partial\Omega} \frac{\partial y}{\partial n}(\Omega)\frac{\partial p}{\partial n}(\Omega)\langle V(0), n\rangle_{\mathbb{R}^N} d\Gamma \ .$$

We refer the reader to Section 6 for stability analysis of solutions to a class of shape optimization problems.

Let $\overline{S} = \varphi(\overline{\mathcal{O}})$, and let $J(S)$ be a given differentiable shape functional. For any vector field V such that $\mathrm{spt}V(0) \cap \partial S = \emptyset$, the Eulerian derivative $dJ(S; V)$ of the shape functional takes the following form.

Corollary 2.1. *There exists a distribution* $\mathcal{G}_S \in \mathcal{D}^{-k}(S)$ *such that*

$$dJ(S; V) = \langle \mathcal{G}_S, w\rangle_{\mathcal{D}^{-k}(S)\times\mathcal{D}^k(S)} \ , \tag{2.18}$$

where, as in Section 2.4, w is the normal speed of the surface S_t at $t = 0$, ie. $w(x) = w(\varphi(\xi)) = \langle V(0, \varphi(\xi)), \mathbf{a}_3(\xi)\rangle_{\mathbb{R}^3}$, for $x = \varphi(\xi) \in S$ and $\xi = \varphi^{-1}(x) \in \mathcal{O}$.

Remark. 2.6. If $\mathrm{spt}V(0) \cap \partial S \neq \emptyset$, then in the above formula for $dJ(S; V)$ an additional term related to the boundary ∂S may appear. The term takes the form

$$\langle \mathcal{G}_{\partial S}, V_\ell(0)\rangle_{\mathcal{D}^{-k}(\partial S;\mathbb{R}^3)\times\mathcal{D}^k(\partial S;\mathbb{R}^3)} \ , \tag{2.19}$$

where $V_\ell(0)$ is a component of the tangent vector field $\gamma_{\partial S}V_\tau(0)$ orthogonal to the tangent vector $\mathbf{b}$ on C^1 curve ∂S, ie. $V_\ell(0, \varphi(\xi)) = \langle V(0, \varphi(\xi)), \mathbf{l}(\xi)\rangle_{\mathbb{R}^3}\mathbf{l}(\xi)$ for $\xi \in \partial\mathcal{O}$. $\gamma_{\partial S}V_\tau(0)$ denotes the trace of $V_\tau(0) = V - \langle V, n\rangle_{\mathbb{R}^3} n$ on ∂S.

We recall the formulae for the derivatives of integrals. Given a family of shape differentiable functions $y(\Omega_t)$, $z(\Gamma_t)$, $\Omega_t = T_t(\Omega)$, then

$$\frac{d}{dt}\left[\int_{\Omega_t} y(\Omega_t)dx\right]_{|t=0} = \int_\Omega y'(\Omega; V)dx + \int_{\partial\Omega} y(\Omega)\langle V(0), n\rangle_{\mathbb{R}^N} d\Gamma, \tag{2.20}$$

$$\frac{d}{dt}\left[\int_{\Gamma_t} z(\Gamma_t)dx\right]_{|t=0} = \int_\Gamma z'(\Gamma; V) - 2\kappa z(\Gamma)\langle V(0), n\rangle_{\mathbb{R}^N} d\Gamma, \tag{2.21}$$

where $\kappa = -\frac{1}{2}\mathrm{div}_\Gamma n$ is the mean curvature on $\partial\Omega$.

Let us consider a surface integral $J(S) = \int_S G d\Gamma$. For $S = \varphi(\mathcal{O}) \subset \mathbb{R}^3$ it follows that $J(S) = \int_{\mathcal{O}} G \circ \varphi dS$, $dS = a^{\frac{1}{2}} d\xi$, where $a = \det [a_{\alpha\beta}]$.

Given vector field V and the transformation $T_t = T_t(V) : S \mapsto S_t$, we denote $\varphi_t = T_t \circ \varphi \, : \, \mathcal{O} \mapsto S_t$, $V(t,x) = \left[\frac{\partial T_t}{\partial t} \circ T_t^{-1}\right](x)$ for $x \in S_t$. Consider the shape functional defined on S_t,

$$J(S_t) = \int_{S_t} F(t,x) d\Gamma = \int_{S_t} z(S_t) d\Gamma = j(t) = \int_{\mathcal{O}} F_t dS_t \tag{2.22}$$

where $F(\cdot,\cdot) : \mathbb{R} \times \mathbb{R}^3 \mapsto \mathbb{R}$ is a given sufficiently smooth function and the following notation is used,

$$\begin{aligned} &z(S_t) = F(t,\cdot)_{|S_t} \, , \\ &F_t(\xi) = F(t, \varphi_t(\xi)), \quad \xi \in \mathcal{O}, \\ &dS_t = \det([DT_t] \circ \varphi) \| \left[{}^*DT_t^{-1} \circ \varphi\right] \cdot \mathbf{a}_{t3} \|_{\mathbb{R}^3} dS \, . \end{aligned}$$

We are going to evaluate the second order shape derivative of the shape functional $J(S_t)$ in the direction of vector fields V, W. For the first order shape derivative we have the following representations

$$\begin{aligned} dJ(S_t; V(t)) &= \int_{S_t} z'(S_t; V(t)) - 2\kappa(t) z(S_t) \langle V(t), n(t) \rangle_{\mathbb{R}^3} d\Gamma_t \qquad (2.23) \\ &= \int_{S_t} \frac{\partial F}{\partial t}(t) + \left(\frac{\partial F}{\partial n}(t) - 2\kappa(t) F(t)\right) \langle V(t), n(t) \rangle_{\mathbb{R}^3} d\Gamma_t \\ &= \frac{dj}{dt}(t) = \int_{\mathcal{O}} \left[\frac{\delta F_t}{\delta t} - 2\kappa_t F_t w_t\right] dS_t \end{aligned}$$

where we denote

$$\begin{aligned} &\kappa(t,x) \text{ is the mean curvature at } x \in S_t \, , \\ &n(t) = \mathbf{a}_{t3} \circ \varphi_t^{-1} \text{ the normal vector at } x = \varphi_t^{-1}(\xi) \in S_t \, , \xi \in \mathcal{O}, \\ &\frac{\partial F}{\partial n}(t) = \langle \nabla F(t), n(t) \rangle_{\mathbb{R}^3} \, , \\ &\kappa_t(\xi) = \kappa(t, \varphi_t(\xi)), \\ &w_t(\xi) = \left\langle \frac{\delta \varphi_t}{\delta t}(\xi), \mathbf{a}_{t3}(\xi) \right\rangle_{\mathbb{R}^3} \, , \; \xi \in \mathcal{O} \, . \end{aligned}$$

Remark. 2.7. In general, we have the following formula for derivative of a surface integral, we refer the reader to (Sokołowski, Zolesio, 1992) for the proof.

Let Σ be a connected surface contained in Γ with C^2 boundary $\partial\Sigma$, $\mathbf{l}$ denote the unit normal vector to $\partial\Sigma$ that is perpendicular to the surface normal n to Γ

and directed into the exterior of Σ. For $\Sigma_t = T_t(\Sigma) \subset \Gamma_t$ it follows that,

$$
\begin{aligned}
&\frac{d}{dt}\left[\int_{\Sigma_t} z(\Gamma_t)d\Gamma_t\right]_{|t=0} = \frac{d}{dt}\left[\int_{\Sigma} z(\Gamma_t)\circ T_t(V)\omega(t)d\Gamma\right]_{|t=0} \\
&= \int_{\Sigma} \dot{z}(\Gamma;V) + \dot{\omega}(\Gamma;V)z(\Gamma)d\Gamma \\
&= \int_{\Sigma} z'(\Gamma;V) + \nabla_\Gamma z(\Gamma)\cdot V_\tau(0) - 2\kappa z(\Gamma)V(0)\cdot n + z(\Gamma)\mathrm{div}_\Gamma V_\tau(0)d\Gamma \\
&= \int_{\Sigma} z'(\Gamma;V) - 2\kappa z(\Gamma)V(0)\cdot n d\Gamma + \int_{\Sigma} \mathrm{div}_\Gamma\left(z(\Gamma)V_\tau(0)\right)d\Gamma \\
&= \int_{\Sigma} z'(\Gamma;V) - 2\kappa z(\Gamma)V(0)\cdot n d\Gamma + \int_{\partial\Sigma} z(\Gamma)\langle V_\tau(0), \mathbf{l}\rangle_{\mathbb{R}^3} dl,
\end{aligned}
$$

where $\omega(t) = \gamma(t)\|{}^*DT_t^{-1}\cdot n(t)\|_{\mathbb{R}^3}$ and $\dot{\omega} = -2\kappa V(0)\cdot n + \mathrm{div}_\Gamma V_\tau(0)$ in the notation of (Sokołowski, Zolesio, 1992).

To differentiate the shape functional

$$
I(S) = \mathfrak{D}J(S;V(0))
$$

in the direction of a vector field $W(\cdot,\cdot)$ we need the following notation, $r\in(-\epsilon,\epsilon)$ is a parameter,

$$
\begin{aligned}
&S_r = T_r(S), \text{ where } T_r = T_r(W) : S \mapsto S_r\ , \\
&V(r) = V\left(r, \varphi_r(\xi)\right), \\
&\varphi_r = T_r(W)\circ\varphi, \\
&w_r(\xi) = \left\langle \frac{\delta\varphi_r}{\delta r}(\xi), \mathbf{a}_{r3}(\xi)\right\rangle_{\mathbb{R}^3}\ , \\
&F_r(\xi) = F(r, \varphi_r(\xi)), \\
&\kappa_r(\xi) = \kappa(r, \varphi_r(\xi)), \\
&\frac{\delta F_r}{\delta t}(\xi) = \frac{\partial F}{\partial t}(r, \varphi_r(\xi)) + \langle \nabla_x F(r, \varphi_r(\xi)), \mathbf{a}_{r3}(\xi)\rangle_{\mathbb{R}^3},\ \xi\in\mathcal{O}\ .
\end{aligned}
$$

With the notation we have

$$
\begin{aligned}
I(S_r) &= dJ(S_r;V(r)) \\
&= \int_{S_r} z'(S_r;V(r)) - 2\kappa(r)z(S_r)\langle V(r), n(r)\rangle_{\mathbb{R}^3} d\Gamma \\
&= \int_{S_r} \frac{\partial F}{\partial t}(r) + \left(\frac{\partial F}{\partial n}(r) - 2\kappa(r)F(r)\right)\langle V(r), n(r)\rangle_{\mathbb{R}^3} d\Gamma \\
&= \frac{dj}{dt}(r) = \int_{\mathcal{O}} \left[\frac{\delta F_r}{\delta t} - 2\kappa_r F_r w_r\right] dS_r
\end{aligned}
$$

For $u(S_r) = z'(S_r; V(r)) - 2\kappa(r)z(S_r)\langle V(r), n(r)\rangle_{\mathbb{R}^3}$ and $u_r = u(S_r) \circ \varphi_r = \frac{\delta F_r}{\delta t} - 2\kappa_r F_r w_r$ it follows that

$$\begin{aligned} \mathfrak{D}I(S; W(0)) &= \mathfrak{D}^2 J(S; V(0), W(0)) \qquad (2.26) \\ &= \int_S u'(S; W) - 2\kappa u(S)\langle W(0), n\rangle_{\mathbb{R}^3} d\Gamma \\ &= \left[\frac{d}{dr}\frac{dj}{dt}(r)\right]_{|r=0} = \int_{\mathcal{O}} \frac{\delta u_r}{\delta r}_{|r=0} - 2\left(\kappa_r u_r w_r\right)_{|r=0} dS \ . \end{aligned}$$

In Section 8.5 second order derivatives of shape functionals are evaluated for the fields $W = V = \frac{\partial \varphi_t}{\partial t} \circ \varphi_t^{-1}$.

Above, as well as in Sections 6, 8, second order derivatives of shape functionals are used.

Let us recall that the second order derivative $d^2 J(\Omega; V, W)$ of the shape functional $J(\Omega)$ in the direction of vector fields V, W is defined as follows

$$d^2 J(\Omega; V, W) = \lim_{s\downarrow 0} \frac{1}{s}[dJ(\Omega_s; V(s)) - dJ(\Omega; V)] \ ,$$

where $\Omega_s = T_s(W)(\Omega)$, $V(s) = V(s, , x)$ with $x = T_s(W)(X)$ for $X \in \Omega$.

It can be shown, that we have the following representation of the second order shape derivative, if the shape derivative exists,

$$\begin{aligned} d^2 J(\Omega; V, W) &= \mathfrak{d}^2 J(\Omega; V(0), W(0)) + \mathfrak{D}J(\Omega; \dot{V}(0)) \ , \qquad (2.27) \\ \mathfrak{D}J(\Omega; \dot{V}(0)) &= \mathfrak{D}J(\Omega; \partial_t V(0)) + \mathfrak{D}J(\Omega; [DV \cdot W](0)) \end{aligned}$$

where $\dot{V}(0)$ denotes the material derivative of the velocity field V in the direction of the field W,

$$\dot{V}(0, \cdot) = \frac{\partial V}{\partial t}(0, \cdot) + [DV \cdot W](0, \cdot) = \partial_t V(0) + [DV \cdot W](0) \ . \qquad (2.28)$$

The second order Eulerian derivative

$$\mathfrak{D}^2 J(\Omega; V(0), W(0)) \equiv \mathfrak{d}^2 J(\Omega; V(0), W(0)) + \mathfrak{D}J(\Omega; [DV \cdot W](0))$$

can be evaluated by taking the shape derivative of the shape functional $I(\Omega) = \mathfrak{D}J(\Omega; V(0))$ in the direction of a vector field W, we refer the reader to eg. (Delfour, Zolesio, 1992) for related results.

In Section 6.2 we use the derivative $\mathfrak{d}^2 J(\Omega; \cdot, \cdot)$ in stability analysis of the solutions to shape optimization problems. In Section 8.5 the Eulerian derivative $\mathfrak{D}^2 J(\Omega; \cdot, \cdot)$ is evaluated for a class of shape functionals for thin shells.

Example. Consider $J(\Omega) = \int_\Omega dx$, then

$$dJ(\Omega; V) = \int_\Omega \mathrm{div} V(0) dx = \int_{\partial\Omega} \langle V(0), n\rangle_{\mathbb{R}^N} d\Gamma \,. \tag{2.29}$$

Therefore

$$d^2 J(\Omega; V, W) = \int_\Omega \mathrm{div}[V(0)\mathrm{div} W(0) - DW(0)\cdot V(0) + \dot{V}(0)]dx \;, \tag{2.30}$$

hence

$$\begin{aligned}\mathfrak{d}^2 J(\Omega; V(0), W(0)) &= \int_\Omega \mathrm{div}[V(0)\mathrm{div} W(0) - DW(0)\cdot V(0)]dx \qquad (2.31)\\ &= \int_{\partial\Omega} [\langle V(0), n\rangle_{\mathbb{R}^N} \mathrm{div} W(0) - n\cdot DW(0)\cdot V(0)]d\Gamma\\ &= \int_{\partial\Omega} [\langle W(0), n\rangle_{\mathbb{R}^N} \mathrm{div} V(0) - n\cdot DV(0)\cdot W(0)]d\Gamma\\ &= \int_{\partial\Omega} [\kappa\langle V(0), n\rangle_{\mathbb{R}^N} \langle W(0), n\rangle_{\mathbb{R}^N} - \nabla_\Gamma(\langle V(0), n\rangle_{\mathbb{R}^N})\cdot W(0)\\ &\quad - \nabla_\Gamma(\langle W(0), n\rangle_{\mathbb{R}^N})\cdot V(0)]d\Gamma \,.\end{aligned}$$

Remark. 2.8. For $T_t = I + t\Theta$, the associated velocity field takes the form $V = \frac{\partial T_t}{\partial t}\circ T_t^{-1} = \Theta\circ(I+t\Theta)^{-1}$, whence the material derivative $\dot{V}(0) = \frac{d\Theta}{dt} = 0$ in any direction W.

Hence for vector fiels $V(t,x) = [\Theta\circ(I+t\Theta)^{-1}](x)$, $W(t.x) = [\Psi\circ(I+t\Psi)^{-1}](x)$, we have

$$\begin{aligned} &d^2 J(\Omega; V, W) = \mathfrak{d}^2 J(\Omega; V(0), W(0)) = \mathfrak{d}^2 J(\Omega; \Theta, \Psi), \qquad (2.32)\\ &\mathfrak{d}^2 J(\Omega; \Theta, \Psi) = \mathfrak{d}^2 J(\Omega; \Psi, \Theta) \quad \text{for all admissible vector fields } \Psi, \Theta\end{aligned}$$

and therefore the second order derivative $\mathfrak{d}^2 J(\Omega; V(0), W(0))$ is symmetric with respect to directions $V(0), W(0)$ whenever the second order shape derivative $d^2 J(\Omega; V, W)$ exists, see eg. (Novruzi, Roche, 1994) for applications to the Newton method.

Results on second order differentiability of shape functionals were derived by several authors, eg. by M. Delfour, K. Dems, N. Fujii, P. Guillaume, M. Masmoudi, Z. Mróz, H. Petryk, O. Pironneau, J. Simon and J.-P. Zolesio.

3 Unilateral problems in $H_0^2(\Omega)$

We examine the differentiability of the metric projection in the Sobolev space $H_0^2(\Omega)$ onto the shifted nonnegative cone.

We prove directional differentiability of the metric projection in the Sobolev space $H_0^2(\Omega)$ onto the convex set $K = \{f \in H_0^2(\Omega) \mid f \geq \psi\}$, where $\Omega \subset \mathbb{R}^N$ is an open, bounded domain. For any element $f \in K$ we give an explicit formula for the tangent cone $T_K(f)$. Using similar arguments we derive also explicit formulae for the tangent cones of the convex set

$$K = \{f \in H_0^m(\Omega) \mid f \geq \psi\}, \quad m = 2, 3, \ldots$$

where $\psi \in H^m(\Omega)$, $\psi < 0$ on $\partial\Omega$. Differentiability of the metric projection in the Sobolev space $H_0^1(\Omega)$ onto the nonnegative cone was investigated in (Mignot, 1976), where the so-called conical differential of the metric projection is given. However, the proof presented therein is based on potential theory in Dirichlet spaces, and cannot be directly applied to the Sobolev space $H_0^2(\Omega)$.

The importance of the concept of polyhedricity can be seen from Corollary 1.1, and formula (1.16). Namely, the polyhedricity of K implies directional differentiability of the metric projection onto K with an explicit formula for the differential. In Section 3.2 we give sufficient conditions for a set K to be polyhedric at a given point $f \in K$.

Applications of the results on differentiability of the metric projection are shown in Section 3.3. In Section 3.3.1 we give a formula for the shape derivative of solutions to an obstacle problem. In Section 3.3.2 we obtain necessary optimality conditions for an optimal design problem for the Kirchhoff plate. In Section 3.3.3 we examine problems of optimization of the obstacle shape. Finally, in Section 4 the form of the tangent cone at any element of the set defined by nonpenetration conditions in the case of model of simply supported Kirchhoff plate in the Sobolev space $H^2(\Omega) \cap H_0^1(\Omega)$ is derived.

We recall some properties of Sobolev spaces and the notion of capacity (Ziemer, 1989). The Sobolev spaces $H_0^1(\Omega)$ and $H_0^2(\Omega)$ are the completions of $C_0^\infty(\Omega)$ with the norms

$$\|\varphi\|^2_{H_0^1(\Omega)} = \int_\Omega |\nabla\varphi|^2 dx \ ,$$

$$\|\varphi\|^2_{H_0^2(\Omega)} = \int_\Omega |\Delta\varphi|^2 dx \ ,$$

respectively. By definition, if $\varphi \in H_0^2(\Omega)$, then $\frac{\partial\varphi}{\partial x_i} \in H_0^1(\Omega)$ for $i = 1, 2, \ldots, n$. Functions belonging to $H_0^1(\Omega)$ are defined quasi everywhere and are quasi continuous in the sense precised below.

The C_1-capacity of a compact set F is defined as

$$C_1(F) = \inf\{\int |\nabla\varphi|^2 dx \ : \ \varphi \geq 1 \text{ on } F, \ \ 0 \leq \varphi \in C_0^\infty(\Omega)\} \ .$$

Similarly, C_2-capacity is defined as

$$C_2(F) = \inf\{\int |\Delta\varphi|^2 dx \ : \ \varphi \geq 1 \text{ on } F, \ \ 0 \leq \varphi \in C_0^\infty(\Omega)\} \ .$$

The capacity of a Borel set is defined as the supremum over the capacities of its compact subsets. We say that a property holds C_i-q.e., $i = 1, 2$, if it holds except for a set of C_i-capacity zero, $i = 1, 2$. With this definition we have the following results. For further details we refer the reader to (Ziemer, 1989).

(i) Let $\varphi \in H_0^1(\Omega)$, and $\{\varphi_n\} \subset C_0^\infty(\Omega)$ converge to φ in $H_0^1(\Omega)$. Then a subsequence of $\{\varphi_n\}$ converge C_1-q.e. and this is a representative of φ.
(ii) Let $\varphi \in H_0^1(\Omega)$. Then φ has a quasicontinuous representative. There is a representative $\overline{\varphi}$ such that given $\varepsilon > 0$, there is an open set $U(\varepsilon)$ of C_1-capacity less than ε such that the restriction of $\overline{\varphi}$ to the complement of $U(\varepsilon)$ is continuous.
(iii) Any two quasicontinuous representatives of $\varphi \in H_0^1(\Omega)$ agree C_1-q.e.
(iv) Every set of positive Lebesgue measure has positive C_1-capacity.

The same holds for the Sobolev space $H_0^2(\Omega)$ with respect to C_2-capacity.

3.1 The tangent cone

We shall consider the metric projection onto the following convex set

$$K = \{f \in H_0^2(\Omega) \mid f(x) \geq \psi(x), x \in \Omega\} \tag{3.1}$$

with respect to the scalar product

$$(y, z) = \int_\Omega \Delta y(x) \Delta z(x) dx \ . \tag{3.2}$$

We assume that $\psi \in H^2(\Omega)$, $\psi(x) < 0$ on $\partial\Omega$. Therefore, the set K given by (3.1) is nonempty. The metric projection $z = P_K(y)$, $y \in H_0^2(\Omega)$, is a unique solution to the variational inequality

$$z \in K : \int_\Omega \Delta z(x) \Delta(\varphi - z)(x) dx \geq \int_\Omega \Delta y(x) \Delta(\varphi - z)(x) dx, \quad \forall \varphi \in K \ . \tag{3.3}$$

We denote

$$C_K(z) = \{\varphi \in H_0^2(\Omega) \mid \exists t > 0 \text{ such that } z + t\varphi \in K\} \ . \tag{3.4}$$

For any element z from the convex set K we derive an explicit formula for the tangent cone $T_K(z) = \mathrm{cl}(C_K(z))$, cl stands for the closure.

Theorem 3.1. *For any element $z \in K$, the tangent cone $T_K(z)$ takes the form*

$$T_K(z) = \{\varphi \in H_0^2(\Omega) \mid \varphi(x) \geq 0 \ , \ C_2 - q.e. \ on \ \Xi\} \tag{3.5}$$

where $\Xi = \{x \in \Omega \mid z(x) = \psi(x)\} \subset \Omega$.

Proof. Note that $C_K(z)$ and hence also $T_K(z)$ are convex cones containing all nonnegative elements of $H_0^2(\Omega)$.

Take any $V \in H_0^2(\Omega)$, $V \geq 0$ C_2-q.e. on Ξ. There exists a unique element $\phi_0 \in T_K(z)$ such that

$$\|V - \phi_0\|^2_{H_0^2(\Omega)} = \inf \{\|V - \phi\|^2_{H_0^2(\Omega)} \mid \phi \in C_K(z)\} . \tag{3.6}$$

It is easy to see that $\phi_0 + t\phi \in T_K(z)$ for any $H_0^2(\Omega) \ni \phi \geq 0$, $t \geq 0$. In a standard way (3.6) implies that

$$(V - \phi_0, \phi)_{H_0^2(\Omega)} \leq 0 , \ 0 \leq \phi \in H_0^2(\Omega) . \tag{3.7}$$

Hence, there exists a nonnegative Radon measure μ on Ω such that

$$(V - \phi_0, \phi)_{H_0^2(\Omega)} = - \int \phi d\mu , \ \phi \in C_0^\infty(\Omega) . \tag{3.8}$$

It follows that for $\phi \geq 0$

$$\int \phi d\mu = -(V - \phi_0, \phi)_{H_0^2(\Omega)} \leq \|V - \phi_0\|_{H_0^2(\Omega)} \|\phi\|_{H_0^2(\Omega)} .$$

So, by the definition of C_2-capacity, μ cannot charge sets of zero C_2-capacity. Since the measure may be large near the boundary it is not evident that (3.8) holds for all $\phi \in H_0^2(\Omega)$. We can clarify this by repeated use of a result by (Hedberg, 1981, Theorem 3.1).

We claim that (3.8) holds for any bounded and nonnegative $\phi \in H_0^2(\Omega)$ with the compact support. Indeed, for suitable ϱ_n , $\phi \star \varrho_n \in C_0^\infty(\Omega)$ has the compact support and tends boundedly pointwise C_2-q.e. in $H_0^2(\Omega)$ to ϕ. Since μ is a Radon measure, by the Lebesgue dominated convergence, we get the claim.

In the general case when $0 \leq \phi \in H_0^2(\Omega)$, by the above cited theorem of Hedberg, we can select $0 \leq w_k \leq 1$, $k = 1, 2, \ldots$ such that $w_k\phi$ has the compact support and approximates ϕ in the norm of L^∞. In particular, $w_k\phi$ tends to ϕ C_2-q.e. By (3.8),

$$\int w_k \phi d\mu = -(V - \phi_0, w_k\phi)_{H_0^2(\Omega)}$$

is bounded, so by the Fatou lemma, $\phi \in L^1(\mu)$. On the other hand, $w_k\phi \leq \phi$, so the dominated convergence applies and

$$- \int \phi d\mu = (V - \phi_0, \phi)_{H_0^2(\Omega)} , \ 0 \leq \phi \in H_0^2(\Omega) . \tag{3.9}$$

Let $\phi \in C_0^\infty(\Omega)$, $0 \leq \phi \leq 1$. Then $\phi(z - \psi) \in H_0^2(\Omega)$. We show that

$$\phi_0 + t\phi(z - \psi) \in T_K(z), \ \ -1 \leq t \leq 1 .$$

It is sufficient to show that $\varphi + t\phi(z-\psi) \in C_K(z)$ for any $\varphi \in C_K(z)$. Now $\varepsilon\varphi + z - \psi \geq$ in Ω for some $\varepsilon > 0$, hence for $s > 0$, $\frac{s}{1-s} < \varepsilon$ we have

$$s[\varphi + t\phi(z-\psi)] + z - \psi \geq 0, \text{ in } \Omega$$

since $(1 + st\phi)(z-\psi) \geq (1-s)(z-\psi)$.

Using this in (3.6) with ϕ replaced by $\phi_0 + t\phi(z-\psi)$ we obtain

$$(V - \phi_0, \phi(z-\psi))_{H_0^2(\Omega)} = 0$$

which, because $\phi(z-\psi)$ has compact support and belongs to $H_0^2(\Omega)$ means that

$$\int \phi(z-\psi)d\mu = 0 \ .$$

Hence

$$\mu(x \ : \ z > \psi) = 0 \ ,$$

ie. μ is concentrated on Ξ.

Our next step is to show that $\phi_0 = 0$ μ-a.e. To this end, using the fact that $T_K(z)$ is a cone and replacing ϕ with $t\phi_0$ in (3.6) we get

$$(V - \phi_0, \phi_0)_{H_0^2(\Omega)} = 0 \ . \tag{3.10}$$

Now we use Hedberg's result once more. Choose w_k , $0 \leq w_k \leq 1$ such that $w_k\phi_0$ has the compact support and converges to ϕ_0 in $H_0^2(\Omega)$. Since $\phi_0 \geq 0$ on Ξ and μ is concentrated on Ξ, $w_k\phi_0 \leq \phi_0$ μ-a.e. So using the same argument as above we get

$$0 = (V - \phi_0, \phi_0)_{H_0^2(\Omega)} = -\int \phi_0 d\mu$$

ie. $\phi_0 = 0$ μ-a.e.

Finally, since $\phi_0 = 0$ μ-a.e and $V \geq 0$ C_2-q.e. on Ξ we can repeat the above argument to get

$$(V - \phi_0, V - \phi_0)_{H_0^2(\Omega)} = -\int (V - \phi_0)d\mu = -\int V d\mu \ .$$

But the right-hand side is ≤ 0 because $V \geq 0$. Thus, $V = \phi_0$. □

Remark. 3.1. For $N = 1, 2, 3$ the proof of Theorem 3.1 simplifies since, by the Sobolev embedding theorem, $H_0^2(\Omega) \subset C(\overline{\Omega})$. It is clear that

$$T_K(u) \subset \{\varphi \in H_0^2(Q) \mid \varphi(x) \geq 0\,, \text{ on } \Xi\}\ .$$

Therefore, it is enough to show that any element $V \geq 0$ on Ξ actually belongs to $T_K(u)$. Since Ξ is compact, there exists $0 \leq \eta \in C_0^\infty(\Omega)$, $\eta \equiv 1$ on Ξ. By the Sobolev embedding theorem, $u, \psi, V \in C(\overline{\Omega})$. Therefore, for any $\varepsilon > 0$ there exists $t > 0$ such that

$$t(V + \varepsilon\eta) + u - \psi \geq 0, \quad \text{in } \Omega\ .$$

Thus,

$$V + \varepsilon\eta \in C_K(u), \quad \forall\varepsilon > 0$$

and

$$V + \varepsilon\eta \to V \quad \text{in } H_0^2(Q) \quad \text{strongly as } \varepsilon \downarrow 0\ .$$

Hence $V \in \mathrm{cl}(C_K(u)) = T_K(u)$.

3.2 Differentiability of metric projections

We now investigate the differentiability of the metric projection P_K in the Hilbert space $H = H_0^2(\Omega)$ onto the convex closed set $K \subset H$ given by (3.1). Let $T_K(f)$ be the tangent cone to K at $f \in K$. Recall that $T_K(f)$ is the closure in the space $H_0^2(\Omega)$ of the convex cone

$$C_K(f) = \{v \in H_0^2(\Omega) \mid \exists t > 0 \text{ such that } f(x) + tv(x) \geq \psi(x) \text{ in } \Omega\}\ . \tag{3.11}$$

For a given element $g \in H_0^2(\Omega)$, such that $f = P_K(g)$ we define the following convex cone in the space $H_0^2(\Omega)$

$$S = T_K(f) \cap [P_K(g) - g]^\perp = T_K(f) \cap [f - g]^\perp\ . \tag{3.12}$$

According to Definition 1.3, a set $K \subset H_0^2(\Omega)$ is polyhedric at $f \in K$, if

$$T_K(f) \cap [f - g]^\perp = \ \mathrm{cl}(C_K(f) \cap [f - g]^\perp) \tag{3.13}$$

for any $g \in H_0^2(\Omega)$ such that $f = P_K(g)$.

If, for given $(f, g) \in H_0^2(\Omega) \times H_0^2(\Omega)$, $f = P_K(g)$, condition (3.13) is satisfied, then for all $h \in H_0^2(\Omega)$ and for $t > 0$, t small enough

$$P_K(g + th) = P_K(g) + tP_S(h) + o(t)\ . \tag{3.14}$$

This means that the metric projection P_K is conically differentiable at $g \in H_0^2(\Omega)$. It turns out that condition (3.13) is satisfied if the support of the non-negative Radon measure defined by (3.15) is admissible in the following sense.

Definition 3.1. A compact set F is *admissible* if for any element $\varphi \in H_0^2(\Omega)$ the condition $\varphi = 0$ on F implies that $\varphi \in H_0^2(\Omega \setminus F)$.

We denote by $B(x,r)$, $x \in \mathbb{R}^N$, $r > 0$, the ball of radius r and centre x, $|A|$ denotes the Lebesgue measure of any set $A \subset \mathbb{R}^N$.

Proposition 3.1. *Let $F \subset \Omega$ be compact. If*

$$|F \cap B(x,r)| > 0 \quad ,$$

for C_1-quasi every $x \in F$ and for all $r > 0$, r small enough, then F is admissible.

Proof. By (Hedberg, 1981, Theorem 1.1), it is enough to show that if $\varphi \in H_0^2(\Omega)$ and $\varphi = 0$ C_2-q.e. on F, then $\nabla\varphi = 0$ C_1-q.e. on F. Now $\varphi \in H_0^1(\Omega)$ so by a standard result, $\nabla\varphi = 0$ a.e. on F. Since $\varphi \in H_0^2(\Omega)$, each component of $\nabla\varphi$ belongs to $H_0^1(\Omega)$ and hence has a finely continuous version (Ziemer, 1981). If, for $x \in F$, $|\nabla\varphi|(x) > 0$, then in a fine neighborhood of x the same inequality will obtain. Since finely open sets have positive measure, and since $\nabla\varphi = 0$ a.e. on F, this violates our condition on F. Thus, $\nabla\varphi = 0$ C_1-q.e. on F. □

Let $\nu \geq 0$ be a Radon measure defined as

$$-\int \varphi d\nu = \int_\Omega \Delta(g-f)\Delta\varphi dx, \ \forall\varphi \in C_0^\infty(\Omega) \tag{3.15}$$

and let $\mathcal{S}_K(f)$ be the following convex cone,

$$\mathcal{S}_K(f) = \{\varphi \in H_0^2(\Omega \setminus F) \mid \varphi \geq 0 \ \ C_2\text{-q.e. on } \Xi \setminus F\}, \tag{3.16}$$

where $F = \text{spt}\nu \subset \Xi$ is compact, $\text{spt}\nu$ denotes the support of the Radon measure ν.

Theorem 3.2. *For any $f \in H_0^2(\Omega)$,*

$$\mathcal{S}_K(f) \subset \ \text{cl}(C_K(f) \cap [f-g]^\perp) \ . \tag{3.17}$$

Proof. The proof is similar to that of Theorem 3.1. We repeat it here for the convenience of the reader.

The cone $C_K(f) \cap [f-g]^\perp$ and hence also $\text{cl}(C_K(f) \cap [f-g]^\perp)$ are convex cones containing all nonnegative elements of $H_0^2(\Omega \setminus F)$. Let $V \in \mathcal{S}_K(f)$. In view of (3.16), we have $V \in H_0^2(\Omega \setminus F)$ and $V \geq 0$ on Ξ.

Suppose that $V \notin \text{cl}(C_K(f) \cap [f-g]^\perp)$. There exists a unique element $\phi_0 \in \text{cl}(C_K(f) \cap [f-g]^\perp)$ such that

$$\|V - \phi_0\|^2_{H_0^2(\Omega\setminus F)} = \inf\{\|V-\phi\|^2_{H_0^2(\Omega\setminus F)} \mid \phi \in C_K(f) \cap [f-g]^\perp\} \ , \tag{3.18}$$

i.e. ϕ_0 is the metric projection of V onto $\text{cl}(C_K(f) \cap [f-g]^\perp)$.

It follows that

$$(V - \phi_0, \phi)_{H_0^2(\Omega\setminus F)} \le 0 \, , \; 0 \le \phi \in H_0^2(\Omega \setminus F) \, ,$$

and hence there exists a nonnegative Radon measure μ on $\Omega \setminus F$ such that

$$(V - \phi_0, \phi)_{H_0^2(\Omega\setminus F)} = -\int \phi d\mu \, , \; \phi \in C_0^\infty(\Omega \setminus F) \, . \tag{3.19}$$

It follows that for $\phi \ge 0$

$$\int \phi d\mu = -(V - \phi_0, \phi)_{H_0^2(\Omega\setminus F)} \le \|V - \phi_0\|_{H_0^2(\Omega\setminus F)} \|\phi\|_{H_0^2(\Omega\setminus F)} \, .$$

Hence, μ cannot charge sets of zero capacity. To show that (3.19) holds for all $\phi \in H_0^2(\Omega \setminus F)$, we first show that (3.19) holds for any nonnegative $\phi \in H_0^2(\Omega \setminus F)$ with the compact support. We can choose ϱ_n , such that $\phi \star \varrho_n \in C_0^\infty(\Omega \setminus F)$, has the compact support and tends boundedly pointwise C_2-q.e. in $H_0^2(\Omega \setminus F)$ to ϕ. Since μ is a Radon measure, by the Lebesgue dominated convergence,

$$\lim_{n\to\infty} \int \phi \star \varrho_n d\mu = \int \phi d\mu \, .$$

In the general case when $0 \le \phi \in H_0^2(\Omega\setminus F)$, we can select $0 \le w_k \le 1, \; k = 1, 2, \ldots$ such that each $w_k\phi$ has the compact support and in L^∞ approximates ϕ. By (3.19),

$$\int w_k \phi d\mu = -(V - \phi_0, w_k\phi)_{H_0^2(\Omega\setminus F)} \tag{3.20}$$

which yields the boundedness, so by Fatou Lemma, $\phi \in L^1(\mu)$. On the other hand, $w_k\phi \le \phi$ and, by the dominated convergence,

$$-\int \phi d\mu = (V - \phi_0, \phi)_{H_0^2(\Omega\setminus F)} \, , \; 0 \le \phi \in H_0^2(\Omega \setminus F) \, . \tag{3.21}$$

Now let $\varphi \in C_0^\infty(\Omega \setminus \Xi) \, , \; 0 \le \varphi \le 1, \; -1 \le t \le 1$. We have $\phi_0 + t\varphi \in \operatorname{cl}(C_K(f) \cap [f - g]^\perp)$. Using this in (3.18) with $\phi = \phi_0 + t\varphi$ we obtain

$$(V - \phi_0, \varphi)_{H_0^2(\Omega\setminus F)} = 0$$

which means

$$\int \varphi d\mu = 0$$

and hence

$$\mu(x \notin \Xi) = 0$$

i.e. μ is concentrated on Ξ.

We claim that $\phi_0 = 0$ μ-a.e. Using the fact that $\operatorname{cl}(C_K(f) \cap [f - g]^\perp)$ is a cone and replacing $t\phi_0$ with ϕ in (3.18) we get

$$(V - \phi_0, \phi_0)_{H_0^2(\Omega\setminus F)} = 0 \, . \tag{3.22}$$

By the same arguments as in the proof of Theorem 3.1,

$$0 = (V - \phi_0, \phi_0)_{H_0^2(\Omega \setminus F)} = -\int \phi_0 d\mu$$

i.e. $\phi_0 = 0$ μ-a.e. Finally, since $\phi_0 = 0$ μ-a.e and $V \geq 0$ on Ξ we get

$$(V - \phi_0, V - \phi_0)_{H_0^2(\Omega \setminus F)} = -\int (V - \phi_0) d\mu = -\int V d\mu \, .$$

But the right-hand side is ≤ 0 because $V \geq 0$, and thus it must be $V = \phi_0$. □

Proposition 3.2. *Let $f \in K \subset H_0^2(\Omega)$, where K is defined by (3.1) and $\Xi = \{x \in \Omega \mid f(x) = \psi(x)\}$.*

For any Radon measure $\nu \in H^{-2}(\Omega)$, $\nu \geq 0$, $\operatorname{spt}\nu \subset \Xi$, there exists $g \in H_0^2(\Omega)$, $f = P_K(g)$, such that

$$\int_\Omega \Delta(f-g)\Delta\varphi dx = \int \varphi d\nu, \quad \forall \varphi \in H_0^2(\Omega) \; . \tag{3.23}$$

Proof. Choose $g \in H_0^2(\Omega)$ such that

$$\int_\Omega \Delta g \Delta \varphi dx = \int_\Omega \Delta f \Delta \varphi dx - \int \varphi d\nu, \quad \forall \varphi \in H_0^2(\Omega) \; .$$

We claim that $f = P_K(g)$. To see this we first note that

$$\int \varphi d\nu \geq 0, \quad \forall \varphi \in T_K(f) \; . \tag{3.24}$$

Since $\eta - f \in T_K(f)$, $\forall \eta \in K$, we have

$$\int (\eta - f) d\nu \geq 0, \quad \forall \eta \in K \tag{3.25}$$

and hence

$$\int (\eta - f) d\nu = \int_\Omega \Delta(f-g)\Delta(\eta - f) dx \geq 0, \quad \forall \eta \in K \tag{3.26}$$

which shows that $f = P_K(g)$. □

Corollary 3.1. *If* $F = \text{spt}\nu$ *is admissible, then (3.13) and (3.14) hold, where the cone* S *is defined by (3.16).*

Therefore, condition (3.13) can be satisfied if $C_1(\Xi) = 0$. The set K is polyhedric at f if and only if the set $\Xi = \{x | f(x) = \psi(x)\}$ is of null C_1 capacity.

3.3 Applications to optimal design

We apply our differentiability results to optimization problems involving fourth order elliptic variational inequalities. We concentrate on variational inequalities for the Kirchhoff model of an elastic plate with an obstacle. In Section 3.3.1 we derive the shape derivative of solutions to obstacle problems. In Section 3.3.2 we formulate first order necessary optimality conditions for optimal design problems. Finally, in Section 3.3.3. an optimization problem related to an optimal choice of the obstacle is analysed.

3.3.1. Shape sensitivity analysis of obstacle problems. Let $\Omega \subset \mathbb{R}^2$ be a domain with smooth boundary $\Gamma = \partial\Omega$. Our aim is to derive the shape derivative $w'(\Omega; V)$ of the solution $w(\Omega)$ to an obstacle problem for the Kirchhoff plate in the direction of a vector field V. By $\{\Omega_t\} \subset \mathbb{R}^2, t \in [0, \delta)$, we denote a family of domains such that $\Omega_t = T_t(V)(\Omega)$.

Let $\psi \in H^3(\mathbb{R}^2)$ and $\psi(x) \leq 0$ in an open neighborhood in $\mathbb{R}^2$ of $\partial\Omega \subset \mathbb{R}^2$. For $t > 0, t$ small enough the set

$$K(\Omega_t) = \{\varphi \in H_0^2(\Omega_t) \mid \varphi(x) \geq \psi(x), \text{ in } \Omega_t\} \tag{3.27}$$

is a nonempty closed convex subset of the Sobolev space $H_0^2(\Omega_t)$. The restriction of $\psi \in H^3(\mathbb{R}^2)$ to the domain $\Omega_t, t \in [0, \delta)$ is again denoted by ψ.

Let $f \in H^2(\mathbb{R}^2)$. By $w_t \in H^2(\Omega)$, $t \in [0, \delta)$, we denote a unique solution to the variational inequality

$$\begin{aligned} &w_t \in K(\Omega_t) \ : \\ &\int_{\Omega_t} \Delta w_t \Delta(\varphi - w_t) dx \geq \int_{\Omega_t} f(\varphi - w_t) dx, \quad \forall \varphi \in K(\Omega_t) \ . \end{aligned} \tag{3.28}$$

For $t = 0$, we put $w(x) = w_0(x)$, $x \in \Omega$. Let $\nu \geq 0$ be a Radon measure,

$$\int \varphi d\nu = \int_\Omega \{\Delta w \Delta\varphi - \varphi f\} dx \ .$$

An extension $\widetilde{w_t}(x)$, $x \in \mathbb{R}^2$, of the element $w_t(x)$, $x \in \Omega$, $t \in [0, \delta)$ is defined as

$$\widetilde{w_t}(x) = \begin{cases} w_t(x), & x \in \Omega_t, \ t \in [0, \delta) \\ 0, & x \in \mathbb{R}^2 \setminus \Omega_t, \ t \in [0, \delta). \end{cases} \tag{3.29}$$

Using the material derivative method combined with Theorem 3.1 we shall prove that $\widetilde{w_t}_{|\Omega} \in H^2(\Omega)$ is right-differentiable with respect to t, at $t = 0$. Put $V(0) = V(0, \cdot)$.

Theorem 3.3. *Assume that* sptν *is admissible in the sense of Definition 3.1 and* $\nabla w \cdot V(0) \in H^2(\Omega)$. *Then for* $t > 0, t$ *small enough*

$$\widetilde{w_{t|\Omega}} = w + tw' + o(t), \quad in\ H^2(\Omega) \tag{3.30}$$

where $\|o(t)\|_{H^2(\Omega)}/t \to 0$ *as* $t \downarrow 0$.
The shape derivative $w' \in H^2(\Omega)$ *uniquely solves the variational inequality*

$$\begin{aligned} w' \in S_v =&\{\varphi \in H^2(\Omega) \cap H_0^1(\Omega) \mid \frac{\partial \varphi}{\partial n} = -v\frac{\partial^2 w}{\partial n^2} \quad on\ \partial\Omega, \\ &\varphi(x) \geq 0\ on\ \Xi^+, \quad \varphi(x) = 0\ on\ \Xi^0\} \end{aligned} \tag{3.31}$$

$$\int_\Omega \Delta w' \Delta(\varphi - w')dx \geq 0, \quad \forall \varphi \in S_v(\Omega)\ , \tag{3.32}$$

where $v(x) = \langle V(0,x), n(x)\rangle_{\mathbb{R}^2}$, $x \in \partial\Omega$; $n(x)$, $x \in \partial\Omega$ *is the unit outward normal vector on* $\partial\Omega$ *and* $\Xi^0 =$ sptν,

$$\Xi^+ = \Xi \setminus \Xi^0 \tag{3.33}$$

$$\Xi = \{x \in \Omega \mid w(x) = \psi(x)\}\ . \tag{3.34}$$

Proof. Recall that the shape derivative $w' = w'(\Omega; V)$ of the solution $w = w(\Omega)$ to variational inequality

$$w \in K(\Omega): \quad \int_\Omega \Delta w \Delta(\varphi - w)dx \geq \int_\Omega f(\varphi - w)dx, \quad \forall \varphi \in K(\Omega) \tag{3.35}$$

in the direction of a vector field V is given as

$$w' = \dot{w} - \nabla w \cdot V(0) \tag{3.36}$$

where

$$\dot{w} = \lim_{t\downarrow 0}(w_t \circ T_t - w)/t \tag{3.37}$$

and $w_t \in H_0^2(\Omega)$ is a unique solution to variational inequality (3.28).

We first characterize the material derivative $\dot{w} = \dot{w}(\Omega; V)$. To this end we transport variational inequality (3.28) to the fixed domain Ω using the mapping $T_t : X \mapsto x = x(t)$ defined by (2.3). The element

$$w^t = w \circ T_t \in H_0^2(\Omega), \ t \in [0, \delta) \tag{3.38}$$

satisfies

$$\begin{aligned} &w^t \in K^t = \{\varphi \in H_0^2(\Omega) \mid \varphi \geq \psi^t \text{ in } \Omega\} \\ &\mathcal{Q}^t(w^t, \varphi - w^t) \geq \int_\Omega f^t(\varphi - w^t)dx, \quad \forall \varphi \in K^t\ , \end{aligned} \tag{3.39}$$

where

$$\psi^t = \psi \circ T_t, \quad f^t = \gamma_t f \circ T_t, \quad \gamma_t = \det(DT_t),$$

DT_t is the Jacobian of the mapping $T_t : \mathbb{R}^2 \mapsto \mathbb{R}^2$,

$$\mathcal{Q}^t(z,\varphi) = \int_\Omega \gamma_t^{-1} \text{div}(A_t \cdot \nabla z)\text{div}(A_t \cdot \nabla \varphi)dx, \quad \forall z,\varphi \in H_0^2(\Omega) \ , \tag{3.40}$$

$$\text{with } A_t = \gamma_t DT_t^{-1} \cdot (DT_t^{-1})^\star \ . \tag{3.41}$$

For

$$z^t = w^t - \psi^t, \quad \xi^t(\varphi) = \text{div}(A_t \cdot \nabla\varphi) \tag{3.42}$$

we get

$$\begin{aligned} &z^t \in K_0 = \{\varphi \in H_0^2(\Omega) \mid \varphi \geq 0 \text{ in } \Omega\} \\ &\mathcal{Q}^t(z^t, \varphi - z^t) \geq \int_\Omega f^t(\varphi - z^t)dx - \mathcal{Q}^t(\psi^t, \varphi - z^t), \ \forall\varphi \in K_0 \ . \end{aligned} \tag{3.43}$$

By Corollary 3.1 and Theorem 1.2, for $t > 0$, t small enough

$$z^t = z^0 + t\dot{z} + o(t), \quad \text{in } H_0^2(\Omega) \ , \tag{3.44}$$

where $\dot{z} \in H_0^2(\Omega)$ solves the variational inequality

$$\begin{aligned} &\dot{z} \in S(\Omega) = \{\varphi \in H_0^2(\Omega) \mid \varphi \geq 0 \text{ on } \Xi^+, \ \varphi = 0 \text{ on } \Xi^0\} \ : \\ &\int_\Omega \Delta\dot{z}\Delta(\varphi - \dot{z})dx \geq \int_\Omega \dot{f}(\varphi - \dot{z})dx - \mathcal{Q}'(w, \varphi - \dot{z}) - \mathcal{Q}^0(\dot{\psi}, \varphi - \dot{z}), \ \forall\varphi \in S(\Omega) \ . \end{aligned} \tag{3.45}$$

Since

$$\dot{z} = \dot{w} - \dot{\psi} = \dot{w} - \nabla\psi \cdot V \tag{3.46}$$

we obtain

$$\begin{aligned} &\dot{w} \in S(\Omega) + \nabla\psi.V \ : \\ &\int_\Omega \Delta\dot{w}\Delta(\varphi - \dot{w})dx \geq \int_\Omega \dot{f}(\varphi - \dot{w})dx - \mathcal{Q}'(w, \varphi - \dot{w}) \ , \\ &\qquad\qquad \forall\varphi \in S(\Omega) + \nabla\psi \cdot V \end{aligned} \tag{3.47}$$

where

$$\begin{aligned} &\dot{\psi} = \nabla\psi \cdot V \in H^2(\Omega), \quad \dot{f} = \text{div}(fV), \\ &\mathcal{Q}'(z,\varphi) = \int_\Omega \{-\text{div}V \Delta z \Delta\varphi + \dot{\xi}(z)\Delta\varphi + \Delta z\dot{\xi}(\varphi)\}dx, \quad \forall z,\varphi \in H^2(\Omega), \end{aligned}$$

and $\dot{\xi}(\varphi) = \text{div}(A' \cdot \nabla\varphi)$, $A' = \text{div}VI - DV - (DV)^\star$.

The shape derivative depends actually on the normal component $v = \langle V(0), n\rangle_{\mathbb{R}^2}$ of the vector field $V(0,\cdot)$ on $\Gamma = \partial\Omega$. Hence,

$$\dot{w} = \nabla w \cdot V(0)$$

for any vector field V such that $v(x) = 0$ on $\partial\Omega$ and from (3.47) we obtain the following Green formula

$$0 = -\int_\Omega \Delta(\nabla w \cdot V(0))\Delta\varphi dx + \int_\Omega \dot{f}\varphi dx - \mathcal{Q}'(w,\varphi), \quad \forall\varphi \in \{S(\Omega) - S(\Omega)\} \quad (3.48)$$

which holds for any vector field V such that $v(x) = 0$ on $\partial\Omega$. For any vector field V and a test function φ smooth enough, integrating by parts (3.47), and using (3.36),(3.48), we get

$$\int_\Omega \Delta w' \Delta(\varphi - w')dx \geq 0 \ . \quad (3.49)$$

Furthermore,

$$w' \in \{\zeta \mid \zeta = \varphi - \nabla w \cdot V(0), \quad \varphi \in S(\Omega)\} \equiv S_v(\Omega)$$

for we can select $V(0,\cdot)$ with the support in a small open neighborhood of $\partial\Omega$. This completes the proof of Theorem 3.3. □

3.3.2. Optimal design problem. Now we formulate necessary optimality conditions for an optimal design problem related to the Kirchhoff plate with an obstacle. Let

$$a(h;\cdot,\cdot):\ H_0^2(\Omega) \times H_0^2(\Omega) \mapsto \mathbb{R} \quad (3.50)$$

be the bilinear form associated with the Kirchhoff plate,

$$a(h;y,\varphi) = \int_\Omega h^3(x) b_{ijkl} \frac{\partial^2 y}{\partial x_i \partial x_j}(x) \frac{\partial^2 \varphi}{\partial x_k \partial x_l}(x) dx, \quad \forall y,\varphi \in H_0^2(\Omega) \ . \quad (3.51)$$

We use the summation convention over repeated indices $i,j,k,l = 1,2$. Assume that

$$h \in U_{ad} = \{h \in L^\infty(\Omega) \mid 0 < h_{min} \leq h(x) \leq h_{max}, \quad \text{for a.e. } x \in \Omega\} \quad (3.52)$$

and

$$b_{ijkl} = b_{jikl} = b_{klij}, \quad i,j,k,l = 1,2 \quad (3.53)$$

$$b_{ijkl}\xi_{ij}\xi_{kl} \geq c\xi_{ij}\xi_{ij}, \quad c > 0, \quad (3.54)$$

for all symmetric matrices $[\xi_{ij}]_{2\times 2}$.

We consider boundary value problems with homogeneous boundary conditions. Analogous results for nonhomogeneous boudary conditions can also be derived.

By asumptions (3.52)–(3.54), bilinear form (3.51) is continuous, symmetric, and $H_0^2(\Omega)$-elliptic, i.e.,

$$a(h;y,y) \geq \alpha \|y\|^2_{H^2(\Omega)}, \quad \alpha > 0, \quad \forall y \in H_0^2(\Omega)\ . \tag{3.55}$$

Denote

$$K = \{\varphi \in H_0^2(\Omega) \mid \varphi(x) \geq \psi(x) \text{ in } \Omega\}\ , \tag{3.56}$$

where $\psi \in H^2(\Omega) \subset C(\bar{\Omega})$ is such that the set (3.56) is nonempty.

For any $h \in U_{ad}$, by $w = w(h;x),\ x \in \Omega$, we denote a unique solution to the variational inequality

$$w \in K\ :\quad a(h;w,\varphi - w) \geq \int_\Omega f(\varphi - w)dx, \quad \forall \varphi \in K\ , \tag{3.57}$$

where $f \in H^{-2}(\Omega)$ is a given element, $H^{-2}(\Omega)$ is the dual of $H_0^2(\Omega)$.

Now, using Corollary 3.1 combined with Theorem 1.2, we give a formula for the right-differential of the nonlinear mapping

$$L^\infty(\Omega) \ni h \mapsto w(h;.) \in H_0^2(\Omega) \tag{3.58}$$

at a given $h \in U_{ad}$.

Lemma 3.1. *Assume that μ is a Radon measure such that*

$$\int \varphi d\mu = a(h;w,\varphi) - \int_\Omega f\varphi dx, \quad \forall \varphi \in H_0^2(\Omega)$$

and the support of μ is admissible in the sense of Definition 3.1.

Then for $\varepsilon > 0, \varepsilon$ small enough

$$\forall v \in L^\infty(\Omega):\quad w(h+\varepsilon v) = w(h) + \varepsilon q(v) + o(\varepsilon)\ , \tag{3.59}$$

where $\|o(\varepsilon)\|_{H_0^2(\Omega)}/\varepsilon \to /0$ as $\varepsilon \downarrow 0$, and $q = q(v) \in H_0^2(\Omega), v \in L^\infty(\Omega)$, uniquely solves the variational inequality

$$q \in S\ :\quad a(h;q,\varphi - q) + a'_v(h;w(h),\varphi - q) \geq 0,\ \forall \varphi \in S\ , \tag{3.60}$$

where

$$a'_v(h;y,\varphi) = \int_\Omega 3h^2(x)v(x)b_{ijkl}\frac{\partial^2 y}{\partial x_i \partial x_j}(x)\frac{\partial^2 \varphi}{\partial x_k \partial x_l}(x)dx, \quad \forall y,\varphi \in H_0^2(\Omega), \tag{3.61}$$

$$S = \{\varphi \in H_0^2(\Omega) \mid \varphi = 0 \textit{ on } \operatorname{spt}\mu,\ \varphi \geq 0 \textit{ on } \Xi \setminus \operatorname{spt}\mu\}, \tag{3.62}$$

$$\Xi = \{x \in \Omega \mid w(h;x) = \psi(x)\} \quad \textit{is compact.} \tag{3.63}$$

Now we consider an optimal design problem for the Kirchhoff plate in the following form.

Problem (P)**:** Find an element $h \in U_{ad}$ which minimizes the functional

$$J(h) = \max_{x \in \Omega} |w(h;x)|$$

over the set U_{ad}.

Assuming that there exists an optimal solution $h \in U_{ad}$ to problem (P) we derive necessary optimality conditions. In general, however, the existence of optimal solutions is not guaranteed. This leads to the notion of generalized solutions to problem (P) (Myslinski, Sokołowski, 1985). For generalized solutions necessary optimality conditions of the same type can be given.

Theorem 3.4. *For any optimal solution $h \in U_{ad}$ to problem (P)*

$$\max_{x \in \Omega^\star(h^\star)} \; sign\,\{w(h^\star;x)\} q(v - h^\star;x) \geq 0, \quad \forall v \in U_{ad}\ , \tag{3.64}$$

where

$$\Omega^\star(h) = \{x \in \Omega \mid J(h) = |w(h;x)|\}, \quad \forall h \in U_{ad}\ .$$

The proof of Theorem 3.4 follows directly from Theorem 3.1 combined with Lemma 3.1 and therefore we omit it here.

Remark. 3.2. An optimal design problem for the Kirchhoff plate with a finite number of pointwise obstacles was investigated in (Bendsøe, Sokolowski, 1988).

3.3.3. Shape optimization of obstacles. In this section we assume that the thickness h of the plate is fixed. Let $\Psi_{ad} \subset H^2(\Omega)$ be a closed and convex set such that there exist elements $a \in H^{\frac{3}{2}}(\Gamma)$, $a(x) \leq a_0 < 0$ for $x \in \Gamma$, $b \in H^{\frac{1}{2}}(\Gamma)$.

$$\psi_{|\Gamma} = a, \quad \frac{\partial \psi}{\partial n}_{|\Gamma} = b \quad \forall \psi \in \Psi_{ad}\ .$$

Denote

$$\mathcal{K}_\psi = \{\varphi \in H_0^2(\Omega) | \varphi \geq \psi \quad \text{in } \Omega\},$$

where $w = w_\psi$, $\psi \in \Psi_{ad}$ solves the variational inequality

$$w \in \mathcal{K}_\psi \ : \quad a(h; w, \varphi - w) \geq \int_\Omega f(\varphi - w)dx \quad \forall \varphi \in \mathcal{K}_\psi\ . \tag{3.65}$$

Consider the nonsmooth shape optimization problem.

Problem (P): find an element $\psi^* \in \Psi_{ad}$ which minimizes the functional

$$\mathcal{J}(\psi) = \max_{x \in \Omega} |w_\psi(h;x)|$$

over the set Ψ_{ad}.

Theorem 3.5. *There exists an optimal solution $\psi^* \in \Psi_{ad}$ to the above problem.*

Assume that a Radon measure ν given as

$$\int \varphi d\nu = a(h; w_{\psi^*}, \varphi) - \int_\Omega f\varphi dx \quad \forall \varphi \in H_0^2(\Omega)$$

satisfies the condition that for any $\varphi \in H_0^2(\Omega)$, $\varphi = 0$ on $F = \text{spt}\nu$, we have $\varphi \in H_0^2(\Omega \setminus F)$.

Then an optimal solution $\psi^* \in \Psi_{ad}$ satisfies the first order optimality condition

$$\max_{x \in \Omega^*(\psi^*)} \text{sign}\,\{w_{\psi^*}(h;x)\} p_{\psi-\psi^*}(h;x) \geq 0, \quad \forall \psi \in \Psi_{ad}\ , \tag{3.66}$$

where

$$\Omega^*(\psi) = \{x \in \Omega | \mathcal{J}(\psi) = w_\psi(h;x)\} \quad \forall \psi \in \Psi_{ad}$$

and $p^* = p_{\psi-\psi^*}$, $\psi \in \Psi_{ad}$, solves uniquely the variational inequality

$$p^* \in \mathcal{S}_{\psi-\psi^*} : \quad a(h; p^*, \varphi - p^*) \geq 0 \quad \forall \varphi \in \mathcal{S}_{\psi-\psi^*}\ .$$

The convex cone $\mathcal{S}_{\psi-\psi^*}$ takes the form

$$\mathcal{S}_{\psi-\psi^*} = \{\varphi \in H_0^2(\Omega) | \varphi = \psi - \psi^* \quad \text{on spt}\nu \quad \varphi \geq \psi - \psi^* \quad \text{on } \Xi \setminus \text{spt}\nu\}\ ,$$

where

$$\Xi = \{x \in \Omega | w_{\psi^*}(h;x) = \psi^*(x)\}\ .$$

The above theorem can be proved in the following way. Let $\chi \in H^2(\Omega)$ be an element such that

$$\chi_{|\Gamma} = a, \quad \frac{\partial \chi}{\partial n}_{|\Gamma} = b\ .$$

Then $z = w_\psi + \chi - \psi \in H_0^2(\Omega)$ is a unique solution to the variational inequality

$$\chi \leq z \in H_0^2(\Omega) : \quad a(h; z, \varphi - z) \geq \int_\Omega f(\varphi - z)dx - a(h; \chi + \psi, \varphi - z)$$
$$\chi \leq \varphi \in H_0^2(\Omega)\ .$$

Under our assumptions the affine mapping

$$H^2(\Omega) \ni \psi \mapsto z(\psi) \in H_0^2(\Omega)$$

is conically differentiable, which yields the first order optimality conditions for the optimization problem under consideration. Here we make use of optimality conditions for composite cost functionals with max type functions.

4 Unilateral problems in $H^2(\Omega) \cap H_0^1(\Omega)$

Directional differentiability of solutions with respect to design in obstacle problems for the simply supported Kirchhoff plate can be investigated in the same framework as in Section 3 for the clamped plate. We derive here an explicit formula for the tangent cone at any element of the convex set defined by nonpenetration condition. The method of the proof differs from that used in the proof of Theorem 3.1. Directional differentiability of solutions to obstacle problems with respect to the design can be established in the same way as before.

4.1 Obstacle problem for the simply supported Kirchhoff plate

Consider the space $H = H^2(\Omega) \cap H_0^1(\Omega)$ equipped with the scalar product

$$(y, z)_H = \int_\Omega \Delta y \Delta z dx, \quad \forall y, z \in H . \tag{4.1}$$

Let K be a convex set,

$$K = \{\varphi \in H \mid \varphi \geq \psi \quad \text{in } \Omega\} . \tag{4.2}$$

For $u \in K$, denote $\Xi = \{x \in \Omega \mid u(x) = \psi(x)\}$.

In this section "q.e." means C_2-q.e. with respect to the capacity defined by formulae (4.6) and (4.7).

Theorem 4.1. *Assume that $\psi \in H$ and Ξ is compact. Then*

$$T_K(u) = \{\varphi \in H \mid \varphi \geq 0 \ \ q.\ e.\ on\ \Xi\} . \tag{4.3}$$

Before we proceed let us establish the framework. It is not difficult to see that

$$H = \{\mathcal{G}f \mid f \in L^2(\Omega)\} , \tag{4.4}$$

where $\mathcal{G}$ is the Green function of Ω, i.e. $\mathcal{G} = (-\Delta)^{-1}$ with homogeneous Dirichlet boundary conditions. We introduce the inner product in H by the formula

$$(\mathcal{G}f, \mathcal{G}g) = \int_\Omega f(x)g(x)dx . \tag{4.5}$$

Note that the corresponding topology is that inherited from $H^2(\Omega)$. For the purpose of this section the C_2-capacity $C_2(F)$ of a compact set $F \subset \Omega$ is defined as

$$C_2(F) = \inf\{\|f\|^2_{L^2(\Omega)} \mid \mathcal{G}f \geq 1 \text{ on } F\} . \tag{4.6}$$

For any Borel set B,

$$C_2(B) = \sup\{C_2(F) \mid \text{ compact } F \subset B\} . \tag{4.7}$$

Proof of Theorem 4.1. We start with an observation that $C_K(u)$ (and its closure) has the following properties.

(i) $C_K(u)$ contains all nonnegative elements of H.

(ii) If $\varphi_i \in C_K(u), a_i \geq 0$, then $\sum_i a_i\varphi_i \in C_K(u)$.

(iii) If $\varphi \in C_K(u), 0 \leq \xi \in C_0^\infty(\Omega)$, then $\xi\varphi \in C_K(u)$.

(iv) If $\varphi = 0$ in a neighborhood of Ξ, then $\varphi \in C_K(u)$.

These properties are direct consequences of the definition of the tangent cone. Since convergence in the space $H = H^2(\Omega) \cap H_0^1(\Omega)$ implies q.e. convergence, it is clear that

$$T_K(u) \subset \{\varphi \in H \mid \varphi \geq 0 \text{ q. e. on } \Xi\} .$$

Let $\phi \in H^2(\Omega) \cap H_0^1(\Omega)$, $\phi \geq 0$ q.e. on $\Xi = \{u = \psi\}$. We show that $\phi \in T_K(u)$. To see this, consider a unique element ϕ_0 of $T_K(u)$ such that

$$\|\phi - \phi_0\|_H = \inf_{\varphi \in T_K(u)} \|\phi - \varphi\|_H . \tag{4.8}$$

This implies that

$$(\phi_0 - \phi, \varphi)_H \geq 0, \quad \forall \varphi \in T_K(u) . \tag{4.9}$$

Introduce a linear map

$$L\varphi = (\phi_0 - \phi, \varphi)_H, \quad \forall \varphi \in H^2(\Omega) \cap H_0^1(\Omega) . \tag{4.10}$$

Choose $f_0 \in L^2(\Omega)$ such that

$$\phi_0 - \phi = \mathcal{G} f_0 . \tag{4.11}$$

If $g \geq 0$, then $\varphi = \mathcal{G} g \geq 0$ and φ belongs to $T_K(u)$. By (4.9), $\int f_0 g \geq 0$ which means that $f_0 \geq 0$ a.e. If $0 \leq \varphi \in C_0^\infty(\Omega)$, then again, by (4.9),

$$\int f_0 \Delta\varphi \leq 0, \quad 0 \leq \varphi \in C_0^\infty(\Omega)$$

i.e. f_0 is superharmonic. By the Riesz decomposition,

$$f_0 = \mathcal{G}\mu + h_0 \tag{4.12}$$

where μ is a positive Radon measure and h_0 is nonnegative harmonic in Ω. For clarity we break up the proof into small steps.

Step 1:

For all $\varphi \in H = H^2(\Omega) \cap H_0^1(\Omega)$,

$$\int |\varphi| d\mu \leq \|L\| \|\varphi\|_H . \tag{4.13}$$

Indeed, let $0 \leq f \in L^2(\Omega)$. There is a sequence of nonnegative elements of $C_0^\infty(\Omega)$ pointwise increasing to $\mathcal{G} f$,

$$\mathcal{G} f = \lim_{k \to \infty} \varphi_k, \quad 0 \leq \varphi_k \in C_0^\infty(\Omega) .$$

From (4.12) and (4.9) it follows that

$$L(\mathcal{G}f) \geq L(\varphi_k) = \int \varphi_k d\mu \ .$$

By monotone convergence we get

$$\int (\mathcal{G}f) d\mu = \lim_{k\to\infty} \int \varphi_k d\mu \leq L(\mathcal{G}f) \ .$$

Now, if $\varphi = \mathcal{G}f$, then

$$\int |\varphi| d\mu \leq \int (\mathcal{G}|f|) d\mu \leq L(\mathcal{G}|f|) \leq \|L\| \|f\|_{L^2(\Omega)} = \|L\| \|\varphi\|_{H^2(\Omega)\cap H_0^1(\Omega)} \quad (4.14)$$

In particular, by (4.14), if φ_k converges to φ in $H^2(\Omega) \cap H_0^1(\Omega)$. it also converges in $L^1(\mu)$.

Step 2:
If $\varphi \in H^2(\Omega) \cap H_0^1(\Omega)$ has the compact support, then

$$\int \varphi d\mu = L(\varphi) \ . \quad (4.15)$$

Indeed, for such a φ, there is a sequence $\{\varphi_k\} \subset C_0^\infty(\Omega)$ converging to φ in $H^2(\Omega) \cap H_0^1(\Omega)$.

In virtue of Step 1, $\{\varphi_k\}$ converges in $L^1(\mu)$ to φ and L agrees with μ on $C_0^\infty(\Omega)$. Thus, (4.15) holds.

Step 3:
For $\varphi \in T_K(u)$,

$$0 \leq \int \varphi d\mu \leq L(\varphi) \ . \quad (4.16)$$

Indeed, let $0 \leq \xi \leq 1$, $\xi \in C_0^\infty(\Omega)$. Then $\xi\varphi \in T_K(u)$, and $(1-\xi)\varphi \in T_K(u)$. Thus,

$$\int \xi\varphi d\mu = L(\xi\varphi) \leq L(\varphi)$$

because $L(\varphi) = L(\xi\varphi) + L((1-\xi)\varphi))$ with the second term nonnegative. Now let ξ increase to 1 on Ω.

Step 4: μ is concentrated on Ξ.
Indeed, since $u - \psi \geq 0$,

$$u - \psi + t\varphi(u-\psi) \geq 0$$

for $0 \leq \varphi \leq 1$, $\varphi \in C_0^\infty(\Omega)$ and $-1 \leq t \leq 1$. In other words, $t\varphi(u-\psi) \in C_K(u)$. Since

$$\int t\varphi(u-\psi) d\mu \geq 0, \quad -1 \leq t \leq 1 \ , \quad (4.17)$$

$\int \varphi(u-\psi)d\mu = 0$ and $u > \psi$ off Ξ, therefore μ is concentrated on Ξ.

Step 5: $\mu = 0$.

To show this note first that

$$\int \phi_0 d\mu = 0 \ . \tag{4.18}$$

Indeed, we know that $L(\phi_0) = 0$. Since $\phi_0 \geq 0$ on Ξ, by (4.17), we get (4.18). Now, $\phi_0 - \phi = \mathcal{G} f_0$ and $f_0 \geq 0$. Hence, $\phi_0 - \phi$ is nonnegative and superharmonic, and, in consequence, either identically equal zero or strictly positive everywhere. Since $\int(\phi_0 - \phi)d\mu = 0$, we must have $\mu = 0$.

Step 6: We claim that $h_0 = 0$.

To prove this we use property (iv) of $T_K(u)$. Let D be a relatively compact open set containing Ξ. By Proposition 3.3 below, there is an $0 \leq f \in L^2(\Omega)$ such that $\mathcal{G}f \equiv 1$ on D. Let $\varphi \in C_0^\infty(\Omega)$ and $\varphi \equiv 1$ on D. Then $\varphi - \mathcal{G}f$ vanishes on D, and hence $\varphi - \mathcal{G}f \in C_K(u)$. Hence, $L(\varphi - \mathcal{G}f) = \int f_0 \Delta(\varphi - \mathcal{G}f) = \int h_0(\Delta\varphi + f) = \int h_0 f$ because h_0 is harmonic. Since $f \geq 0$ we get $h_0 \equiv 0$. Thus, $L = 0$ or $\phi \in T_K(u)$, which completes the proof. □

Proposition 3.3. *Let F be a compact subset of a bounded domain Ω. Let $\mathcal{G}$ be the Green function of Ω. Then there exists an element $f \in L^\infty(\Omega)$, $f \geq 0$ such that $\mathcal{G}f \equiv 1$ on F.*

Proof. Let D be open and relatively compact, $F \subset D$. Then there exists a finite measure μ on ∂D such that $\mathcal{G}\mu \equiv 1$ on D. Choose $2\delta =$ dist$(F, \partial D)$ and a function φ, $\varphi \in C^\infty$, which is radial and vanishes off $B(0,\delta)$, and $\int \varphi dx = 1$.

Let $x \in F$ be fixed and $y \in \partial D$. Then $\mathcal{G}(x,z)$ is harmonic in $B(y,\delta)$. So, for all $x \in F$ and $y \in \partial D$

$$\mathcal{G}(x,y) = \int \mathcal{G}(x,z)\varphi(y-z)dz \ .$$

Intergrating relative to μ we get

$$1 = \int \mathcal{G}(x,y)\mu(dy) = \int \mathcal{G}(x,z) \int \varphi(y-z)\mu(dy)$$

for all $x \in F$ and $\int \varphi(y-z)\mu(dy)$ is C^∞ with compact support in Ω. □

5 Systems with unilateral conditions

In this section we investigate shape sensitivity of solutions to variational inequalities defined on convex subsets of the Sobolev space $H_0^2(\Omega) \times H_0^1(\Omega) \times H_0^1(\Omega)$. For the sake of simplicity we shall consider a simplified model with unilateral constraints. The same results are valid for a shallow shell with an obstacle.

Consider the convex set

$$\mathcal{K} = \{ \varphi = (\varphi_1, \varphi_2, \varphi_3) \in \mathcal{H} \mid \mathcal{R}\varphi \geq \psi \text{ in } \Omega \} , \tag{5.1}$$

where $\mathcal{H} = H_0^2(\Omega) \times H_0^1(\Omega) \times H_0^1(\Omega)$, $\psi(x)$, $x \in \Omega$, denotes the obstacle, and $\mathcal{R}$ is a linear mapping of the form

$$\mathcal{R}\varphi = \varphi_1 - a_2\varphi_2 - a_3\varphi_3 \tag{5.2}$$

where $a_2 = \frac{\partial \psi}{\partial x_1}$, $a_3 = \frac{\partial \psi}{\partial x_2}$.

For the sake of simplicity we assume that the obstacle is sufficiently smooth, and hence

$$\mathcal{R}\varphi \in H_0^1(\Omega), \quad \forall \varphi \in \mathcal{H} . \tag{5.3}$$

Moreover, we assume that the set

$$\{x \in \Omega \; | a_1(x) = a_2(x) = 0\}$$

is of null C_1 capacity.

Let

$$H^1_{\mathcal{R}}(\Omega) = \{ \phi \in H_0^1(\Omega) \mid \phi = \mathcal{R}\varphi, \text{ for some } \varphi \in \mathcal{H} \} \tag{5.4}$$

be the image in $H_0^1(\Omega)$ of $\mathcal{H}$ under the mapping $\mathcal{R}$.

Consider the metric projection $P_{\mathcal{K}}$ in $\mathcal{H}$ onto the convex set $\mathcal{K}$ with respect to the scalar product

$$(\varphi, \phi)_{\mathcal{H}} = \int_\Omega (\Delta\varphi_1 \Delta\phi_1 + \nabla\varphi_2 \cdot \nabla\phi_2 + \nabla\varphi_3 \cdot \nabla\phi_3) dx$$

or, equivalently, the following variational inequality

$$u \in \mathcal{K} : \quad (u, \varphi - u)_{\mathcal{H}} \geq \langle f, \varphi - u \rangle \quad \forall \varphi \in \mathcal{K} , \tag{5.5}$$

where

$$\langle f, \phi \rangle = \int_\Omega (f_1\phi_1 + f_2\phi_2 + f_3\phi_3) dx$$

and $f_i \in H^1(\mathbb{R}^2)$, $i = 1, 2, 3$, are given elements.

In this section q.e. means "quasi everywhere" with respect to the capacity of the space $H^1_{\mathcal{R}}(\Omega)$ equipped with the smallest norm for which the mapping $\mathcal{R} : \mathcal{H} \to H^1_{\mathcal{R}}(\Omega)$ is continuous.

Theorem 5.1. *Let $\nabla u(\Omega) \cdot V \in H^2(\Omega) \times H^1(\Omega) \times H^1(\Omega)$. The shape derivative of a solution $u(\Omega)$ to variational inequality (5.5) in the direction of the vector field V is a unique solution to the following variational inequality*

$$u' \in S_v = \{\varphi \in (H^2(\Omega) \cap H_0^1(\Omega)) \times H^1(\Omega) \times H^1(\Omega) | \frac{\partial \varphi_1}{\partial n} = -v \frac{\partial^2 u_1}{\partial n^2} \text{ on } \partial\Omega,$$

$$\varphi_2 = -v\frac{\partial u_2}{\partial n}, \; \varphi_3 = -v\frac{\partial u_3}{\partial n} \text{ on } \partial\Omega, \; \mathcal{R}\varphi(x) \geq 0 \text{ q.e. on } \Xi, \quad (u, \varphi)_{\mathcal{H}} = \langle f, \varphi \rangle\}$$

$$(u', \varphi - u')_{\mathcal{H}} \geq 0 \quad \forall \varphi \in S_v(\Omega), \tag{5.6}$$

where $\Xi = \{x \in \Omega | \mathcal{R}u(\Omega)(x) = 0\}$.

Using this theorem we can derive necessary optimality conditions of optimal shape design problems for free boundary problems described by variational inequality (5.5). In this case we get the same result as for the Dirichlet space in the scalar case.

6 Shape estimation problems

In this section we investigate differentiability of solutions to a class of shape optimization problems for systems governed by elliptic partial differential equations. In our analysis, we make an extensive use of the fact that the differentiability of solutions to nonconvex parametric optimization problems can be proved with the help of the Hadamard derivatives of metric projections onto sets of admissible parameters. The underlying observation we exploit is that local solutions to regularized parametric optimization problems are fixed points of metric projections onto sets of admissible parameters.

We now describe the framework of our approach. Let K be a closed and convex subset of a Hilbert space H, $J_{\alpha,\varepsilon}(u) = J_\varepsilon(u) + \frac{\alpha}{2}\|u\|_H^2$ a C^2 functional, $\alpha > 0$, $\varepsilon \in [0, \delta)$ parameters. By $\mathcal{G}_\varepsilon(u)$, and $\mathcal{H}_\varepsilon(u)$ we denote the gradient and the Hessian of $J_\varepsilon(u)$, respectively.

Consider the optimization problem

$$\text{Minimize } \; J_{\alpha,\varepsilon}(u) \; \text{ over } \; K \; .$$

For this problem necessary optimality conditions take the form

$$u_\varepsilon \in K : \quad (\mathcal{G}_\varepsilon(u_\varepsilon), v - u_\varepsilon)_H + \alpha(u_\varepsilon, v - u_\varepsilon)_H \geq 0 \quad \forall v \in K \; . \tag{6.1}$$

Hence,

$$u_\varepsilon = P_K(-\alpha^{-1}\mathcal{G}_\varepsilon(u_\varepsilon)) \tag{6.2}$$

ie. u_ε is a fixed point, P_K is the metric projection in H onto K.

Denoting $\mathcal{F}_\varepsilon(v) = -\alpha^{-1}\mathcal{G}_\varepsilon(v)$ we get

$$u_\varepsilon = P_K(\mathcal{F}_\varepsilon(u_\varepsilon)) \; . \tag{6.3}$$

If, for a fixed $\alpha > 0$, the nonlinear C^1 mapping $\mathcal{F}_0$ is a contraction, ie.

$$\|\mathcal{F}_0(v) - \mathcal{F}_0(u)\|_H \leq L\|v - u\|_H \tag{6.4}$$

for some $0 < L < 1$, and the mapping $\varepsilon \to \mathcal{F}_\varepsilon(v)$ is Lipschitz continuous,

$$\|\mathcal{F}_\varepsilon(v) - \mathcal{F}_0(v)\|_H \leq C_1\varepsilon\|v\|_H \; , \tag{6.5}$$

then the solution u_ε to (6.1) is unique, and Lipschitz continuous with respect to ε,

$$\|u_\varepsilon - u_0\|_H \leq \frac{C_1\varepsilon}{1-L}\|u_\varepsilon\|_H \leq C\varepsilon \tag{6.6}$$

since the norm $\|u_\varepsilon\|_H$ of any solution to (6.1) is bounded. We have the boundedness when eg. $J_\varepsilon(v) \geq 0$ for all $v \in K$.

To obtain the differentiability of u_ε with respect to ε at 0^+, we need the differentiability with respect to ε of $f_\varepsilon = \mathcal{G}_\varepsilon(u_\varepsilon)$. By Lipschitz continuity (6.6), there exists an element $q \in H$, in general non-unique, such that for $\varepsilon > 0$, ε small enough,

$$u_\varepsilon = u_0 + \varepsilon q + r(\varepsilon) \ . \tag{6.7}$$

By assuming compactness of $\mathcal{G}_\varepsilon(u_\varepsilon) : H \to H$ for $\varepsilon \geq 0$, we get

$$\mathcal{G}_\varepsilon(u_\varepsilon) = \mathcal{G}_0(u_0) + \varepsilon[\mathcal{H}_0(u_0)q + \partial\mathcal{G}(u_0)] + o(\varepsilon), \tag{6.8}$$

where $\partial\mathcal{G}$ denotes the directional derivative of $\mathcal{G}_\varepsilon$ with respect to ε at $\varepsilon = 0^+$. We can use (6.8) and differentiate (6.2) provided that the metric projection P_K is directionally differentiable in the sense of Hadamard at $-\alpha^{-1}\mathcal{G}_0(u_0)$. Suppose that there exists a mapping $\mathcal{Q} : H \mapsto H$ such that for all $h \in H$ and $\tau > 0$, τ small enough,

$$P_K(-\alpha^{-1}\mathcal{G}_0(u_0) + \tau h) = P_K(-\alpha^{-1}\mathcal{G}_0(u_0)) + \tau\mathcal{Q}(h) + o(\tau) \quad \text{in } H, \tag{6.9}$$

where $\|o(\tau)\|_H/\tau \to 0$ as $\tau \downarrow 0$, uniformly with respect to h on compact subsets of H. Then

$$q = \mathcal{Q}(-\alpha^{-1}\mathcal{H}_0(u_0)q - \alpha^{-1}\partial\mathcal{G}(u_0)), \tag{6.10}$$

ie. the directional derivative q in (6.7) is a fixed point of the nonlinear mapping $\mathcal{Q}$.
For some special cases the mapping $\mathcal{Q}$ is given in Section 3. In particular, (6.9) holds for any ball in a Hilbert space, see Proposition 1.2 in Section 1 of this chapter. In Sobolev spaces, for convex sets described by local constraints (see Corollary 6.1) we can prove that (under some assumptions) $\mathcal{Q} = P_S$, where S is a convex cone of the form

$$S = T_K(u_0) \cap [\alpha^{-1}\mathcal{G}_0(u_0) + u_0]^\perp \tag{6.11}$$

$T_K(u_0)$ is the tangent cone to K at $u_0 \in K$.

If $\mathcal{Q} = P_S$, then a local solution u_ε to (6.1) is Lipschitz continuous and directionally differentiable with respect to ε at 0^+ provided that

$$\exists\beta > 0 : \quad d^2 J_\alpha(u_0; v, v) \geq \beta\|v\|_H^2 \quad \forall v \in \{S - S\} \tag{6.12}$$

(see Proposition 6.1). Condition (6.12) implies uniqueness of the fixed point q defined by (6.10).

Finally, it can be shown shown that the metric projection P_K in H onto K is directionally differentiable in the sense of Hadamard at $-\alpha^{-1}\mathcal{G}(u_0)$.

By applying this framework we establish differential stability of local solutions to shape optimization problems (in general nonconvex) for second order elliptic equations.

The proposed method can be also applied to systems governed by linear partial differential equations of parabolic and hyperbolic types as well as to some nonlinear partial differential equations. In subsequent sections we prove the Lipschitz continuity of local optimal solutions to necessary optimality conditions with respect to the parameter, and exploiting the structure of the optimality system, we derive differential stability results.

As we have already mentioned, the key observation exploited in our approach is that local solutions to regularized optimization problems can be represented as fixed points of metric projections onto sets of admissible parameters. This allows us to apply results on the Hadamard differentiability of the metric projection onto sets of admissible parameters (graphs) in the Sobolev spaces $H^s(Q)$ and $H_0^s(Q)$, where $Q \subset \mathbb{R}^{N-1}$ is a domain, $s > \frac{N+1}{2}$. An admissible domain $\Omega \subset \mathbb{R}^N$ is defined by (6.13) below as the graph of a given function $f(x')$, $x' \in Q \subset \mathbb{R}^{N-1}$, f belongs to the Sobolev space $H^s(Q)$. Sets of admissible graphs with norm constraints in the Sobolev space $H^s(Q)$ are considered in Section 6.1. In Section 6.2 admissible graphs satisfy local inequality constraints.

6.1 Admissible domains with norm constraints on graphs

We shall now consider a shape optimization problem for a second order elliptic equation.

Let $Q \subset \mathbb{R}^{N-1}$, $N \geq 2$, be a given domain with the smooth boundary ∂Q. By $\Omega = \Omega_f \subset \mathbb{R}^N$ we denote a domain of the form

$$\Omega = \{x = (x', x_N) |\, 0 < x_N < f(x')\,, \quad x' \in Q\}, \tag{6.13}$$

where $x' = (x_1, \ldots, x_{N-1}) \in \mathbb{R}^{N-1}$, and f is a given function which depends on an auxiliary functional parameter $v \in H^s(Q)$. The function f is given as

$$f(x') = \frac{1}{2}[v(x')]^2 + \psi(x')\,, \quad x' \in Q, \tag{6.14}$$

where $\psi \in H^s(Q)$, $\psi(x') \geq c > 0$, and v belongs to the set K of admissible parameters,

$$K = \{v \in H^s(Q) \mid \ \|v\|_{H^s(Q)} \leq 1\}\ . \tag{6.15}$$

For $v \in H^s(Q)$, $s > \frac{N+1}{2}$, by the Sobolev imbedding theorem,

$$v, v_{,i} = \frac{\partial v}{\partial x_i} \in C(\overline{Q})\ , \ i = 1, \ldots, N-1\ . \tag{6.16}$$

The domain $\Omega(v) = \Omega_{f(v)}$ which depends on $v \in K$ is given as

$$\Omega(v) = \{x \mid 0 < x_N < f(v; x') = \frac{1}{2}[v(x')]^2 + \psi(x')\,, \quad x' \in Q\}\ . \tag{6.17}$$

The boundary $\partial\Omega(v)$ is divided into two parts, the graph of $f(v)(\cdot)$ is denoted by $\Gamma(v) \subset \partial\Omega(v)$,

$$\Gamma(v) = \{x = (x', x_N) |\, x_N = \frac{1}{2}[v(x')]^2 + \psi(x')\,, \quad x' \in Q\}\ . \tag{6.18}$$

Let

$$D = \{x |\ x' \in Q\,, \quad 0 < x_N < M\} \tag{6.19}$$

with M chosen so as to have $\Omega(v) \subset D$ for all $v \in K$, and let $F \in H^1(D)$ be given. We shall consider the following shape optimization problem.

Problem (P) **:** Find an element $v \in K$ which minimizes the cost functional

$$J(v) = \frac{1}{2}\|y(v) - z\|^2_{L^2(\Omega(v))}\,, \tag{6.20}$$

where $z \in H^1(D)$ is given, $y(v,.) \in H^1_0(\Omega(v))$ is a unique solution to a state equation given as an elliptic boundary value problem

$$\begin{aligned} -\mathrm{div}(\varrho\nabla y) &= F \quad \text{in } \Omega(v), \\ y &= 0 \quad \text{on } \partial\Omega(v), \end{aligned} \tag{6.21}$$

where $\varrho \in W^{1,\infty}(D)$, $\varrho(x) \geq c_2 > 0$.

The problem is to reconstruct a partially unknown domain on the basis of the observation z of the solution to the state equation. We can ensure the existence of a solution to problem (P) by applying a regularization technique.

Consider the following regularization of problem (P).

Problem (P_α) **:** Find an element $v \in K$ which minimizes the cost functional

$$J_\alpha(v) = \frac{1}{2}\|y(v) - z\|^2_{L^2(\Omega(v))} + \frac{\alpha}{2}\|v - \overline{v}\|^2_{H^s(Q)}\,, \tag{6.22}$$

where $\alpha > 0$ is a constant, $\overline{v} \in H^s(Q)$ is given. Below we assume for simplicity that $\overline{v} = 0$.

We shall investigate the differentiability of solutions to problem (P_α) with respect to perturbations of the observation z. Let $\varepsilon \geq 0$ be a parameter. Denote

$$z_\varepsilon = z + \varepsilon\vartheta, \tag{6.23}$$

where $\vartheta \in H^1(D)$ is given.

Let $v^\star_\varepsilon$ be a solution to the problem

Problem (P^ε_α) **:** Find an element $v \in K$ which minimizes the cost functional

$$J_{\alpha,\varepsilon}(v) = \frac{1}{2}\|y(v) - z_\varepsilon\|^2_{L^2(\Omega(v))} + \frac{\alpha}{2}\|v\|^2_{H^s(Q)}\,. \tag{6.24}$$

An optimal solution to problem (P^ε_α), though in general not unique, exists provided $\alpha > 0$. Below we shall prove that there exists $\alpha_0 > 0$ such that for $\alpha > \alpha_0$ an optimal solution $v^\star_\varepsilon$ is locally unique.

Since in this section we consider the fixed domain formulation of shape optimization problems, the problem (P_α^ε) is transported to the fixed domain Ω by an appropriate change of variables.

Transport to the fixed domain. Let

$$\Omega = \Omega_1 = \{x|\ 0 < x_N < 1 \quad x' \in Q\} \tag{6.25}$$

be a fixed domain in $\mathbb{R}^N$. We transport problem (P_α^ε) defined in the domain $\Omega_f = \Omega(v), f = f(v), v \in K$, to the fixed domain Ω. To this end we introduce a transformation of variables. Namely, for any $\zeta = (\zeta', \zeta_N) \in \Omega_f$ we put $x' = \zeta'$ and $x_N = \zeta_N / f(\zeta')$. Consequently, $\zeta' = x'$ and $\zeta_N = x_N f(x')$.

To transport the state equation to the fixed domain we introduce a new function $u(x)$ defined as

$$u(x) = y(x', x_N f(x')) \ . \tag{6.26}$$

Hence

$$\frac{\partial y}{\partial x_i} = y_{,i} = u_{,i} - \frac{\zeta_N}{[f(\zeta')]^2} f_{,i} u_{,N} = u_{,i} - \frac{x_N}{f(x')} f_{,i} u_{,N}$$

$i = 1, \ldots, N-1$, and

$$y_{,N} = \frac{1}{f(x')} u_{,N} \ .$$

In view of (6.26)

$$\nabla_\zeta y(\zeta) = A(x) \cdot \nabla_x u(x), \tag{6.27}$$

where $A = [a_{ij}]$, $a_{ii} = 1$ for $i = 1, \ldots, N-1$, $a_{NN} = 1/f(x')$, $a_{iN} = (-x_N f_{,i}(x'))/f(x')$, and $a_{ij} = 0$ otherwise.

Simple calculations show that state equation (6.21) in the fixed domain Ω takes the form of an integral identity,

$$u \in H_0^1(\Omega) \ : \ \int_\Omega \langle B(v) \cdot \nabla u, \nabla \varphi \rangle_{\mathbb{R}^N} dx = \int_\Omega F(v) \varphi dx \ \ \forall \varphi \in H_0^1(\Omega), \tag{6.28}$$

where

$$B(v)(x) = B(v; x) = \varrho(x', x_N f(x')) f(x') A^\star(x) \cdot A(x), \tag{6.29}$$

$A^\star$ is the transpose of A,

$$F(v)(x) = F(v; x) = f(x') F(x', x_N f(x')), \tag{6.30}$$

$$f(x') = f(v)(x') = f(v; x') = \frac{1}{2}[v(x')]^2 + \psi(x') \ . \tag{6.31}$$

We denote

$$z(v)(x) = z(v; x) = z(x', x_N f(v)(x')), \tag{6.32}$$

$$\vartheta(v)(x) = \vartheta(v; x) = \vartheta(x', x_N f(v)(x')),$$

and

$$z^\varepsilon(x) = z(v; x) + \varepsilon \vartheta(v; x) \ . \tag{6.33}$$

Problem (P_α^ε) is equivalent to the following problem defined in the fixed domain Ω.

Problem ($\Pi^{\varepsilon}_{\alpha}$) :

Find an element $v \in K$ which minimizes the cost functional

$$I_{\alpha,\varepsilon}(v) = \frac{1}{2}\|(u(v) - z^{\varepsilon})(f(v))^{\frac{1}{2}}\|^2_{L^2(\Omega)} + \frac{\alpha}{2}\|v\|^2_{H^s(Q)} \cdot \tag{6.34}$$

Recall that any local solution to problem ($\Pi^{\varepsilon}_{\alpha}$) satisfies the following necessary optimality conditions.

Find $(u^\star_\varepsilon, p^\star_\varepsilon, v^\star_\varepsilon)$ satisfying the following nonlinear system.

State equation:

$$-\mathrm{div}(B^\star_\varepsilon \cdot \nabla u^\star_\varepsilon) = F^\star_\varepsilon \quad \text{in } \Omega, \tag{6.35}$$

$$u^\star_\varepsilon = 0 \quad \text{on } \partial\Omega\ . \tag{6.36}$$

Adjoint state equation:

$$\mathrm{div}(B^\star_\varepsilon \cdot \nabla p^\star_\varepsilon) = f(v^\star_\varepsilon)(u^\star_\varepsilon - z^\varepsilon) \quad \text{in } \Omega, \tag{6.37}$$

$$p^\star_\varepsilon = 0 \quad \text{on } \partial\Omega\ . \tag{6.38}$$

Optimality conditions:

$$v^\star_\varepsilon \in K\ : \quad \int_\Omega \langle B'(v^\star_\varepsilon; v - v^\star_\varepsilon) \cdot \nabla u^\star_\varepsilon, \nabla p^\star_\varepsilon\rangle_{\mathbb{R}^N} dx \tag{6.39}$$

$$- \int_\Omega f(v^\star_\varepsilon)(u^\star_\varepsilon - z^\varepsilon)z'_\varepsilon(v^\star_\varepsilon; v - v^\star_\varepsilon)dx - \int_\Omega F'(v^\star_\varepsilon; v - v^\star_\varepsilon)p^\star_\varepsilon dx$$

$$+ \frac{1}{2}\int_\Omega (u^\star_\varepsilon - z^\varepsilon)^2 v^\star_\varepsilon (v - v^\star_\varepsilon)dx + \alpha(v - v^\star_\varepsilon, v^\star_\varepsilon)_{H^s(Q)} \geq 0 \quad \forall v \in K\ ,$$

where $B^\star_\varepsilon = B(v^\star_\varepsilon)$, $F^\star_\varepsilon = F(v^\star_\varepsilon)$ and $f^\star_\varepsilon = f(v^\star_\varepsilon) = \frac{1}{2}[v^\star_\varepsilon(x')]^2 + \psi(x')$. Here we assume the existence of the strong limits

$$B'(v; a) = \lim_{t \downarrow 0} \frac{1}{t}(B(v + ta) - B(v)) \quad \text{in } L^\infty(\Omega; \mathbb{R}^{N^2}), \tag{6.40}$$

$$F'(v; a) = \lim_{t \downarrow 0} \frac{1}{t}(F(v + ta) - F(v)) \quad \text{in } H^{-1}(\Omega), \tag{6.41}$$

$$z'_\varepsilon(v; a) = \lim_{t \downarrow 0} \frac{1}{t}(z^\varepsilon(v + ta) - z^\varepsilon(v)) \quad \text{in } L^2(\Omega)\ . \tag{6.42}$$

In view of optimality conditions (6.39), any local solution $v^\star_\varepsilon$ to (6.35)–(6.39) is a fixed point,

$$v^\star_\varepsilon = P_K(g^\star_\varepsilon), \tag{6.43}$$

where $g_\varepsilon^\star$ is a unique solution to the variational equation

$$\alpha(g_\varepsilon^\star,\varphi)_{H^s(Q)} = -\int_\Omega \langle B'(v_\varepsilon^\star;\varphi)\cdot\nabla u_\varepsilon^\star, \nabla p_\varepsilon^\star\rangle_{\mathbb{R}^N} dx \tag{6.44}$$
$$-\frac{1}{2}\int_\Omega (u_\varepsilon^\star - z^\varepsilon)^2 v_\varepsilon^\star \varphi dx + \int_\Omega f(v_\varepsilon^\star)(u_\varepsilon^\star - z^\varepsilon) z_\varepsilon'(v_\varepsilon^\star;\varphi)dx$$
$$+\int_\Omega F'(v_\varepsilon^\star;\varphi)p_\varepsilon^\star dx \quad \forall\varphi\in H^s(Q) .$$

In deriving optimality system (6.35)–(6.39) we use the directional derivative of the cost functional $I_{\alpha,\varepsilon}(v)$ in a direction $a\in H^s(Q)$,

$$dI_{\alpha,\varepsilon}(v_\varepsilon^\star;a) = -\int_\Omega \langle B'(v_\varepsilon^\star;a)\cdot\nabla u_\varepsilon^\star, \nabla p_\varepsilon^\star\rangle_{\mathbb{R}^N} dx + \frac{1}{2}\int_\Omega (u_\varepsilon^\star - z^\varepsilon)^2 v_\varepsilon^\star a dx$$
$$-\int_\Omega f(v_\varepsilon^\star)(u_\varepsilon^\star - z^\varepsilon) z_\varepsilon'(v_\varepsilon^\star;a)dx + \int_\Omega F'(v_\varepsilon^\star;a)p_\varepsilon^\star dx + \alpha(v_\varepsilon^\star,a)_{H^s(Q)} . \tag{6.45}$$

It is also useful to derive the second derivative of the functional $I_{\alpha,\varepsilon}(v)$. Namely, we differentiate the first derivative $dI_{\alpha,\varepsilon}(v_\varepsilon^\star;a)$ with respect to $v_\varepsilon^\star$ in a direction b and we obtain

$$d^2 I_{\alpha,\varepsilon}(v_\varepsilon^\star;a,b) \tag{6.46}$$
$$= \frac{1}{2}\int_\Omega (u_\varepsilon^\star - z^\varepsilon)^2 ab dx + \int_\Omega (u_\varepsilon^\star - z^\varepsilon)(\xi_\varepsilon^\star - z_\varepsilon'(v_\varepsilon^\star;b))v_\varepsilon^\star a dx$$
$$-\int_\Omega (u_\varepsilon^\star - z^\varepsilon) z_\varepsilon'(v_\varepsilon^\star;b) v_\varepsilon^\star a dx - \int_\Omega f(v_\varepsilon^\star)(\xi_\varepsilon^\star - z_\varepsilon'(v_\varepsilon^\star;b)) z_\varepsilon'(v_\varepsilon^\star;a)dx$$
$$-\int_\Omega f(v_\varepsilon^\star)(u_\varepsilon^\star - z^\varepsilon) z_\varepsilon''(v_\varepsilon^\star;a,b)dx + \int_\Omega \langle B''(v_\varepsilon^\star;a,b)\cdot\nabla u_\varepsilon^\star, \nabla p_\varepsilon^\star\rangle_{\mathbb{R}^N} dx$$
$$+\int_\Omega [\langle B'(v_\varepsilon^\star;a)\cdot\nabla\xi_\varepsilon^\star, \nabla p_\varepsilon^\star\rangle_{\mathbb{R}^N} + \langle B'(v_\varepsilon^\star;a)\cdot\nabla u_\varepsilon^\star, \nabla\eta_\varepsilon^\star\rangle_{\mathbb{R}^N}]dx$$
$$-\int_\Omega [F''(v_\varepsilon^\star;a,b)p_\varepsilon^\star + F'(v_\varepsilon^\star;a)\eta_\varepsilon^\star]dx + \alpha(a,b)_{H^s(Q)},$$

where $\xi_\varepsilon^\star, \eta_\varepsilon^\star$, respectively, are unique solutions to the linearized equations,

Linearized state equation:

$$-\mathrm{div}(B_\varepsilon^\star\cdot\nabla\xi_\varepsilon^\star) - \mathrm{div}(B'(v_\varepsilon^\star;b)\cdot\nabla u_\varepsilon^\star) = F'(v_\varepsilon^\star;b) \quad \text{in } \Omega,$$
$$\xi_\varepsilon^\star = 0 \quad \text{on } \partial\Omega ,$$

Linearized adjoint state equation:

$$\mathrm{div}(B_\varepsilon^\star\cdot\nabla\eta_\varepsilon^\star) + \mathrm{div}(B'(v_\varepsilon^\star;b)\cdot\nabla p_\varepsilon^\star)$$
$$= f(v_\varepsilon^\star)(\xi_\varepsilon^\star - z_\varepsilon'(v_\varepsilon^\star;b)) + (u_\varepsilon^\star - z^\varepsilon)v_\varepsilon^\star b \quad \text{in } \Omega,$$
$$\eta_\varepsilon^\star = 0 \quad \text{on } \partial\Omega .$$

Let

$$\mathcal{F}_\varepsilon : v \to g \tag{6.47}$$

be a nonlinear mapping depending on z^ε such that for $v \in K$ the element $g = \mathcal{F}_\varepsilon(v) \in H^s(Q)$ satisfies

$$\begin{aligned}\alpha(g, \varphi)_{H^s(Q)} = &- \int_\Omega \langle B'(v;\varphi) \cdot \nabla u, \nabla p\rangle_{\mathbb{R}^N} dx \\ &- \frac{1}{2}\int_\Omega (u - z^\varepsilon)^2 v\varphi dx \\ &+ \int_\Omega f(v)(u - z^\varepsilon) z'_\varepsilon(v;\varphi) dx \\ &+ \int_\Omega F'(v;\varphi) p dx \quad \forall \varphi \in H^s(Q).\end{aligned} \tag{6.48}$$

and u, p are unique solutions to the following elliptic problems.

State equation:

$$-\text{div}(B(v) \cdot \nabla u) = F(v) \quad \text{in } \Omega, \tag{6.49}$$
$$u = 0 \quad \text{on } \partial\Omega\,. \tag{6.50}$$

Adjoint state equation:

$$\text{div}(B(v) \cdot \nabla p) = f(v)(u - z^\varepsilon) \quad \text{in } \Omega, \tag{6.51}$$
$$p = 0 \quad \text{on } \partial\Omega\,. \tag{6.52}$$

There exists $\alpha_0 > 0$ such that for $\alpha > \alpha_0$ and for any $\varepsilon \in [0, \delta)$ the mapping $\mathcal{F}_\varepsilon$ is a contraction in an open neighborhood of $v^\star_\varepsilon$. Henceforth, for $\alpha > \alpha_0$ a fixed point $v^\star_\varepsilon$ is locally unique and Lipschitz continuous with respect to ε.

To get differentiability of the element $v^\star_\varepsilon$ with respect to ε at $\varepsilon = 0^+$ we shall use Proposition 1.2 from Section 1 of this chapter. By Proposition 1.2, the differentiability of fixed point (6.43) with respect to ε follows from the implicit function theorem, whenever $\|g^\star_\varepsilon\|_{H^s(Q)} > 1$. This condition, in turn, is satisfied when we choose the parameter α so that the Frechet derivative of the nonlinear mapping $I - P_K\mathcal{F}_\varepsilon$ at $v^\star_\varepsilon$ is an isomorphism. When $\|g^\star_\varepsilon\|_{H^s(Q)} = 1$ the projection P_K is merely directionally differentiable and we cannot proceed in the same way (see also Theorem 6.2).

Denote

$$\Phi(a) = D\mathcal{F}_0(v^\star_0; a), \tag{6.53}$$
$$\Psi(\vartheta) = \partial\mathcal{F}(v^\star_0) = \partial_\varepsilon\mathcal{F}_\varepsilon(v^\star_\varepsilon)_{|\varepsilon=0}\,. \tag{6.54}$$

We have

$$\Phi(a) \in H^s(Q): \quad (\Phi(a), \varphi)_{H^s(Q)} = \frac{1}{2}\int_\Omega (u_0^\star - z^0) a\varphi dx \tag{6.55}$$
$$+ \int_\Omega (u_0^\star - z^0)(\xi_0^\star - z_0'(v_0^\star;\varphi))v_0^\star a dx - \int_\Omega (u_0^\star - z^0) z_0'(v_0^\star; a) v_0^\star \varphi dx$$
$$- \int_\Omega f(v_0^\star)(\xi_0^\star - z_0'(v_0^\star;\varphi)) z_0'(v_0^\star; a) dx - \int_\Omega f(v_0^\star)(u_0^\star - z^0) z_0''(v_0^\star; a, \varphi) dx$$
$$+ \int_\Omega [\langle B''(v_0^\star; a, \varphi)\cdot \nabla u_0^\star, \nabla p_0^\star\rangle_{\mathbb{R}^N} + \langle B'(v_0^\star; a)\cdot \nabla \xi_0^\star, \nabla p_0^\star\rangle_{\mathbb{R}^N}]$$
$$+ \int_\Omega \langle B'(v_0^\star; a)\cdot \nabla u_0^\star, \nabla \eta_0^\star\rangle_{\mathbb{R}^N} dx$$
$$- \int_\Omega [F''(v_0^\star; a, \varphi) p_0^\star + F'(v_0^\star; a)\eta_0^\star] dx \quad \forall \varphi \in H^s(Q) .$$

The partial derivative of $\mathcal{F}_\varepsilon$ with respect to ε is a unique solution to the equation

$$\Psi(a) \in H^s(Q): \quad (\Psi(\vartheta), \varphi)_{H^s(Q)} = \int_\Omega \langle B'(v_0^\star;\varphi)\cdot \nabla u_0^\star, \nabla \phi\rangle_{\mathbb{R}^N} dx \tag{6.56}$$
$$- \int_\Omega (u_0^\star - z(v_0^\star))^2 \vartheta(v_0^\star) v_0^\star \varphi dx - \int_\Omega f(v_0^\star)\vartheta(v_0^\star) z_0'(v_0^\star;\varphi) dx \quad \forall \varphi \in H^s(Q),$$

where

$$\operatorname{div}(B_0^\star \cdot \nabla \phi) = -f(v_0^\star)\vartheta(v_0^\star) \quad \text{in } \Omega, \tag{6.57}$$
$$\phi = 0 \quad \text{on } \partial\Omega . \tag{6.58}$$

Theorem 6.1. *Assume that there exists $\beta > 0$ such that*

$$\|a - \chi\Phi(a)\|_{H^s(Q)} \geq \beta \|a\|_{H^s(Q)} \quad \forall a \in \{S - S\}, \tag{6.59}$$

where $\chi = \frac{1}{\alpha\|g_0^\star\|_{H^s(Q)}}$.

(i) If

$$\|g_0^\star\|_{H^s(Q)} > 1 \ , \tag{6.60}$$

then there exists $\delta > 0$ such that for $|\varepsilon| < \delta$

$$v_\varepsilon^\star = v_0^\star + \varepsilon q + o(\varepsilon) \quad \text{in } H^s(Q), \tag{6.61}$$

where q is a unique solution to the linear problem

$$q \in S = \{\varphi \in H^s(Q) \mid (g_0^\star, \varphi)_{H^s(Q)} = 0\}, \tag{6.62}$$
$$\alpha\|g_0^\star\|_{H^s(Q)}(q,\varphi)_{H^s(Q)} = (\Phi(q),\varphi)_{H^s(Q)} + (\Psi(\vartheta),\varphi)_{H^s(Q)} \quad \forall \varphi \in H^s(Q) .$$

(ii) If $\|g_0^\star\|_{H^s(Q)} = 1$, then for $\varepsilon > 0$, ε small enough,

$$v_\varepsilon^\star = v_0^\star + \varepsilon q + o(\varepsilon) \quad \text{in } H^s(Q), \tag{6.63}$$

where q is a unique solution to the variational inequality

$$q \in S = \{\varphi \in H^s(Q) \mid (g_0^\star, \varphi)_{H^s(Q)} \le 0\}, \tag{6.64}$$
$$\alpha \|g_0^\star\|_{H^s(Q)} (q, \varphi - q)_{H^s(Q)} \ge (\Phi(q), \varphi - q)_{H^s(Q)} + (\Psi(\vartheta), \varphi - q)_{H^s(Q)} \quad \forall \varphi \in S .$$

Proof. We make use of Proposition 1.2.

(i) In view of (1.22) and (6.59), we can use the implicit function theorem in order to differentiate (6.43) with respect to ε, at $\varepsilon = 0$, which yields the required result.

(ii) By the same argument as in Proposition 6.1, in view of (1.20), for any $\varepsilon > 0$ there exists a local solution to the necessary optimality conditions $v_\alpha^\star$ such that

$$\|v_\varepsilon^\star - v_0^\star\|_{H^s(Q)} \le c\varepsilon$$

Therefore, we can proceed in the same way as in the proof of Theorem 6.2 in order to obtain (6.63). □

6.2 Admissible domains with local constraints on graphs

We assume that an admissible domain $\Omega = \Omega_f$ is the graph of f, ie. Ω is of the form (6.13), where f belongs to the set

$$K = \{f \in H_0^s(Q) | 0 < \psi_1(x') \le f(x') \le \psi_2(x') \quad \text{for all } x' \in Q\}$$

and $\psi_i \in H_0^s(Q)$, $i = 1, 2$, are given elements such that the set K is nonempty. In this section we shall not use an auxiliary variable v.

We put

$$D = \{x = (x', x_N) | 0 < x_N < \psi_2(x'), \ x' \in Q\} .$$

Therefore, $\Omega_f \subset D$ for all $f \in K$.

Shape sensitivity analysis in a variable domain. We now prove some results concerning shape sensitivity analysis for an elliptic state equation and an associated shape functional in the variable domain setting. We use shape derivatives of solutions to elliptic equations. Such an approach is in some sense more straightforward as compared to the fixed domain approach used in Section 6.1.

Consider an elliptic equation

$$-\Delta y = F \quad \text{in } \Omega_f , \tag{6.65}$$
$$y = 0 \quad \text{on } \partial\Omega_f . \tag{6.66}$$

Denote

$$\Omega_t = \{x = (x', x_N) | 0 < x_N < f(x') + th(x'), \ x' \in Q\}, \tag{6.67}$$

where $h \in H_0^s(Q)$ is given, t is small.

We use the following notation.

$$\begin{aligned}
&T_t(V) \,:\, \Omega \mapsto \Omega_t = \Omega_t(V), \quad \Omega \equiv \Omega_f = \Omega_{t|t=0}\,,\\
&V(t,x) = (0,\dots,0, x_N h(x')(f(x')+th(x'))^{-1})\,,\\
&V(0,x) = (0,\dots,0,h(x'))\,,\\
&h_\nu(x') = \langle V(0,x), n(x')\rangle_{\mathbb{R}^N}\,,\\
&n(t,x) = \left(1+|\nabla(f+th)(x')|^2\right)^{-\frac{1}{2}} \begin{pmatrix} -\nabla f(x') - t\nabla h(x') \\ 1 \end{pmatrix},\\
&n(x') = n(0,x)\,,\\
&J_\alpha(f) = J(\Omega_f) + \frac{\alpha}{2}\|f\|^2\,,\\
&\mathfrak{D}J(\Omega_f; V(0)) = (\mathcal{G}_0(f), h_\nu)\,,\\
&\mathfrak{d}^2 J(\Omega_f; V(0), W(0)) = (\mathcal{H}_0(f)h, v)\,,\\
&W(t,x) = (0,\dots,0,x_N v(x')(f(x')+tv(x'))^{-1})\,,\\
&[V\circ T_t(V)\circ T_t^{-1}(W)](x) = (0,\dots,0,x_N h(x')(f(x')+tv(x'))^{-1})\,,
\end{aligned}$$

and for $y(t) = y_t(\Omega_t)$ the shape derivative in the direction of a field V defined above is denoted $y'(h_\nu) = y'(\Omega; V)$.

With the notation,

$$h_\nu(x') = \frac{h(x')}{(1+|\nabla f(x')|^2)^{\frac{1}{2}}}$$

is the normal component of the vector field

$$\frac{\partial T_t}{\partial t}(V) = V \circ T_t = (0,\dots,0,h(x')) \in \mathbb{R}^N$$

on $\Gamma(f)$ and

$$-\Delta y(t) = F \quad \text{in } \Omega_t\,, \tag{6.68}$$

$$y(t) = 0 \quad \text{on } \partial\Omega_t\,. \tag{6.69}$$

For $t = 0$, a solution to (6.68), (6.69) is denoted $y(f) = y(t)_{|t=0} \in H_0^1(\Omega_f)$, ie. $-\Delta y(f;x) = F(x)$ in $\Omega = \Omega_f$.

The shape derivative $y' = y'(h_\nu)$ can be defined as

$$y'(x) = \frac{\partial Y}{\partial t}(0,x), \tag{6.70}$$

where

$$Y(t,x) = \begin{cases} y(t,x) & \text{for } x \in \Omega_t, t \geq 0 \\ 0 & \text{for } x \notin \Omega_t, t \geq 0\,. \end{cases}$$

The shape derivative is a unique solution to the elliptic equation

$$-\Delta y' = 0 \quad \text{in } \Omega_f \,, \tag{6.71}$$

$$y' = -h_\nu \frac{\partial y}{\partial n} \quad \text{on } \Gamma(f), \tag{6.72}$$

$$y' = 0 \quad \text{on } \partial\Omega \setminus \Gamma(f) \,. \tag{6.73}$$

For the shape functional

$$J_\alpha(f) = \frac{1}{2}\int_{\Omega_f} (y(f;x) - z(x))^2 dx + \frac{\alpha}{2}\|f\|^2_{H_0^s(Q)} \tag{6.74}$$

the directional derivative in a direction h takes the form

$$\begin{aligned} dJ_\alpha(f;h) &= \mathfrak{D}J(\Omega_f; V(0)) + \alpha(f,h)_{H_0^s(Q)} \\ &= (\mathcal{G}_0(f), h)_{H_0^s(Q)} + \alpha(f,h)_{H_0^s(Q)} \\ &= \int_{\Omega_f} (y(f;x) - z(x)) y'(h_\nu) dx + \alpha(f,h)_{H_0^s(Q)} \,, \end{aligned} \tag{6.75}$$

where we assume $z = 0$ on $\Gamma(f)$.

The adjoint state is defined in the standard way,

$$-\Delta p = y - z \quad \text{in } \Omega_f \,, \tag{6.76}$$

$$p = 0 \quad \text{on } \partial\Omega_f \,, \tag{6.77}$$

and consequently,

$$dJ_\alpha(f;h) = \int_{\Gamma(f)} h_\nu \frac{\partial p}{\partial n}\frac{\partial y}{\partial n} d\Gamma + \alpha(f,h)_{H_0^s(Q)} \,. \tag{6.78}$$

Finally, the second derivative takes the form

$$\begin{aligned} d^2 J_\alpha(f;h,v) &= \lim_{t\downarrow 0} \frac{1}{t}(dJ_\alpha(f+tv;h) - dJ_\alpha(f;h)) \\ &= (\mathcal{H}_0(f)h, v)_{H_0^s(Q)} + \alpha(h,v)_{H_0^s(Q)} \,. \end{aligned} \tag{6.79}$$

where

$$\begin{aligned} &\mathfrak{d}^2 J(\Omega_f; V(0), W(0)) = (\mathcal{H}_0(f)h, v)_{H_0^s(Q)} \\ &= -\int_{\Gamma(f)} \left(h_\nu \frac{\langle \nabla f(x'), \nabla v(x')\rangle_{\mathbb{R}^3}}{(1+|\nabla f(x')|^2)} + v_\nu \frac{\langle \nabla f(x'), \nabla h(x')\rangle_{\mathbb{R}^3}}{(1+|\nabla f(x')|^2)} + 2\kappa v_\nu h_\nu \right) \frac{\partial p}{\partial n}\frac{\partial y}{\partial n} d\Gamma \\ &+ \int_{\Gamma(f)} h_\nu v_\nu \frac{\partial}{\partial n}\left(\frac{\partial y}{\partial n}\frac{\partial p}{\partial n}\right) d\Gamma + \int_{\Gamma(f)} h_\nu \left(\frac{\partial p'}{\partial n}\frac{\partial y}{\partial n} + \frac{\partial p}{\partial n}\frac{\partial y'}{\partial n}\right) d\Gamma \,. \end{aligned}$$

Here κ is the mean curvature on $\Gamma(f)$.

The shape derivatives $y'(v_\nu) = y'(\Omega_f; W)$ and $p'(v_\nu) = p'(\Omega_f; W)$ satisfy the elliptic equation

$$-\Delta y' = 0 \quad \text{in } \Omega_f \,, \tag{6.80}$$
$$y' = -v_\nu \frac{\partial y}{\partial n} \quad \text{on } \Gamma(f) \,, \tag{6.81}$$
$$y' = 0 \quad \text{on } \partial\Omega \setminus \Gamma(f) \,, \tag{6.82}$$

and

$$-\Delta p' = y' - z' \quad \text{in } \Omega_f \,, \tag{6.83}$$
$$p' = -v_\nu \frac{\partial y}{\partial n} \quad \text{on } \Gamma(f) \,, \tag{6.84}$$
$$p' = 0 \quad \text{on } \partial\Omega \setminus \Gamma(f), \tag{6.85}$$

here $z' = z'(\Omega_f; W)$ is the shape derivative of the observation z, and we can assume $z' = 0$.

To obtain (6.79) we evaluate the shape derivative $\mathfrak{D}I(\Omega; W(0))$ of the shape functional

$$I(\Omega) = I(\Omega_f) = \int_{\Gamma(f)} h_\nu \frac{\partial p}{\partial n} \frac{\partial y}{\partial n} d\Gamma$$

in the direction of the vector field $W = W(t, x)$.

The functional I in domain $\Omega_t = \Omega_t(W)$ takes the following form

$$\begin{aligned} I(\Omega_t) &= \int_{\Gamma(f+tv)} h(x') \left(1 + |\nabla(f+tv)(x')|^2\right)^{-\frac{1}{2}} \nabla y(t) \cdot n(t) \; \nabla p(t) \cdot n(t) d\Gamma_t \\ &= \int_{\Gamma(f+tv)} \psi(t, x) d\Gamma_t \,, \end{aligned}$$

where

$$\begin{aligned} &y(t) = y(\Omega_t(W)) = y(f+tv), \\ &p(t) = p(\Omega_t(W)) = p(f+tv), \\ &n(t, x) = \left(1 + |\nabla(f+tv)(x')|^2\right)^{-\frac{1}{2}} \begin{pmatrix} -\nabla f(x') - t\nabla v(x') \\ 1 \end{pmatrix} . \end{aligned}$$

We have the following formulae for the derivatives at $t = 0$.

$$\begin{aligned} &\frac{\partial}{\partial t}\left(\nabla y(t) \cdot n(t) \; \nabla p(t) \cdot n(t)\right) = \frac{\partial p'}{\partial n}(v_\nu) \frac{\partial y}{\partial n} + \frac{\partial p}{\partial n} \frac{\partial y'}{\partial n}(v_\nu) \,, \\ &\frac{\partial}{\partial t}\left(h(x') \left(1 + |\nabla(f+tv)(x')|\right)^{-\frac{1}{2}}\right) = -h_\nu \frac{\langle \nabla f(x'), \nabla v(x') \rangle_{\mathbb{R}^3}}{(1 + |\nabla f(x')|^2)} \,, \\ &\frac{\partial \psi}{\partial n} W \cdot n = -v_\nu \frac{\langle \nabla f(x'), \nabla h(x') \rangle_{\mathbb{R}^3}}{(1 + |\nabla f(x')|^2)} \frac{\partial y}{\partial n} \frac{\partial p}{\partial n} + v_\nu h_\nu \frac{\partial}{\partial n}\left(\frac{\partial y}{\partial n} \frac{\partial p}{\partial n}\right) \,, \end{aligned}$$

therefore, with the formula for derivative of a surface integral,

$$\mathfrak{d}^2 J(\Omega_f; V(0), W(0)) = \mathfrak{D}I(\Omega; W(0))$$
$$= \int_{\Gamma(f)} \frac{\partial \psi}{\partial t}(0,x) + \left[\frac{\partial \psi}{\partial n}(0,x) - 2\kappa(x)\psi(0,x)\right] W(0,x)\cdot n(x')d\Gamma$$

we obtain second order shape derivative.

Remark. 6.1. Since $y'(h_\nu) = -h_\nu \frac{\partial y}{\partial n}, p'(h_\nu) = -h_\nu \frac{\partial p}{\partial n}$, on $\Gamma(f)$ we have

$$\int_{\Gamma(f)} h_\nu \left(\frac{\partial p'}{\partial n}\frac{\partial y}{\partial n} + \frac{\partial p}{\partial n}\frac{\partial y'}{\partial n}\right) d\Gamma \tag{6.86}$$
$$= -\int_{\Gamma(f)} y'(h_\nu)\frac{\partial p'}{\partial n}(v_\nu) + p'(h_\nu)\frac{\partial y'}{\partial n}(v_\nu) d\Gamma \ .$$

Suppose that

$$z_\varepsilon = z + \varepsilon\vartheta \ .$$

Then the derivative η of the adjoint state p with respect to ε is a unique solution to the elliptic equation

$$-\Delta\eta = -\vartheta \quad \text{in } \Omega_f \ ,$$
$$\eta = 0 \quad \text{on } \partial\Omega_f \ .$$

For any $\alpha > 0$, there exists an element $f_\varepsilon^\star$ which minimizes the perturbed cost functional

$$J_{\alpha,\varepsilon}(f) = \frac{1}{2}\int_{\Omega_f} (y(f) - z_\varepsilon)^2 dx + \frac{\alpha}{2}\|f\|^2_{H_0^s(Q)}$$

over the set $K \subset H_0^s(Q)$, $s > \frac{N+1}{2}$. We assume that $z = 0$ on $\Gamma(f_0^\star)$.

Theorem 6.2. *Assume that there exists $\beta > 0$ such that*

$$d^2 J_\alpha(f_0^\star; v, v) \geq \beta\|v\|_{H_0^s(Q)} \quad \textit{for all } v \in \{S - S\} \tag{6.87}$$

and condition (1.25) is satisfied for $f = f_0^\star = P_K(-\alpha^{-1}\mathcal{G}_0(f_0^\star))$ and $g = -\alpha^{-1}\mathcal{G}_0(f_0^\star)$, ie. the set K is polyhedric at (g, f).
Then for $\varepsilon > 0$, ε small enough,

$$f_\varepsilon^\star = f_0^\star + \varepsilon q + o(\varepsilon) \quad \textit{in } H_0^s(Q), \tag{6.88}$$

where $\|o(\varepsilon)\|_{H_0^s(Q)}/\varepsilon \downarrow 0$ with $\varepsilon \downarrow 0$ and the element q uniquely solves the optimality system (we denote $f = f_0^\star$):

Find (y', p', q) such that the following system is satisfied

State equation:

$$-\Delta y' = 0 \quad \textit{in } \Omega_f \ , \tag{6.89}$$
$$y' = -q_\nu \frac{\partial y}{\partial n} \quad \textit{on } \Gamma(f), \tag{6.90}$$
$$y' = 0 \quad \textit{on } \partial\Omega \setminus \Gamma(f) \ . \tag{6.91}$$

Adjoint state equation:

$$-\Delta p' = y' \quad in\ \Omega_f\ , \tag{6.92}$$

$$p' = -q_\nu \frac{\partial p}{\partial n} \quad on\ \Gamma(f), \quad p' = 0 \quad on\ \partial\Omega \setminus \Gamma(f)\ . \tag{6.93}$$

Optimality conditions:

$$q \in S = T_K(f) \cap [f + \frac{1}{\alpha}\mathcal{G}_0(f)]^\perp\ : \tag{6.94}$$

$$- \int_{\Gamma(f)} \left((h-q)_\nu \frac{\langle \nabla f(x'), \nabla q(x') \rangle_{\mathbb{R}^3}}{(1+|\nabla f(x')|^2)} + q_\nu \frac{\langle \nabla f(x'), \nabla (h-q)(x') \rangle_{\mathbb{R}^3}}{(1+|\nabla f(x')|^2)} \right) \frac{\partial p}{\partial n}\frac{\partial y}{\partial n} d\Gamma$$

$$+ 2 \int_{\Gamma(f)} \kappa q_\nu (h-q)_\nu \frac{\partial p}{\partial n}\frac{\partial y}{\partial n} d\Gamma$$

$$+ \int_{\Gamma(f)} (h-q)_\nu q_\nu \frac{\partial}{\partial n}\left(\frac{\partial y}{\partial n}\frac{\partial p}{\partial n}\right) d\Gamma + \int_{\Gamma(f)} (h-q)_\nu \left(\frac{\partial p'}{\partial n}\frac{\partial y}{\partial n} + \frac{\partial p}{\partial n}\frac{\partial y'}{\partial n}\right) d\Gamma$$

$$+ \alpha((h-q), q)_{H_0^s(Q)} + \int_{\Gamma(f)} (h-q)_\nu \frac{\partial \eta}{\partial n}\frac{\partial y}{\partial n} d\Gamma \ \geq 0 \quad \forall h \in S\ .$$

Here $T_K(v)$ denotes the tangent cone to K at $v \in K$, $[f-v]^\perp$ is the hyperplane orthogonal in $H_0^s(Q)$ to $f-v$. The cone S takes the form

$$S = \{\varphi \in H_0^s(Q) | \varphi \geq 0 \quad \text{on } \Xi_1\ ,\ \varphi \leq 0 \ \text{ on } \Xi_2\ ,$$
$$(f_0^\star + \frac{1}{\alpha}\mathcal{G}_0(f_0^\star), \varphi)_{H_0^s(Q)} = 0\},$$

where

$$\Xi_i = \{x \in Q | f_0^\star(x) = \psi_i(x)\}\ ,\ i = 1, 2\ .$$

Proof of Theorem 6.2. Let $\{\varepsilon_n\}$ be a sequence such that $\varepsilon_n \downarrow 0$ as $n \to \infty$. Denote $\varepsilon = \varepsilon_n$ and let $\mathcal{F}_\varepsilon = -\frac{1}{\alpha}\mathcal{G}_\varepsilon\ ,\ \mathcal{F}'_\varepsilon = -\frac{1}{\alpha}\mathcal{H}_\varepsilon$. By Proposition 6.1 below we have

$$\|f_\varepsilon^\star - f_0^\star\|_{H_0^s(Q)} \leq c\varepsilon\ . \tag{6.95}$$

From (6.95) it follows that there exists an element $q \in H_0^s(Q)$, such that for $\varepsilon > 0$, ε small enough

$$f_\varepsilon^\star = f_0^\star + \varepsilon q + r(\varepsilon) \quad \text{in } H_0^s(Q), \tag{6.96}$$

where $r(\varepsilon)/\varepsilon \to 0$ weakly in $H_0^s(Q)$, with $\varepsilon \downarrow 0$. Therefore,

$$\mathcal{F}_\varepsilon = \mathcal{F}_0 + \varepsilon(\mathcal{F}'_0 q + \partial\mathcal{F}_0) + o(\varepsilon) \quad \text{in } H_0^s(Q)\ .$$

By Corollary 6.1, Corollary 1.1 and (6.96) it follows that

$$f_\varepsilon^\star = P_K(\mathcal{F}_\varepsilon) = P_K(\mathcal{F}_0 + \varepsilon(\mathcal{F}'_0 q + \partial\mathcal{F}_0) + o(\varepsilon))$$
$$= P_K(\mathcal{F}_0) + \varepsilon P_S(\mathcal{F}'_0 q + \partial\mathcal{F}_0) + o(\varepsilon) \quad \text{in } H_0^s(Q)\ , \tag{6.97}$$
$$q = P_S(\mathcal{F}'_0 q + \partial\mathcal{F}_0) \tag{6.98}$$

which completes the proof of Theorem 6.2 since, by (6.87), q is unique. □

It remains to show that under our assumptions (6.95) holds.

Proposition 6.1. *Assume that*
(i) there exists $\beta > 0$ *such that*

$$d^2 J_\alpha(f_0^\star; v, v) \geq \beta \|v\|_{H_0^s(Q)} \quad \forall v \in S \; .$$

(ii) P_K *is directionally differentiable at* $f_0^\star = P_K(-\alpha^{-1}\mathcal{G}_0(f_0^\star))$ *in the sense of Hadamard ie., for all* $h \in H_0^s(Q)$ *and* $\tau > 0$, τ *small enough,*

$$P_K(f_0^\star + \tau h) = P_K(f_0^\star) + \tau P_S(h) + o(\tau) \quad \text{in } H_0^s(Q),$$

where $\|o(\tau)\|_{H_0^s(Q)}/\tau \to 0$ *with* $\tau \downarrow 0$ *uniformly with respect to* h *on compact subsets of* $H_0^s(Q)$.
Then (6.95) holds for any sequence $\varepsilon_n \downarrow 0$.

Proof. We have $f_\varepsilon^\star = P_K(\mathcal{F}_\varepsilon(f_\varepsilon^\star))$, and

$$\|f_\varepsilon^\star\|_{H_0^s(Q)} \leq c.$$

Hence, there exists an element $v \in H_0^s(Q)$ such that for a subsequence, still denoted $\varepsilon_n \downarrow 0$,

$$f_{\varepsilon_n} \to v \text{ weakly in } H_0^s(Q) \; . \tag{6.99}$$

Since the mapping

$$H_0^s(Q) \ni f \to \mathcal{G}_\varepsilon(f) \in H_0^s(Q)$$

is compact it follows that

$$\mathcal{F}_{\varepsilon_n}(f_{\varepsilon_n}^\star) \to \mathcal{F}_0(v) \text{ strongly in } H_0^s(Q), \tag{6.100}$$

and hence

$$v = P_K(\mathcal{F}_0(v)), \tag{6.101}$$

ie. $v = f_0^\star$ is a local solution to the shape optimization problem and, by (6.100), $f_{\varepsilon_n}^\star \to f_0^\star$ strongly in $H_0^s(Q)$. Therefore, the set $\{f_{\varepsilon_n}^\star\}_{n=1}^\infty \cup f_0^\star$ is compact in $H_0^s(Q)$. By the Hadamard differentiability of P_K,

$$\begin{aligned} f_{\varepsilon_n}^\star - f_0^\star =& P_K(\mathcal{F}_{\varepsilon_n}(f_{\varepsilon_n}^\star)) - P_K(\mathcal{F}_0(f_{\varepsilon_n}^\star)) + P_S(\mathcal{F}_0'(f_0^\star)(f_{\varepsilon_n}^\star - f_0^\star)) \\ &+ o(\|f_{\varepsilon_n}^\star - f_0^\star\|_{H_0^s(Q)}), \end{aligned} \tag{6.102}$$

where, in view of assumption (ii), $o(\|f_{\varepsilon_n}^\star - f_0^\star\|_{H_0^s(Q)})$ is uniform on the compact set $\{f_{\varepsilon_n}^\star\}_{n=1}^\infty \cup f_0^\star$ of $H_0^s(Q)$.

We show that (6.102) implies (6.95). To this end let us observe that

$$\|P_K(\mathcal{F}_{\varepsilon_n}(f_{\varepsilon_n}^\star)) - P_K(\mathcal{F}_0(f_{\varepsilon_n}^\star))\|_{H_0^s(Q)} \leq c\varepsilon_n \tag{6.103}$$

On the other hand, S is a closed and convex cone. Denoting by $S^\star$ the polar cone we have

$$P_S + P_{S^\star} = I \; . \tag{6.104}$$

By taking the metric projection of (6.102) onto $S^\star$ we get

$$\begin{aligned}\|P_{S^\star}(f^\star_{\varepsilon_n} - f^\star_0)\|_{H^s_0(Q)} &\le c\varepsilon_n + \|P_{S^\star} o(\|f^\star_{\varepsilon_n} - f^\star_0\|_{H^s_0(Q)})\|_{H^s_0(Q)} \\ &\le c\varepsilon_n + \|o(\|f^\star_{\varepsilon_n} - f^\star_0\|_{H^s_0(Q)})\|_{H^s_0(Q)} \,.\end{aligned} \tag{6.105}$$

From (6.102) it follows

$$\begin{aligned}&\|f^\star_{\varepsilon_n} - f^\star_0\|_{H^s_0(Q)} \le \|P_K(\mathcal{F}_{\varepsilon_n}(f^\star_{\varepsilon_n})) - P_K(\mathcal{F}_0(f^\star_{\varepsilon_n}))\|_{H^s_0(Q)} \\ &+ \|P_S(\mathcal{F}'_0(f^\star_0)(f^\star_{\varepsilon_n} - f^\star_0))\|_{H^s_0(Q)} + \|o(\|f^\star_{\varepsilon_n} - f^\star_0\|_{H^s_0(Q)})\|_{H^s_0(Q)} \\ &\le \|P_S(\mathcal{F}'_0(f^\star_0)P_S(f^\star_{\varepsilon_n} - f^\star_0))\|_{H^s_0(Q)} + c\varepsilon_n + \|o(\|f^\star_{\varepsilon_n} - f^\star_0\|_{H^s_0(Q)})\|_{H^s_0(Q)} \,.\end{aligned} \tag{6.106}$$

By our assumption (i)

$$-(\mathcal{F}'_0 v, v)_{H^s_0(Q)} + (v, v)_{H^s_0(Q)} \ge \frac{\beta}{\alpha}(v, v)_{H^s_0(Q)} \quad \forall v \in S,$$

and hence

$$\begin{aligned}&\|P_S(\mathcal{F}'{}_0(f^\star_0)P_S(f^\star_{\varepsilon_n} - f^\star_0))\|_{H^s_0(Q)} \\ &\le (1 - \frac{\beta}{\alpha})\|P_S(f^\star_{\varepsilon_n} - f^\star_0)\|_{H^s_0(Q)} \le (1 - \frac{\beta}{\alpha})\|f^\star_{\varepsilon_n} - f^\star_0\|_{H^s_0(Q)} \,.\end{aligned} \tag{6.107}$$

Thus,

$$-\|P_S(\mathcal{F}'_0(f^\star_0)P_S(f^\star_{\varepsilon_n} - f^\star_0))\|_{H^s_0(Q)} \ge (\frac{\beta}{\alpha} - 1)\|f^\star_{\varepsilon_n} - f^\star_0\|_{H^s_0(Q)}$$

which, in view of (6.106), completes the proof of Proposition 6.1. □

6.3 Differentiability of the metric projection

Consider the set K as defined in Section 6.2,

$$K = \{f \in H^m_0(Q) | \psi_2(x) \ge f(x) \ge \psi_1(x), x \in Q\}, \tag{6.108}$$

where $\psi_2(x) > \psi_1(x)$, $x \in Q$, are given functions in $H^m_0(Q)$, $Q \subset \mathbb{R}^2$ and $m \ge 2$.

For $g \in H^m_0(Q)$, such that $f = P_K(g)$ we consider the following convex cone

$$S = T_K(f) \cap [g - P(g)]^\perp = T_K(f) \cap [f - g]^\perp \,. \tag{6.109}$$

Remark. 6.2. In general

$$S \ne \mathrm{cl}(C_K(f) \cap [f - g]^\perp) \tag{6.110}$$

in the Sobolev spaces $H^m_0(\Omega)$, $m = 2, 3, \dots$

Denote

$$\Xi_i = \{x \in Q | f(x) = \psi_i(x)\}, \quad i = 1, 2, \tag{6.111}$$

and assume that Ξ_i is compact for $i = 1, 2,$. Let $\mu_i \geq 0$ be Radon measures defined as

$$(g - f, \varphi)_{H_0^m(Q)} = \int \varphi d\mu_1 - \int \varphi d\mu_2 \quad 0 \leq \varphi \in C_0^\infty(Q) \ . \tag{6.112}$$

By sptμ_i we denote the support of μ_i, $i = 1, 2$. Note that spt$\mu_i \subset \Xi_i$ are compact for $i = 1, 2$.

Definition 6.1. A compact $F = \text{spt}\mu$ is admissible if

$$S = \text{cl}(C_K(f) \cap [f - g]^\perp) \ .$$

Corollary 6.1. *If a compact* $F = \text{spt}\mu$ *satisfies the condition that*
$$\varphi \in H_0^m(\Omega),\ \varphi = 0 \ \text{on } F \ \text{implies } \varphi \in H_0^m(Q \setminus F) \ ,$$
then sptμ *is admissible, and*

$$\begin{aligned} S &= T_K(f) \cap [g(0) - f]^\perp \\ &= \{\varphi \in H_0^m(Q) | \varphi(x) \geq 0 \ \text{on } \Xi_1 \ , \ \varphi(x) \leq 0 \ \text{on } \Xi_2 \ , \int \varphi d\mu = 0\} \ . \end{aligned} \tag{6.113}$$

6.4 Shape estimation problem for the wave equation

In the same way as before we can consider the shape estimation problem for the wave equation

$$y_{tt} - \Delta y = F \quad \text{in } \Omega_f \times (0, T), \tag{6.114}$$
$$y = 0 \quad \text{on } \partial\Omega_f \times (0, T), \tag{6.115}$$
$$y(0) = y_0 \ , \ y_t(0) = y_1 \quad \text{in } \Omega_f \ , \tag{6.116}$$

where $y_{tt} = \frac{\partial^2 y}{\partial t^2}$ and we denote $y = y(f)$.

We assume that F, z are defined in $D \times (0, T)$, y_0, y_1 are defined in D, and all data are sufficiently smooth.

For the functional

$$\begin{aligned} J_\alpha(f) &= J(\Omega_f) + \frac{\alpha}{2} \|f\|^2_{H_0^s(Q)} \\ &= \frac{1}{2} \int_0^T \int_{\Omega_f} (y(f) - z)^2 dx dt + \frac{\alpha}{2} \|f\|^2_{H_0^s(Q)} \end{aligned} \tag{6.117}$$

the directional derivative is given by

$$dJ_\alpha(f; h) = \int_0^T \int_{\Gamma(f)} h_\nu \frac{\partial p}{\partial n} \frac{\partial y}{\partial n} d\Gamma dt + \alpha(f, h)_{H_0^s(Q)} \ . \tag{6.118}$$

Here we assume $z = 0$ on $\Gamma(f) \times (0,T)$. The adjoint state p satisfies the following equation

$$p_{tt} - \Delta p = y - z \quad \text{in } \Omega_f \times (0,T), \tag{6.119}$$
$$p = 0 \quad \text{on } \partial\Omega_f \times (0,T), \tag{6.120}$$
$$p(T) = 0 \; , \; p_t(T) = 0 \quad \text{in } \Omega_f \; . \tag{6.121}$$

We can evaluate the shape derivative $y'(h_\nu)$, which satisfies the following wave equation

$$y'_{tt} - \Delta y' = 0 \quad \text{in } \Omega_f \times (0,T), \tag{6.122}$$
$$y' = -h_\nu \frac{\partial y}{\partial n} \quad \text{on } \partial\Omega_f \times (0,T), \tag{6.123}$$
$$y'(0) = 0 \; , \; y_t(0) = 0 \quad \text{in } \Omega_f \; . \tag{6.124}$$

As before we can evaluate the second derivative of the cost functional

$$\begin{aligned}
&d^2 J_\alpha(f; h, v) = \\
&-\int_0^T \int_{\Gamma(f)} \left(h_\nu \frac{\langle \nabla f(x'), \nabla v(x') \rangle_{\mathbb{R}^3}}{(1 + |\nabla f(x')|^2)} + v_\nu \frac{\langle \nabla f(x'), \nabla h(x') \rangle_{\mathbb{R}^3}}{(1 + |\nabla f(x')|^2)} \right) \frac{\partial p}{\partial n} \frac{\partial y}{\partial n} d\Gamma dt \\
&- 2 \int_0^T \int_{\Gamma(f)} \kappa v_\nu h_\nu \frac{\partial p}{\partial n} \frac{\partial y}{\partial n} d\Gamma dt + \int_0^T \int_{\Gamma(f)} h_\nu v_\nu \frac{\partial}{\partial n} \left(\frac{\partial y}{\partial n} \frac{\partial p}{\partial n} \right) d\Gamma dt \\
&+ \int_0^T \int_{\Gamma(f)} h_\nu \left(\frac{\partial p'}{\partial n} \frac{\partial y}{\partial n} + \frac{\partial p}{\partial n} \frac{\partial y'}{\partial n} \right) d\Gamma dt + \alpha(h, v)_{H_0^s(Q)} \; ,
\end{aligned}$$

where κ is the mean curvature on $\Gamma(f)$, and the shape derivative $y'(v_\nu)$ satisfies (6.122)–(6.124) with h_ν replaced by v_ν. The shape derivative $p'(v_\nu)$ satisfies the following wave equation

$$p'_{tt} - \Delta p' = y' - z' \quad \text{in } \Omega_f \times (0,T), \tag{6.125}$$
$$p' = -v_\nu \frac{\partial p}{\partial n} \quad \text{on } \partial\Omega_f \times (0,T), \tag{6.126}$$
$$p'(T) = 0 \; , \; p'_t(T) = 0 \quad \text{in } \Omega_f \; , \tag{6.127}$$

where $z'(v_\nu)$ denotes the shape derivative of the observation z.

Stability results for shape optimization problems corresponding to the wave equation can be obtained in much the same way as it was done for the Laplace equation in Section 6.3. We refer the reader to (Lasiecka, Sokolowski, 1991) for sensitivity analysis of convex control problems for the wave equation.

7 Domain optimization problem for parabolic equations

In this section we study the heat equation and related domain optimization problems. The problem is to select in an optimal way two C^1 curves being interfaces between subsets of the lateral boundary of a parabolic cylinder with Neumann and Dirichlet boundary conditions. We prove the existence of an optimal solution and derive first order necessary optimality conditions.

Let $Q = \Omega \times (0,T)$, $T > 0$, be a cylinder in $\mathbb{R}^3$, where $\Omega \subset \mathbb{R}^2$ is a bounded domain with smooth boundary $\partial\Omega$. Assume we are given two compact simply connected subsets $\mathcal{K}_i \subset \partial\Omega$, $i = 1, 2$, $\mathcal{K}_1 \cap \mathcal{K}_2 = \emptyset$, and two sufficiently smooth curves $\mathcal{X}_i : [0,T] \to \mathcal{K}_i \subset \mathbb{R}^3$ on $\overline{\Sigma} = \partial\Omega \times [0,T]$. The set $\overline{\Sigma}$ is composed of two subsets $\overline{\Sigma}_i$, $i = 1, 2$, with

$$\overline{\Sigma}_1 \cap \overline{\Sigma}_2 = \bigcup_{t\in[0,T]} \{\mathcal{X}_1(t) \times t\} \cup \{\mathcal{X}_2(t) \times t\} \ .$$

We assume that $\mathcal{X}_i \in H^2(0,T;\mathbb{R}^2)$, $i = 1, 2$, and denote $\mathcal{X} = \{\mathcal{X}_1, \mathcal{X}_2\} \in \mathcal{U} = H^2(0,T;\mathbb{R}^4)$. The set of admissible curves $\mathcal{U}_{ad} = \{\mathcal{X} \in H^2(0,T;\mathbb{R}^4) | \mathcal{X}(t) \in \mathcal{K}_1 \times \mathcal{K}_2\}$ is obviously nonconvex.

The domain optimization problem can be formulated as

$$\inf_{\mathcal{X}\in\mathcal{U}_{ad}} J(\mathcal{X})$$

with

$$J(\mathcal{X}) = \frac{1}{2}\int_0^T \int_\Omega (y(\mathcal{X};x,t) - y_d(x,t))^2 dxdt + \frac{\alpha}{2}\|\mathcal{X}\|^2_{\mathcal{U}} \ ,$$

where $y_d \in L^2(Q)$ is given, $\alpha \geq 0$ is a regularization parameter.

For a given function $F \in L^2(Q)$, the function $y(\mathcal{X})(x,t)$, $\mathcal{X} \in \mathcal{U}_{ad}$, $(x,t) \in \Omega \times (0,T)$ satisfies the heat equation

$$\frac{\partial y}{\partial t} - \Delta y = F, \qquad \text{in } Q = \Omega \times (0,T)$$

with an initial condition in Ω and mixed boundary conditions imposed on the lateral boundary Σ of Q.

Boundary conditions imposed on $y = y(\mathcal{X})$ are in the form of a nonhomegeneous Dirichlet condition on Σ_2 and a nonhomogeneous Neumann condition on Σ_1. Interfaces $\{\mathcal{X}_1, \mathcal{X}_2\} \equiv \overline{\Sigma}_1 \cap \overline{\Sigma}_2$ are given as admissible curves $\mathcal{X} = \{\mathcal{X}_1, \mathcal{X}_2\} \in \mathcal{U}_{ad}$.

The existence of an optimal domain, ie. the existence of an element $\mathcal{X}^\star \in \mathcal{U}_{ad}$ such that $J(\mathcal{X}^\star) \leq J(\mathcal{X})$ for all $\mathcal{X} \in \mathcal{U}_{ad}$, is guaranteed under the assumption that there exists a family of admissible domains which is compact in an appropriate sense. For our problem it is enough to assume that for any minimizing sequence of domains $\{Q_m\}$ there exists a subsequence, still denoted by $\{Q_m\}$, such that the sequence of characteristic functions,

$$\chi_m \equiv \text{ characteristic function of } \Sigma_2^m,$$

converges in $L^2(\Sigma)$ to a characteristic function $\overline{\chi}$.

To prove the existence of an optimal domain we use the standard technique, namely, a regularizing term is introduced for $\alpha > 0$.

7.1 Parabolic equation in a variable domain

Consider the following parabolic equation

$$\begin{cases} \frac{\partial y}{\partial t} - \Delta y = F & \text{in } Q = \Omega \times (0,T) \\ \frac{\partial y}{\partial n} = f & \text{on } \Sigma_1 \\ y = g & \text{on } \Sigma_2 \\ y(x,0) = y_0(x) & \text{in } \Omega \end{cases} \tag{7.1}$$

where $f, g \in L^2(\Sigma)$ are given,

$$\Sigma_i = \{ \ (x,t) \in \Gamma_i(t) \times \{t\}, \ t \in (0,T) \ \}, \quad i = 1,2 \tag{7.2}$$

$$\text{and } \overline{\Gamma_1(t)} \cap \overline{\Gamma_2(t)} = \mathcal{X}_1(t) \cup \mathcal{X}_1(t) \text{ for all } t \in [0,T] \ . \tag{7.3}$$

Here Ω is a given domain, $\overline{\partial\Omega} = \Gamma_1(0) \cup \Gamma_2(0) \cup \{\mathcal{X}_1(0)\} \cup \{\mathcal{X}_2(0)\}$, and $\Gamma_0 \equiv \Gamma_0(0)$, $\Gamma_1 \equiv \Gamma_1(0)$.

To derive first order necessary optimality conditions for the optimization problem defined in a variable domain setting, we assume that

$$\mathcal{X}_1(t) = T_t(V)(\mathcal{X}_1(0)), \ \mathcal{X}_2(t) = T_t(V)(\mathcal{X}_2(0)), \ \forall t \in [0,T]$$

for a given vector field $V \in C(0, T+\delta_1; C^1(\mathbb{R}^2, \mathbb{R}^2))$, $\delta_1 > 0$, V is given with the support in a compact neighbourhood of $\Sigma = \partial\Omega \times (0,T) \subset \mathbb{R}^3$.

We investigate differentiability of a solution to system (7.1) with respect to perturbations of curves $\mathcal{X}_1(t), \mathcal{X}_2(t) \in \partial\Omega \subset \mathbb{R}^2, t \in [0,T]$.

Let $\mathcal{X}_1^\varepsilon, \mathcal{X}_2^\varepsilon$ be perturbed curves defined below with $\varepsilon \in [0,\delta)$ as a parameter. Denote by y_ε a solution to the parabolic equation in the perturbed domain,

$$\begin{cases} \frac{\partial y_\varepsilon}{\partial t} - \Delta y_\varepsilon = F & \text{in } Q = \Omega \times (0,T) \\ \frac{\partial y_\varepsilon}{\partial n} = f & \text{on } \Sigma_1^\varepsilon \\ y_\varepsilon = g & \text{on } \Sigma_0^\varepsilon \\ y_\varepsilon(x,0) = y_0(x) & \text{in } \Omega \end{cases} \tag{7.4}$$

where $f = f_\varepsilon$, $g = g_\varepsilon$ is the restriction of the function f (resp. g) defined on Σ to Σ_2^ε (resp. Σ_1^ε). We assume that for sufficiently small t the curves $\mathcal{X}_1(\cdot), \mathcal{X}_2(\cdot)$ are not perturbed, ie.,

$$\mathcal{X}_1^\varepsilon(t) = \mathcal{X}_1(t) \ , \mathcal{X}_2^\varepsilon(t) = \mathcal{X}_2(t), \ t \in [0,\delta_1), \ \delta_1 > 0 \ . \tag{7.5}$$

Furthermore, we assume that $\mathcal{X}_1^\varepsilon(t),\ \mathcal{X}_2^\varepsilon(t) \in \partial\Omega,\ \mathcal{X}_1^\varepsilon(t) \neq \mathcal{X}_2^\varepsilon(t),\ \forall t \in [0,T]$, $\forall \varepsilon \in [0,\delta)$ and there exist the limits in $H^2(0,T;\mathbb{R}^2)$,

$$h_i = \lim_{\varepsilon \downarrow 0} \frac{1}{\varepsilon}(\mathcal{X}_i^\varepsilon - \mathcal{X}_i^0),$$

for $i = 1,2$, where $h_1(t),\ h_2(t),\ t \in (0,T)$, are the tangent vectors on $\Sigma = \partial\Omega \times (0,T)$.

We derive the derivative y' of a solution $y_\varepsilon = y(\mathcal{X}_\varepsilon) \in L^2(Q)$ to (7.4), with respect to ε, at $\varepsilon = 0$, assuming that the derivative exists in the space $L^2(Q)$. By the transposition method, system (7.4) is equivalent to the following integral identity

$$\begin{aligned} y_\varepsilon \in L^2(Q): \quad & \int_0^T \int_\Omega y_\varepsilon \Big(-\frac{\partial \varphi}{\partial t} - \Delta\varphi\Big) dx dt = \int_0^T \int_\Omega F\varphi dx dt \\ & + \int_{\Sigma_2^\varepsilon} g \frac{\partial \varphi}{\partial n} d\Sigma + \int_{\Sigma_1^\varepsilon} f\varphi d\Sigma + \int_\Omega y_0(x)\varphi(0,x)dx, \\ & \forall \varphi \in H^{2,1}(Q),\ \varphi(T) = 0 \text{ in } \Omega,\ \varphi = 0 \text{ on } \Sigma_2^\varepsilon \,. \end{aligned} \tag{7.6}$$

Let φ be a sufficiently smooth test function independent of the parameter ε, ie. $\varphi = 0$ in a small neighbourhood of $\Sigma_2^\varepsilon \subset \mathbb{R}^3$. If y_ε is differentiable with respect to ε, at $\varepsilon = 0$, the derivative satisfies the following integral identity

$$\begin{aligned} y' \in L^2(Q): \quad & \int_0^T \int_\Omega y'(-\frac{\partial \varphi}{\partial t} - \Delta\varphi) dx dt \\ & = \int_{\partial\Sigma_2} g \frac{\partial \varphi}{\partial n} \mathcal{L}(h) d\ell + \int_{\partial\Sigma_1} f\varphi\mathcal{L}(h) d\ell \\ & \forall \varphi \in H^{2,1}(Q),\ \varphi(T) = 0 \text{ in } \Omega,\ \varphi = 0 \text{ on } \Sigma_2^\varepsilon \,, \end{aligned} \tag{7.7}$$

where

$$h = \begin{cases} h_1 = \lim_{\varepsilon\downarrow 0} \frac{1}{\varepsilon}(\mathcal{X}_1^\varepsilon - \mathcal{X}_0), & \text{on } \mathcal{X}_1^0 \equiv \mathcal{X}_1 \\ h_2 = \lim_{\varepsilon\downarrow 0} \frac{1}{\varepsilon}(\mathcal{X}_2^\varepsilon - \mathcal{X}_2^0), & \text{on } \mathcal{X}_2^0 \equiv \mathcal{X}_2 \end{cases} \tag{7.8}$$

$$dl = (1 + \Big|\frac{d\mathcal{X}_i}{dt}(t)\Big|^2)^{1/2} dt, \qquad \text{on } \mathcal{X}_i \,, \quad i = 1,2 \,,$$

$$\mathcal{L}(h)(t) = \frac{1}{\alpha(t)}[h_i(t) - (h_i(t), V(t,\mathcal{X}_i(t))_{\mathbb{R}^2} / \|V(t,\mathcal{X}_i(t))\|_{\mathbb{R}^2}], \text{ on } \mathcal{X}_i \,, \quad i = 1,2 \,.$$

Remark. 7.1. If φ is not smooth enough, the right-hand side of (7.7) is not well defined. Then, under the assumption that $\varepsilon \to y_\varepsilon$ is differentiable, the corresponding result can be stated as follows.

Proposition 7.1. *There exist distributions $\mathcal{G}_0$, $\mathcal{G}_1$ supported on $\partial\Sigma_2$, $\partial\Sigma_1$, respectively, such that*

$$\begin{aligned} y' \in L^2(Q) : \int_0^T \int_\Omega y'(-\frac{\partial \varphi}{\partial t} - \Delta\varphi) dx dt \qquad (7.9) \\ = < \mathcal{G}_0(\varphi), h_1 >_{\partial\Sigma_2} + < \mathcal{G}_1(\varphi), h_2 >_{\partial\Sigma_1} \\ \forall \varphi \in H^{2,1}(Q),\ \varphi(T) = 0 \text{ in } \Omega,\ \varphi = 0 \text{ on } \Sigma_2 = \Sigma_2^0 . \end{aligned}$$

In view of (7.7) we can identify the distributions $\mathcal{G}_0$, $\mathcal{G}_1$ for sufficiently smooth test functions.

7.2 Differentiability of the cost functional

Consider the following cost functional depending on ε, $\mathcal{X}_\varepsilon = (\mathcal{X}_1^\varepsilon, \mathcal{X}_2^\varepsilon)$,

$$\begin{aligned} j(\varepsilon) = J(\mathcal{X}_\varepsilon) = \frac{1}{2} \int_0^T \int_\Omega (y_\varepsilon - y_d)^2 dx dt \qquad (7.10) \\ + \frac{\alpha}{2} \|\mathcal{X}_\varepsilon\|_{\mathcal{U}}^2 + \frac{\beta_1}{2} \int_{\Sigma_2^\varepsilon} g^2 d\Sigma + \frac{\beta_2}{2} \int_{\Sigma_1^\varepsilon} f^2 d\Sigma \end{aligned}$$

where $\varepsilon \in [0,\delta)$, $\alpha > 0$, $\beta_1, \beta_2 \geq 0$ are given, $\mathcal{U} = H^2(0,T;\mathbb{R}^4)$ and $\mathcal{X}_1^\varepsilon(t), \mathcal{X}_2^\varepsilon(t) \in \partial\Omega$, $\forall t \in [0,T]$, $\forall \varepsilon \in [0,\delta)$.

We obtain the derivative

$$j'(0) = \lim_{\varepsilon \to 0} \frac{1}{\varepsilon}(j(\varepsilon) - j(0))$$

which is linear with respect to the direction $h = (h_1, h_2)$. To this end we differentiate subsequent terms in (7.10). It follows that

$$\frac{d}{d\varepsilon}\left[\frac{1}{2}\int_0^T \int_\Omega (y_\varepsilon - y_d)^2 dx dt\right]_{|\varepsilon=0} = \int_0^T \int_\Omega (y_\varepsilon - y_d) y' dx dt\,, \qquad (7.11)$$

$$\frac{d}{d\varepsilon}\left[\frac{\alpha}{2}\|\mathcal{X}_\varepsilon\|_{\mathcal{U}}^2\right]_{|\varepsilon=0} = \alpha \Sigma_{i=1}^2 (\mathcal{X}_i, h_i)_{H^2(0,T;\mathbb{R}^2)}\,, \qquad (7.12)$$

$$\frac{d}{d\varepsilon}\left[\frac{\beta_1}{2}\int_{\Sigma_2^\varepsilon} g^2 d\Sigma\right]_{|\varepsilon=0} = \beta_1 \int_{\partial\Sigma_2} g^2 \mathcal{L}(h_1) dl\,, \qquad (7.13)$$

$$\frac{d}{d\varepsilon}\left[\frac{\beta_2}{2}\int_{\Sigma_1^\varepsilon} f^2 d\Sigma\right]_{|\varepsilon=0} = \beta_2 \int_{\partial\Sigma_1} f^2 \mathcal{L}(h_2) dl\,. \qquad (7.14)$$

The adjoint state $p \in W(0,T)$ satisfies by definition the following system

$$\begin{cases} -\frac{\partial p}{\partial t} - \Delta p = y - y_d & \text{in } Q = \Omega \times (0,T) \\ \frac{\partial p}{\partial n} = 0 & \text{on } \Sigma_1 \\ y = 0 & \text{on } \Sigma_2 \\ p(x,T) = 0 & \text{in } \Omega\,. \end{cases} \qquad (7.15)$$

Then, in view of (7.15), (7.9) we can rewrite (7.11) as

$$\int_0^T \int_\Omega (y - y_d) y' dx dt = \int_0^T \int_\Omega y' (-\frac{\partial p}{\partial t} - \Delta p) dx dt$$
$$= < \mathcal{G}_0(p), h_1 >_{\partial \Sigma_2} + < \mathcal{G}_1(p), h_2 >_{\partial \Sigma_1}$$
$$= \int_{\partial \Sigma_2} g \frac{\partial p}{\partial n} \mathcal{L}(h) d\Sigma + \int_{\partial \Sigma_1} f p \mathcal{L}(h) d\Sigma, \qquad \forall h = (h_1, h_2)$$

provided that the adjoint state is sufficiently smooth.

Now we are going to show the existence of the shape derivative y'. Therefore, we evaluate the material derivative.

7.3 Shape sensitivity analysis

Consider the cylinder $\mathcal{Q} = \Omega_\xi \times (0, T)$, $(\xi, s) \in \mathcal{Q}$, where $\partial \Omega_\xi = \Gamma_1 \cup \Gamma_2$. Assume we are given points on $\partial \Omega_\xi$, $\xi_i = \mathcal{X}_i(0)$, $i = 1, 2$, such that $\overline{\Gamma}_1 \cap \overline{\Gamma}_2 = \{\xi_1\} \cup \{\xi_2\}$. The lateral boundary of $\mathcal{Q}$ takes the form $\Sigma_\xi = \Sigma_1^\xi \cup \Sigma_2^\xi$, where $\Sigma_i^\xi = \Gamma_i \times (0, T)$ for $i = 1, 2$. The interfaces between Σ_1^ξ and Σ_2^ξ are two segments $\{\xi_i\} \times (0, T)$ for $i = 1, 2$.

We define a transformation $\mathcal{T}_\varepsilon$ of a neighbourhood of the cylinder $\mathcal{Q}$ onto a neighbourhood of the cylinder Q, $(x, t) \in Q$, such that for interfaces we have

$$\mathcal{T}_\varepsilon : \mathbb{R}^3 \supset \{\xi_i\} \times (0, T) \mapsto \cup_{t \in (0,T)} \{(\mathcal{X}_i(t), t)\} \subset \mathbb{R}^3 \ .$$

The transformation $\mathcal{T}_\varepsilon$ is constructed with the help of the vector fields V, $H \in C^1(0, T + \delta; C^2(D, D))$, $\delta > 0$, where D is an open and bounded set in $\mathbb{R}^3$ such that $D \supset \overline{\Omega}$. We assume that

$$V(t, x) \cdot n(x) = H(t, x) \cdot n(x) = 0, \ x \in \partial \Omega, \ t \in (0, T + \delta)$$
$$V(t, x) = H(t, x) = 0, \ x \in D, \ t \in (0, \delta)$$

where $n(x)$, $x \in \partial \Omega$, is the unit outward normal vector. Moreover, we assume that the cylinder $Q = \Omega_x \times (0, T)$, $x \in \Omega, t \in (0, T)$ is the image of the cylinder $\mathcal{Q} = \Omega_\xi \times (0, T)$, $\xi \in \Omega, s \in (0, T)$, under the transformation $\mathcal{T}_\varepsilon = T_t(V_\varepsilon)$ defined as follows

$$T_t(V_\varepsilon)(\xi) = X_\varepsilon(t) : \qquad X_\varepsilon(0) = \xi \text{ and for } t \in (0, T)$$
$$\frac{dX_\varepsilon}{dt}(t) = V(t, X_\varepsilon(t)) + \varepsilon H(t, X_\varepsilon(t)) \equiv V_\varepsilon(t, X_\varepsilon(t)) \ . \tag{7.16}$$

Here $X_\varepsilon(t) = X_\varepsilon(t, \xi)$ is a solution to (7.16), where ξ stands for the initial condition.

Using the transformation $\mathcal{T}_\varepsilon$, the image of $(\xi, s) \in \mathcal{Q}$ is $(X^\varepsilon(t), t) \in Q$ with $t = s$,

$$\mathcal{T}_\varepsilon : \ \mathcal{Q} \ni (\xi, s) \mapsto (X_\varepsilon(t, \xi), t) \in Q \ .$$

For the initial conditions ξ_i , $i = 1, 2$, we have

$$\mathcal{T}_\varepsilon : \mathcal{Q} \ni (\xi_i, s) \mapsto (\mathcal{X}_i^\varepsilon(t), t) \in Q$$

where $\mathcal{X}_i^\varepsilon(t) \equiv X_\varepsilon(t, \xi_i)$.

For simplicity, consider system (7.1) with homogeneous Dirichlet boundary conditions, ie. for $g \equiv 0$. Denote $y^\varepsilon \equiv y_\varepsilon \circ \mathcal{T}_\varepsilon$. By standard arguments, y^ε satisfies the following parabolic equation.

Proposition 7.2. *For* $\varepsilon > 0$, ε *small enough,* $y^\varepsilon \in W(0,T)$,

$$\begin{aligned}
&y^\varepsilon(s) \in H^1_{\Gamma_0}(\Omega) \ : \\
&\int_\Omega [a_\varepsilon(\xi, s)\frac{\partial y^\varepsilon}{\partial s}(\xi, s) - b_\varepsilon(\xi, s) \cdot \nabla y^\varepsilon(\xi, s)]\varphi(\xi, s) d\xi \\
&+ \int_\Omega (A_\varepsilon(\xi, s) \cdot \nabla y^\varepsilon(\xi, s), \nabla\varphi(\xi, s))_{\mathbb{R}^2} d\xi \\
&= \int_\Omega F_\varepsilon(\xi, s)\varphi(\xi, s) d\xi + \int_{\Gamma_1} \sigma_\varepsilon(\xi, s) f_\varepsilon(\xi, s)\varphi(\xi, s) d\Gamma, \\
&\forall \varphi \in H^1_{\Gamma_0}(\Omega) = \{\ \varphi \in H^1(\Omega) \mid \varphi = 0, \quad on\ \Gamma_0\ \}, \qquad for\ a.e.\ t \in (0,T)
\end{aligned} \tag{7.17}$$

with the initial condition

$$y^\varepsilon(\xi, 0) = y_0(\xi), \qquad in\ \Omega$$

where the coefficients are defined as follows:

$$\begin{aligned}
a_\varepsilon(\xi, s) &= \det(DT_s(V_\varepsilon)) \\
b_\varepsilon(\xi, s) &= \det(DT_s(V_\varepsilon)) DT_s^{-1} \circ T_s(V_\varepsilon) \cdot V_\varepsilon \circ T_s(V_\varepsilon) \\
A_\varepsilon(\xi, s) &= \det(DT_s(V_\varepsilon)) DT_s^{-1} \circ T_s(V_\varepsilon) \cdot {}^*DT_s^{-1} \circ T_s(V_\varepsilon) \\
F_\varepsilon(\xi, s) &= \det(DT_s(V_\varepsilon)) F \circ T_s(V_\varepsilon) \\
\sigma_\varepsilon(\xi, s) &= \det(DT_s(V_\varepsilon)) \|{}^*DT_s^{-1} \circ T_s(V_\varepsilon) \cdot n\|_{\mathbb{R}^2} \\
f_\varepsilon(\xi, s) &= f \circ T_s(V_\varepsilon) = (f \circ T_s(V_\varepsilon))(\xi, s)
\end{aligned}$$

Here $DT_s(V_\varepsilon)$ *denotes the Jacobian of the mapping* $T_s(V_\varepsilon)$ *and* ${}^*DT_s(V_\varepsilon)$ *is the transpose of* $DT_s(V_\varepsilon)$.

On the lateral boundary Σ we have

$$\mathcal{X}_i^\varepsilon(t) \equiv T_t(V_\varepsilon)(\mathcal{X}_i(0)),\ \forall t \in [0,T],\ i = 1, 2\,,$$

and for $\varepsilon = 0$, $\mathcal{X}_i(t) = T_t(V)(\mathcal{X}_i(0)) \equiv T_t(V)(\xi_i)$, $\forall t \in [0,T]$, $i = 1, 2$.

Proposition 7.3. *There exist limits*

$$h_i = \lim_{\varepsilon \downarrow 0} \frac{1}{\varepsilon}(\mathcal{X}_i^\varepsilon - \mathcal{X}_i), \quad i = 1, 2,$$

in $C([0,T])$, *where* $h_i(\cdot)$, $i = 1,2$, *are solutions to the following linearized systems of ordinary differential equations*

$$\begin{aligned}
&\frac{dh_i}{dt}(t) = DV(t, \mathcal{X}_i(t)) \cdot h_1(t) + H(t, \mathcal{X}_i(t)), \quad t \in (0,T),\\
&h_i(0) = 0.
\end{aligned}$$

Now we give results on differentiability of solutions to system (7.17) with respect to the parameter ε.

Let y^ε be a solution to (7.17) defined on the cylinder $\mathcal{Q}$. The following lemma establishes the existence of the strong material derivative $\dot{y} \in W(0,T)$.

Lemma 7.1. *For* $\varepsilon > 0$, ε *small enough*

$$y^\varepsilon = y^0 + \varepsilon \dot{y} + o(\varepsilon) \qquad \text{in } W(0,T),$$

where $\|o(\varepsilon)\|_{W(0,T)}/\varepsilon \to 0$ *with* $\varepsilon \to 0$, *and the material derivative* $\dot{y}$ *satisfies the following parabolic equation*

$$\begin{aligned}
&\dot{y}(.,s) \in H^1_{\Gamma_0}(\Omega) \ :\\
&\int_\Omega [a_0(\xi,s)\frac{\partial \dot{y}}{\partial s}(\xi,s) - b_0(\xi,s)\cdot\nabla\dot{y}(\xi,s)]\varphi(\xi,s)d\xi\\
&+\int_\Omega (A_0(\xi,s)\cdot\nabla\dot{y}(\xi,s), \nabla\varphi(\xi,s))_{\mathbb{R}^2}d\xi \qquad\qquad (7.18)\\
&+\int_\Omega [a_0'(\xi,s)\frac{\partial y^0}{\partial s}(\xi,s) - b_0'(\xi,s)\cdot\nabla y^0(\xi,s)]\varphi(\xi,s)d\xi\\
&+\int_\Omega (A_0'(\xi,s)\cdot\nabla y^0(\xi,s), \nabla\varphi(\xi,s))_{\mathbb{R}^2}d\xi\\
&=\int_\Omega F_0'(\xi,s)\varphi(\xi,s)d\xi + \int_{\Gamma_1} \sigma_2'(\xi,s)f_0(\xi,s)\varphi(\xi,s)d\Gamma\\
&+\int_{\Gamma_1} \sigma_2(\xi,s)f_0'(\xi,s)\varphi(\xi,s)d\Gamma\\
&\forall\varphi \in H^1_{\Gamma_0}(\Omega) = \{\varphi \in H^1(\Omega) | \varphi = 0, \text{ on } \Gamma_0\}, \text{ for a.e. } s \in (0,T),\\
&\text{with the initial conditions}\\
&\dot{y}(\xi,0) = 0, \qquad \text{in } \Omega
\end{aligned}$$

where

$$a_0'(\xi,s) = \lim_{\varepsilon\to 0}\frac{1}{\varepsilon}(a_\varepsilon(\xi,s) - a_0(\xi,s)) \qquad \text{in } C^1(0,T;C(\overline{\Omega})).$$

Since $y \in L^2(0,T;H^1(\Omega))$, in view of Lemma 7.1, the existence of the shape derivative y' follows by standard arguments.

Corollary 7.1. *There exists the shape derivative y' in $L^2(Q)$ of the following form*

$$y' \equiv \frac{\partial y_\varepsilon}{\partial \varepsilon}_{|\varepsilon=0} = \dot{y} \circ T_t^{-1}(V) - \nabla y \cdot \frac{\partial T_\varepsilon}{\partial \varepsilon}_{|\varepsilon=0} \circ T_t^{-1}(V) \in L^2(Q) \ .$$

7.4 Optimization problem

Now we derive necessary optimality conditions for optimization problems of the form:

minimize the cost functional

$$\begin{aligned} J(\mathcal{X}) =& \frac{1}{2}\int_0^T \int_\Omega (y_\varepsilon - y_d)^2 dx dt \\ &+ \frac{\alpha}{2}\|\mathcal{X}\|_{\mathcal{U}} + \frac{\beta_1}{2}\int_{\Sigma_2} g^2 d\Sigma + \frac{\beta_2}{2}\int_{\Sigma_1} f^2 d\Sigma \end{aligned} \tag{7.19}$$

subject to $\mathcal{X}(t) \in \mathcal{U}_{ad}$, where $\alpha > 0,\ \beta_1, \beta_2 \geq 0$ are given.

Proposition 7.4. *There exists an optimal solution $\overline{\mathcal{X}}$ to the above problem.*

Proof. Let $\{\mathcal{X}_m\}$ be a minimizing sequence. Then the sequence of solutions $\{y_m\}$ to (7.6) is bounded in $L^2(Q)$, and thus, there exists an element $\overline{y} \in L^2(Q)$ such that for a subsequence, still denoted by $\{y_m\}$,

$$y_m \rightharpoonup \overline{y}, \text{ weakly in } L^2(Q) \ . \tag{7.20}$$

For any $m = 1, 2, \ldots$, the function y_m satisfies the integral identity

$$\begin{aligned} y_m \in L^2(Q) \ : \ & \int_0^T \int_\Omega y_m\Big(-\frac{\partial \varphi}{\partial t} - \Delta\varphi\Big) dx dt = \int_0^T \int_\Omega F\varphi dx dt \\ &+ \int_{\Sigma_2^m} g\frac{\partial \varphi}{\partial n} d\Sigma + \int_{\Sigma_1^m} f\varphi d\Sigma + \int_\Omega y_0(x)\varphi(0,x) dx \\ &\forall \varphi \in H^{2,1}(Q),\ \varphi(T) = 0,\ \varphi = 0 \ \text{ on } \Sigma_2^m \ . \end{aligned} \tag{7.21}$$

This identity can be rewritten as

$$\begin{aligned} y_m \in L^2(Q) \ : \ & \int_0^T \int_\Omega y_m\Big(-\frac{\partial \varphi}{\partial t} - \Delta\varphi\Big) dx dt = \int_0^T \int_\Omega F\varphi dx dt \\ &+ \int_\Sigma [\chi_m g\frac{\partial \varphi}{\partial n} + (1 - \chi_m) f\varphi] d\Sigma + \int_\Omega y_0(x)\varphi(0,x) dx, \end{aligned} \tag{7.22}$$

where χ_m is the characteristic function of Σ_2^m.

Since the sequence $\{\mathcal{X}_m\}$ is bounded in $\mathcal{U}$, there exist an element $\overline{\mathcal{X}}$ such that passing to a subsequence, if necessary, we have

$$\mathcal{X}_m \rightharpoonup \overline{\mathcal{X}} \quad \text{weakly in } H^2(0,T;\mathbb{R}^4)\ . \tag{7.23}$$

Furthermore, $\overline{\mathcal{X}}_1(t) \in \mathcal{K}_1$, $\overline{\mathcal{X}}_2(t) \in \mathcal{K}_2$, $\forall t \in [0,T]$, thus $\overline{\mathcal{X}} \in \mathcal{U}_{ad}$. In view of (7.23),

$$\chi_m \to \overline{\chi} \quad \text{in } L^2(Q) \tag{7.24}$$

where $\overline{\chi}$ is the characteristic function of the subset of Σ defined by the curves $\overline{\mathcal{X}}_1, \overline{\mathcal{X}}_2$. We can pass to the limit in (7.22) which completes the proof. □

In order to derive first order necessary optimality conditions we assume that there exists a vector field $\overline{V}$ such that

$$\overline{\mathcal{X}}_i(t) \equiv T_t(\overline{V})(\mathcal{X}_i(0)) \qquad \forall t \in [0,T], \quad i = 1,2\ .$$

A vector field H defines an admissible perturbation of the optimal solution $\overline{\mathcal{X}}$ provided that for $\varepsilon > 0$, ε small enough, $T_t(V_\varepsilon)(\mathcal{X}_1(0)) \in \mathcal{K}_1$, $T_t(V_\varepsilon)(\mathcal{X}_2(0)) \in \mathcal{K}_2$, $\forall t \in [0,T]$, where $V_\varepsilon = \overline{V} + \varepsilon H$.

For simplicity, we denote an optimal solution to the optimization problem by $\mathcal{X}$.

Theorem 7.1. *An optimal solution $\mathcal{X} \in \mathcal{U}_{ad}$ satisfies the following optimality system*

State equation:

$$\begin{cases} \frac{\partial y}{\partial t} - \Delta y = F & \text{in } Q = \Omega \times (0,T) \\ \frac{\partial y}{\partial n} = f & \text{on } \Sigma_1 \\ y = g & \text{on } \Sigma_2 \\ y(x,0) = y_0(x) & \text{in } \Omega \end{cases} \tag{7.25}$$

Adjoint state equation:

$$\begin{cases} -\frac{\partial p}{\partial t} - \Delta p = y - y_d & \text{in } Q = \Omega \times (0,T) \\ \frac{\partial p}{\partial n} = 0 & \text{on } \Sigma_1 \\ y = 0 & \text{on } \Sigma_2 \\ p(x,T) = 0 & \text{in } \Omega \end{cases} \tag{7.26}$$

Optimality conditions:

$$\begin{aligned} &\int_{\partial\Sigma_2} g\frac{\partial p}{\partial n}\mathcal{L}(h)d\Sigma + \int_{\partial\Sigma_1} fp\mathcal{L}(h)d\Sigma + \alpha\Sigma_{i=1}^2(\mathcal{X}_i, h_i)_{\mathcal{U}} \\ &+ \beta_1\int_{\partial\Sigma_2} g^2\mathcal{L}(h_1)dl + \beta_2\int_{\partial\Sigma_1} f^2\mathcal{L}(h_2)dl \geq 0, \qquad \forall h = (h_1,h_2) \end{aligned} \tag{7.27}$$

for any admissible vector field H, where $\mathcal{L}(h)$ is defined by (7.8) and h_1, h_2 are given by Proposition 2.

Remark. 7.2. The same arguments apply to the shape functional $\mathcal{J}_1$,

$$\mathcal{J}_1(Q) = \min_{(f,g)\in L^2(\Sigma)} \mathcal{J}(Q) .$$

8 Shape sensitivity analysis for thin shells

In this section we present the shape sensitivity analysis of an elastic thin shell. The displacement derivatives are used to derive the shape derivatives of shape functionals for thin shells.

8.1 Thin shells

The geometry of a general thin shell can be characterised by two different mappings:

i) the mapping φ which defines the middle surface S of the shell as the image of the closure of a bounded domain $\mathcal{O}$ of the plane;
ii) the mapping e which defines the thickness of the shell at any point of the middle surface along the normal of this surface.

The shape optimization problem for such a shell consists in finding the geometry of the shell (middle surface and thickness) which minimizes a given functional (for example, the weight of the shell) and satisfies some constraints (for example, bounds on the thickness, on the strain energy, on the displacements).

Example. We can consider the following shape functional as an example,

$$J(S) = \int_S \mathfrak{F}(x, \mathbf{u}(S)(x), e(S)(x))d\Gamma(x)$$

where $\mathbf{u}(S)$ is the displacement field of the shell.

We denote by $\mathfrak{J}(\varphi, e; \mathbf{u}) = J(S)$ the integral functional defined on the reference domain $\mathcal{O}$,

$$\mathfrak{J}(\varphi, e; \mathbf{u}) = \int_{\mathcal{O}} \mathfrak{F}(\varphi(\xi), \mathcal{L}\cdot\mathbf{u}(\xi), e(\xi))dS(\xi)$$

with

$$\mathcal{L}\cdot\mathbf{u}(\xi) \equiv \mathbf{u}(S)(\varphi(\xi)),$$
$$e(\xi) \equiv e(S)(\varphi(\xi)) \quad \xi \in \mathcal{O},$$

where $\mathbf{u}(\xi) = \text{col}(u_1(\xi), u_2(\xi), u_3(\xi))$ denotes covariant components of the displacement field $\mathbf{u}(S) \circ \varphi$ of the middle surface $\overline{S}$, and linear mapping $\mathcal{L}$ is defined in terms of contravariant basis on S, ie. $\mathcal{L} \cdot \mathbf{u} \equiv u_i \mathbf{a}^i$.

Computational algorithms usually require that we can compute the derivatives of shape functionals with respect to the geometry, ie., with respect to φ and e. This is a difficult problem since in general shape functionals depend on φ and e not only explicitely but also implicitly through the dependence of the displacement field $\mathbf{u}(S)$ on φ and e. To circumvent this difficulty we make use of the classical adjoint state method.

We use the shell model presented in (Koiter, 1970). A detailed discussion related to numerical analysis of such problems can be found in (Bernadou, Ciarlet, 1976), (Bernadou, Boisserie, 1982).

Geometrical definition of the thin shell. Let $\mathcal{O}$ be a bounded domain of a plane $\mathbb{R}^2$ with boundary $\Gamma = \partial\mathcal{O}$. We assume that the middle surface $\overline{S}$ of the shell is the image of the set $\overline{\mathcal{O}}$ under a regular mapping φ, ie.,

$$\begin{cases} \varphi: \ (\xi^1, \xi^2) \in \overline{\mathcal{O}} \subset \mathbb{R}^2 \mapsto \varphi(\xi^1, \xi^2) \in \overline{S} \subset \mathbb{R}^3 \\ \overline{S} = \varphi(\overline{\mathcal{O}}) \ . \end{cases} \tag{8.1}$$

On $\overline{S}$ we define two local bases (see Section 2.4 of this chapter): the covariant basis $(\mathbf{a}_i, \ i = 1, 2, 3)$ and the contravariant basis $(\mathbf{a}^i, \ i = 1, 2, 3)$.

With the covariant and contravariant bases we associate the first $(a_{\alpha\beta}, a^{\alpha\beta})$ and second $(b_{\alpha\beta}, b^{\beta}_{\alpha}, b^{\alpha\beta})$ fundamental forms of the middle surface, respectively.

The thickness of the shell can be defined as a regular mapping

$$e : (\xi^1, \xi^2) \in \overline{\mathcal{O}} \subset \mathbb{R}^2 \mapsto e(\xi_1, \xi_2) \in \{ \ x \in R : \ x > 0 \ \} \ . \tag{8.2}$$

Then the shell $\mathcal{S}$ is the set

$$\begin{aligned} \mathcal{S} = \{ & M \in \mathbb{R}^3 \ : \ \overline{OM} = \varphi(\xi^1, \xi^2) + \xi^3 \mathbf{a}_3(\xi^1, \xi^2), \ (\xi^1, \xi^2) \in \overline{\mathcal{O}}, \\ & -\frac{1}{2} e(\xi^1, \xi^2) \leq \xi^3 \leq \frac{1}{2} e(\xi^1, \xi^2) \} \ . \end{aligned} \tag{8.3}$$

Remark. 8.1. We assume that $e = \tilde{e} \circ \varphi^{-1}$, where $\tilde{e}(x)$, $x \in \mathbb{R}^3$, is a given function restricted to S. This means that the thickness $e \circ \varphi$ is independent of the parametrization φ of the surface S.

Deformation of a thin shell. For simplicity we shall assume that

i) the shell is clamped on its boundary ∂S;
ii) the shell is loaded by a distribution of forces whose resultant has density $\mathbf{p}$ on S,
iii) the shell is elastic, homogenuous and isotropic.

Moreover, according to (Koiter, 1970), we assume that

iv) the normals to the middle surface remain normals to the deformed middle surface;

v) the stresses are approximatively plane and parallel to the tangent plane to the middle surface.

Under these assumptions, the deformations of the shell can be entirely described by the displacement field $\mathbf{u}(S) \circ \varphi = \mathcal{L} \cdot \mathbf{u} = u_i \mathbf{a}^i$. Thus, the unknowns of the problem are the covariant components of the displacement $\mathcal{L} \cdot \mathbf{u}$ of the middle surface $\overline{S}$, ie. the three functions

$$u_i : (\xi^1, \xi^2) \in \overline{\mathcal{O}} \subset \mathbb{R}^2 \mapsto u_i(\xi^1, \xi^2) \in R, \quad i = 1, 2, 3. \tag{8.4}$$

For a surface S_t which depends on a parameter $t \in [0, \varepsilon)$, $\varepsilon > 0$, we shall use the following notation for the displacement field

$$\begin{aligned} \mathcal{L}_t \cdot \mathbf{u}_t &= u_{ti} \mathbf{a}_t^i \ , \\ \mathbf{u}_t &= \mathrm{col}(u_{t1}, u_{t2}, u_{t3}) \ . \end{aligned}$$

Remark. 8.2. Using the mapping φ_t, $t \geq 0$, we are able to associate to any function g defined on $\overline{S}_t$, a function $h = g \circ \varphi_t$ which is defined on $\overline{\mathcal{O}}$ and we have $g = h \circ {\varphi_t}^{-1}$. We use the following notation, given $g(t, x)$, we denote $g_t(\xi) = g(t, \varphi_t(\xi))$, ie. $g_t = g \circ \varphi_t$.

Variational formulation of the thin shell problem. Under the assumptions i) to v) above, the problem takes the following variational form.

$$\begin{aligned} &\text{Find } \mathbf{u} = (u_1, u_2, u_3) \in \mathcal{H} = (H_0^1(\mathcal{O}))^2 \times H_0^2(\mathcal{O}) \text{ such that} \\ &a(\mathbf{u}, \mathbf{v}) = f(\mathbf{v}) \ \forall \mathbf{v} \in \mathcal{H} \ , \end{aligned} \tag{8.5}$$

where

$$a(\mathbf{u}, \mathbf{v}) = \int_{\mathcal{O}} e E^{\alpha\beta\lambda\iota} \{\gamma_{\alpha\beta}(\mathbf{u}) \gamma_{\lambda\iota}(\mathbf{v}) + \frac{e^2}{12} \rho_{\alpha\beta}(\mathbf{u}) \rho_{\lambda\iota}(\mathbf{v})\} dS, \tag{8.6}$$

$$f(\mathbf{v}) = \int_{\mathcal{O}} \mathbf{p} \cdot \mathbf{v} dS, \tag{8.7}$$

$$E^{\alpha\beta\lambda\iota} = \frac{E}{2(1+\mu)} [a^{\alpha\lambda} a^{\beta\iota} + a^{\alpha\iota} a^{\beta\lambda} + \frac{2\mu}{1-\mu} a^{\alpha\beta} a^{\lambda\iota}], \tag{8.8}$$

$$dS = \sqrt{\det(a_{\alpha\beta})} d\xi^1 d\xi^2 \ ,$$

$E =$ Young's modulus ; $\nu =$ Poisson's coefficient.

$$\gamma_{\alpha\beta}(\mathbf{u}) = \frac{1}{2}(u_{\alpha|\beta} + u_{\beta|\alpha}) - b_{\alpha\beta} u_3 \tag{8.9}$$

$$\overline{\rho}_{\alpha\beta}(\mathbf{u}) = u_{3|\alpha\beta} - b_\alpha^\lambda b_{\lambda\beta} u_3 + b_\alpha^\lambda|_\beta u_\lambda + b_\alpha^\lambda u_\lambda|_\beta + b_\beta^\lambda u_\lambda|_\alpha \ . \tag{8.10}$$

These expressions can be simplified by taking into account deformations of general thin shallow shells. This allows to keep $\gamma_{\alpha\beta}$ unchanged and to replace $\overline{\rho}_{\alpha\beta}(\mathbf{u})$ by

$$\rho_{\alpha\beta}(\mathbf{u}) = u_{3|\alpha\beta} \tag{8.11}$$

Theorem 8.1. *Problem (8.5) has a unique solution.*

The proof of Theorem 8.1 is given in (Bernardou, Ciarlet, 1976).

Theorem 8.2. *Problem (8.5) with $\overline{\rho}_{\alpha\beta}(\mathbf{u})$ replaced by $\rho_{\alpha\beta}(\mathbf{u})$ (see (8.11)) has a unique solution.*

The proof of Theorem 8.2 is given in (Bernardou, Lalanne, 1986).

8.2 Displacement derivatives

The basic assumption we make in this section is that the shape functional $J(S) = \mathfrak{J}(\varphi, e; \mathbf{u})$ under consideration depends only on $S = \varphi(\mathcal{O})$ and is independent of the parametrization φ for a fixed surface S and the given reference domain $\mathcal{O}$.

Under this assumption, by Corollary 3.1 in Section 2.5, we can obtain the form of shape gradient of the shape functional $J(S) = \mathfrak{J}(\varphi, e; \mathbf{u})$. In particular, we can use displacement derivatives of solutions to shell equations to evaluate the Eulerian derivative $dJ(S; V)$ in a direction V.

The family of surfaces S_t. Let t be a real parameter which belongs to $[0, \delta]$, $\delta > 0$. Let S_t be a family of surfaces which are images of the reference plane domain $\mathcal{O}$ under a regular mapping φ_t:

$$\begin{aligned} &\varphi_t \;:\; (\xi^1, \xi^2) \in \overline{\mathcal{O}} \subset \mathbb{R}^2 \mapsto \varphi_t(\xi^1, \xi^2) \in \overline{S} \subset \mathbb{R}^3 \\ &\overline{S_t} = \varphi_t(\overline{\mathcal{O}}) \ . \end{aligned} \tag{8.12}$$

When $t = 0$ we recover the surface S, or, in the other words,

$$\varphi_0 = \varphi, \qquad \text{and} \qquad S_0 = S \ .$$

According to the results obtained by (Kosinski, 1986, formulae (1.13),(1.16) and (1.17)) we have

$$\frac{\partial \varphi_t}{\partial t} = c_t^\alpha \mathbf{a}_{t\alpha} + w_t \mathbf{a}_{t3} = c_t^\alpha \mathbf{a}_{t\alpha} + \frac{\delta \varphi_t}{\delta t} \ . \tag{8.13}$$

We use the following notation.

For a given function $F(t, x), t \in \mathbb{R}, x \in \mathbb{R}^3$, we denote $F_t(\xi) = F(t, \varphi_t(\xi))$ for $x = \varphi(\xi) \in S_t, \xi \in \mathcal{O}$. The mapping $T_t(V) = \varphi_t \circ \varphi^{-1} \ : \ S \mapsto S_t$ is defined for the vector field $V(t, x) = \left[\frac{\partial \varphi_t}{\partial t} \circ \varphi_t^{-1}\right](x)$. The normal component on S_t of the vector field V is given by $\langle \frac{\delta \varphi_t}{\delta t} \circ \varphi_t^{-1} = w_t \circ \varphi_t^{-1} \cdot \mathbf{a}_{3t} \circ \varphi_t^{-1} \rangle_{\mathbb{R}^3}$.

For $t = 0$, we denote $\partial_t \varphi \equiv \frac{\delta \varphi_t}{\delta t}_{|t=0} = V(0, \varphi(\cdot))$.

Variational formulation of the thin shell problem over S_t. As in (8.5), we can formulate the problem over the middle surface S_t. The problem is to

$$\text{find } \mathbf{u}_t = (u_{t1}, u_{t2}, u_{t3}) \in \mathcal{H} \equiv (H_0^1(\mathcal{O}))^2 \times H_0^2(\mathcal{O}) \text{ such that} \tag{8.14}$$
$$a_t(\mathbf{u}_t, \mathbf{v}) = f_t(\mathbf{v}) \ \forall \mathbf{v} \in \mathcal{H},$$

where

$$a_t(\mathbf{u}_t, \mathbf{v}) = \int_{\mathcal{O}} e_t E_t^{\alpha\beta\lambda\iota}\{\gamma_{t\alpha\beta}(\mathbf{u}_t)\gamma_{t\lambda\iota}(\mathbf{v}) + \frac{e_t^2}{12}\rho_{t\alpha\beta}(\mathbf{u}_t)\rho_{t\lambda\iota}(\mathbf{v})\}dS_t \ ,$$
$$f_t(\mathbf{v}) = \int_{\mathcal{O}} \mathbf{p}_t \cdot \mathbf{v} dS_t \ ,$$
$$dS_t = \sqrt{\det(a_{t\alpha\beta})}d\xi^1\delta\xi^2 \ .$$

The index t in e_t, $E_t^{\alpha\beta\lambda\iota}$, $\gamma_{t\alpha\beta}$ and $\rho_{t\alpha\beta}$ is used to indicate that these quantities depend on φ_t, and consequently also on t. For example,

$$E_t^{\alpha\beta\lambda\iota} = \frac{E}{2(1+\nu)}[a_t^{\alpha\lambda}a_t^{\beta\iota} + a_t^{\alpha\iota}a_t^{\beta\lambda} + \frac{2\nu}{1-\nu}a_t^{\alpha\beta}a_t^{\lambda\iota}], \tag{8.15}$$
$$\gamma_{t\alpha\beta}(\mathbf{v}) = \frac{1}{2}(v_{\alpha,\beta} + v_{\beta,\alpha}) - \Gamma_{t\alpha\beta}^{\lambda}v_\lambda - b_{t\alpha\beta}v_3 \ , \tag{8.16}$$
$$\overline{\rho}_{t\alpha\beta}(\mathbf{v}) = v_{3\alpha\beta} - \Gamma_{t\alpha\beta}^{\lambda}v_{3,\lambda} \ . \tag{8.17}$$

In order to determine the displacement derivative $\frac{\delta \mathbf{u}_t}{\delta t}$ of a solution to (8.14) we differentiate (8.14) with respect to t at $t = 0$,

$$\frac{d}{dt}[a_t(\mathbf{u}_t, \mathbf{v})] = \frac{d}{dt}[f_t(\mathbf{v})].$$

Here we assume that system (8.14) is defined for $t \in (-\varepsilon, \varepsilon)$. Then

$$\frac{d}{dt}[a_t(\mathbf{u}_t, \mathbf{v})] = \frac{\partial a_t}{\partial t}(\mathbf{u}_t, \mathbf{v}) + a_t(\frac{\partial \mathbf{u}_t}{\partial t}, \mathbf{v}) + a_t(\mathbf{u}_t, \frac{\partial \mathbf{v}}{\partial t}) \ ,$$
$$\frac{\partial a_t}{\partial t}(\mathbf{u}_t, \mathbf{v}) = \frac{\partial a_t}{\partial \varphi_t}(\mathbf{u}_t, \mathbf{v})\frac{\partial \varphi_t}{\partial t} + \frac{\partial a_t}{\partial e_t}(\mathbf{u}_t, \mathbf{v})\frac{\partial e_t}{\partial t} \ ,$$
$$\frac{d}{dt}[f_t(\mathbf{v})] = \frac{\partial f_t}{\partial t}(\mathbf{v}) + f_t(\frac{\partial \mathbf{v}}{\partial t}) \ .$$

Remark. 8.3. We have $\frac{d}{dt}\left[\int_{S_t} G_t dS_t\right] = \int_{S_t} \frac{\delta G_t}{\delta t} - 2\kappa_t w_t G_t dS_t$, therefore

$$\frac{d}{dt}[a_t(\mathbf{u}, \mathbf{v})] = \frac{\partial a_t}{\partial \varphi_t}(\mathbf{u}, \mathbf{v})\frac{\delta \varphi_t}{\delta t}_{|t=0} + \frac{\partial a_t}{\partial e_t}(\mathbf{u}, \mathbf{v})\frac{\delta e_t}{\delta t}_{|t=0} \ .$$

here we assume that $\frac{\delta \mathbf{u}}{\delta t} = \frac{\delta \mathbf{v}}{\delta t} = 0$.

Remark. 8.4. For vector function $\mathbf{u}(S_t) = [\mathcal{L}_t \cdot \mathbf{u}_t] \circ \varphi_t^{-1}$ we have the following formula for the material derivative,

$$\dot{\mathbf{u}}(S;V) = \mathbf{u}'(S;V) + D_\Gamma \mathbf{u}(S) \cdot V \ .$$

From the latter formula nonhomogenuous Dirichlet boundary conditions for $\frac{\delta \mathbf{u}_t}{\delta t}$ at $t = 0$ can be obtained. Under our assumptions on the mapping φ_t we have however the homogenuous boundary conditions for $\frac{\delta \mathbf{u}_t}{\delta t} \in \mathcal{H}$.

Now we are going to evaluate the derivatives with respect to t of integrals over $\mathcal{O}$.

By using the displacement derivative as in (Kosinski, 1986, Sect. 2), we obtain

$$\frac{d}{dt}\Big[\int_{\mathcal{O}} F_t dS_t\Big] = \int_{\mathcal{O}} \frac{\delta}{\delta t}(F_t dS_t) = \int_{\mathcal{O}} \frac{\delta F_t}{\delta t} dS + \int_{\mathcal{O}} F \frac{\delta dS_t}{\delta t}$$

but

$$\frac{\delta dS_t}{\delta t} = \frac{\delta \sqrt{a}_t}{\delta t} d\xi^1 d\xi^2 = \frac{1}{2\sqrt{a}_t} \frac{\delta a_t}{\delta t} d\xi^1 d\xi^2$$

and, by (Kosinski, 1986, (2.21)), we obtain

$$\frac{\delta dS_t}{\delta t} = \frac{1}{2\sqrt{a}_t}(-2a_t w_t b^\alpha_{t\alpha}) d\xi^1 d\xi^2 = -w_t b^\alpha_{t\alpha} \sqrt{a}_t d\xi^1 d\xi^2$$

so that

$$\frac{d}{dt}\Big[\int_{\mathcal{O}} F_t dS_t\Big] = \int_{\mathcal{O}} \Big[\Big(\frac{\delta F_t}{\delta t}\Big) - F_t w_t b^\alpha_{t\alpha}\Big] dS_t \ . \tag{8.18}$$

If $F_t(.)$ is the restriction to S_t of a given function $F(t,x)$, $x \in \mathbb{R}^3$. then by (Kosinski, 1986, (2.8))

$$\frac{\delta F_t}{\delta t}(\xi) = \frac{\partial F}{\partial t}(t, \varphi_t(\xi)) + w_t(\xi) \langle \nabla_x F(t, \varphi_t(\xi)), \mathbf{a}_{t3}(\xi) \rangle_{\mathbb{R}^3} \ . \tag{8.19}$$

For the displacement field $\mathcal{L}_t \cdot \mathbf{u}_t$ we have

$$\frac{\delta(\mathcal{L}_t \cdot \mathbf{u}_t)}{\delta t} = \frac{\delta}{\delta t}(u_{ti}\mathbf{a}^i_t) = \frac{\delta u_{ti}}{\delta t}\mathbf{a}^i_t + u_{ti}\frac{\delta \mathbf{a}^i_t}{\delta t} = \mathcal{L}_t \cdot \frac{\delta \mathbf{u}_t}{\delta t} + \frac{\delta \mathcal{L}_t}{\delta t} \cdot \mathbf{u}_t \ ,$$

where

$$\frac{\delta \mathbf{u}_t}{\delta t} = \operatorname{col}\Big(\frac{\delta u_{t1}}{\delta t}, \frac{\delta u_{t2}}{\delta t}, \frac{\delta u_{t3}}{\delta t}\Big).$$

Differentiating system (8.14) with respect to t, at $t = 0$ we get the system of equations to which the vector function $\frac{\delta \mathbf{u}}{\delta t}$ is a solution.

Keeping this in mind, it remains to compute all the necessary derivatives for our general thin shell problem. This is done in the next paragraph.

8.3 Shape sensitivity analysis of thin shells

First, we differentiate the right-hand side of (8.14). By (8.18),

$$\frac{d}{dt}[f_t(\mathbf{v})] = \int_{\mathcal{O}} \left\{ \frac{\delta \mathbf{p}_t \cdot \mathbf{v}}{\delta t} - 2w_t\kappa_t \mathbf{p}_t \cdot \mathbf{v} \right\} dS_t$$

where, in view of (8.19),

$$\begin{aligned} \frac{\delta(\mathbf{p}_t \cdot \mathbf{v})}{\delta t} &= \frac{\delta \mathbf{p}_t}{\delta t} \cdot \mathbf{v} + \mathbf{p}_t \cdot \frac{\delta \mathbf{v}}{\delta t} \\ &= \frac{\partial \mathbf{p}_t}{\partial t} \cdot \mathbf{v} + w_t \frac{\partial \mathbf{p}_t}{\partial n} \cdot \mathbf{v} + \mathbf{p}_t \cdot \frac{\partial \mathbf{v}}{\partial t} + w \mathbf{p}_t \cdot \frac{\partial \mathbf{v}}{\partial n} \ . \end{aligned} \tag{8.20}$$

Here $\mathbf{v} = \widetilde{\mathbf{v}} \circ \varphi_t \in \mathcal{H}$ is a smooth test function, $\widetilde{\mathbf{v}}$ is a function defined in an open neighbourhood of S. We assume that the function $\widetilde{\mathbf{v}}$ satisfies $\nabla \widetilde{v}_i \cdot \mathbf{a}_3 = 0$, $i = 1, 2, 3$. Hence,

$$\frac{\delta \mathbf{v}}{\delta t}_{|t=0} = \frac{\delta \widetilde{\mathbf{v}} \circ \varphi_t}{\delta t}_{|t=0} = 0 \ . \tag{8.21}$$

Therefore,

$$\frac{d}{dt}[f_t(\mathbf{v})]_{|t=0} = \int_{\mathcal{O}} [\frac{\partial \mathbf{p}_t}{\partial t}_{|t=0} \cdot \mathbf{v} + w_t \frac{\partial \mathbf{p}_t}{\partial n}_{|t=0} \cdot \mathbf{v} - 2w\kappa \mathbf{p}_{t|t=0} \cdot \mathbf{v}] dS \ . \tag{8.22}$$

Remark. 8.5. For some kinds of loading, $\mathbf{p}_t$ is independent of t. For instance, this is the case for weight loadings as opposed to pressure loadings. When $\mathbf{p}_t$ is independent of t, relation (8.22) is simpler and takes the form

$$\frac{d}{dt}[f_t(\mathbf{v})]_{|t=0} = \int_{\mathcal{O}} [w_t \frac{\partial \mathbf{p}_t}{\partial n}_{|t=0} \cdot \mathbf{v} - 2w\kappa \mathbf{p}_{t|t=0} \cdot \mathbf{v}] dS \ . \tag{8.23}$$

Next, we differentiate the left-hand side of (8.14). Again by (8.18),

$$\begin{aligned} &\frac{d}{dt}[a_t(\mathbf{u}_t, \mathbf{v})]_{|t=0} \\ &= \frac{d}{dt}\Big[\int_{\mathcal{O}} e_t E_t^{\alpha\beta\lambda\iota} \{\gamma_{t\alpha\beta}(\mathbf{u}_t)\gamma_{t\lambda\iota}(\mathbf{v}) + \frac{e_t^2}{12}\rho_{t\alpha\beta}(\mathbf{u}_t)\rho_{t\lambda\iota}(\mathbf{v})\} dS_t \Big]_{|t=0} \\ &= \int_{\mathcal{O}} \frac{\delta}{\delta t}[e_t E_t^{\alpha\beta\lambda\iota} \{\gamma_{t\alpha\beta}(\mathbf{u}_t)\gamma_{t\lambda\iota}(\mathbf{v}) + \frac{e_t^2}{12}\rho_{t\alpha\beta}(\mathbf{u}_t)\rho_{t\lambda\iota}(\mathbf{v})\}]_{|t=0} \\ &\quad - 2w\kappa[e E^{\alpha\beta\lambda\iota} \{\gamma_{\alpha\beta}(\mathbf{u})\gamma_{\lambda\iota}(\mathbf{v}) + \frac{e^2}{12}\rho_{\alpha\beta}(\mathbf{u})\rho_{\lambda\iota}(\mathbf{v})]\} dS \ . \end{aligned} \tag{8.24}$$

To compute the first term of the right-hand side of (8.24), we have to evaluate the following derivatives

$$\frac{\delta e_t}{\delta t}, \ \frac{\delta E_t^{\alpha\beta\lambda\iota}}{\delta t}, \ \frac{\delta}{\delta t}\gamma_{t\alpha\beta}(\mathbf{u}_t), \ \frac{\delta}{\delta t}\rho_{t\alpha\beta}(\mathbf{u}_t) \ .$$

Recall that all these functions depend only on (ξ^1, ξ^2) and are independent of the normal coordinate ξ^3. Thus, for these functions, relation (8.19) implies that for the convected parametrization φ, ie. such that $c_t^\alpha \equiv 0$ as in (8.13), $\frac{\delta}{\delta t}$ is just reduced to $\frac{\partial}{\partial t}$ since $\frac{\partial}{\partial n_t}$ is identically zero. In general, we have to use the displacement derivatives in the case of an arbitrary parametrization.

Computations of $\frac{\delta e_t}{\delta t}$. The derivative $\frac{\delta e_t}{\delta t}$ can be defined in the following way.

For a given function $\Psi(t,\cdot)$, $t \in (-\varepsilon, \varepsilon)$, defined in an open neighbourhood of the surface $S_t \subset \mathbb{R}^3$, we denote

$$e_t(\xi) = \Psi(t, \varphi_t(\xi)), \text{ thus}$$
$$\frac{\delta e_t}{\delta t}(\xi) = \frac{\partial \Psi}{\partial t}(t, \varphi_t(\xi)) + \nabla_x \Psi(t, \varphi_t(\xi)) \cdot \frac{\delta \varphi_t}{\delta t}(\xi) \ . \tag{8.25}$$

Computations of $\frac{\delta E_t^{\alpha\beta\lambda\iota}}{\delta t}$. Recall that

$$E_t^{\alpha\beta\lambda\iota} = \frac{E}{2(1+\nu)}[a_t^{\alpha\lambda}a_t^{\beta\iota} + a_t^{\alpha\iota}a_t^{\beta\lambda} + \frac{2\nu}{1-\nu}a_t^{\alpha\beta}a_t^{\lambda\iota}] \ .$$

So we have to compute $\frac{\partial}{\partial t}(a_t^{\alpha\beta})$. According to (Kosinski, 1986, (2.50)), we obtain

$$\frac{\delta}{\delta t}(a_t^{\alpha\beta}) = 2w_t b_t^{\alpha\beta} \tag{8.26}$$

so that, $w_t = w$ for $t = 0$,

$$\frac{\delta E_t^{\alpha\beta\lambda\iota}}{\delta t} \tag{8.27}$$
$$= \frac{Ew}{1+\nu}[a_t^{\alpha\lambda}b_t^{\beta\iota} + b_t^{\alpha\lambda}a_t^{\beta\iota} + a_t^{\alpha\iota}b_t^{\beta\lambda} + b_t^{\alpha\iota}a_t^{\beta\lambda} + \frac{2\nu}{1-\nu}(a_t^{\alpha\beta}b_t^{\lambda\iota} + b_t^{\alpha\beta}a_t^{\lambda\iota})] \ .$$

Remark. 8.6. We could also consider the evolving mechanical characteristics E and ν depending on t.

Computations of $\frac{\delta}{\delta t}\gamma_{t\alpha\beta}(\mathbf{u}_t)$. We have

$$\gamma_{t\alpha\beta}(\mathbf{u}_t) = \frac{1}{2}(u_{t\alpha,\beta} + u_{t\beta,\alpha}) - \Gamma_{t\alpha\beta}^{\lambda}u_{t\lambda} - b_{t\alpha\beta}u_{t3} \ .$$

so that

$$\frac{\delta}{\delta t}\gamma_{t\alpha\beta}(\mathbf{u}_t) = \frac{1}{2}\Big[\Big(\frac{\delta u_{t\alpha}}{\delta t}\Big)_{,\beta} + \Big(\frac{\delta u_{t\beta}}{\delta t}\Big)_{,\alpha}\Big] \tag{8.28}$$
$$- \Gamma_{t\alpha\beta}^{\lambda}\frac{\delta u_{t\alpha}}{\delta t} - b_{t\alpha\beta}\frac{\delta u_{t3}}{\delta t} - u_{t\lambda}\frac{\delta}{\delta t}(\Gamma_{t\alpha\beta}^{\lambda}) - u_{t3}\frac{\delta}{\delta t}\Big(b_{t\alpha\beta}\Big)$$
$$= \frac{1}{2}\Big[\frac{\delta u_{t\alpha}}{\delta t}_{|\beta} + \frac{\delta u_{t\beta}}{\delta t}_{|\alpha}\Big] - b_{t\alpha\beta}\frac{\delta u_{t3}}{\delta t} - u_{t\lambda}\frac{\delta}{\delta t}(\Gamma_{t\alpha\beta}^{\lambda}) - u_{t3}\frac{\delta}{\delta t}\Big(b_{t\alpha\beta}\Big) \ .$$

By (Kosinski, 1986, (2.24)),

$$\frac{\delta b_{t\alpha\beta}}{\delta t} = w_{t,\alpha\beta} - w_{t,\lambda}\Gamma^{\lambda}_{t\alpha\beta} - w_t b^{\lambda}_{t\alpha} b_{t\lambda\beta} \ . \tag{8.29}$$

Moreover, using the relations (Gauss and Weingarten relations)

$$\mathbf{a}_{t\alpha,\beta} = \Gamma^{\lambda}_{t\alpha\beta}\mathbf{a}_{t\lambda} + b_{t\alpha\beta}\mathbf{a}_{t3} \ ; \qquad \mathbf{a}_{t3,\alpha} = -b^{\lambda}_{t\alpha}\mathbf{a}_{t\lambda} \ ; \qquad \mathbf{a}^{\alpha}_{t,\beta} = -\Gamma^{\alpha}_{t\beta\lambda}\mathbf{a}^{\lambda}_t + b^{\alpha}_{t\beta}\mathbf{a}_{t3} \ ,$$

so that $\Gamma^{\lambda}_{t\alpha\beta} = \mathbf{a}_{t\alpha,\beta} \cdot \mathbf{a}^{\lambda}_t$. Then

$$\frac{\delta}{\delta t}\left(\Gamma^{\lambda}_{t\alpha\beta}\right) = \left(\frac{\delta}{\delta t}\left(\mathbf{a}_{t\alpha,\beta}\right)\right) \cdot \mathbf{a}^{\lambda}_t + \mathbf{a}_{t\alpha,\beta} \cdot \frac{\delta}{\delta t}\left(\mathbf{a}^{\lambda}_t\right) \ . \tag{8.30}$$

Comparing with (8.19),(8.30) and (Kosinski, 1986, (2.18)), we get

$$\begin{aligned}
\left(\frac{\delta}{\delta t}\left(\mathbf{a}_{t\alpha,\beta}\right)\right) \cdot \mathbf{a}^{\iota}_t &= \left(\frac{\delta\varphi_t}{\delta t}\right)_{,\alpha\beta} \cdot \mathbf{a}^{\iota}_t = \left(w_t\mathbf{a}_{t3}\right)_{,\alpha\beta} \cdot \mathbf{a}^{\iota}_t \\
&= \left(w_{t,\alpha}\mathbf{a}_{t3,\beta} + w_{t,\beta}\mathbf{a}_{t3,\alpha} + w_t\mathbf{a}_{t3,\alpha\beta}\right) \cdot \mathbf{a}^{\iota}_t \\
&= -w_{t,\alpha}b^{\iota}_{t\beta} - w_{t,\beta}b^{\iota}_{t\alpha} - w_t\left(b^{\iota}_{t\alpha,\beta} + b^{\lambda}_{t\alpha}\Gamma^{\iota}_{t\alpha\beta}\right), \\
\mathbf{a}_{t\alpha,\beta} \cdot \frac{\delta}{\delta t}\left(\mathbf{a}^{\iota}_t\right) &= \left(\Gamma^{\iota}_{t\alpha\beta}\mathbf{a}^{\lambda}_t + b_{t\alpha\beta}\mathbf{a}_{t3}\right) \cdot \left(w_{t,\omega}a^{\omega\iota}_t\mathbf{a}_{t3} + w_t b^{\iota}_{t\omega}\mathbf{a}^{\omega}_t\right)
\end{aligned}$$

$$= w_{t,\omega}a^{\omega\iota}_t b_{t\alpha\beta} + w_t b^{\iota}_{t\omega}\Gamma^{\omega}_{t\alpha\beta} \ .$$

Finally, since

$$b^{\iota}_{t\alpha|\beta} = b^{\iota}_{t\alpha,\beta} + \Gamma^{\iota}_{t\alpha\beta}b^{\omega}_{t\alpha} - \Gamma^{\omega}_{t\alpha\beta}b^{\iota}_{t\omega}$$

we get

$$\frac{\delta}{\delta t}\left(\Gamma^{\iota}_{t\alpha\beta}\right) = -w_t b^{\iota}_{t\alpha|\beta} + w_{t,\omega}a^{\omega\iota}_t b_{t\alpha\beta} - w_{t,\alpha}b^{\iota}_{t\beta} - w_{t,\beta}b^{\iota}_{t\alpha} \ . \tag{8.31}$$

In a similar way, since $\frac{\delta v_i}{\delta t} = 0$ we obtain

$$\frac{\delta}{\delta t}\gamma_{t\alpha\beta}(\mathbf{v}) = -v_{\lambda}\frac{\delta}{\delta t}\left(\Gamma^{\lambda}_{t\alpha\beta}\right) - v_3\frac{\delta}{\delta t}\left(b_{t\alpha\beta}\right) \ . \tag{8.32}$$

Computations of $\frac{\delta}{\delta t}\left[\overline{\rho}_{t\alpha\beta}(\mathbf{u}_t)\right]$. We have

$$\overline{\rho}_{t\alpha\beta}(\mathbf{u}_t) = u_{t3}|_{\alpha\beta} - b^{\lambda}_{t\alpha}b_{t\lambda\beta}u_{t3} + b^{\lambda}_{t\alpha}\big|_{\beta}\, u_{t\lambda} + b^{\lambda}_{t\alpha}\ u_{t\lambda}|_{\beta} + b^{\lambda}_{t\beta}\ u_{t\lambda}|_{\alpha} \ .$$

Thus, we obtain

$$\begin{aligned}
\frac{\delta}{\delta t}\left(u_{t3}|_{\alpha\beta}\right) &= \frac{\delta}{\delta t}\left(u_{t3,\alpha\beta}\right) - \Gamma^{\lambda}_{t\alpha\beta}\frac{\delta}{\delta t}\left(u_{t3,\lambda}\right) - u_{t3,\lambda}\frac{\delta}{\delta t}\left(\Gamma^{\lambda}_{t\alpha\beta}\right) \\
&= \left.\frac{\delta u_{t3}}{\delta t}\right|_{\alpha\beta} - u_{t3,\lambda}\frac{\delta}{\delta t}\left(\Gamma^{\lambda}_{t\alpha\beta}\right) \\
\frac{\delta}{\delta t}\left(u_{t\lambda}|_{\beta}\right) &= \left.\frac{\delta u_{t\lambda}}{\delta t}\right|_{\beta} - u_{t3,\lambda}\frac{\delta}{\delta t}\left(\Gamma^{\iota}_{t\lambda\beta}\right) \ ,
\end{aligned}$$

so that

$$
\begin{aligned}
\frac{\delta}{\delta t}\left[\overline{\rho}_{t\alpha\beta}(\mathbf{u}_t)\right] = & \left.\frac{\delta u_{t3}}{\delta t}\right|_{\alpha\beta} - u_{t3,\lambda}\frac{\delta}{\delta t}\left(\Gamma^{\lambda}_{t\alpha\beta}\right) - b^{\lambda}_{t\alpha} b_{t\lambda\beta}\frac{\delta u_{t3}}{\delta t} \\
& - u_{t3}\frac{\delta}{\delta t}\left(b^{\lambda}_{t\alpha} b_{t\lambda\beta}\right) + b^{\lambda}_{t\alpha}\big|_{\beta}\frac{\delta u_{t\lambda}}{\delta t} + u_{t\lambda}\frac{\delta}{\delta t}\left(b^{\lambda}_{t\alpha}\big|_{\beta}\right) \\
& + b^{\lambda}_{t\alpha}\left.\frac{\delta u_{t\lambda}}{\delta t}\right|_{\beta} - u_{t\nu} b^{\lambda}_{t\alpha}\frac{\delta}{\delta t}\left(\Gamma^{\nu}_{t\lambda\beta}\right) + u_{t\lambda}\big|_{\beta}\frac{\delta}{\delta t}\left(b^{\lambda}_{t\alpha}\right) \\
& + b^{\lambda}_{t\beta}\left.\frac{\delta u_{t\lambda}}{\delta t}\right|_{\alpha} - u_{t\nu} b^{\lambda}_{t\beta}\frac{\delta}{\delta t}\left(\Gamma^{\nu}_{t\lambda\alpha}\right) + u_{t\lambda}\big|_{\alpha}\frac{\delta}{\delta t}\left(b^{\lambda}_{t\beta}\right)
\end{aligned}
$$

or, equivalently

$$
\begin{aligned}
\frac{\delta}{\delta t}\left[\overline{\rho}_{t\alpha\beta}(\mathbf{u}_t)\right] = & \overline{\rho}_{t\alpha\beta}\left(\frac{\delta \mathbf{u}_t}{\delta t}\right) - u_{t3,\lambda}\frac{\delta}{\delta t}\left(\Gamma^{\lambda}_{t\alpha\beta}\right) - u_{t3}\frac{\delta}{\delta t}\left(b^{\lambda}_{t\alpha} b_{t\lambda\beta}\right) \\
& + u_{t\lambda}\big|_{\beta}\frac{\delta}{\delta t}\left(b^{\lambda}_{t\alpha}\right) + u_{t\lambda}\big|_{\alpha}\frac{\delta}{\delta t}\left(b^{\lambda}_{t\beta}\right) \\
& + u_{t\nu}\left[\frac{\delta}{\delta t}\left(b^{\nu}_{t\alpha}\big|_{\beta}\right) - b^{\lambda}_{t\alpha}\frac{\delta}{\delta t}\left(\Gamma^{\nu}_{t\lambda\beta}\right) - b^{\lambda}_{t\beta}\frac{\delta}{\delta t}\left(\Gamma^{\nu}_{t\lambda\alpha}\right)\right] .
\end{aligned}
\tag{8.33}
$$

It remains to compute derivatives appearing in the right-hand side of (8.33).

First, $\frac{\delta}{\delta t}\left(b_{t\alpha\beta}\right)$ and $\frac{\delta}{\delta t}\left(\Gamma^{\lambda}_{t\alpha\beta}\right)$ are given by relations (8.29) and (8.31). By (Kosinski, 1986, (2.51)),

$$
\frac{\delta}{\delta t}\left(b^{\beta}_{t\alpha}\right) = w_t\big|_{\alpha\lambda} a^{\lambda\beta}_t + w_t b^{\lambda}_{t\alpha} b^{\beta}_{t\lambda} . \tag{8.34}
$$

Then

$$
\frac{\delta}{\delta t}\left(b^{\lambda}_{t\alpha} b_{t\lambda\beta}\right) = w_t\big|_{\alpha\iota} b^{\iota}_{t\beta} + w_t\big|_{\iota\beta} b^{\iota}_{t\alpha} . \tag{8.35}
$$

Finally it remains to compute $\frac{\delta}{\delta t}\left(b^{\iota}_{t\alpha}\big|_{\beta}\right)$. Observe that

$$
b^{\iota}_{t\alpha} = a^{\iota\lambda}_t b_{t\alpha\lambda}
$$

so that $\left(a^{\iota\lambda}_t\big|_{\beta} \equiv 0\right)$

$$
b^{\iota}_{t\alpha}\big|_{\beta} = a^{\iota\lambda}_t \, b_{t\alpha\lambda}\big|_{\beta} .
$$

In view of (2.10) from Section 2, and (8.30) we obtain

$$
\begin{aligned}
b_{t\alpha\lambda}\big|_{\beta} &= b_{t\alpha\lambda,\beta} - \Gamma^{\omega}_{t\alpha\beta} b_{t\omega\alpha} - \Gamma^{\omega}_{t\alpha\beta} b_{t\omega\alpha} \\
&= \left(\mathbf{a}_{t3}\cdot\mathbf{a}_{t\alpha,\lambda}\right)_{,\beta} - \Gamma^{\omega}_{t\alpha\beta} b_{t\omega\alpha} - \Gamma^{\omega}_{t\alpha\beta} b_{t\omega\alpha} \\
&= \mathbf{a}_{t3,\beta}\cdot\varphi_{t,\alpha\lambda} + \mathbf{a}_{t3}\cdot\varphi_{t,\alpha\lambda\beta} - \Gamma^{\omega}_{t\alpha\beta} b_{t\omega\alpha} - \Gamma^{\omega}_{t\alpha\beta} b_{t\omega\alpha} \\
&= -b^{\omega}_{t\beta}\mathbf{a}_{t\omega}\cdot\left(\Gamma^{\iota}_{t\alpha\lambda}\mathbf{a}_{t\iota} + b_{t\alpha\lambda}\mathbf{a}_{t3}\right) + \mathbf{a}_{t3}\cdot\varphi_{t,\alpha\lambda\beta} - \Gamma^{\omega}_{t\alpha\beta} b_{t\omega\alpha} - \Gamma^{\omega}_{t\alpha\beta} b_{t\omega\alpha} \\
&= \mathbf{a}_{t3}\cdot\varphi_{t,\alpha\lambda\beta} - \Gamma^{\omega}_{t\alpha\lambda} b_{t\beta\omega} - \Gamma^{\omega}_{t\alpha\beta} b_{t\omega\alpha} - \Gamma^{\omega}_{t\alpha\beta} b_{t\omega\alpha} .
\end{aligned}
$$

It follows that

$$\frac{\delta}{\delta t}\left(b^{\iota}_{t\alpha}|_{\beta}\right) = b_{t\alpha\lambda}|_{\beta}\frac{\delta}{\delta t}\left(a^{\iota\lambda}_{t}\right) + a^{\iota\lambda}_{t}\frac{\delta}{\delta t}\left(b_{t\alpha\lambda}|_{\beta}\right) .$$

Moreover, by (Kosinski, 1986, (2.15)), we have

$$\frac{\delta \mathbf{a}_{t3}}{\delta t} = -w_{t,\alpha}\mathbf{a}^{\alpha}_{t}$$

so with (3.2), (4.8) and (8.31) we obtain

$$\begin{aligned}
&\frac{\delta}{\delta t}\left(b^{\iota}_{t\alpha}|_{\beta}\right)\\
&= 2w_t b^{\iota\lambda}_{t}\, b_{t\alpha\lambda}|_{\beta} + a^{\iota\lambda}_{t}\Big\{-w_{t,\omega}\mathbf{a}^{\omega}_{t}\cdot\varphi_{t,\alpha\lambda\beta} + \mathbf{a}_{t3}\cdot\left(w_t\mathbf{a}_{t3}\right)_{,\alpha\lambda\beta}\\
&+ b_{t\beta\omega}\left(w_t(b^{\omega}_{t\alpha}|_{\lambda} - w_{t,\theta}a^{\theta\omega}_{t}b_{t\alpha\lambda} + w_{t,\alpha}b^{\omega}_{t\lambda} + w_{t,\lambda}b^{\omega}_{t\alpha}\right) - \Gamma^{\omega}_{t\alpha\lambda}\left(w_t|_{\beta\omega} - w_t b^{\iota}_{t\beta}b_{t\iota\beta}\right)\\
&+ b_{t\omega\lambda}\left(w_t(b^{\omega}_{t\alpha}|_{\beta} - w_{t,\iota}a^{\iota\omega}_{t}b_{t\alpha\beta} + w_{t,\alpha}b^{\omega}_{t\beta} + w_{t,\beta}b^{\omega}_{t\alpha}\right) - \Gamma^{\omega}_{t\alpha\beta}\left(w_t|_{\omega\lambda} - w_t b^{\iota}_{t\omega}b_{t\iota\lambda}\right)\\
&+ b_{t\omega\alpha}\left(w_t(b^{\omega}_{t\lambda}|_{\beta} - w_{t,\iota}a^{\iota\omega}_{t}b_{t\lambda\beta} + w_{t,\lambda}b^{\omega}_{t\beta} + w_{t,\beta}b^{\omega}_{t\lambda}\right) - \Gamma^{\omega}_{t\lambda\beta}\left(w_t|_{\omega\alpha} - w_t b^{\iota}_{t\omega}b_{t\iota\alpha}\right)\Big\} .
\end{aligned}$$

But

$$\begin{aligned}
\mathbf{a}_{t3}\cdot\left(w_t\mathbf{a}_{t3}\right)_{,\alpha\beta\lambda} &= w_{t,\alpha\beta\lambda} - w_{t,\alpha}b^{\iota}_{t\beta}b_{t\lambda\iota} - w_{t,\beta}b^{\iota}_{t\lambda}b_{t\alpha\iota} - w_{t,\lambda}b^{\iota}_{t\alpha}b_{t\iota\beta}\\
&\quad - w_t b^{\iota}_{t\alpha,\beta}b_{t\iota\lambda} - w_t b^{\iota}_{t\alpha}b_{t\iota\lambda,\beta} - w_t b^{\iota}_{t\alpha,\lambda}b_{t\iota\beta} - w_t b^{\iota}_{t\alpha}b_{t\omega\beta}\Gamma^{\omega}_{t\iota\lambda}
\end{aligned}$$

and

$$-w_{t,\omega}a^{\iota\lambda}_{t}\mathbf{a}^{\omega}_{t}\cdot\varphi_{t,\alpha\lambda\beta} = -w_{t,\omega}a^{\iota\lambda}_{t}\left[\Gamma^{\omega}_{t\alpha\lambda,\beta} + \Gamma^{\iota}_{t\alpha\lambda}\Gamma^{\omega}_{t\iota\beta} - b_{t\alpha\lambda}b^{\omega}_{t\beta}\right] .$$

Then

$$\begin{aligned}
\frac{\delta}{\delta t}\left(b^{\iota}_{t\alpha}|_{\beta}\right) &= w_t\Big\{2b^{\iota\lambda}_{t}b_{t\alpha\lambda}|_{\beta} + a^{\iota\lambda}_{t}\Big[-b^{\omega}_{t\alpha,\beta}b_{t\iota\lambda} - b^{\omega}_{t\alpha}b_{t\omega\lambda,\beta} - b^{\omega}_{t\alpha,\lambda}b_{t\omega\beta} - b^{\theta}_{t\alpha}b_{t\omega\beta}\Gamma^{\omega}_{t\theta\lambda}\\
&\quad + b_{t\beta\omega}\, b^{\omega}_{t\alpha}|_{\lambda} + b^{\theta}_{t\beta}b_{t\theta\omega}\Gamma^{\omega}_{t\alpha\lambda} + b_{t\omega\lambda}\, b^{\omega}_{t\alpha}|_{\beta} + b^{\omega}_{t\theta}b_{t\omega\lambda}\Gamma^{\theta}_{t\alpha\beta} + b_{t\omega\alpha}\, b^{\omega}_{t\lambda}|_{\beta}\\
&\quad + b^{\omega}_{t\theta}b_{t\omega\alpha}\Gamma^{\theta}_{t\lambda\beta}\Big]\Big\} - w_{t,\omega}a^{\iota\lambda}_{t}\Big\{\Gamma^{\omega}_{t\alpha\lambda,\beta} + \Gamma^{\iota}_{t\alpha\lambda}\Gamma^{\omega}_{t\iota\beta} - b_{t\alpha\lambda}b^{\omega}_{t\beta}\Big\}\\
&\quad - a^{\iota\lambda}_{t}\Big\{w_{t,\alpha}b^{\omega}_{t\beta}b_{t\lambda\omega} + w_{t,\beta}b^{\omega}_{t\lambda}b_{t\alpha\omega} + w_{t,\lambda}b^{\omega}_{t\alpha}b_{t\omega\beta}\Big\}\\
&\quad + a^{\iota\lambda}_{t}b_{t\beta\omega}\Big\{w_{t,\alpha}b^{\omega}_{t\lambda} + w_{t,\lambda}b^{\omega}_{t\alpha} - w_{t,\theta}a^{\theta\omega}_{t}b_{t\alpha\lambda}\Big\}\\
&\quad + a^{\iota\lambda}_{t}b_{t\omega\lambda}\Big\{w_{t,\alpha}b^{\omega}_{t\omega} + w_{t,\beta}b^{\omega}_{t\alpha} - w_{t,\theta}a^{\theta\omega}_{t}b_{t\alpha\beta}\Big\}\\
&\quad + a^{\iota\lambda}_{t}b_{t\omega\alpha}\Big\{w_{t,\lambda}b^{\omega}_{t\omega} + w_{t,\beta}b^{\omega}_{t\lambda} - w_{t,\theta}a^{\theta\omega}_{t}b_{t\lambda\beta}\Big\}\\
&\quad - a^{\iota\lambda}_{t}\Big\{\Gamma^{\omega}_{t\alpha\lambda}\, w_t|_{\beta\omega} + \Gamma^{\omega}_{t\alpha\beta}\, w_t|_{\omega\lambda} + \Gamma^{\omega}_{t\lambda\beta}\, w_t|_{\omega\alpha}\Big\} + w_{t,\alpha\beta\lambda}
\end{aligned}$$

$$= w_t\Big\{ 2b_t^{\iota\lambda}b_{t\alpha\lambda}\big|_\beta + b_{t\omega}^{\iota}b_{t\alpha}^{\theta}\Gamma_{t\beta\theta}^{\omega} + a_t^{\iota\lambda}b_{t\omega\alpha}b_{t\lambda,\beta}^{\omega}b_{t\omega\alpha}b_t^{\theta\iota}\Gamma_{t\beta\theta}^{\omega} - a_t^{\iota\lambda}b_{t\alpha}^{\omega}b_{t\omega\lambda,\beta}\Big\}$$
$$+ w_{t,\alpha}b_{t\omega}^{\iota}b_{t\beta}^{\omega} + w_{t,\beta}b_{t\omega}^{\iota}b_{t\alpha}^{\omega} + w_{t,\lambda}a_t^{\iota\lambda}b_{t\omega\alpha}b_{t\beta}^{\omega}$$
$$- w_{t,\theta}\Big\{b_t^{\theta\iota}b_{t\alpha\beta} + b_{t\alpha}^{\theta}b_{\theta\beta}^{\iota} + a_t^{\iota\lambda}\Gamma_{t\alpha\lambda,\beta}^{\theta} + a_t^{\iota\lambda}\Gamma_{t\alpha\lambda}^{\nu}\Gamma_{t\nu\beta}^{\theta}\Big\}$$
$$- a_t^{\iota\lambda}\Big\{\Gamma_{t\alpha\lambda}^{\omega}\, w_t\big|_{\beta\omega} + \Gamma_{t\alpha\beta}^{\omega}\, w_t\big|_{\omega\lambda} + \Gamma_{t\lambda\beta}^{\omega}\, w_t\big|_{\omega\alpha}\Big\} + w_{t,\alpha\beta\lambda}\,.$$

Finally, we get

$$\frac{\delta}{\delta t}\left(b_{t\alpha}^{\iota}\big|_\beta\right) = 2w_t\; 2b_t^{\iota\lambda}b_{t\alpha\lambda}\big|_\beta + w_{t,\alpha}b_{t\omega}^{\iota}b_{t\beta}^{\omega} + w_{t,\beta}b_{t\omega}^{\iota}b_{t\alpha}^{\omega} \tag{8.36}$$
$$- w_{t,\omega}\Big\{b_t^{\omega\iota}b_{t\alpha\beta} + b_{t\alpha}^{\omega}b_{t\beta}^{\iota} + a_t^{\iota\lambda}\Gamma_{t\alpha\lambda,\beta}^{\omega} + a_t^{\iota\lambda}\Gamma_{t\alpha\lambda}^{\theta}\Gamma_{t\theta\beta}^{\omega} - a_t^{\iota\omega}b_{\lambda\alpha}b_{t\beta}^{\lambda}\Big\}$$
$$- a_t^{\iota\lambda}\Big\{\Gamma_{t\alpha\lambda}^{\omega}\, w_t\big|_{\beta\omega} + \Gamma_{t\alpha\beta}^{\omega}\, w_t\big|_{\omega\lambda} + \Gamma_{t\lambda\beta}^{\omega}\, w_t\big|_{\omega\alpha}\Big\}$$
$$+ w_{t,\alpha\beta\lambda}\,.$$

It remains to substitute relations (8.31),(8.34) to (8.36) into expression (8.33) in order to obtain the displacement derivative $\frac{\delta}{\delta t}\left[\overline{\rho}_{t\alpha\beta}(\mathbf{u}_t)\right]$.

In a similar way and since the v_i are independent of t we get

$$\frac{\delta}{\delta t}\left[\overline{\rho}_{t\alpha\beta}(\mathbf{u}_t)\right] = -\,v_{3,\lambda}\frac{\delta}{\delta t}\left(\Gamma_{t\alpha\beta}^{\lambda}\right) - v_3\frac{\delta}{\delta t}\left(b_{t\alpha}^{\lambda}b_{t\lambda\beta}\right) + v_\lambda\big|_\beta\frac{\delta}{\delta t}\left(b_{t\alpha}^{\lambda}\right) \tag{8.37}$$
$$+ v_\lambda\big|_\alpha\frac{\delta}{\delta t}\left(b_{t\beta}^{\lambda}\right) + v_\iota\left[\frac{\delta}{\delta t}\left(b_{t\alpha}^{\iota}\big|_\beta\right) - b_{t\alpha}^{\lambda}\frac{\delta}{\delta t}\left(\Gamma_{t\lambda\beta}^{\iota}\right) - b_{t\beta}^{\lambda}\frac{\delta}{\delta t}\left(\Gamma_{t\lambda\alpha}^{\iota}\right)\right]$$

Subsequently, since the variables subjected to optimization are φ_t and e_t, we decompose the variations of $a_t(\cdot,\cdot)$ into three parts. More precisely, by (8.24)–(8.27) and (8.31)–(8.33), we get

$$\frac{d}{dt}\left[a_t(\mathbf{u}_t,\mathbf{v})\right] = \frac{\partial a_t}{\partial u_t}\Big(\frac{\delta \mathbf{u}_t}{\delta t},\mathbf{v}\Big) + \frac{\partial a_t}{\partial \varphi_t}(\mathbf{u}_t,\mathbf{v})\frac{\delta\varphi_t}{\delta t} + \frac{\partial a_t}{\partial e_t}(\mathbf{u}_t,\mathbf{v})\frac{\delta e_t}{\delta t}\,, \tag{8.38}$$

where

$$\frac{\partial a_t}{\partial u_t}\Big(\frac{\delta \mathbf{u}_t}{\delta t},\mathbf{v}\Big) = \int_{\mathcal{O}} e_t E_t^{\alpha\beta\lambda\mu}\Big\{\gamma_{t\alpha\beta}\left(\frac{\delta\mathbf{u}_t}{\delta t}\right)\gamma_{t\lambda\mu}(\mathbf{v}) \tag{8.39}$$
$$+ \frac{e_t^2}{12}\overline{\rho}_{t\alpha\beta}\left(\frac{\delta\mathbf{u}_t}{\delta t}\right)\overline{\rho}_{t\lambda\mu}(\mathbf{v})\Big\}dS_t$$

$$\frac{\partial a_t}{\partial \varphi_t}(\mathbf{u}_t,\mathbf{v})\frac{\delta\varphi_t}{\delta t} \tag{8.40}$$
$$= \int_{\mathcal{O}} 2e_t w_t\Big\{\frac{E}{1+\nu}\left[a_t^{\alpha\lambda}b_t^{\beta\mu} + a_t^{\alpha\mu}b_t^{\beta\lambda} + \frac{\nu}{1+\nu}\left(a_t^{\alpha\beta}b_t^{\lambda\mu} + a_t^{\lambda\mu}b_t^{\alpha\beta}\right)\right]$$
$$- \kappa_{tm}E_t^{\alpha\beta\lambda\mu}\Big\}\Big\{\gamma_{t\alpha\beta}(\mathbf{u}_t)\,\gamma_{t\lambda\mu}(\mathbf{v}) + \frac{e_t^2}{12}\overline{\rho}_{t\alpha\beta}(\mathbf{u}_t)\,\overline{\rho}_{t\lambda\mu}(\mathbf{v})\Big\}$$

$$
\begin{aligned}
& - e_t E_t^{\alpha\beta\lambda\mu} \left\{ \left[u_{t\iota} \frac{\delta}{\delta t} \left(\gamma_{t\alpha\beta}^{\iota} \right) + u_{t3} \frac{\delta}{\delta t} \left(b_{t\alpha\beta} \right) \right] \gamma_{t\lambda\mu} (\mathbf{v}) \right. \\
& + \gamma_{t\alpha\beta} (\mathbf{u}_t) \left[v_\iota \frac{\delta}{\delta t} \left(\Gamma_{t\lambda\mu}^{\iota} \right) + v_3 \frac{\delta}{\delta t} \left(b_{t\lambda\mu} \right) \right] \\
& + \frac{e_t^2}{12} \left[u_{t3,\iota} \frac{\delta}{\delta t} \left(\Gamma_{t\alpha\beta}^{\iota} \right) + u_{t3} \frac{\delta}{\delta t} \left(b_{t\alpha}^{\omega} b_{t\omega\beta} \right) - u_{t\omega} \frac{\delta}{\delta t} \left(b_{t\alpha}^{\omega}|_\beta \right) - u_{t\omega,\beta} \frac{\delta}{\delta t} \left(b_{t\alpha}^{\omega} \right) \right. \\
& \left. + u_{t\iota} \frac{\delta}{\delta t} \left(b_{t\alpha}^{\omega} \Gamma_{t\omega\beta}^{\iota} \right) - u_{t\omega,\alpha} \frac{\delta}{\delta t} \left(b_{t\beta}^{\omega} \right) + u_{t\iota} \frac{\delta}{\delta t} \left(b_{t\beta}^{\omega} \Gamma_{t\omega\alpha}^{\iota} \right) \right] \overline{\rho}_{t\lambda\mu} (\mathbf{v}) \\
& + \frac{e_t^2}{12} \overline{\rho}_{t\alpha\beta} (\mathbf{u}_t) \left[v_{3,\iota} \frac{\delta}{\delta t} \left(\Gamma_{t\lambda\mu}^{\iota} \right) + v_3 \frac{\delta}{\delta t} \left(b_{t\lambda}^{\omega} b_{t\omega\mu} \right) - v_\omega \frac{\delta}{\delta t} \left(b_{t\lambda}^{\omega}|_\mu \right) - v_{\omega,\mu} \frac{\delta}{\delta t} \left(b_{t\lambda}^{\omega} \right) \right. \\
& \left. \left. + v_\iota \frac{\delta}{\delta t} \left(b_{t\lambda}^{\omega} \Gamma_{t\omega\mu}^{\iota} \right) - v_{\omega,\lambda} \frac{\delta}{\delta t} \left(b_{t\mu}^{\omega} \right) + v_\iota \frac{\delta}{\delta t} \left(b_{t\mu}^{\omega} \Gamma_{t\omega\lambda}^{\iota} \right) \right] \right\} dS_t
\end{aligned}
$$

$$
\begin{aligned}
& \frac{\partial a_t}{\partial e_t} (\mathbf{u}_t, \mathbf{v}) \frac{\delta e_t}{\delta t} \qquad (8.41) \\
& = \int_{\mathcal{O}} E_t^{\alpha\beta\lambda\iota} \frac{\delta e_t}{\delta t} \{ \gamma_{t\alpha\beta}(\mathbf{u}_t) \gamma_{t\lambda\iota}(\mathbf{v}) + \frac{e_t^2}{4} \overline{\rho}_{t\alpha\beta}(\mathbf{u}_t) \overline{\rho}_{t\lambda\iota}(\mathbf{v}) \} dS_t \ .
\end{aligned}
$$

In (8.40) we left for the reader to make substitutions of relations (8.29), (8.31), (8.34) to (8.36). The following result holds true.

Theorem 8.3. *Assume we are given the derivatives* $\frac{\delta e_t}{\delta t}$ *and* $\frac{\delta \varphi_t}{\delta t}$.

The derivative $\mathbf{u}' \equiv \frac{\delta \mathbf{u}_t}{\delta t}_{|t=0}$ *of a solution to equation (8.14) at* $t = 0$ *is a unique solution to the problem*

find $\mathbf{u}' \in \mathcal{H} = (H_0^1(\mathcal{O}))^2 \times H_0^2(\mathcal{O})$ *such that*

$$
\begin{aligned}
& a(\mathbf{u}', \mathbf{v}) + \frac{\partial a_t}{\partial \varphi_t} (\mathbf{u}, \mathbf{v}) \frac{\delta \varphi_t}{\delta t}_{|t=0} + \frac{\partial a_t}{\partial e_t} (\mathbf{u}, \mathbf{v}) \frac{\delta e_t}{\delta t}_{|t=0} = \frac{\delta f_t}{\delta t} (\mathbf{v})_{|t=0} \qquad (8.42) \\
& \forall \mathbf{v} \in (H_0^1(\mathcal{O}))^2 \times H_0^2(\mathcal{O})
\end{aligned}
$$

where the subsequent terms are given respectively by (8.39), (8.40), (8.41) and (8.22).

8.4 Computation of derivatives of cost functionals

Let $\mathfrak{J}(\varphi_t, e_t; \mathbf{u}_t)$ be a functional that we are going to optimize. We assume that

$$
\mathfrak{J}(\varphi_t, e_t; \mathbf{u}_t) = J(S_t), \text{ where } S_t = \varphi_t(\mathcal{O}) \qquad (8.43)
$$

and $J(S_t)$ is a shape functional.

Put

$$
j(t) = \mathfrak{J}(\varphi_t, e_t; \mathbf{u}_t) \ .
$$

Under the assumption that data are sufficiently smooth and $\mathfrak{J}$ is differentiable, we can compute

$$\begin{aligned}\frac{dj}{dt}(0) &= \lim_{t\to 0}\frac{j(t)-j(0)}{t} \qquad (8.44)\\ &= \left[\frac{\partial \mathfrak{J}}{\partial \varphi_t}(\varphi, e_t; \mathbf{u}_t)\frac{\partial \varphi_t}{\partial t} + \frac{\partial \mathfrak{J}}{\partial e_t}(\varphi, e_t; \mathbf{u}_t)\frac{\partial e_t}{\partial t} + \frac{\partial \mathfrak{J}}{\partial u_t}(\varphi, e_t; \mathbf{u}_t)\frac{\partial \mathbf{u}_t}{\partial t}\right]_{t=0} .\end{aligned}$$

Denote

$$\frac{\delta j}{\delta t}(0) = \left[\frac{\partial \mathfrak{J}}{\partial \varphi_t}(\varphi, e_t; \mathbf{u}_t)\frac{\delta \varphi_t}{\delta t} + \frac{\partial \mathfrak{J}}{\partial e_t}(\varphi, e_t; \mathbf{u}_t)\frac{\delta e_t}{\delta t} + \frac{\partial \mathfrak{J}}{\partial u_t}(\varphi, e_t; \mathbf{u}_t)\frac{\delta \mathbf{u}_t}{\delta t}\right]_{t=0} \qquad (8.45)$$

By applying the structure theorem,

$$\frac{dj}{dt}(0) = \frac{\delta j}{\delta t}(0) .$$

Since, in general, we are not able to compute $\frac{\delta \mathbf{u}_t}{\delta t}$ for any $\frac{\delta \varphi_t}{\delta t}$, we introduce the adjoint state equation

Find $\mathbf{q} \in (H_0^1(\mathcal{O}))^2 \times H_0^2(\mathcal{O})$ such that

$$a(\mathbf{q}, \mathbf{v}) = \left[\frac{\partial \mathfrak{J}}{\partial u_t}(\varphi_t, e_t; \mathbf{u}_t)\right]_{|t=0} \mathbf{v}, \quad \forall \mathbf{v} \in (H_0^1(\mathcal{O}))^2 \times H_0^2(\mathcal{O}) . \qquad (8.46)$$

Clearly, equation (8.46) is uniquely solvable. By combining (8.13), (8.42) with (8.46), and in view of (8.45), expression (8.44) takes the form

$$\begin{aligned}\frac{dj}{dt}(0) = &- \left[\frac{\partial a_t}{\partial u_t}(\mathbf{u}_t, \mathbf{q})\frac{\delta \varphi_t}{\delta t}\right]_{|t=0} - \left[\frac{\partial a_t}{\partial e_t}(\mathbf{u}_t, \mathbf{q})\frac{\delta e_t}{\delta t}\right]_{|t=0}\\ &+ \left[\frac{\delta f_t}{\delta t}(\mathbf{q})\right]_{|t=0} + \left[\frac{\partial \mathfrak{J}}{\partial \varphi_t}(\varphi_t, e_t; \mathbf{u}_t)\frac{\delta \varphi_t}{\delta t}\right]_{|t=0} + \left[\frac{\partial \mathfrak{J}}{\partial e_t}(\varphi_t, e_t; \mathbf{u}_t)\frac{\delta e_t}{\delta t}\right]_{|t=0} ,\end{aligned}$$

where the subsequent terms of the right-hand side are given by relations (8.22) (8.40) (8.41) and (8.46), respectively. One should note that the right-hand side is actually a linear mapping with respect to $(w, \varepsilon = \frac{\delta e_t}{\delta t}(0))$.

8.5 Computation of the second derivative

For any $t \geq 0$, (8.44) can be written as

$$\frac{dj}{dt}(t) = \frac{\partial \mathfrak{J}}{\partial \varphi_t}(\varphi, e_t; \mathbf{u}_t)\frac{\delta \varphi_t}{\delta t} + \frac{\partial \mathfrak{J}}{\partial e_t}(\varphi, e_t; \mathbf{u}_t)\frac{\delta e_t}{\delta t} + \frac{\partial \mathfrak{J}}{\partial u_t}(\varphi, e_t; \mathbf{u}_t)\frac{\delta \mathbf{u}_t}{\delta t} \qquad (8.47)$$

and for any $t \geq 0$ we can define the following adjoint state equation.

Find $\mathbf{q}_t \in (H_0^1(\mathcal{O}))^2 \times H_0^2(\mathcal{O})$ such that

$$a_t(\mathbf{q}_t, \mathbf{v}) = \frac{\partial \mathfrak{J}}{\partial u_t}(\varphi, e_t; \mathbf{u}_t)\mathbf{v}, \quad \forall \mathbf{v} \in (H_0^1(\mathcal{O}))^2 \times H_0^2(\mathcal{O}) . \qquad (8.48)$$

As in Theorem 8.3, we get the following result.

Theorem 8.4. *The displacement derivative $\frac{\delta \mathbf{u}_t}{\delta t}$ of a solution to (8.14) is a unique solution to the following problem*
given $\left(\frac{\delta e_t}{\delta t}, \frac{\delta \varphi_t}{\delta t}\right)$, find $\frac{\delta \mathbf{u}_t}{\delta t} \in (H_0^1(\mathcal{O}))^2 \times H_0^2(\mathcal{O})$ such that

$$a(\frac{\delta \mathbf{u}_t}{\delta t}, \mathbf{v}) + \frac{\partial a_t}{\partial \varphi_t}(\mathbf{u}_t, \mathbf{v}) \frac{\delta \varphi_t}{\delta t} + \frac{\partial a_t}{\partial e_t}(\mathbf{u}_t, \mathbf{v}) \frac{\delta e_t}{\delta t} = \frac{\delta f_t}{\delta t}(\mathbf{v}) \quad (8.49)$$
$$\forall \mathbf{v} \in (H_0^1(\mathcal{O}))^2 \times H_0^2(\mathcal{O}) ,$$

where the subsequent terms are respectively given by (8.22) and (8.39) to (8.42).

Furthermore, the displacement derivative $\frac{\delta \mathbf{q}_t}{\delta t}$ of a solution to (8.48) is given as a unique solution to the problem
find $\frac{\delta \mathbf{q}_t}{\delta t} \in (H_0^1(S))^2 \times H_0^2(S)$ such that

$$a(\frac{\delta \mathbf{q}_t}{\delta t}, \mathbf{v}) + \frac{\partial a_t}{\partial \varphi_t}(\mathbf{q}_t, \mathbf{v}) \frac{\delta \varphi_t}{\delta t} + \frac{\partial a_t}{\partial e_t}(\mathbf{q}_t, \mathbf{v}) \frac{\delta e_t}{\delta t} \quad (8.50)$$
$$= + \frac{\partial^2 J}{(\partial u_t)^2}(\varphi_t, e_t; \frac{\delta \mathbf{u}_t}{\delta t})\mathbf{v} + \frac{\partial^2 J}{\partial u_t \partial \varphi_t}(\varphi_t, e_t; \mathbf{u}_t)\left(\mathbf{v}, \frac{\delta \varphi_t}{\delta t}\right)$$
$$+ \frac{\partial^2 J}{\partial u_t \partial e_t}(\varphi_t, e_t; \mathbf{u}_t)\left(\mathbf{v}, \frac{\delta e_t}{\delta t}\right)$$
$$\forall \mathbf{v} \in (H_0^1(\mathcal{O}))^2 \times H_0^2(\mathcal{O}) .$$

Therefore, by relations (8.47)–(8.49), we obtain

$$\frac{dj}{dt}(t) = \int_{\mathcal{O}} G_t dS_t = -\frac{\partial a_t}{\partial \varphi_t}(\mathbf{u}_t, \mathbf{q}_t)\frac{\delta \varphi_t}{\delta t} - \frac{\partial a_t}{\partial e_t}(\mathbf{u}_t, \mathbf{q}_t)\frac{\delta e_t}{\delta t} \quad (8.51)$$
$$+ \frac{\delta f_t}{\delta t}(\mathbf{q}_t) + \frac{\partial \mathfrak{J}}{\partial \varphi_t}(\varphi, e_t; \mathbf{u}_t)\frac{\delta \varphi_t}{\delta t} + \frac{\partial \mathfrak{J}}{\partial e_t}(\varphi, e_t; \mathbf{u}_t)\frac{\delta e_t}{\delta t} .$$

Using (8.43) it follows that

$$\frac{dj}{dt}(t) = dJ(S_t; V(t)), \quad V(t) = \frac{\partial \varphi_t}{\partial t} \circ \varphi_t^{-1}$$

therefore,

Corollary 8.1. *The second derivative of shape functional (8.43) is given by*

$$\mathfrak{D}^2 J(S; V(0), V(0)) = \frac{d^2 j}{dt^2}(0) = \int_{\mathcal{O}} \frac{\delta G_t}{\delta t}_{|t=0} - 2w\kappa G dS$$

where $G = G_{t|t=0}$, $w = w_{t|t=0}$, *and the first part of the latter integral takes the following form*

$$\int_O \frac{\delta G_t}{\delta t}_{|t=0} dS = \frac{\partial^2 J}{(\partial \varphi_t)^2}(\varphi, e_t; \mathbf{u}_t)\left(\frac{\delta \varphi_t}{\delta t}\right)^2 + 2\frac{\partial^2 J}{\partial \varphi_t \partial e_t}(\varphi, e_t; \mathbf{u}_t)\left(\frac{\delta \varphi_t}{\delta t}, \frac{\delta e_t}{\delta t}\right)$$
$$+ \frac{\partial^2 J}{\partial \varphi_t \partial u_t}(\varphi, e_t; \mathbf{u}_t)\left(\frac{\delta \varphi_t}{\delta t}, \frac{\delta \mathbf{u}_t}{\delta t}\right) + \frac{\partial^2 J}{(\partial e_t)^2}(\varphi, e_t; \mathbf{u}_t)\left(\frac{\delta e_t}{\delta t}\right)^2$$
$$+ \frac{\partial^2 J}{\partial e_t \partial u_t}(\varphi, e_t; \mathbf{u}_t)\left(\frac{\delta e_t}{\delta t}, \frac{\delta \mathbf{u}_t}{\delta t}\right)$$
$$+ \frac{\partial \mathfrak{J}}{\partial \varphi_t}(\varphi, e_t; \mathbf{u}_t)\frac{\delta^2 \varphi}{\delta t^2} + \frac{\partial \mathfrak{J}}{\partial e_t}(\varphi, e_t; \mathbf{u}_t)\frac{\delta^2 e_t}{\delta t^2}$$
$$+ \frac{d^2 f_t}{dt^2}(\mathbf{q}_t) + \frac{d}{dt} f_t\left(\frac{\delta \mathbf{q}_t}{\delta t}\right)$$
$$- \frac{\partial^2 a_t}{(\partial \varphi_t)^2}(\mathbf{u}_t, \mathbf{q}_t)\left(\frac{\delta \varphi_t}{\delta t}\right) - 2\frac{\partial^2 a_t}{\partial \varphi_t \partial e_t}(\mathbf{u}_t, \mathbf{q}_t)\left(\frac{\delta \varphi_t}{\delta t}, \frac{\delta e_t}{\delta t}\right)$$
$$- \frac{\partial a_t}{\partial \varphi_t}\left(\frac{\delta \mathbf{u}_t}{\delta t}, \mathbf{q}_t\right)\frac{\delta \varphi_t}{\delta t} - \frac{\partial a_t}{\partial \varphi_t}\left(\mathbf{u}_t, \frac{\delta \mathbf{q}_t}{\delta t}\right)\frac{\delta \varphi_t}{\delta t}$$
$$- \frac{\partial^2 a_t}{(\partial e_t)^2}(\mathbf{u}_t, \mathbf{q}_t)\left(\frac{\delta e_t}{\delta t}\right)^2 - \frac{\partial a_t}{\partial e_t}\left(\frac{\delta \mathbf{u}_t}{\delta t}, \mathbf{q}_t\right)\frac{\delta e_t}{\delta t} - \frac{\partial a_t}{\partial e_t}\left(\mathbf{u}_t, \frac{\delta \mathbf{q}_t}{\delta t}\right)\frac{\delta e_t}{\delta t}$$

The second order derivative $\mathfrak{d}^2 J(s; \cdot, \cdot)$ can be used to obtain second order sufficient optimality conditions as well as stability results for local solutions to a class of shape optimization problems.

References

Adams R.A. (1975): Sobolev Spaces. Academic Press, New York.

Annin B.D., Khludnev A.M. (1979): On the variational inequality for nonlinear plates. 21st Polish Solid Mechanics Conference. Abstracts, p. 2.

Anzellotti G. (1983): On the existence of the rates of stress and displacement for Prandtl-Reuss plasticity. Quart. Appl. Math. **41** (2), pp. 181–208.

Anzellotti G., Giaquinta M. (1980): Existence of the displacements fields for an elasto-plastic body subject to Hencky's law and Von Mises yield condition. Manuscripta Math. **32**, pp. 101–136.

Anzellotti G., Giaquinta M. (1982): On the existence of the fields of stresses and displacements for an elasto-perfectly plastic body in static equilibrium. J. Math. Pure Appl. **61** (3), pp. 219–244.

Arkhipova A.A. (1984): Regularity of solutions to systems of variational inequalities with constraints in $\mathbb{R}^N$. Transactions of Leningrad State University. Series math., mechanics, astronomy **13**, pp. 5–9 (in Russian).

Arutunyan N.H., Shoihet B.A. (1981): The asymptotic behaviour of solutions in creep theory. Proceedings of the USSR Academy of Sciences, Mechanics of Solids **3**, pp. 31–48 (in Russian).

Athanasopoulos I. (1981): Stability of the coincidence set for the Signorini problem. Indiana Univ. Math. J. **30** (2), pp. 235–247.

Attouch H., Picard C. (1983): Variational inequalities with varying obstacles: The general form of the limit problem. J. Funct. Anal. **50**, (3), pp. 329–386.

Bamberger A., Schatzman M. (1983): New results on the vibrating string with a continuous obstacle. SIAM J. Math. Anal. **14** (3), pp. 560–595.

Banichuk N.V. (1970): The small parameter method in a definition of curvilinear shape crack. Transactions of the USSR Academy of Sciences, Mechanics of Solids **2**, pp. 130–137 (in Russian).

Banichuk N.V. (1980): Shape optimization of elastic bodies. Nauka, Moscow (in Russian).

Barbu V. (1984): Optimal Control of Variational Inequalities. Research Notes in Mathematics Vol. 100, Pitman, London.

Bauer J., Sokołowski J. (1984): On the unilateral problem of a rigid punch pressed on an elastic arch. Bulletin of the Polish Academy of Sciences, Technical Sciences **11–12**(32), pp. 717–721.

Ben Abda A. (1993): Sur quelques problèmes inverses géométriques via des équations de conduction elliptiques. Etude théorique et numérique. Thesis, École Nationale d'Ingénieurs de Tunis.

Bendsoe M.P., Olhoff N., Sokołowski J. (1985): Sensitivity analysis of problems of elasticity with unilateral constraints. J. Struct. Mech. **2**(13), pp. 201–222.

Bendsoe M.P., Sokołowski J. (1987a): Sensitivity analysis and optimization of elastic-plastic structures. Journal of Engineering Optimization, ASI Special ISSUE, pp. 31–38.

Bendsoe M.P., Sokołowski J. (1988): Design sensitivity analysis of elastic-plastic analysis problem. Mechanics of Structures and Machines, **1**(16), pp. 81–102.

Bendsoe M.P., Sokołowski J. (1987b): Sensitivity analysis and optimal design of elastic plates with unilateral point supports. Mechanics of Structures and Machines **15**(3), pp. 383–393.

Bendsoe M.P., Sokołowski J. (1993): Shape sensitivity analysis of optimal compliance functionals, Report No. 477, December 1993, The Danish Center of Applied Mathematics and Mechanics, The Technical University of Denmark, Lyngby, to appear in: Mechanics of Structures and Machines.

Bendsoe M.P., Sokołowski J. (1994): Analiza wrazliwosci ze wzgledu na kształt dla pewnej klasy funkcjonalow w optymalizacji konstrukcji. Zeszyty Naukowe Politechniki Śląskiej, Seria: Mechanika **116**, pp. 37–46.

Bendsoe M.P., Sokołowski J. (1994): Shape sensitivity analysis of nonsmooth shape functionals, to appear in: Proceedings of the seventh French-German Conference on Optimization, Lecture Notes in Economics and Mathematical Systems, R. Durier, C. Michelot (Eds).

Benedict R., Sokołowski J., Zolesio J.P. (1984): Shape optimization for contact problems. In: System Modelling and Optimization, P. Thoft-Christensen (Ed.), Lecture Notes in Control and Information Sciences, Vol. 59, Springer, Berlin, pp. 790–799.

Bernadou M., Ciarlet P.E. (1976): Sur l'ellipticité du modèle linéaire de coques de W.T. Koiter. In Computing Methods in Applied Sciences and Engineering, R. Glowinski and J.L. Lions (Eds.), Lectures Notes in Economics and Mathematical Systems, Vol. 134, Springer, Berlin, pp. 89–136.

Bernadou M., Boisserie J.M. (1982): The Finite Element Method in Thin Shell Theory: Application to Arch Dam Simulations. Birkhäuser, Boston.

Bernadou M., Lalanne B. (1986): On the approximation of thin shell by "B-spline and Finite Element Methods". In: Innovative Numerical Methods in Engineering, R.P. Shaw, J. Périaux, A. Chaudouet, J. Wu, C. Marino and C.A. Brebbia (Eds.), Springer, Berlin, pp. 585–592.

Bernadou M., Palma F.J., Rousselet, B. (1991): Shape optimization of an elastic thin shell under various. Structural Optimization **3**, pp. 7–21.

Biancardi M. (1984): Su una sisequazione variazionale nello spazio $H^2(\Omega) \times H^1(\Omega)$ relativa ad un operatore del quarto ordine. Boll. Unione Mat. Ital., B. **3** (3), pp. 779–795.

Biroli M. (1979): A de Giorgi-Nash Moser result for a variational inequality. Boll. Unione Mat. Ital., A **16** (3), pp. 598–605.

Bock I., Lovišek J. (1987): Optimal control problems for variational inequalities with controls in coefficients and in unilateral constraints. Appl. Mat. **32** (4), pp. 301–314.

Bourbaki N.: Éléments de mathématique. Integration. Livre VI, Paris.

Brezis H., Stampacchia G. (1977): Remarks on some fourth order variational inequalities. Ann. Scuola Norm. Sup. Pisa Cl. sci., **4** (2), pp. 363–371.

Caffarelli L.A., Friedman A. (1979): The obstacle problem for the biharmonic operator. Ann. Scuola Norm. Sup. Pisa, ser. IV, **6** (1), pp. 151–184.

Caffarelli L.A., Friedman A., Torelli A. (1982): The two-obstacle problem for the biharmonic operator. Pacific J. Math., **103** (3), pp. 325–335.

Cea J. (1978): Optimization, Theory and Algorithms. Springer, Berlin.

Chenais D. (1987): Optimal design of midsurface of shells: differentiability proof and sensitivity computations. Appl. Math. Optim. **16**, pp. 93–133.

Chenais D. (1994): Shape optimization of shells. Control and Cybernetics, **3** (23), pp. 351–382.

Cimetiére A., Léger A. (1993): Un résultat de différentiabilité dans un problème d'obstacle pour des poutres en flexion. C.R. Acad. Sci. Paris, **316** Série I, pp. 749–754.

Cimatti G. (1973): The constrained elastic beam. Meccanica, **8** (2), pp. 119–124.

Citrini C. (1979): Discontinuous solutions of a nonlinear hyperbolic equations with unilateral constraints. Manuscripta Math. **29**, pp. 323–352.

Delfour M.C., Zolesio J.P. (1988a): Shape sensitivity analysis via Min Max differentiability, SIAM J. Control and Optimization **26**, pp. 834–862.

Delfour M.C., Zolesio J.P. (1988b): Shape sensitivity analysis via penalization method Annali di Matematica, Bologna, pp. 1449–1483.

Delfour M.C., Zolesio J.P. (1992): Structure of shape derivatives for nonsmooth domains. Journal of Functional Analysis, **104**, pp. 1–33.

Demengel F. (1983): Problems variationneles en plasticitè parfaite des plaques, **6** (1), pp. 73–119.

Demengel F. (1984): Fonctions a hessien borné. Ann. Inst. Fourier **34** (2), pp. 155–190.

Demkowicz L., Oden J.T. (1982): On some existence and uniqueness results in contact problems with nonlocal friction. Nonlinear Anal. Theory, Meth. and Appl. **6** (10), pp. 1072–1092.

Dems K., Mróz Z. (1984): Variational approach by means of adjoint systems to structural optimization and sensitivity analysis. II.Structure shape variation. Int. J. Solids Struct. **6**(20), pp. 527–552.

Duvaut G., Lions J.-L. (1972): Les inéquations en mécanique et en physique. Dunod, Paris.

Eisen G. (1983): On obstacle problem with a volume constraint. Manuscr. Math. **43** (1), pp. 73–83.

Ekeland I., Temam R. (1976): Convex analysis and variational problems. North-Holland Publishing Company - Amsterdam, Oxford. American Elsevier Publishing Company, INC., New York.

Erkhov M.I. (1978): Theory of perfectly plastic bodies and structures. Nauka, Moscow, (in Russian).

Fichera G. (1972): Boundary value problems of elasticity with unilateral constraints. In: Handbuch der Physik, Band 6a/2. Springer, Berlin, Heidelberg, New York.

Frehse J. (1972): Two dimensional variational problems with thin obstacles. Math. Z. **143**, pp. 279–288.

Frehse J. (1972): On the regularity of the solution of the biharmonic variational inequality. Manuscr. Math. **9** (1), pp. 91–103.

Friedman A. (1982): Variational principles and free-boundary problems. New York.

Friedman A., Vogelius M. (1989): Determining cracks by boundary measurements. Indiana Univ. Math. J. **38** (3), pp. 527–556.

Fujii N. (1990): Second order necessary conditions in a domain optimization problem. JOTA **65**(2), pp. 223–244.

Fujii N. (1994): Sufficient conditions for optimality in shape optimizations. Control and Cybernetics **3** (23), pp. 393–406.

Fuchs M., Weigner M. (1989): The regularity of minima of variational problems with graph obstacles. Arch. Math. **53** (1), pp. 75–81.

Galin L.A. (1980): Contact problems in elasticity and viscoelasticity. Nauka, Moscow (in Russian).

Gałka A., Telega J.J., Bielski W.R. (1994): Contribution to dual variational principles for nonlinear elastic beams. Control and Cybernetics **3** (23), pp. 641–670.

Guillaume P., Masmoudi M. (1992): Dérivées d'ordre supérieur en optimisation de domaines, C.R. Acad. Sci. Paris **315**, série I, pp. 859–862.

Glowinski R., Lions J.L., Tremolieres R. (1981): Numerical Analysis of Variational Inequalities. North-Holland, Amsterdam.

Grigolyuk E.I., Tolkachev V.M. (1980): Contact problems in the theory of plates and shells. Mashinostroenie, Moscow (in Russian).

Grigolyuk E.I., Kulikov G.M. (1981): Simplified methods of solving nonlinear problems in theory of elastic plates and shells. Some applied problems of a plate and shell theory. Moscow University, pp. 94–121 (in Russian).

Haraux A. (1977): How to differentiate the projection on a convex set in Hilbert space. Some applictions to variational inequalities. J. Math. Soc. Japan **29**(4), pp. 615–631.

Haslinger J., Neittaanmäki P., Tiihonen T. (1986): Shape optimization in contact problems based on penalization of the state inequality. Apl. Mat. **31** (1), pp. 54–77.

Haslinger J., Neittaanmäki P. (1988): Finite element approximation for optimal shape design. Theory and applications. John Wiley & Sons Ltd., Chichester.

Haug E.J., Cea J. (Eds.)(1981): Optimization of Distributed Parameter Structures Vols. I and II. Sijthoff and Noordhoff, Alpen aan den Rijn.

Hayman W.K., Kennedy P.B. (1976): Subharmonic functions, Vol. 1, Academic Press, London, New York, San Francisco.

Hedberg, L.I. (1981): Spectral Synthesis in Sobolev Spaces, and Uniqueness of Solutions of Dirichlet Problem. Acta Math. **147**, pp. 237–264.

Henrot A. (1994): Continuity with respect to the domain for the Laplacian: a survey. Control and Cybernetics **3** (23), pp. 427–444.

Hlaváček I., Bock I., Lovišek J. (1984): Optimal control of a variational inequality with applications to structural analysis. Part I. Optimal design of a beam with unilateral supports. Appl. Math. Optim. **11**, pp. 111–142.

Hlaváček I., Bock I., Lovišek J. (1985): Optimal control of a variational inequality with applications to structural analysis. Part II. Local optimization of the stress of a beam. Part III. Optimal design of an elastic plate. Appl. Math. Optim. **13**, pp. 117–136.

Hlavaček I., Nečas J. (1970): On inequalities of Korn's type. Arch. Rat. Mech. Anal. **36** (4), pp. 305–334.

Hlaváček I., Nečas J. (1982): Optimization of the domain in elliptic unilateral boundary value problems by finite element method, RAIRO. Num. Anal. **16**, pp. 351–373.

Hlaváček I., Haslinger J., Nečas J., Lovišek J. (1983): Riešenie variačnych nerovnosti v mechanike. Praha.

Hoffmann K.H., Sokołowski J. (1994): Interface optimization problems for parabolic equations. Control and Cybernetics **3** (23), 445–452.

Ikonnikov A.K. (1988): Optimal control of elastic beams. Transactions of the 26th student scientific conference. Mathematics, Novosibirsk State University, pp. 41–45 (in Russian).

Ivanov G.V. (1969): Elasto-plastic flow of shells under von Mises yield condition. Proceedings of the USSR Academy of Sciences, Mechanics of Solids, pp. 85–90 (in Russian).

Jensen R. (1980): Boundary regularity for variational inequalities. Ind. Univ. Math. J. **29** (4), pp. 495–504.

Jensen R. (1983): Regularity for elasto-plastic type variational inequalities. Ind. Univ. Math. J. **32** (3), pp. 407–423.

John O. (1977): On Signorini problem for von Karman equation. Apl. Mat. **22** (1), pp. 52–68.

Johnson C. (1976): Existence theorems for plasticity problems. J. Math. Pures et Appl. **55**, pp. 431–444.

Kantorovich L.V., Akilov G.P. (1984): Functional analysis. Nauka, Moscow (in Russian).

Katyshev V.V. (1984): A unilateral problem for hyperbolic equations with restriction imposed upon the solution. Transactions of Sobolev seminar, Novosibirsk **2**, pp. 61–64 (in Russian).

Kazhikhov A.V. (1979): Some questions in Navier-Stokes theory of compressible fluids. Dynamics of the Continuum. Novosibirsk **38**, pp. 33–47 (in Russian).

Khludnev A.M. (1978): A unilateral problem for a weakly nonlinear hyperbolic operator. Siberian Math. J. **19** (2), pp. 412–417 (in Russian).

Khludnev A.M. (1981): A variational approach to contact problems for shallow shells. Transactions of the Sobolev seminar. Novosibirsk **2**, pp. 109–114 (in Russian).

Khludnev A.M. (1982): A variational approach to a contact problem for two plates. Reports of the USSR Academy of Sciences **262** (4), pp. 851–852 (in Russian).

Khludnev A.M. (1982): A remark on the regularity of solutions to a fourth-order variational inequality. Dynamics of Continuum. Novosibirsk **55**, pp. 107–112 (in Russian).

Khludnev A.M. (1982): On a variational inequality for the elastic beam. Appl. Math. and Mech. **46** (3), pp. 456–460 (in Russian).

Khludnev A.M. (1983): A contact problem of a linear elastic body and a rigid punch (variational approach). Appl. Math. and Mech. **47** (6), pp. 999–1005 (in Russian).

Khludnev A.M. (1983): Existence of solutions in elastoplastic one-dimensional problems. J. of Appl. Mech. and Tech. Phys. **2**, pp. 150–156 (in Russian).

Khludnev A.M. (1983): Variational inequalities in a shallow shell theory. Differential Equations **19** (4), pp. 673–679 (in Russian).

Khludnev A.M. (1983): A contact of two elastic plates. Appl. Math. and Mech. **47** (1), pp. 140–146 (in Russian).

Khludnev A.M. (1984): On the existence of solutions in Hencky theory. J. of Appl. Mech. and Tech. Phys. **1**, pp. 174–177 (in Russian).

Khludnev A.M. (1984): Existence and regularity of solutions to unilateral boundary value problems for shallow shells. Differential Equations **20** (11), pp. 1968–1975 (in Russian).

Khludnev A.M. (1984): On solutions of boundary value problems for creep equations with boundary constraints. Appl. Math. and Mech. **48** (1), pp. 44–49 (in Russian).

Khludnev A.M. (1984): Existence of solutions in quasistatic elastoplastic problems for shells. Siberian Math. J. **25** (5), pp. 168–176 (in Russian).

Khludnev A.M. (1986): Regularity of solutions to the fourth-order variational inequalities. Dynamics of Continuum, Novosibirsk **77**, pp. 137–141 (in Russian).

Khludnev A.M. (1987): A variational inequality for a shallow shell operator with restrictions imposed on the boundary. Appl. Math. and Mech. **51** (2), pp. 345–348 (in Russian).

Khludnev A.M. (1988): On variational inequalities in contact plastic problems. Differential Equations **24** (9), pp. 1622–1628 (in Russian).

Khludnev A.M. (1989): The limiting transitions in optimal control problems for the fourth-order operator. Differential Equations **25** (8), pp. 1427–1435 (in Russian).

Khludnev A.M. (1989): Optimal control of punch shape in contact problem for a plate. Dynamics of Continuum, Novosibirsk **93–94**, pp. 163–175 (in Russian).

Khludnev A.M. (1989): Extreme crack shapes in a plate contacting with a rigid punch. In: Boundary problems for nonclassical equations of mathematical physics, Novosibirsk, pp. 60–67 (in Russian).

Khludnev A.M. (1990): Optimal control of a plate with obstacles. Siberian Math. J. **31** (1), pp. 172–178 (in Russian).

Khludnev A.M. (1990): Variational inequalities in contact problems. In: Free boundary problems. Theory and applications, K.-H. Hoffmann, J. Sprekels (Eds.), Vol. 1, Scientific & Longman Technical, pp. 312–314.

Khludnev A.M. (1991): On the convergence of solutions in problems of optimal control of variational inequality of the fourth-order. In: Numerical methods for free boundary problems, P. Neittaanmäki (Ed.). Birkhäuser, Basel, pp. 225–233.

Khludnev A.M. (1991): Optimal control in one-dimensional elastoplastic models. J. of Appl. Mech. and Tech. Phys. **5**, pp. 112–115 (in Russian).

Khludnev A.M. (1992): Contact viscoelastoplastic problem for a beam. In: Free boundary problems in continuum mechanics, S.N.Antontsev, K.-H. Hoffmann, A.M. Khludnev (Eds.). Birkhäuser, Basel, pp. 159–166.

Khludnev A.M. (1992): An optimal control problem for a fourth-order variational inequality. In: Partial Differential Equations. Banach Center Publications, Vol. 27. Institute of Mathematics Polish Academy of Sciences, Warszawa, pp. 225–231.

Khludnev A.M. (1992): On extreme crack shapes in a plate. Transactions of Russian Academy of Sciences. Mechanics of Solid **1**, pp. 170–176 (in Russian).

Khludnev A.M., Hoffmann K.-H. (1992): A variational inequality in a contact elastoplastic problem for a bar. Advances in Mathematical Sciences and Applications, Vol. 1, No. 1, pp. 127–136.

Kikuchi N., Oden J.T. (1987): Contact Problems in Elasticity. SIAM, Philadelphia.

Kinderlehrer D. (1981): Remarks about Signorini's problems in linear elasticity. Ann. Scuola Norm. Sup. Pisa **8** (4), ser. IV, pp. 605–645.

Kinderlehrer D., Stampacchia G. (1980): An introduction to variational inequalities and their applications. New York.

Kiwiel K.C. (1985): Methods of descent for nondifferentiable optimization. Lecture Notes in Mathematics Vol. 1133. Springer, New York.

Koiter V.T. (1960): General theorems of elastic plastic solids. In: Progress in solid mechanics, Vol. 1, Ch. IV. North-Holland, Amsterdam, pp. 165–221.

Koiter W.T. (1970): On the foundations of the linear theory of thin elastic shells, Proc. Kon. Ned. Akad. **B73**, pp. 169–195.

Kosinski W. (1986): Introduction to field singularities and wave analysis PWN – Ellis Horwood.

Kravchuk A.S. (1977): The Hertz problem for linear and nonlinear elastic bodies of bounded sizes. Appl. Math. and Mech. **41** (2), pp. 329–337 (in Russian).

Kravchuk A.S. (1978): Contact problems for deformable bodies as problems of nonlinear programming. Appl. Math. and Mech. **42** (3), pp. 466–474 (in Russian).

Kuz'menko V.I. (1979): A variational approach in the theory of laminated bodies. Appl. Math. and Mech. **43** (5), pp. 893–901 (in Russian).

Landkof N.S. (1966): Foundations of modern theory of a potential. Nauka, Moscow (in Russian).

Lar'kin N.A. (1984): A unilateral problem for nonlocal hyperbolic equation in elasticity theory. Reports of the USSR Academy of Sciences **274** (6), pp. 1341–1344 (in Russian).

Lasiecka I., Sokołowski J. (1988): Regularity and strong convergence of a variational approximation to nonhomogeneous Dirichlet hyperbolic boundary value problems. SIAM Journal on Mathematical Analysis **19** (3), pp. 528–540.

Lasiecka I., Sokołowski J. (1989): Semidiscrete approximations of nonhomogeneous Dirichlet hyperbolic boundary value problems. SIAM Journal on Mathematical Analysis **20** (6), pp. 1366–1387.

Lasiecka I., Sokołowski J., Neittaanmäki P.(1990): Regularization and finite element approximation of the wave equation with Dirichlet boundary data. In: Banach Center Publications, Vol. 24, pp. 329–354.

Lasiecka I., Sokołowski J. (1991): Sensitivity analysis of constrained optimal control problem for wave equation. SIAM Journal on Control and Optimization **29** (5), pp. 1128–1149.

Lewy H., Stampacchia G. (1969): On the regularity of the solution of a variational inequality. Comm. Pure Appl. Math. **22**, pp. 155–188.

Lions J.-L., Magenes E. (1968): Problèmes aux limites non homogénes et applications, Vol. 1, Dunod Gauthier-Villars, Paris.

Lions J.-L. (1969): Quelques methodes de resolution des problemes aux limites non linéaires. Dunod Gauthier-Villars, Paris.

Lions J.L. (1973): Perturbations Singulières dans les Problèmes aux Limités et en Contrôle Optimal. Lecture Notes in Mathematics, Springer, Berlin.

Litvinov V.G. (1987): An optimization in elliptic boundary problems with applications to mechanics. Nauka, Moscow (in Russian).

L'vov G.I. (1982): Variational approaches to contact problems for linear and nonlinear shallow shells. Appl. Math. and Mech. **46** (5), pp. 841–846 (in Russian).

Lukasiewicz S. (1979): Local loads in plates and shells. PWN-Polish Scientific Publishers. Warszawa Noordhoff International Publishing, Leyden.

Malanowski K. (1987): Stability of Solutions to Convex Problems of Optimization. Lecture Notes in Control and Information Sciences Vol. 93, Springer, Berlin.

Mignot F. (1976): Contrôle dans les inequations variationelles elliptiques. J. Funct. Anal. **22**, pp. 25–39.

Mignot F., Murat F., Puel J.P. (1979): Variation d'un point de rétournemont par rapport au domaine. Comm. PDE **41**(4), pp. 1263–1297.

Mignot F., Puel J.P. (1984): Optimal control of some variational inequalities. SIAM J. Control and Optimization **22**, pp. 466–478.

Mikhailov V.P. (1976): Differential equations with partial derivatives. Nauka, Moscow (in Russian).

Moet H.J.K. (1979): Singular perturbation methods in a one-dimensional free boundary problems. Lecture Notes in Mathematics, Vol. 711, Springer, Berlin, pp. 63–75.

Moet H.J.K. (1982): Asymptotic analysis of the boundary in singularly perturbed elliptic variational inequalities. Lect. Notes Math., 942, pp. 1–17.

Morozov N.F. (1984): Mathematial foundations of a crack theory. Nauka, Moscow (in Russian).

Mosco U. (1969): Convergence of convex sets and of solutions of variational inequalities. Adv. Math. **3** (4), pp. 510–585.

Mossino J., Zolesio J.P. (1977): Formulation variationelle d'un probléme issu de la physique de plasmas. C.R. Acad. Sc. Paris **285**, pp. 1033.

Mota Soares, C.A. (Ed.)(1986): Computer Aided Optimal Design. NATO/NASA/NSF-/USAF Advanced Study Institute, Center of Mechanics and Materials of the Technical University of Lisbon, Troia, Portugal, June 29–July 11, 1986.

Mróz Z. (1963): Limit analysis of plastic structures subject to boundary variations. Arch. Mech. Stos. **15**, pp. 63–76.

Mróz Z., Sokołowski J. (Eds.), (1990): Proceedings of the 2nd Symposium on *Optimal Design and Control of Structures* in Jablonna, Poland, June 4–9, 1990, special issue of Control and Cybernetics **19** (3–4).

Murat F., Simon J. (1976): Sur le Contrôle par un Domaine Geometique. Publications du Laboratoire d'Analyse Numerique, Universite de Paris VI.

Myślinski A., Sokołowski J. (1985): Nondifferentiable optimization problems for elliptic systems. SIAM J. Control and Optimization **23**, pp. 632–648.

Naumann J. (1975): On some unilateral boundary value problems for the von Karmàn equations. Apl. Mat. **20** (2), pp. 96–105.

Nečas J. (1967): Les Méthodes Directes en Théorie des Équations Elliptiques. Masson, Paris.

Nečas J. (1975): On regularity of solutions to nonlinear variational inequalities for second-order elliptic systems. Rend. Math. **8**, pp. 481–495.

Nečas J., Hlaváček I. (1981): Mathematical Theory of Elastic and Elasto-Plastic Bodies. Elsevier, Amsterdam.

Neittaanmäki P., Sokołowski J., Zolesio J.P. (1988): Optimization of the domain in elliptic variational inequalities. Appl. Math. and Optimiz. **18**, pp. 85–98.

Novruzi A., Roche J.R. (1994): Newton and Quasi-Newton methods in numerical computations of free surfaces in the electromagnetic shaping of liquid metals. Zeszyty Naukowe Politechniki Śląskiej, Seria: Mechanika **116**, pp. 149–160.

Oleinik O.A., Kondratyev V.A., Kopacek I. (1981): Asymptotic properties of solutions of a biharmonic equation. Differential Equations **17** (10), pp. 1886–1899 (in Russian).

Petryk H., Mróz Z. (1986): Time derivatives of integrals and functionals defined on varying volume and surface domains. Arch. Mech. **38** (5–6), pp. 697–724.

Pierre M., Sokołowski J. (1996): Differentiability of projection and applications, Lecture Notes in Pure and Applied Mathematics, Vol. 174, Marcel Dekker, E. Casas (Ed.), pp. 231–240.

Pironneau O. (1984): Optimal Shape Design for Elliptic Systems. Springer, New York.

Pisarenko G.S., Mozharovsky N.S. (1981): Equations and boundary value problems in plasticity and creep theory. Naukowa Dumka, Kiev (in Russian).

Rabotnov Y.N. (1979): Mechanics of solids. Nauka, Moscow (in Russian).

Rao M., Sokołowski J. (1988): Sensitivity analysis of Kirchhoff plate with obstacle. Rapport de Recherche No. 771, INRIA, Rocquencourt, France.

Rao M., Sokołowski J. (1989): Shape sensitivity analysis of state constrained optimal control problems for distributed parameter systems. In: Proceedings of IFIP Conference on Optimal Control of Systems Governed by Partial Differential Equations, Santiago de Compostela, July 6–9, 1987, Lecture Notes in Control and Information Sciences Vol. 114, Bermudez A. (Ed.) Springer, Berlin, pp. 236–245.

Rao M., Sokołowski J. (1991a): Differential stability of solutions to parametric optimization problems. Mathematical Methods in Applied Sciences. **14**, pp. 281–294.

Rao M., Sokołowski J. (1991b): Sensitivity analysis of shallow shell with obstacles. In: Lecture Notes in Control and Information Sciences, Vol.154, Kurzhanski A., Lasiecka I. (Eds.) Springer, Berlin pp. 135–144.

Rao M., Sokołowski J. (1993a): Sensitivity analysis of unilateral problems in $H_0^2(\Omega)$ and applications. Numerical Functional Analysis and Optimization, 14(1&2), pp. 125–143.

Rao M., Sokołowski J. (1993b): Polyhedricity of convex sets in Sobolev space $H_o^2(\Omega)$. Nagoya Math. J. **130**, pp. 101–110.

Rektorys K. (1980): Variational Methods in Mathematics, Science and Engineering, D. Reidel, Boston.

Reshetnyak Y.G. (1982): Stability theorems in geometry and analysis. Nauka, Novosibirsk (in Russian).

Rousselet B. (1982): Quelques Résultats en Optimisation de Domains. Thèse d'Etat, Université de Nice.

Schatzman M. (1980): A hyperbolic problem of second order with unilateral constraints: the vibration string with a concave obstacle. J. Math. Anal. Appl. **73** (1), pp. 138–191.

Schild B. (1984): A regularity result for polyharmonic variational inequalities with thin obstacles. Ann. Scuola Norm. Sup. Pisa **11** (1), pp. 87–122.

Schuss Z. (1976): Singular perturbations and the transition from thin plate to membrane. Proc. Amer. Math. Soc. **58**, pp. 139–147.

Schwartz L. (1967): Analyse mathematique. Vol. 1, Hermann, Paris.

Seregin G.A. (1984): On correctness of variational problems in mechanics of elasto-plastic medium. Reports of the USSR Academy of Sciences **276** (1), pp. 71–75 (in Russian).

Simon J. (1980): Differentiability with respect to the domain in boundary value problems. Numer. Funct. Anal. and Optimiz. **2**(7&8), pp. 649–687.

Sokołowski J. (1981a): Optimal control in coefficients for weak variational problems in Hilbert space. Applied Mathematics and Optimization, **7**, pp. 283–293.

Sokołowski J. (1981b): Control in coefficients for PDE. In: Theory of Nonlinear Operators.(Abh. der Akad. d. Wiss. d. DDR) Berlin, Akademie Verlag, Jahrgang, pp. 287–295.

Sokołowski J. (1981c): Sensitivity analysis for a class of variational inequalities. In: Optimization of Distributed Parameter Structures, Vol. 2, E.J. Haug and J. Cea (Eds.) Sijthoff and Noordhoff, Rockville, Maryland, pp. 1600–1609.

Sokołowski J. (1983): Optimal control in coefficients of boundary value problems with unilateral constraints. Bulletin of the Polish Academy of Sciences, Technical Sciences, **1–12**(31), pp. 71–81.

Sokołowski J. (1985a): Differential stability of solutions to constrained optimization problems. Appl. Math. Optim. **13**, pp. 97–115.

Sokołowski J. (1985b): Sensitivity Analysis and Parametric Optimization of Optimal Control Problems for Distributed Parameter Systems. Habilitation Thesis, Warsaw Technical University Publications, Prace Naukowe, Elektronika, z. 73 (in Polish).

Sokołowski J. (1986a): Differential stability of projection in Hilbert space onto convex set, Applications to sensitivity analysis of optimal control problems. In: Analysis and Algorithms of Optimization Problems, K. Malanowski and K. Mizukami (Eds.) Lecture Notes in Control and Information Sciences, Vol. 82, Springer, Berlin, pp. 1–37.

Sokołowski J. (1986b): Parametric optimization problems for evolution initial-boundary value problems. In: Analysis and Algorithms of Optimization Problems, K. Malanowski and K. Mizukami (Eds.), Lecture Notes in Control and Information Sciences, Vol. 82, Springer, Berlin pp. 61–87.

Sokołowski J. (1987a): Sensitivity analysis of control constrained optimal control problems for distributed parameter systems. SIAM J. Control and Optimization **25**(6), pp. 1542–1556.

Sokołowski J. (1987b): Sensitivity analysis of Signorini variational inequality. In: Banach Center Publications, B. Bojarski (Ed.), Polish Scientific Publisher. Warsaw, pp. 287–299.

Sokołowski J. (1988a): Shape sensitivity analysis of nonsmooth variational problems. In: Boundary Control and Boundary Variations. Proceedings of the IFIP WG 7.2 Conference, Nice, France, June 10–13, 1986, J.P. Zolesio (Ed.), Lecture Notes in Control and Information Sciences Vol. 100, Springer, Berlin, pp. 265–285.

Sokołowski J. (1988b): Shape sensitivity analysis of boundary optimal control problems for parabolic systems. SIAM Journal on Control and Optimization **26**, pp. 763–787.

Sokołowski J. (1988c): Sensitivity analysis of contact problems with prescribed friction. Appl. Math. Optim. **18**, pp. 99–117.

Sokołowski J. (1990a): Sensitivity analysis of contact problem with friction. In: Free Boundary Problems : Theory and Applications, K.-H. Hoffmann and J. Sprekels (Eds.), Pitman Research Notes in Mathematics **185**, pp. 329–334.

Sokołowski J. (1990b): Sensitivity analysis of shape optimization problems. Control and Cybernetics **19** 3–4, pp. 271–286.

Sokołowski J. (1992): Shape Sensitivity Analysis of Variational Inequalities. In: Shape Optimization and Free Boundaries, M.C. Delfour (Ed.) Proceedings of the NATO ASI and Seminaire de Mathematiques superieure on Shape Optimization and Free Boundaries, Montreal, June 25–July 15, 1990, Kluwer Academic Publishers, Dordrecht, pp. 287–320.

Sokołowski J. (1993a): Sensitivity analysis of shape estimation problems. In the volume dedicated to Jean Cea, Mechanics of Structures and Machines **21**(1), pp. 67–94.

Sokołowski J. (1993b): Nonsmooth optimal design problems for the Kirchhoff plate with unilateral conditions. Kybernetika **29** 3, pp. 284–290.

Sokołowski J. (1993c): Applications of material derivative method. Topology Design of Structures. M.P. Bendsoe and C. Mota-Soares (Eds.), Kluwer Academic Publishers, Dordrecht, pp. 307–312.

Sokołowski J., Editor, (1994a): Proceedings of Minisemester at the Banach Center, Part I, Shape Design and Optimization. Special issue of Control and Cybernetics **23** (3).

Sokołowski J., Editor, (1994b): Proceedings of Minisemester at the Banach Center, Part II, Modelling, Identification, Sensitivity Analysis and Control of Structures. Special issue of Control and Cybernetics **23** (4).

Sokołowski J., Sprekels J. (1991): Dynamical shape control and the stabilization of nonlinear thin rods. Mathematical Methods in Applied Sciences **14**, pp. 63–78.

Sokołowski J., Sprekels J. (1994): Control problems with state constraints for shape memory alloys, Mathematical Methods in Applied Sciences **17**, pp. 943–952.

Sokołowski J., Sprekels J. (1994): Control Problems for Shape Memory Alloys with Constraints on the Shear Strain. In: The Proceedings of the IFIP Workshop on Control of PDE's, Trento, Italy, January 1993, Lecture Notes in Pure and Applied Mathematics, Vol. 165 Marcel Dekker, G. Da Prato, L. Tubaro (Eds), pp. 189–196.

Sokołowski J., Zolesio J.P. (1982a): Shape sensitivity analysis for variational inequalities. In: Systems Modelling and Optimization, Proceedings of the 10th IFIP Conference, New York City, Lecture Notes in Control and Information Sciences, Vol.38, R.F. Drenick and F.Kozin (Eds.), Springer, Berlin, pp. 401–407.

Sokołowski J., Zolesio J.P. (1982b): Dérivation par rapport au domaine dans les problèmes unilatéraux. INRIA, Rapport de Recherche No.132.

Sokołowski J., Zolesio J.P. (1985a): Dériveé par rapport au domaine de la solution d'un problème unilatéral. C.R. Acad. Sc. Paris, **301**(4), Serie I, pp. 103–106.

Sokołowski J., Zolesio J.P. (1985b): Shape sensitivity analysis of an elastic-plastic torsion problem. Bulletin of the Polish Academy of Sciences, Technical Sciences **33**(11–12), pp. 579–586.

Sokołowski J., Zolesio J.P. (1987a): Shape sensitivity analysis of unilateral problems. SIAM J. Math. Anal. **18**(5), pp. 1416–1437.

Sokołowski J., Zolesio J.P. (1987b): Shape design sensitivity analysis of plates and plane elastic solids under unilateral constraints. JOTA **54** (2), pp. 361–382.

Sokołowski J., Zolesio J.P. (1988): Shape sensitivity analysis of contact problems with prescribed friction. Journal of Nonlinear Analysis: Theory, Methods, & Applications **12**(12), pp. 1399–1411.

Sokołowski J., Zolesio J.P. (1990): Shape sensitivity analysis of hyperbolic problems. In: Stabilization of Flexible Structures, J.P. Zolesio (Ed.) Lecture Notes in Information and Control Sciences Vol. 147, Springer, Berlin, pp. 280–297.

Sokołowski J., Zolesio J.P. (1992) Introduction to Shape Optimization. Shape Sensitivity Analysis. Springer, New York.

Stampacchia G. (1975): Su una disequazione variazionale legata al comportamento elasto-plastico delle travi appoggiate agli estremi. Boll. Unione Mat. Ital. **11** (4), pp. 444–454.

Strang G., Temam R. (1980): Functions of bounded deformation. Arch. Rat. Mech. Anal. **75**, pp. 7–21.

Suquet P.M. (1981): Evolution problems for a class of dissipative materials. Quart. Appl. Math. **38** (4), pp. 391–414.

Telega J.J. (1987): Variational methods in contact problems of mechanics. Advances in Mechanics **10** (2), pp. 3–95 (in Russian).

Telega J.J., Lewinski T. (1994): Mathematical aspects of modelling the macroscopic behaviour of cross-ply laminates with intralaminar cracks. Control and Cybernetics, **3** (23), pp. 773–792.

Temam R. (1979): Navier-Stokes equations. Theory and numerical analysis. Horth-Holland, Amsterdam.

Temam R. (1983): Problèmes mathématiques en plasticité. Gauthier-Villars, Paris.

Temam R., Strang G. (1980): Duality and relaxation in the variational problems of plasticity. J. de Mécanique **19** (3), pp. 493–527.

Torelli A. (1979): Some regularity results for a family of variational inequalities. Ann. Scuola Norm. Sup. Pisa **6** (3), pp. 497–510.

Vainberg M.M. (1972): A variational method and monotonous operators method. Nauka, Moscow (in Russian).

Vergara Caffarelli G. (1971): Regolarita di un problema di desequazioni variazionalli relativo a due membrane. Atti Acad. Naz. Lincei Rend. **50**, pp. 659–662.

Vergara Caffarelli G. (1974): Variational inequalities for two surfaces of constant mean curvature. Arch. Rat. Mech. Anal. **56**, pp. 334–347.

Vladimirov V.S. (1981): Equations of mathematical physics. Nauka, Moscow (in Russian).

Vorovich I.I. (1957): Some direct methods in nonlinear vibration theory of shallow shells. Proceedings of the USSR Academy of Sciences. Series math. **21**, pp. 747–784 (in Russian).

Vorovich I.I., Lebedev L.P. (1972): Existence of solutions in nonlinear theory of shallow shells. Appl. Math. and Mech. **36** (4), pp. 691–704 (in Russian).

Yakunina G.V. (1981): Smoothness of solutions of variational inequalities. Partial differential equations. Spectral theory. Leningrad State University **8**, pp. 213–220 (in Russian).

Zarantonello F.H. (1971): Projections on convex sets in Hilbert space and spectral theory. In: Contributions to Nonlinear Functional Analysis, Publ. No.27, Math. Res. Center. Univ. Wisconsin, Madison, Academic Press, New York – London, pp. 237–424.

Ziemer P.W. (1989): Weakly Differentiable Functions. Springer, New York.

Zolesio J.P. (1979): Identification de domains par déformations. Thèse d'Etat, Université de Nice.

Index

International Series of Numerical Mathematics

Edited by
K.–H. Hoffmann, Technische Universität München, Germany
H.D. Mittelmann, Arizona State University, Tempe, CA, USA
J. Todd, California Institute of Technology, Pasadena, CA, USA

International Series of Numerical Mathematics *is open to all aspects of numerical mathematics. Some of the topics of particular interest include free boundary value problems for differential equations, phase transitions, problems of optimal control and optimization, other nonlinear phenomena in analysis, nonlinear partial differential equations, efficient solution methods, bifurcation problems and approximation theory. When possible, the topic of each volume is discussed from three different angles, namely those of mathematical modeling, mathematical analysis, and numerical case studies.*

Titles previously published in the series

ISNM 95 **K.-H. Hoffmann, J. Sprekels (Eds.):** Free Boundary Value Problems, 1990 (3-7643-2474-0)

ISNM 96 **J. Albrecht, L. Collatz, P. Hagedorn, W. Velte (Eds):** Numerical Treatment of Eigenvalue Problems, Vol. 5, 1991 (3-7643-2575-5)

ISNM 97 **R.U. Seydel, F.W. Schneider, T.G. Küpper, H. Troger (Eds):** Bifurcation and Chaos: Analysis, Algorithms, Applications, 1991 (3-7643-2593-3)

ISNM 98 **W. Hackbusch, U. Trottenberg (Eds):** Multigrid Methods III, 1991 (3-7643-2632-8)

ISNM 99 **P. Neittaanmäki (Ed.):** Numerical Methods for Free Boundary Problems, 1991 (3-7643-2641-7)

ISNM 100 **W. Desch, F. Kappel, K. Kunisch (Eds):** Estimation and Control of Distributed Parameter Systems, 1991 (3-7643-2676-X)

ISNM 101 **G. Del Piero, F. Maceri (Eds):** Unilateral Problems in Structural Analysis IV, 1991 (3-7643-2487-2)

ISNM 102 **U. Hornung, P. Kotelenez, G. Papanicolaou (Eds):** Random Partial Differential Equations, 1991 (3-7643-2688-3)

ISNM 103 **W. Walter (Ed.):** General Inequalities 6, 1992 (3-7643-2737-5)

ISNM 104 **E. Allgower, K. Böhmer, M. Golubitsky (Eds):** Bifurcation and Symmetry, 1992 (3-7643-2739-1)

ISNM 105 **D. Braess, L.L. Schumaker (Eds):** Numerical Methods in Approximation Theory, Vol. 9, 1992 (3-7643-2746-4)

ISNM 106 **S.N. Antontsev, K.-H. Hoffmann, A.M. Khludnev (Eds):** Free Boundary Problems in Continuum Mechanics, 1992 (3-7643-2784-7)

ISNM 107 **V. Barbu, F.J. Bonnans, D. Tiba (Eds):** Optimization, Optimal Control and Partial Differential Equations, 1992 (3-7643-2788-X)

ISNM 108 **H. Antes, P.D. Panagiotopoulos:** The Boundary Integral Approach to Static and Dynamic Contact Problems. Equality and Inequality Methods, 1992 (3-7643-2592-5)

ISNM 109 **A.G. Kuz'min:** Non-Classical Equations of Mixed Type and their Applications in Gas Dynamics, 1992 (3-7643-2573-9)

ISNM 110 **H.R.E.M. Hörnlein, K. Schittkowski (Eds):** Software Systems for Structural Optimization, 1992 (3-7643-2836-3)

ISNM 111 **R. Burlisch, A. Miele, J. Stoer, K.H. Well:** Optimal Control, 1993 (3-7643-2887-8)

ISNM 112 **H. Braess, G. Hämmerlin (Eds):** Numerical Integration IV. Proceedings of the Conference at the Mathematical Research Institute at Oberwolfach, November 8-14, 1992, 1993 (3-7643-2922-X)

ISNM 113 **L. Quartapelle:** Numerical Solution of the Incompressible Navier-Stokes Equations, 1993 (3-7643-2935-1)

ISNM 114 **J. Douglas, U. Hornung (Eds):** Flow in Porous Media. Proceedings of the Oberwolfach Conference, June 21-27, 1992 (ISBN 3-7643-2949-1)

ISNM 115 **R. Bulirsch, D. Kraft (Eds):** Computational Optimal Control, 1994 (ISBN 3-7643-5015-6)

ISNM 116 **P.W. Hemker, P. Wesseling (Eds):** Multigrid Methods IV. Proceedings of the Fourth European Multigrid Conference, Amsterdam, July 6-9, 1993 (ISBN 3-7643-5030-X)

ISNM 117 **R.E. Bank, R. Bulirsch, H. Gajewski, K. Merten (Eds):** Mathematical Modelling and Simulation of Electrical Circuits and Semiconductor Devices, 1994 (ISBN 3-7643-5053-9)

ISNM 118 **W. Desch, F. Kappel, K. Kunisch (Eds):** Control and Estimation of Distributed Parameter Systems: Nonlinear Phenomena. International Conference on Control and Estimation of Distributed Parameter Systems, Vorau, July 18-24, 1993, 1994 (ISBN 3-7643-5098-9)

ISNM 119 **R. Zahar (Ed.):** Approximation and Computation: A Festschrift in Honor of Walter Gautschi, 1995 (ISBN 0-8176-3753-2)

ISNM 120 **W. Hackbusch:** Integral Equations: Theory and Numerical Treatment, 1995 (ISBN 3-7643-2871-1)

ISNM 121 **R. Jeltsch, M. Mansour (Eds):** Stability Theory. Hurwitz Centenary Conference, Centro Stefano Franscini, Ascona, 1995, 1996 (ISBN 3-7643-5474-7)

Systems and Control: Foundations and Applications

A series of monographs and advanced graduate texts

Edited by **Christopher Byrnes**, Washington University, St. Louis, MO, USA

Systems and Control *is designed for the publication of research level monographs and advanced graduate textbooks in all areas of systems and control theory and its applications to a wide variety of scientific disciplines.*

Previously published titles:

SC1, *Banks, H.T. / Kunisch, K.*: Estimation Techniques for Distributed Parameter Systems. 1989 (ISBN 3-7643-3433-9)

SC2, *Aubin, J.-P. / Frankowska, H.*: Set-Valued Analysis. 1990 (ISBN 3-7643-3478-9)

SC3, *Kushner, H.J.*: Weak Convergence Methods and Singularly Perturbed Stochastic Control and Filtering Problems. 1990 (ISBN 3-7643-3437-1)

SC4, *Falb, P.*: Methods of Algebraic Geometry in Control Theory. Part I: Scalar Linear Systems and Affine Algebraic Geometry. 1990 (ISBN 3-7643-3454-1)

SC5, *Basar, T. / Bernhard, P.*: H_∞-Optimal Control and Related Minimax Design Problems: A Dynamic Game Approach. 1991. ***2nd printing 1995*** (ISBN 3-7643-3814-8)

Aubin, J.-P.: Viability Theory. 1991 (ISBN 3-7643-3571-8)

Chen, H.F. / Guo, L.: Identification and Stochastic Adaptive Control. 1991 (ISBN 3-7643-3597-1)

Bensoussan, A. / Da Prato, G. / Delfour, M.C. / Mitter, S.K.: Representation and Control of Infinite Dimensional Systems. Vol. 1. 1992 (ISBN 3-7643-3641-2)

Zabczyk, J.: Mathematical Control Theory. An Introduction. 1992. ***2nd printing 1995*** (ISBN 3-7643-3645-5)

Asachenkov, A. / Marchuk, G. / Mohler, R. / Zuev, S.: Disease Dynamics. 1993. (ISBN 3-7643-3692-7)

Bensoussan, A. / Da Prato, G. / Delfour, M.C. / Mitter, S.K.: Representation and Control of Infinite Dimensional Systems. Vol. II. 1993 (ISBN 3-7643-3642-0)

Keulen, B. van: H_∞-Infinity Control for Distributed Parameter Systems: A State-Space-Approach. 1993 (ISBN 3-7643-3709-5)

Krasovskii, A.N. / Krasovskii, N.N.: Control Under Lack of Information. 1994 (ISBN 3-7643-3698-6)

Kuijper, M.: First-order Representations of Linear Systems. 1994 (ISBN 3-7643-3754-0)

Lagnese, J.E. / Leugering, G. / Schmidt, E.J.P.G.: Modeling, Analysis and Control of Dynamic Elastic Multi-Link Structures. 1994 (ISBN 3-7643-3705-2)

Li, X. / Yong, J.: Optimal Control Theory for Infinite Dimensional Systems. 1994 (ISBN 3-7643-3722-2)

Sethi, S.P. / Zhang, Q.: Hierarchical Decision Making in Stochastic Manufacturing Systems. 1994 (ISBN 3-7643-3735-4)

Subbotin, A.I.: Generalized Solutions of First Order PDEs: The Dynamical Optimization Perspective. 1994 (ISBN 3-7643-3740-0)

Zelikin, M.I. / Borisov, V.F.: Theory of Chattering Control with Applications to Astronautics, Robotics, Economics, and Engineering. 1994 (ISBN 3-7643-3618-8)

Basar, T. / Bernhard, P.: H^{∞}–Optimal Control and Related Minimax Design Problems. 1995 (ISBN 3-7643-3554-8)

Dullerud, G.: Control of Uncertain Sampled-Data Systems. 1995 (ISBN 3-7643-3851-2)

Kimura, H.: $H-\infty$ Control and Circuit Theory. 1995. (ISBN 3-7643-3787-7)

Subrahmanyam, M.B.: Finite Horizon $H\infty$ and Related Control Problems. 1995 (ISBN 3-7643-3811-3)

Bardi, M. / Dolcetta, I.C.: Optimal Control & Viscosity Solutions of Hamilton-Jacobi-Bellman Equations. 1996 (ISBN 3-7643-3640-4)

Byrnes, C.I. / Datta, B.N. / Martin, C.F. / Gilliam, D.S. (Eds): Systems and Control in the Twenty-First Century. 1996 (ISBN 3-7643-3881-4)

Feuer, A.: Sampling in Digital Signal Processing and Control. 1996 (ISBN 3-7643-3934-9)

Freeman, R.A.: Robust Nonlinear Control Design: State-Space and Lyapunov Techniques. 1996 (ISBN 3-7643-3930-6)

Kurzhanski, A.B. / Vályi, I.: Ellipsoidal Calculus for Estimation and Control. 1996 (ISBN 3-7643-3699-4)

Mareels, I. / Polderman, J.W.: Adaptive Systems: An Introduction. 1996 (ISBN 3-7643-3877-6)

Peters, M.A. / Iglesias P.: Minimum Entropy Control for Time-Varying Systems. 1996 (ISBN 3-7643-3972-1)